Statistical Genetics: Mapping, Linkage and Analysis

Statistical Genetics: Mapping, Linkage and Analysis

Edited by Ivie Nathaniel

SYRAWOOD
PUBLISHING HOUSE

New York

Published by Syrawood Publishing House,
750 Third Avenue, 9th Floor,
New York, NY 10017, USA
www.syrawoodpublishinghouse.com

Statistical Genetics: Mapping, Linkage and Analysis
Edited by Ivie Nathaniel

Cataloging-in-Publication Data

Statistical genetics : mapping, linkage and analysis / edited by Ivie Nathaniel.
    p. cm.
Includes bibliographical references and index.
ISBN 978-1-68286-722-8
1. Genetics--Statistical methods. 2. Gene mapping--Statistical methods.
3. Linkage (Genetics)--Statistical methods. I. Nathaniel, Ivie.
QH438.4.S73 S73 2019
576.5--dc23

TABLE OF CONTENTS

Preface VII

Chapter 1 **In utero exposure to maternal smoking is associated with DNA methylation alterations and reduced neuronal content in the developing fetal brain** 1
Zac Chatterton, Brigham J. Hartley, Man-Ho Seok, Natalia Mendelev, Sean Chen, Maria Milekic, Gorazd Rosoklija, Aleksandar Stankov, Iskra Trencevsja-Ivanovska, Kristen Brennand, Yongchao Ge, Andrew J. Dwork and Fatemeh Haghighi

Chapter 2 **Asymmetric DNA methylation of CpG dyads is a feature of secondary DMRs associated with the *Dlk1/Gtl2* imprinting cluster in mouse** 12
Megan Guntrum, Ekaterina Vlasova and Tamara L. Davis

Chapter 3 **Genome-wide search for Zelda-like chromatin signatures identifies GAF as a pioneer factor in early fly development** 26
Arbel Moshe and Tommy Kaplan

Chapter 4 **Vitamin C induces specific demethylation of H3K9me2 in mouse embryonic stem cells via Kdm3a/b** 40
Kevin T. Ebata, Kathryn Mesh, Shichong Liu, Misha Bilenky, Alexander Fekete, Michael G. Acker, Martin Hirst, Benjamin A. Garcia and Miguel Ramalho-Santos

Chapter 5 **Regions of common inter-individual DNA methylation differences in human monocytes: genetic basis and potential function** 52
Christopher Schröder, Elsa Leitão, Stefan Wallner, Gerd Schmitz, Ludger Klein-Hitpass, Anupam Sinha, Karl-Heinz Jöckel, Stefanie Heilmann-Heimbach, Per Hoffmann, Markus M. Nöthen, Michael Steffens, Peter Ebert, Sven Rahmann and Bernhard Horsthemke

Chapter 6 **Additional sex combs interacts with enhancer of zeste and trithorax and modulates levels of trimethylation on histone H3K4 and H3K27 during transcription of *hsp70*** 70
Taosui Li, Jacob W. Hodgson, Svetlana Petruk, Alexander Mazo and Hugh W. Brock

Chapter 7 **Hypomethylated domain-enriched DNA motifs prepattern the accessible nucleosome organization in teleosts** 87
Ryohei Nakamura, Ayako Uno, Masahiko Kumagai, Shinichi Morishita and Hiroyuki Takeda

Chapter 8 **Histone isoform H2A1H promotes attainment of distinct physiological states by altering chromatin dynamics** 100
Saikat Bhattacharya, Divya Reddy, Vinod Jani, Nikhil Gadewal, Sanket Shah, Raja Reddy, Kakoli Bose, Uddhavesh Sonavane, Rajendra Joshi and Sanjay Gupta

Chapter 9   **Chromatin organization changes during the establishment and maintenance
of the postmitotic state**..119
Yiqin Ma and Laura Buttitta

Chapter 10   **Site-specific regulation of histone H1 phosphorylation in pluripotent cell
differentiation** 139
Ruiqi Liao and Craig A. Mizzen

Chapter 11   **Differential DNA methylation and lymphocyte proportions in a Costa Rican
high longevity region**..160
Lisa M. McEwen, Alexander M. Morin, Rachel D. Edgar, Julia L. MacIsaac,
Meaghan J. Jones, William H. Dow, Luis Rosero-Bixby, Michael S. Kobor and
David H. Rehkopf

Chapter 12   **Links between DNA methylation and nucleosome occupancy in the human
genome**..174
Clayton K. Collings and John N. Anderson

Chapter 13   **Identification of epigenetic signature associated with alpha
thalassemia/mental retardation X-linked syndrome** 193
Laila C. Schenkel, Kristin D. Kernohan, Arran McBride, Ditta Reina,
Amanda Hodge, Peter J. Ainsworth, David I. Rodenhiser, Guillaume Pare,
Nathalie G. Bérubé, Cindy Skinner, Kym M. Boycott, Charles Schwartz and
Bekim Sadikovic

Permissions

List of Contributors

Index

PREFACE

Statistical genetics is a scientific discipline that is concerned with developing statistical methods for analyzing genetic data. This field is generally associated with human genetics. It explores theories and methodologies to support research in the areas of population genetics, genetic epidemiology and quantitative genetics. This book provides significant information of this discipline to help develop a good understanding of statistical genetics and related fields. The various studies that are constantly contributing towards the evolution of this field are examined in detail. Geneticists, molecular biologists, computational biologists, researchers and students engaged in this field will find this book immensely beneficial.

This book has been the outcome of endless efforts put in by authors and researchers on various issues and topics within the field. The book is a comprehensive collection of significant researches that are addressed in a variety of chapters. It will surely enhance the knowledge of the field among readers across the globe.

It gives us an immense pleasure to thank our researchers and authors for their efforts to submit their piece of writing before the deadlines. Finally in the end, I would like to thank my family and colleagues who have been a great source of inspiration and support.

Editor

In utero exposure to maternal smoking is associated with DNA methylation alterations and reduced neuronal content in the developing fetal brain

Zac Chatterton[1,2,7], Brigham J. Hartley[1,2,3], Man-Ho Seok[1,2,3], Natalia Mendelev[1,2,7], Sean Chen[1,2,7], Maria Milekic[5], Gorazd Rosoklija[5,8,9], Aleksandar Stankov[9], Iskra Trencevsja-Ivanovska[10], Kristen Brennand[1,2,3], Yongchao Ge[4], Andrew J. Dwork[5,6,8] and Fatemeh Haghighi[1,2,7]* (ID)

Abstract

Background: Intrauterine exposure to maternal smoking is linked to impaired executive function and behavioral problems in the offspring. Maternal smoking is associated with reduced fetal brain growth and smaller volume of cortical gray matter in childhood, indicating that prenatal exposure to tobacco may impact cortical development and manifest as behavioral problems. Cellular development is mediated by changes in epigenetic modifications such as DNA methylation, which can be affected by exposure to tobacco.

Results: In this study, we sought to ascertain how maternal smoking during pregnancy affects global DNA methylation profiles of the developing dorsolateral prefrontal cortex (DLPFC) during the second trimester of gestation. When DLPFC methylation profiles (assayed via Illumina, HM450) of smoking-exposed and unexposed fetuses were compared, no differentially methylated regions (DMRs) passed the false discovery correction (FDR ≤ 0.05). However, the most significant DMRs were hypomethylated CpG Islands within the promoter regions of GNA15 and SDHAP3 of smoking-exposed fetuses. Interestingly, the developmental up-regulation of SDHAP3 mRNA was delayed in smoking-exposed fetuses. Interaction analysis between gestational age and smoking exposure identified significant DMRs annotated to SYCE3, C21orf56/LSS, SPAG1 and RNU12/POLDIP3 that passed FDR. Furthermore, utilizing established methods to estimate cell proportions by DNA methylation, we found that exposed DLPFC samples contained a lower proportion of neurons in samples from fetuses exposed to maternal smoking. We also show through in vitro experiments that nicotine impedes the differentiation of neurons independent of cell death.

Conclusions: We found evidence that intrauterine smoking exposure alters the developmental patterning of DNA methylation and gene expression and is associated with reduced mature neuronal content, effects that are likely driven by nicotine.

Keywords: Brain, DNA methylation, Epigenetics, Fetal, Neuron, Nicotine, Neurodevelopment, Prenatal, Smoking, Tobacco

Background

Numerous studies have established that maternal smoking during pregnancy is associated with impaired executive function and behavioral problems in the offspring [1–3]. Maternal smoking is associated with altered fetal brain development [4] and reduced volumes of cortical gray matter in childhood [5], indicating that exposure to tobacco smoke constituents in utero may impact brain development and subsequently result in neurodevelopmental abnormalities. Offspring exposed to smoking

*Correspondence: fatemeh.haghighi@mssm.edu
[2] Department of Neuroscience, Icahn School of Medicine at Mount Sinai, 1425 Madison Ave, Floor 10, Room 10-70D, New York, NY 10029, USA
Full list of author information is available at the end of the article

after birth does not exhibit the same adverse trajectories [6, 7], suggesting biologically mediated mechanisms during gestation. Cigarette smoke is a highly complex mixture of more than 5000 chemicals of which approximately 100 are known to be hazardous [8]. Linking specific compound(s) with defined phenotypes has proven difficult. Indirect biological mechanisms caused by cigarette constituents other than nicotine have been proposed, including hypoxia/ischemia and DNA damage [9]. Exposure to nicotine prenatally has a direct impact on brain development. In rodents, prenatal exposure to nicotine is reported to induce abnormal dendritic morphology and reduced synapse density in the cerebral cortex and nucleus accumbens [10]. Additionally, prenatal nicotine exposure during primate brain development up-regulates nicotinic acetylcholine receptors (nAChR), causes cell death, and alters cell size and neurite outgrowth in a regionally dependent manner [11]. Furthermore, nicotine replacement therapy has been suggested to increase the risk for behavioral impairments (for review see [12–17]).

Prenatal exposure to environmental factors such as alcohol [18] and industrial chemicals (lead, methylmercury, PCBs, reviewed in [19, 20]) often manifests as neurodevelopmental disorders. Epigenetic modifications such as DNA methylation regulate gene activity necessary for cell differentiation [21]. Exposure to tobacco smoke can induce alterations in epigenetic patterning that are associated with a wide spectrum of human diseases including cardiovascular, pulmonary, neurobehavioral disorders and cancer [22–28]. Maternal smoking during pregnancy alters DNA methylation in the blood of newborns [29] and can cause DNA methylation changes that persist into childhood [30]. In relation to neurological function, differences in DNA methylation have been reported between offspring of smokers and non-smokers in the promoters of catechol-O-methyltransferase (*COMT*) and monoamine oxidase A (*MAOA*), genes thought to be involved in nicotine dependence and other neurobehavioral disorders [31, 32]. Further, an increase in the DNA methylation of the brain-derived neurotrophic factor-6 (*BDNF-6*) promoter/5'UTR has been found in adolescents exposed to maternal smoking during pregnancy [33]. To our knowledge, no studies have directly examined the epigenetic changes of the developing human brain exposed in utero to maternal cigarette smoking. Here we interrogate DNA methylation patterns in the developing cortex of human fetuses exposed to maternal smoking on a genome scale.

Methods
Sample selection
Fetuses were from second-trimester elective saline abortions performed for non-medical reasons. Fetal sample groups, exposed ($N = 14$) and unexposed ($N = 10$) to maternal smoking, were balanced for sex and gestational age (weeks since the first day of last normal menstrual period) (Table 1). All mothers of the exposed group were active smokers prior to and during pregnancy, whereas no mothers of the unexposed group were active smokers prior to or during pregnancy (Additional file 4: Table S1). Alcohol is a well-described teratogen that affects neuroanatomical development [18, 34]. No mother reported any alcohol abuse or dependence prior to or during pregnancy. A higher proportion of mothers reported consuming some ("any") alcohol during pregnancy in the unexposed (60%) then exposed (29%) groups, although this difference was not significant (p value = 0.12, Chi-Square, Additional file 4: Table S1). All samples were identified as Caucasians.

Sample dissection and processing
Upon delivery, the products of conception were refrigerated, and within hours, they were moved to a −80 °C freezer. For examination, they were placed at −20 °C overnight. Working quickly over dry ice, the brain was removed without thawing. The cortical plate was sampled in the region that becomes the DLPFC in order to obtain post-migratory NeuN-immunoreactive (NeuN$^+$) neurons, which normally become numerous in cortical layers 4–6 between 14 and 20 weeks gestational age, and in layers 2 and 3 between 20 and 24 weeks (Additional file 1: Figure S1) [35]. This region was chosen because it is involved in decision-making and working memory, and its function is compromised in neurodevelopmental and psychiatric conditions, including autism spectrum disorder (ASD). It is readily identified and accessible in second-trimester fetal brain. During the second trimester, the cerebral hemispheric wall in the frontal region grows

Table 1 Fetal cortical samples dissected from the second trimester (ST) of gestation

	Early ST			Late ST			Total
	N	Age in wpc (mean ± SD)	M/F	N	Age in wpc (mean ± SD)	M/F	
Exposed	9	16.63 ± 0.52	5/4	5	22.4 ± 1.14	3/2	14
Unexposed	6	16.67 ± 0.52	3/3	4	23.25 ± 0.96	2/2	10

Fetal sample groups, exposed and unexposed to maternal smoking, were balanced for gestational age and sex from a thickness of ~2 mm at 12 weeks to ~6–8 mm at 18 weeks and ~18 mm at 26 weeks, with cortical plate thickness of ~0.5–1, ~1.5, and ~2 mm, respectively [36–38]. We obtained tissue from the cortical plate from frozen fetal brains by scraping the dorsal prefrontal region of the left hemisphere to a depth of approximately 0.5 mm for the youngest fetuses, where there was no gross demarcation between plate and subplate. In the older fetuses, we were guided by a change in color at the junction of the cortical plate and subplate at approximately the predicted depth. These sample specimens for DNA methylation and gene expression assays were stored at −80 °C for further processing.

Human induced Neuronal Precursor Cells (hiNPC)

All hiNPC lines were derived as previously described [39]. To match the in vivo data generated from postmortem studies, hiNPC lines (NSB553-3-C, NSB2607-4-1 and NSB690-2-1) used in this study were derived from three Caucasian males, and for full details of the donors of the fibroblasts and validation of the hiPSC and NPC lines, please see [40]. Cell culture; NPCs were maintained at high density, grown on growth factor-reduced Matrigel (BD Biosciences)-coated plates in NPC media (Dulbecco's Modified Eagle Medium/Ham's F12 Nutrient Mixture (ThermoFisher Scientific), 1× N2, 1× B27-RA (ThermoFisher Scientific) and 20 ng/ml^{-1} FGF2 and split 1:3 every week with Accutase (Millipore, Billerica, MA, USA). Neural differentiation; NPCs were dissociated with Accutase and plated at 2.0×10^5 cells per cm^2 in NPC media onto growth factor-reduced Matrigel-coated plates. For neuronal differentiation, medium was changed to neural differentiation medium (DMEM/F12, 1× N2, 1× B27-RA, 20 ng/ml^{-1} BDNF (Peprotech), 20 ng/ml^{-1} GDNF (Peprotech), 1 mM dibutyryl-cyclic AMP (Sigma), 200 nM ascorbic acid (Sigma) 1–2 days later. NPC-derived neurons were differentiated for 3 and 6 weeks before being assayed.

Nicotine treatment

Nicotine (N0267-100MG, Sigma) was diluted at three different concentrations [100 nM (low), 10 µM (med) and 1 mM (high)] in neuronal media and added every 2nd day with a complete media change. Control wells were treated with equal volume of vehicle (ethanol) added to neuronal media.

hiNPC assays
Toxicity

The cell impermeant nuclei dye TO-PRO3® (ThermoFisher Scientific, T3605) was added at week 3 of differentiation. Three plates, each containing triplicates of a hiNPC line, were imaged with an Odyssey® infrared

imaging system (LI-COR). TO-PRO3® fluorescence intensity was normalized to control (vehicle treated) wells.

Immunofluorescence analysis

At 3 and 6 weeks of differentiation, cells were washed once with 1× PBS and then fixed in 4% paraformaldehyde (Electron Microscopy Services) for 15 min. Following 3 washes with 1× PBS, cells were then blocked and permeabilized with 1% v/v BSA Fraction V (BSA, ThermoFisher Scientific) with 0.3% v/v Triton-X 100 (T-100X, Sigma). Primary antibodies (Rb-Ki67, 1:500, Abcam, ab15580 and Ms-TUJ1, 1:1000, Covance, MMS-435P) were added overnight in 1%BSA/0.5%T-100X. Appropriate secondary antibodies (AlexaFluor Dk secondaries, Ms-680 and Rb-800) were incubated for 2.5 h in 1%BSA/0.5%T-100X. Following 3 washes with 1× PBS, plates were imaged with an Odyssey® infrared imaging system (LI-COR). Fluorescence intensity was normalized to control wells. Statistical differences between nicotine-treated and vehicle-treated controls were determined by Student's t test using R Language 3.03 [41].

Illumina Infinium Human Methylation BeadChip sample processing

DNA from fetal brains and hiNPCs were isolated and bisulfite converted (Zymo Research), and CpG methylation was determined using Illumina Infinium Human Methylation BeadChip microarrays (HM450), as described previously [42].

DNA methylation data preprocessing

The analyses were performed using R Language 3.03 [41] an environment for statistical computing and Bioconductor 2.13 [43]. Raw data files (.idat) were processed by minfi package [44]. All samples displayed a mean probe-wise detection call for the 485,512 array probes <0.0005. The data were normalized, background subtracted and further normalized by SWAN [45]. M values were used in feature selection models. Beta values (logistic transformed M values) were used for sample sex determination and DNA methylation reporting. Probes mapping to multiple locations ($N = 19,834$), Infinium type I probes with a SNP at the interrogated CpG ($N = 13,708$) and probes mapping to the X- and Y-chromosomes were removed from analysis ($N = 11,648$), as described [46], leaving 452,930 analyzable probes.

DNA methylation analysis

Differentially methylated probes (DMPs) display a mean difference in DNA methylation of at least 20%, corresponding to a methylation difference detectable by the HM450 with 99% confidence [47]. DMPs were mapped to refSeq gene annotations and analyzed using Ingenuity Pathways Analysis (IPA) software (Ingenuity Systems, www.ingenuity.com). Differentially methylated regions (DMR) were found using the bumphunter algorithm applied to DNA methylation M values [48]. Specifically, for each CpG site, we estimate the difference between the M values for the exposed and unexposed adjusting for gestational age, sex and sample chip assignment. An interaction term was included between smoking exposure and gestational age for interaction DMR analysis. The methylation difference estimates are smoothed based on the predefined CpG clusters where the maximal gap between neighboring CpG sites is 500 bp, while the largest cluster size is set to 1500 bp. The smoothed regional methylation difference estimates were obtained using a predefined threshold to identify the putative DMRs, with associated significance levels obtained empirically based on 1000 permutations. Cell-proportion estimates were performed using the methods described in Jaffe et al. [49]. Briefly, publicly available HM450 data from ESC-derived NPC (H9) [50], adult cortical NeuN$^+$ and NeuN$^-$ cells [51] were quantile normalized together [44] and 227 unique probes that separated the 3 cell types were used in a nonlinear mixed modeling [52] to estimate the proportion of each of the 3 cell types within our HM450 fetal dataset. Cell-proportion estimates were also generated for publicly available HM450 data from dissected postnatal DLPFC aged 4, 6 and 10 months, produced by the BrainSpan Consortium [53].

Gene expression analysis
Total RNA was isolated from the same 24 fetal samples used for DNA methylation analysis (ToTALLY RNA™ Total RNA Isolation Kit, Ambion). Fetal mRNA was analyzed using Nanostring nCounter Elements technology. Gene expression analysis of fetal DLPFC samples was performed for the 2 most significant smoking-DMRs (SDHAP3 and GNA15) and 3 of the 5 genes annotated to the most significant interaction DMRs (C21orf56, POLDIP3 and SYCE3). Housekeeping gene selection: We used the Nanostring nCounter Elements technology and selected 4 housekeeping genes for expression normalization. Previously, Penna et al. [54] investigated the stability of a panel of housekeeping genes for mRNA normalization in human postmortem brain samples. Additionally, Madden et al. [55] described a subset of ubiquitously expressed transcripts ideal for using as housekeeping genes within brain tissue. We selected 4 housekeeping genes, 3 of which were identified by both Penna et al. and Madden et al. (*GAPDH*, *YWHAZ* and *CYC*) and *SDHA*, identified by the former group and that we have

previously used successfully in Nanostring interrogation of rat mRNA [56]. Negative control subtraction and normalization to housekeeping genes was performed using the nSolver Analysis Software. Sample fold-changes (FC) were calculated relative to sample FS5777 (one of the 24 fetal samples analyzed chosen at random) gene expression levels for each gene independently. Any expression values of 3 standard deviations from the group mean were deemed outliers and removed from the analysis. No more than 1 result for any assay was removed.

Results
Cortical sampling
Fetal samples exposed to maternal smoking were matched to unexposed fetal samples by age and sex when available (Table 1). Fetal brain weight and total weight were highly correlated ($R^2 = 0.94$, Additional file 1: Figure S1a), and within either early or late second-trimester samples, no significant difference in brain weight was observed between exposure groups (p value = 0.3 and 0.8, respectively, Student's t test) (Additional file 1: Figure S1b, c). The cortical plate was sampled from the presumptive DLPFC in an effort to obtain post-migratory NeuN$^+$ neurons, which normally become numerous in cortical layers 4–6 between 14 and 20 weeks gestational age, and in layers 2 and 3 between 20 and 24 weeks [57].

Maternal smoking-associated differential DNA methylation in the fetal cortex
To examine DMRs associated with maternal smoking exposure in the fetal cortex, we performed DNA methylation microarray analysis among exposure groups ("Methods" section). Across the 452930 CpG sites examined, no DMRs passed multiple testing correction (family wise error rate, fwer, cutoff ≤ 0.05, Fig. 1a). This was likely due to the small number of fetal brains available for analysis. Notably, smoking exposure DMRs with the highest point-wise significance was found within the gene promoters of *SDHAP3* and G protein subunit Alpha 15 (*GNA15*) (Fig. 1b, c). Both DMRs were hypomethylated in smoking exposed (Fig. 1b, c). Gene expression analysis of SDHAP3 and *GNA15* (Fig. 1d, e) did not reveal exposure-related differences. However, we observed up-regulation of mRNA between early and late second-trimester samples that were restricted to smoking-exposed fetuses for expression of *SDHAP3* and larger for exposed than for unexposed for expression of *GNA15* (p value = 0.005 vs 0.02, Fig. 1e). These observations led us to hypothesize that maternal smoking exposure has temporal effects on gene expression and DNA methylation during fetal cortical development.

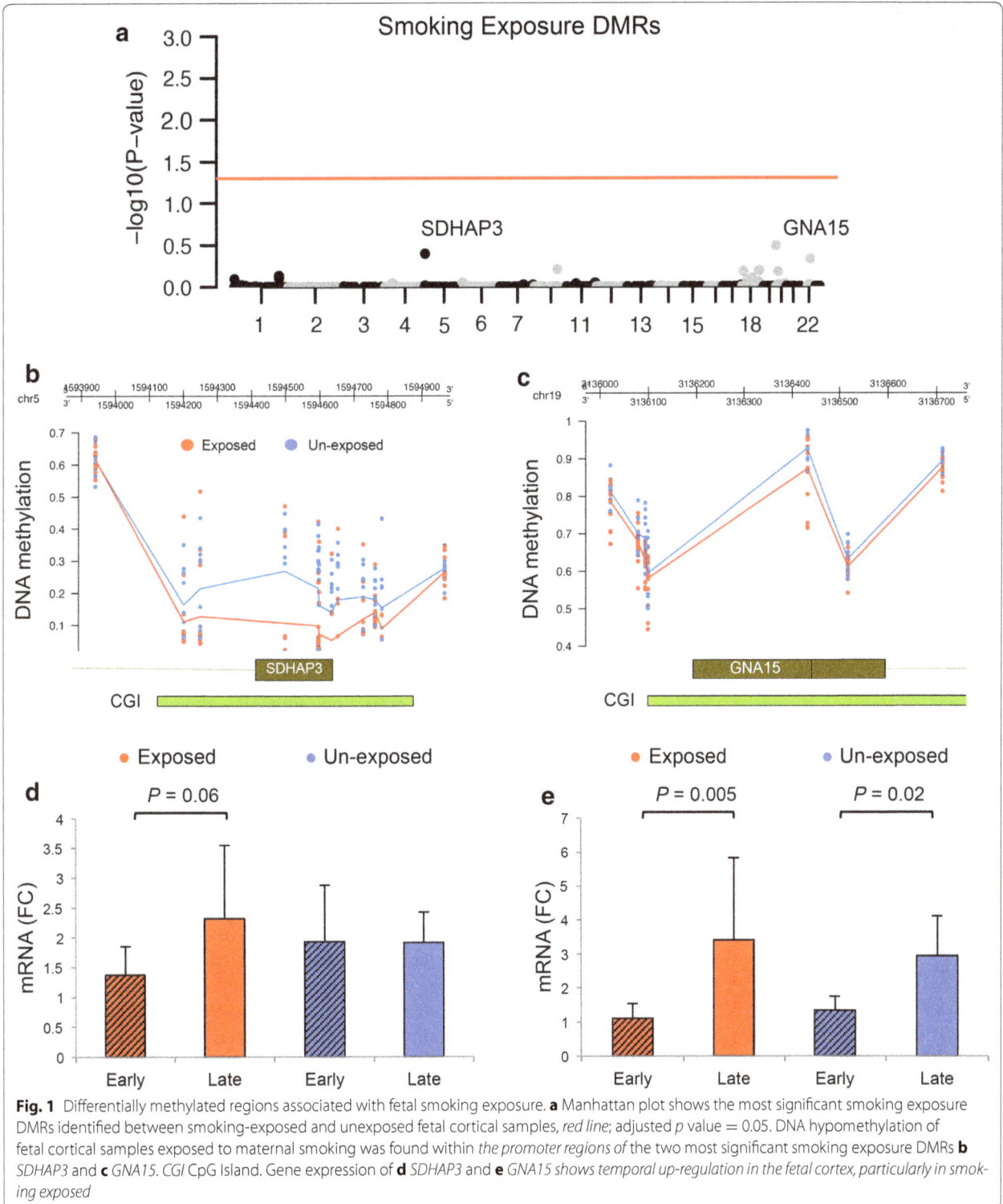

Fig. 1 Differentially methylated regions associated with fetal smoking exposure. **a** Manhattan plot shows the most significant smoking exposure DMRs identified between smoking-exposed and unexposed fetal cortical samples, *red line*; adjusted *p* value = 0.05. DNA hypomethylation of fetal cortical samples exposed to maternal smoking was found within *the promoter regions* of the two most significant smoking exposure DMRs **b** *SDHAP3* and **c** *GNA15*. *CGI* CpG Island. Gene expression of **d** *SDHAP3* and **e** *GNA15 shows temporal up-regulation in the fetal cortex, particularly in smoking exposed*

Developmental interaction with maternal smoking exposure

To explore the effects of smoking exposure on fetal development, we performed an interaction DMR analysis between gestational age and smoking exposure ("Methods" section). Four significant DMRs (fwer *p* value ≤ 0.05) located within the promoters of synaptonemal complex central element protein 3 (*SYCE3*), chromosome

21 open reading frame 56 (*C21orf56/LSS*), sperm-associated antigen 1 (*SPAG1*) and *RNA,* U12 Small Nuclear (*RNU12/POLDIP3*) (Fig. 2a) were identified. Within the promoter region of *C21orf56*, exposed fetal cortices at 22.4 weeks (late) exhibited higher total DNA methylation levels than samples examined at 16.6 weeks (early). Conversely, unexposed fetal cortices displayed higher

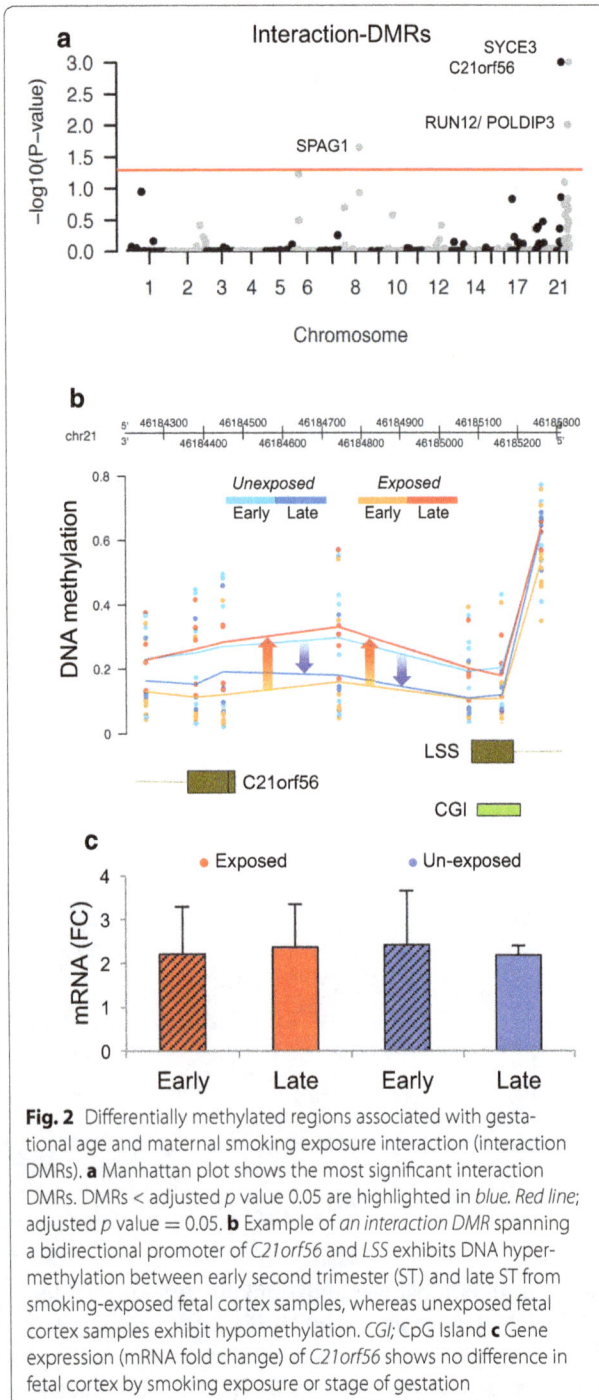

Fig. 2 Differentially methylated regions associated with gestational age and maternal smoking exposure interaction (interaction DMRs). **a** Manhattan plot shows the most significant interaction DMRs. DMRs < adjusted *p* value 0.05 are highlighted in *blue*. *Red line*; adjusted *p* value = 0.05. **b** Example of *an interaction DMR* spanning a bidirectional promoter of *C21orf56* and *LSS* exhibits DNA hyper-methylation between early second trimester (ST) and late ST from smoking-exposed fetal cortex samples, whereas unexposed fetal cortex samples exhibit hypomethylation. *CGI*; CpG Island **c** Gene expression (mRNA fold change) of *C21orf56* shows no difference in fetal cortex by smoking exposure or stage of gestation

DNA methylation levels in early samples compared to late (Fig. 2b). Additionally, we found an intragenic region of *SPAG1* that was conservatively hypomethylated in all samples except a subset of late second-trimester exposed fetal cortices that were hypermethylated (Additional file 2: Figure S2). Nanostring expression analysis of *C21orf56* demonstrated no difference in mRNA levels between exposure/developmental groups (Fig. 2c).

No discernable difference in group-wise DNA methylation patterns by exposure/development was evident in the promoter regions of *RUN12/POLDIP3* or *SYCE3*, regions that had been identified using the bumphunter algorithm as significant (Additional file 3: Figure S3a, b). Gene expression analysis of *SYCE3* revealed low gene expression in early second-trimester smoking exposed (p value = 0.02) (Additional file 3: Figure S3c), analogous to the *SDHAP3* mRNA results indicative of developmental delay in smoking exposed. No difference in *POLDIP3* mRNA was found between exposure/development (Additional file 3: Figure S3d).

Global DNA methylation of fetal cortex exposed to maternal smoking

We examined global DNA methylation changes during cortical development in response to maternal smoking by calculating differentially methylated probes (DMPs) between early and late second-trimester samples in each exposure group (methods). We identified 574/371 hyper/hypomethylated DMPs (Additional file 4: Table S2) in unexposed samples and 399/178 hyper/hypomethylated DMPs (Additional file 4: Table S3) in exposed samples (Fig. 3a). Unsupervised hierarchical clustering of both hypomethylated and hypermethylated DMPs separated samples by gestational age (Fig. 3a). More DMPs were found between early and late second-trimester unexposed samples than between smoking-exposed samples. Notably, a higher proportion of these DMPs were hypermethylated in the unexposed (64%) compared to exposed samples (44%). Hypomethylation is a feature of pluripotent/multipotent stem cells and as cells differentiate, the acquisition of DNA methylation restricts cell lineage and drives cell specification [58]. We postulated that reduced hypermethylated DMPs in smoking-exposed samples could reflect alterations or delay in cell-type specification occurring in the developing cortices in response to smoking exposure. Gene ontology analysis implicated canonical pathways of "Role of NFAT in Cardiac Hypertrophy," "Th2 Pathway" and "Th1 and Th2 Activation Pathway" associated with genes annotated to DMPs between early and late second-trimester unexposed fetal samples (Additional file 4: Table S4). NFAT is a transcription factor that mediates axon growth in developing neurons (reviewed in [59]). Conversely, canonical pathways associated with

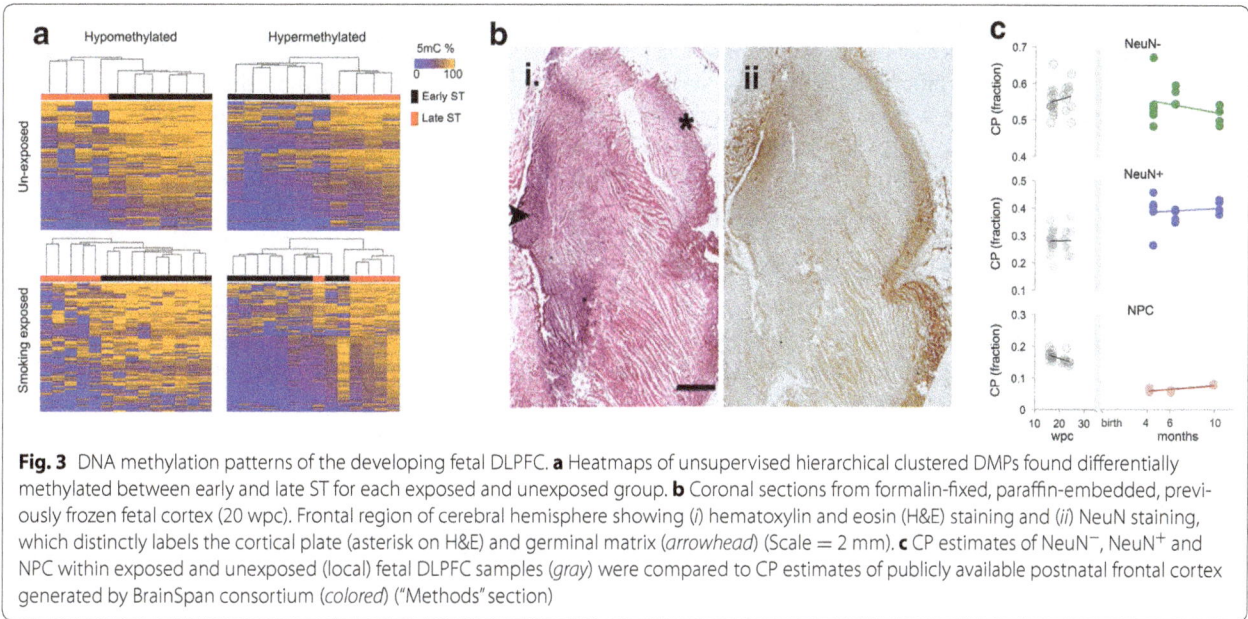

Fig. 3 DNA methylation patterns of the developing fetal DLPFC. **a** Heatmaps of unsupervised hierarchical clustered DMPs found differentially methylated between early and late ST for each exposed and unexposed group. **b** Coronal sections from formalin-fixed, paraffin-embedded, previously frozen fetal cortex (20 wpc). Frontal region of cerebral hemisphere showing (*i*) hematoxylin and eosin (H&E) staining and (*ii*) NeuN staining, which distinctly labels the cortical plate (asterisk on H&E) and germinal matrix (*arrowhead*) (Scale = 2 mm). **c** CP estimates of NeuN⁻, NeuN⁺ and NPC within exposed and unexposed (local) fetal DLPFC samples (*gray*) were compared to CP estimates of publicly available postnatal frontal cortex generated by BrainSpan consortium (*colored*) ("Methods" section)

genes annotated to DMPs between early and late second-trimester exposed fetal samples implicated "T Helper Cell Differentiation," "Epithelial Adherens Junction Signaling" and "Factors Promoting Cardiogenesis in Vertebrates" (Additional file 4: Table S4).

Our cortical plate sectioning of the presumptive DLPFC aimed to enrich for post-migratory NeuN⁺ neurons (Fig. 3b). Other investigators have established cell deconvolution algorithms with remarkable accuracy in estimating NeuN⁺ proportions of whole brain tissues using DNA methylation profiles [51, 60]. Using the DNA methylation profiles generated from the fetal cortices, we were able to estimate the cell proportions (CP estimates) of NeuN⁺, NeuN⁻ and neural precursor cells (NPCs) within the fetal DLPFC sections ("Methods" section). CP estimates revealed our fetal cortical sections contained a high proportion of NPCs (mean = 16%) compared to CP estimates of NPCs of cortical sections (postnatal 4–10 months) produced by the BrainSpan consortium (mean = 7%) [50, 53, 57] (Fig. 3c). These NPCs are presumably undergoing maturation, and thus, our sections provide a rare window into the effects of maternal smoking exposure on neuro-cellular development.

CP estimates of fetal DLPFC revealed significantly fewer NeuN⁺ in smoking-exposed fetal DLPFC (*p* value = 0.04, Fig. 4). Any decrease in CP estimate will be balanced by an increase in the proportion of another cellular population, and indeed, we observed a higher proportion of NeuN⁻ in smoking-exposed fetal samples (*p* value = 0.08, Fig. 4). No difference in NPC proportions was observed between smoking exposure groups (Fig. 4).

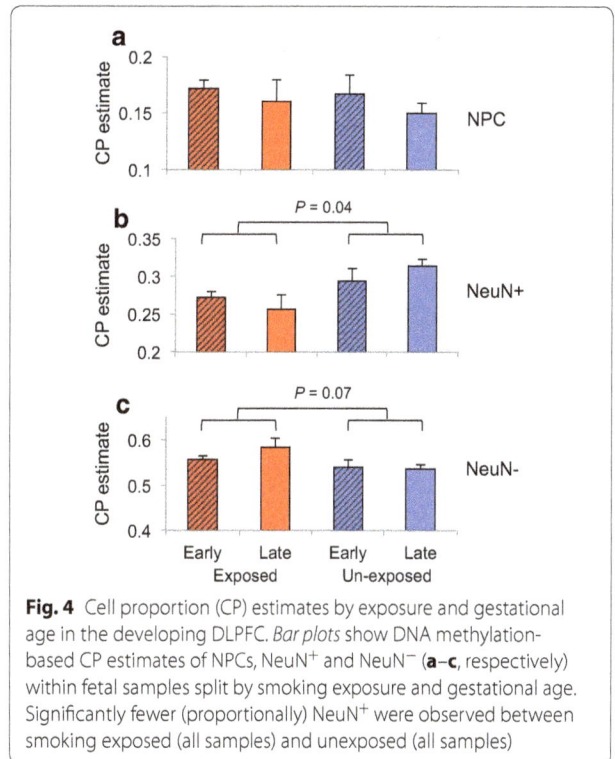

Fig. 4 Cell proportion (CP) estimates by exposure and gestational age in the developing DLPFC. *Bar plots* show DNA methylation-based CP estimates of NPCs, NeuN⁺ and NeuN⁻ (**a–c**, respectively) within fetal samples split by smoking exposure and gestational age. Significantly fewer (proportionally) NeuN⁺ were observed between smoking exposed (all samples) and unexposed (all samples)

The results indicate that tobacco exposure is associated with a reduction in NeuN⁺ in the developing DLPFC. Using ReFACTor [61] method that adjusts for the possible effect of variation in cell-type proportions, we found no difference in methylation associated with smoking exposure or developmental stage interaction. The methylation change associated with smoking exposure reflects differences in rate of cell differentiation during development as observed by the estimated CP changes. To investigate the mechanism of nicotine exposure on neuron development, we turned to an in vitro model of development.

In vitro modeling of neurodevelopment in response to nicotine exposure

To model the effects of nicotine on human neuronal development, we exposed human induced neural precursor cells (hiNPCs) to nicotine over a 6-week period of neuronal differentiation. Following 3 weeks of nicotine exposure, differentiating hiNPCs exhibited a dosage-dependent increase in TO-PRO®-3 fluorescence (live cells are impermeable to the dye, whereas the dye penetrates compromised membranes characteristic of dead cells) (p value <0.05) and an increase in cell proliferation (*Ki67*) that was significant at high dose (p value = 0.01, Fig. 5). We observed a decrease in the staining of βIII Tubulin (*TUJ1*), a marker for mature neurons, and notably, the most significant decrease was found at 0.1 uM (low) nicotine exposure (p value = 0.0003, Fig. 5). No significant differences in cell death or proliferation were detectable by 6 weeks; however, *TUJ1* showed a dosage-dependent increase that was significant at 1000 μM nicotine (p value = 0.03, Fig. 5). These data demonstrate that nicotine elicits dosage-dependent effects on neuronal maturation. Interestingly, the biologically meaningful

nicotine exposure (0.1 μM) restricts early neuron development despite exhibiting a lower toxicity (cell death) than higher exposures. Further, the differences in *TUJ1* staining between 3 and 6 weeks indicate that less differentiated hiNPCs (3 weeks) are more susceptible to the inhibiting effects of nicotine on neuron development (Fig. 5).

Discussion

In this study, we profiled genome-scale DNA methylation of fetal brain development in response to exposure to maternal smoking in utero. Although limited in number, the fetal brain samples have well-characterized maternal health history and exposure data, thus providing a rare opportunity to investigate the impact of nicotine exposure on early human brain development. Notably, our DNA methylation profiling was performed on whole tissue sections from the developing DLPFC that consist of a mixture of neuronal and non-neuronal cell types. Established methods enable the isolation of neuronal nuclei [62]; however, due to the fragility of fetal neuronal nuclei, this technique cannot be applied. Estimating cell proportions using DNA methylation profiles revealed our fetal DLPFC sections contained approximately 16% NPCs that are presumably undergoing maturation, providing a rare window into human neuro-cellular development in response to maternal smoking exposure.

Embryogenesis is a stage of rapid neurological transformation and growth in which epigenomic landscapes undergo dramatic change [63, 64]. In the developing

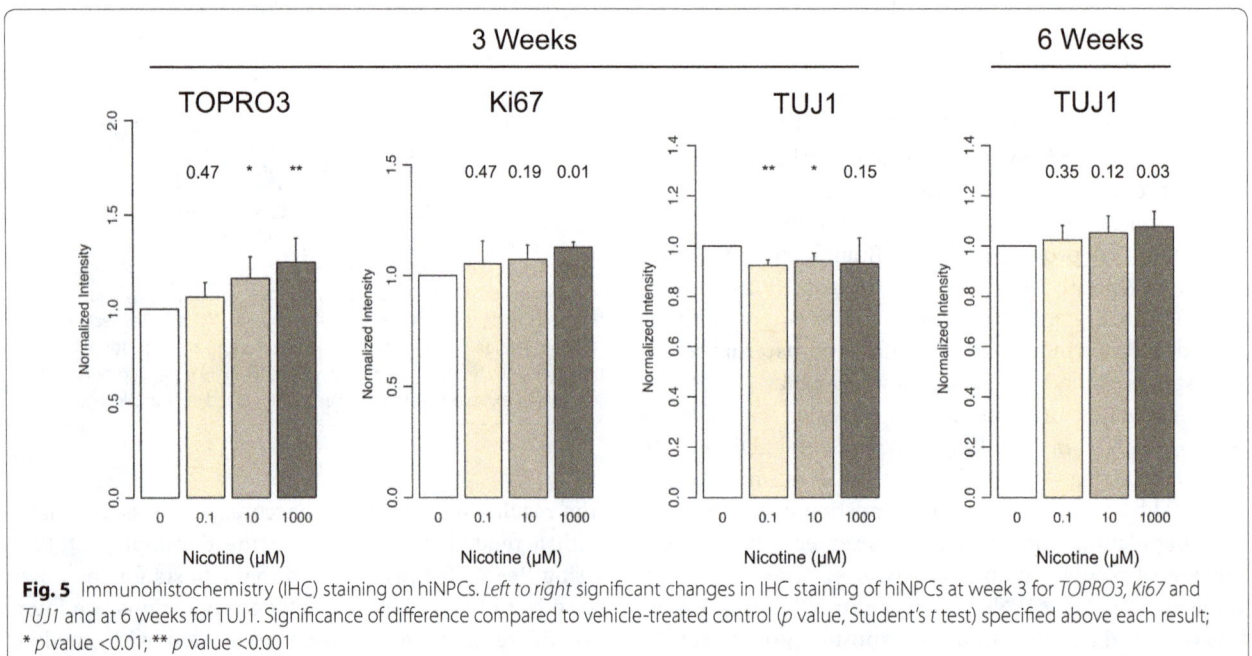

Fig. 5 Immunohistochemistry (IHC) staining on hiNPCs. *Left to right* significant changes in IHC staining of hiNPCs at week 3 for *TOPRO3, Ki67* and *TUJ1* and at 6 weeks for TUJ1. Significance of difference compared to vehicle-treated control (p value, Student's *t* test) specified above each result; * p value <0.01; ** p value <0.001

DLPFC, genome-wide DNA methylation changes that distinguish early and late second-trimester samples were clearly reduced in fetal cortex of smoking exposed indicating alterations in cell-type differentiation. Indeed, our most significant DMRs were identified by interaction analysis between gestational age and maternal smoking exposure. We found a delay in the up-regulation of expression of *SYCE3*, a gene that is conserved among mammals and whose loss leads to a block in synapsis initiation resulting in meiotic arrest [65]. We also identified an interaction DMR in the promoter of *C21orf56* (also known as *SPATC1L*). *C21orf56* is a spermatogenesis and centriole-associated 1-like gene found on chromosome 21 of little-known function. Although we did not observe developmental/exposure dependent changes in gene expression, we observed a gain in promoter methylation in late second-trimester exposed samples that could reduce the transcriptional potential of *C21orf56* later in development.

Genes such as *GNA15* and *SDHAP3* that contained maternal smoking-associated DMRs displayed a developmental delay in mRNA up-regulation in smoking exposed. Notably, *SDHAP3* is a subunit of the succinate dehydrogenase complex located within the mitochondrial membrane and functions in electron transport chain transfer of electrons to coenzyme Q [66]. It has been reported that mutations within succinate dehydrogenase subunits actually increase levels of oxidative stress [67]. Intriguingly, this same DMR was recently found hypermethylated in the cerebellum of patients diagnosed with ASD [68] and differentially methylated in the DLPFC of patients diagnosed with schizophrenia (SCZ) within 3 independent studies [69]. Furthermore, in a separate report, *GNA15* was found to be differentially methylated in the PFC of ASD patients [70]. Both ASD and SCZ probably have prenatal origins [71–73]. Taken together, these results reveal a potential link between maternal smoking-associated DNA methylation perturbation and potential increase risk for neurodevelopmental abnormalities. Notably, *GNA15* is transcriptionally modifiable by acute doses of nicotine in neuroblastoma cell lines [74], indicating nicotine as a potential causative agent.

Cell deconvolution algorithms have shown remarkable accuracy in estimating NeuN$^+$ proportions from DNA methylation profiles from whole brain tissue [51, 60]. CP estimates within our fetal DLPFC revealed a smoking exposure-associated reduction in NeuN$^+$ cells supporting previous observations of reduced gray matter in the cortex of smoking-exposed children [5]. The adverse effects of maternal smoking on fetal development are well described; however, it was estimated that 30% of smokers attempting to quit smoking use cessation aids that contain nicotine [75]. Nicotine is a well-studied substance in tobacco and has been shown to induce oxidative stress in rodent [76, 77] and human neurons [78]. Exposure of hiNPCs to 100 nM nicotine resulted in the lowest amount of toxicity but the greatest suppression of neuronal differentiation (B3-Tubulin). These results recapitulate the reduction in the estimated proportion of NeuN$^+$ cells we observed in human samples and implicate nicotine as a causative agent *in* impeding neuronal development. These data provide direct evidence from primary tissue of in utero exposure to teratogenic agents as found in cigarettes—warranting further investigations of the in utero environment on fetal development and how it impacts offspring health and disease risk through the lifespan.

Conclusions

In summary, we have found evidence that intrauterine smoking exposure alters the developmental patterning of DNA methylation and gene expression and is associated with reduced mature neuronal content, effects that are likely driven by nicotine through mechanisms independent of cell death.

Additional files

Additional file 1: Fig. S1. Fetal brain weight and total body weight of fetus. **a** Correlation between fetal brain weight and total fetal body weight. The change in total body weight (**b**) and brain weight (**c**) in exposed and unexposed fetal samples between early ST and late ST. * *p* value <0.05, ** *p* value >0.1. Error bars = SD.

Additional file 2: Fig. S2. DNA methylation of fetal samples by exposure and gestational age of the significant intergenic interaction DMR found within *SPAG1*. CGI; CpG Island.

Additional file 3: Fig. S3. DNA methylation of fetal DLPFC by exposure and gestational age within interaction DMRs annotated to the promoters of **a** *SYCE3* and **b** *POLDIP3/RNU12*. CGI CpG Island. *Gene expression* analysis revealed temporal up-regulation of c *SYCE3* in smoking *exposed*; however, gene expression analysis of **d** *POLDIP3/RNU12* shows no significant difference between smoking-exposed or unexposed early ST and late ST.

Additional file 4: Table S1. Results of mothers self-reported family history of environmental exposures prior and during pregnancy. **Table S2**. DMPs (late–early) smoking unexposed. **Table S3**. DMPs (late–early) smoking exposed. **Table S4**. Top (significance) canonical pathways associated with genes annotated to DMPs found between late and early ST within smoking-exposed and unexposed fetal samples.

Abbreviations

ASD: autism spectrum disorder; BDNF-6: brain-derived neurotrophic factor-6; C21orf56: chromosome 21 open reading frame 56; COMT: catechol-O-methyltransferase; CP: cell proportion; DLPFC: dorsolateral prefrontal cortex; DMP: differentially methylated positions; DMR: differentially methylated region; FDR: false discovery rate; GNA15: G protein subunit alpha 15; hiPSCs: human induced pluripotent stem cell; HM450: Illumina Infinium Human Methylation Microarray 450K platform; IHC: immunohistochemistry; MAOA: monoamine oxidase A; nAChR: nicotinic acetylcholine receptors; NeuN$^+$: NeuN-immunoreactive; NPC: neural progenitor cell; RNU12: RNA, U12 Small Nuclear; SCZ: schizophrenia; SPAG1: sperm-associated antigen 1; CGI: CpG Island.

Authors' information
Please see title page for author affiliations and contact information.

Authors' contributions
ZC, NM, SC, MM, YG and FH designed and performed DNA methylation and gene expression experiments and analysis. ZC, FH, BH, KB and SMH designed and performed cell culture studies. AD, GR, AS and ITI designed and collected mothers' history and fetal brain samples. AD performed fetal brain dissections. ZC, FH, AD, YG, KB, MM and BH wrote and edited the manuscript. All authors read and approved the final manuscript.

Author details
[1] Friedman Brain Institute, Icahn School of Medicine at Mount Sinai, 1425 Madison Avenue, New York, NY 10029, USA. [2] Department of Neuroscience, Icahn School of Medicine at Mount Sinai, 1425 Madison Ave, Floor 10, Room 10-70D, New York, NY 10029, USA. [3] Department of Psychiatry, Icahn School of Medicine at Mount Sinai, 1425 Madison Avenue, New York, NY 10029, USA. [4] Department of Neurology, Icahn School of Medicine at Mount Sinai, 1425 Madison Avenue, New York, NY 10029, USA. [5] Department of Psychiatry, Columbia University, New York, NY 10032, USA. [6] Department of Pathology and Cell Biology, Columbia University, New York, NY 10032, USA. [7] Medical Epigenetics, James J. Peters VA Medical Center, Bronx, NY 10468, USA. [8] Macedonian Academy of Sciences and Arts, Skopje, Macedonia. [9] School of Medicine, Skopje, Macedonia. [10] Psychiatric Hospital Skopje, Skopje, Macedonia.

Acknowledgements
This work was supported in part through the computational resources and staff expertise provided by Scientific Computing at the Icahn School of Medicine at Mount Sinai.

Competing interests
The authors declare that they have no competing interests.

Funding
The Haghighi Laboratory is supported by the National Institute of Health (NIH) Grant R01MH094774. ZC is supported by a NIDA T32 training grant in Drug Abuse Research from the NIH, USA. Kristen J. Brennand is a New York Stem Cell Foundation—Robertson Investigator. The Brennand Laboratory is supported by the Brain and Behavior Research Foundation, NIH Grants R01 MH101454 and R01 MH106056, and the New York Stem Cell Foundation. Research reported in this paper was supported by the Office of Research Infrastructure of the National Institutes of Health under award number S10OD018522. The content is solely the responsibility of the authors and does not necessarily represent the official views of the National Institutes of Health.

References
1. Cornelius MD, et al. Effects of prenatal cigarette smoke exposure on neurobehavioral outcomes in 10-year-old children of adolescent mothers. Neurotoxicol Teratol. 2011;33(1):137–44.
2. Huijbregts SC, et al. Hot and cool forms of inhibitory control and externalizing behavior in children of mothers who smoked during pregnancy: an exploratory study. J Abnorm Child Psychol. 2008;36(3):323–33.
3. Robinson M, et al. Smoking cessation in pregnancy and the risk of child behavioural problems: a longitudinal prospective cohort study. J Epidemiol Community Health. 2010;64(7):622–9.
4. Roza SJ, et al. Effects of maternal smoking in pregnancy on prenatal brain development. The Generation R Study. Eur J Neurosci. 2007;25(3):611–7.
5. El Marroun H, et al. Prenatal tobacco exposure and brain morphology: a prospective study in young children. Neuropsychopharmacology. 2014;39(4):792–800.
6. Eskenazi B, Bergmann JJ. Passive and active maternal smoking during pregnancy, as measured by serum cotinine, and postnatal smoke exposure. I. Effects on physical growth at age 5 years. Am J Epidemiol. 1995;142(9 Suppl):S10-8.
7. Wakschlag LS, et al. Maternal smoking during pregnancy and severe antisocial behavior in offspring: a review. Am J Public Health. 2002;92(6):966–74.
8. Talhout R, et al. Hazardous compounds in tobacco smoke. Int J Environ Res Public Health. 2011;8(2):613–28.
9. Li Y, et al. Perinatal nicotine exposure increases vulnerability of hypoxic–ischemic brain injury in neonatal rats: role of angiotensin II receptors. Stroke J Cereb Circ. 2012;43(9):2483–90.
10. Mychasiuk R, et al. Long-term alterations to dendritic morphology and spine density associated with prenatal exposure to nicotine. Brain Res. 2013;1499:53–60.
11. Slotkin TA, et al. Effects of prenatal nicotine exposure on primate brain development and attempted amelioration with supplemental choline or vitamin C: neurotransmitter receptors, cell signaling and cell development biomarkers in fetal brain regions of rhesus monkeys. Neuropsychopharmacology. 2005;30(1):129–44.
12. Bruin JE, Gerstein HC, Holloway AC. Long-term consequences of fetal and neonatal nicotine exposure: a critical review. Toxicol Sci Off J Soc Toxicol. 2010;116(2):364–74.
13. Dwyer JB, McQuown SC, Leslie FM. The dynamic effects of nicotine on the developing brain. Pharmacol Ther. 2009;122(2):125–39.
14. Button TM, Maughan B, McGuffin P. The relationship of maternal smoking to psychological problems in the offspring. Early Human Dev. 2007;83(11):727–32.
15. Lim R, Sobey CG. Maternal nicotine exposure and fetal programming of vascular oxidative stress in adult offspring. Br J Pharmacol. 2011;164(5):1397–9.
16. Mahar I, et al. Developmental hippocampal neuroplasticity in a model of nicotine replacement therapy during pregnancy and breastfeeding. PLoS ONE. 2012;7(5):e37219.
17. Pauly JR, Slotkin TA. Maternal tobacco smoking, nicotine replacement and neurobehavioural development. Acta Paediatr. 2008;97(10):1331–7.
18. Jones KL, Smith DW. Recognition of the fetal alcohol syndrome in early infancy. Lancet. 1973;302(7836):999–1001.
19. Grandjean P, Landrigan PJ. Developmental neurotoxicity of industrial chemicals. Lancet. 2006;368(9553):2167–78.
20. Julvez J, Grandjean P. Neurodevelopmental toxicity risks due to occupational exposure to industrial chemicals during pregnancy. Ind Health. 2009;47(5):459–68.
21. Gidekel S, Bergman Y. A unique developmental pattern of Oct-3/4 DNA methylation is controlled by a cis-demodification element. J Biol Chem. 2002;277(37):34521–30.
22. Adcock IM, et al. Epigenetic regulation of airway inflammation. Curr Opin Immunol. 2007;19(6):694–700.
23. Furniss CS, et al. Line region hypomethylation is associated with lifestyle and differs by human papillomavirus status in head and neck squamous cell carcinomas. Cancer Epidemiol Biomarkers Prev. 2008;17(4):966–71.
24. Kanai Y. Alterations of DNA methylation and clinicopathological diversity of human cancers. Pathol Int. 2008;58(9):544–58.
25. Lawrence J, et al. Foetal nicotine exposure causes PKCepsilon gene repression by promoter methylation in rat hearts. Cardiovasc Res. 2011;89(1):89–97.
26. Prescott SL, Clifton V. Asthma and pregnancy: emerging evidence of epigenetic interactions in utero. Curr Opin Allergy Clin Immunol. 2009;9(5):417–26.
27. Satta R, et al. Nicotine decreases DNA methyltransferase 1 expression and glutamic acid decarboxylase 67 promoter methylation in GABAergic interneurons. Proc Natl Acad Sci USA. 2008;105(42):16356–61.
28. Zochbauer-Muller S, et al. Aberrant methylation of multiple genes in the upper aerodigestive tract epithelium of heavy smokers. Int J Cancer. 2003;107(4):612–6.

29. Joubert BR, et al. DNA methylation in newborns and maternal smoking in pregnancy: genome-wide consortium meta-analysis. Am J Hum Genet. 2016;98(4):680–96.

30. Ladd-Acosta C, et al. Presence of an epigenetic signature of prenatal cigarette smoke exposure in childhood. Environ Res. 2016;144(Pt A):139–48.

31. Philibert RA, et al. MAOA methylation is associated with nicotine and alcohol dependence in women. Am J Med Genet B Neuropsychiatr Genet. 2008;147B(5):565–70.

32. Xu Q, et al. Determination of methylated CpG sites in the promoter region of catechol-O-methyltransferase (COMT) and their involvement in the etiology of tobacco smoking. Front Psychiatry. 2010;1:16.

33. Toledo-Rodriguez M, et al. Maternal smoking during pregnancy is associated with epigenetic modifications of the Brain-derived neurotrophic factor-6 exon in adolescent offspring. Am J Med Genet B Neuropsychiatr Genet. 2010;153B(7):1350–4.

34. Roebuck TM, Mattson SN, Riley EP. A review of the neuroanatomical findings in children with fetal alcohol syndrome or prenatal exposure to alcohol. Alcohol Clin Exp Res. 1998;22(2):339–44.

35. Sarnat HB, Nochlin D, Born DE. Neuronal nuclear antigen (NeuN): a marker of neuronal maturation in early human fetal nervous system. Brain Dev. 1998;20(2):88–94.

36. Kostovic I, et al. Laminar organization of the human fetal cerebrum revealed by histochemical markers and magnetic resonance imaging. Cereb Cortex. 2002;12(5):536–44.

37. Rados M, Judas M, Kostovic I. In vitro MRI of brain development. Eur J Radiol. 2006;57(2):187–98.

38. Shepard TH, et al. Organ weight standards for human fetuses. Pediatr Pathol. 1988;8(5):513–24.

39. Lee IS, et al. Characterization of molecular and cellular phenotypes associated with a heterozygous CNTNAP2 deletion using patient-derived hiPSC neural cells. Npj Schizophr. 2015;1:15019.

40. Topol A, et al. Dysregulation of miRNA-9 in a subset of schizophrenia patient-derived neural progenitor cells. Cell Rep. 2016;15(5):1024–36.

41. R Development Core Team. R: a language and environment for statistical computing. R Foundation for Statistical Computing. 2014; http://www.r-project.org/.

42. Bibikova M, et al. High-throughput DNA methylation profiling using universal bead arrays. Genome Res. 2006;16(3):383–93.

43. Gentleman RC, et al. Bioconductor: open software development for computational biology and bioinformatics. Genome Biol. 2004;5(10):R80.

44. Aryee MJ, et al. Minfi: a flexible and comprehensive Bioconductor package for the analysis of Infinium DNA methylation microarrays. Bioinformatics. 2014;30(10):1363–9.

45. Maksimovic J, Gordon L, Oshlack A. SWAN: subset-quantile within array normalization for illumina infinium HumanMethylation450 BeadChips. Genome Biol. 2012;13(6):R44.

46. Naeem H, et al. Reducing the risk of false discovery enabling identification of biologically significant genome-wide methylation status using the HumanMethylation450 array. BMC Genom. 2014;15:51.

47. Bibikova M, et al. High density DNA methylation array with single CpG site resolution. Genomics. 2011;98(4):288–95.

48. Jaffe AE, et al. Bump hunting to identify differentially methylated regions in epigenetic epidemiology studies. Int J Epidemiol. 2012;41(1):200–9.

49. Jaffe AE, et al. Developmental regulation of human cortex transcription and its clinical relevance at single base resolution. Nat Neurosci. 2015;18(1):154–61.

50. Kim M, et al. Dynamic changes in DNA methylation and hydroxymethylation when hES cells undergo differentiation toward a neuronal lineage. Hum Mol Genet. 2014;23(3):657–67.

51. Guintivano J, Aryee MJ, Kaminsky ZA. A cell epigenotype specific model for the correction of brain cellular heterogeneity bias and its application to age, brain region and major depression. Epigenetics. 2013;8(3):290–302.

52. Houseman EA, et al. DNA methylation arrays as surrogate measures of cell mixture distribution. BMC Bioinform. 2012;13:86.

53. BrainSpan: Atlas of the developing human brain [Internet]. Funded by ARRA Awards 1RC2MH089921-01, R.M.-., and 1RC2MH089929-01. © 2011. http://developinghumanbrain.org.

54. Penna I, et al. Selection of candidate housekeeping genes for normalization in human postmortem brain samples. Int J Mol Sci. 2011;12(9):5461–70.

55. Velculescu VE, et al. Analysis of human transcriptomes. Nat Genet. 1999;23(4):387–8.

56. Haghighi F, et al. Neuronal DNA methylation profiling of blast-related traumatic brain injury. J Neurotrauma. 2015;32(16):1200–9.

57. Sasaki T, et al. Type of feeding during infancy and later development of schizophrenia. Schizophr Res. 2000;42:79–82.

58. Mohn F, et al. Lineage-specific polycomb targets and de novo DNA methylation define restriction and potential of neuronal progenitors. Mol Cell. 2008;30(6):755–66.

59. Nguyen T, Di Giovanni S. NFAT signaling in neural development and axon growth. Int J Dev Neurosci. 2008;26(2):141–5.

60. Montano CM, et al. Measuring cell-type specific differential methylation in human brain tissue. Genome Biol. 2013;14(8):R94.

61. Rahmani E, et al. Sparse PCA corrects for cell type heterogeneity in epigenome-wide association studies. Nat Methods. 2016;13(5):443–5.

62. Jiang Y, et al. Isolation of neuronal chromatin from brain tissue. BMC Neurosci. 2008;9:42.

63. Lister R, et al. Global epigenomic reconfiguration during mammalian brain development. Science. 2013;341(6146):1237905.

64. Numata S, et al. DNA methylation signatures in development and aging of the human prefrontal cortex. Am J Hum Genet. 2012;90(2):260–72.

65. Schramm S, et al. A novel mouse synaptonemal complex protein is essential for loading of central element proteins, recombination, and fertility. PLoS Genet. 2011;7(5):e1002088.

66. Yankovskaya V, et al. Architecture of succinate dehydrogenase and reactive oxygen species generation. Science. 2003;299(5607):700–4.

67. Slane BG, et al. Mutation of succinate dehydrogenase subunit C results in increased O^{2-}, oxidative stress, and genomic instability. Cancer Res. 2006;66(15):7615–20.

68. Ladd-Acosta C, et al. Common DNA methylation alterations in multiple brain regions in autism. Mol Psychiatry. 2014;19(8):862–71.

69. Wockner LF, et al. Brain-specific epigenetic markers of schizophrenia. Transl Psychiatry. 2015;5:e680.

70. Nardone S, et al. DNA methylation analysis of the autistic brain reveals multiple dysregulated biological pathways. Transl Psychiatry. 2014;4:e433.

71. Christensen J, et al. Prenatal valproate exposure and risk of autism spectrum disorders and childhood autism. JAMA. 2013;309(16):1696–703.

72. Pidsley R, et al. Methylomic profiling of human brain tissue supports a neurodevelopmental origin for schizophrenia. Genome Biol. 2014;15(10):483.

73. Jaffe AE, et al. Mapping DNA methylation across development, genotype and schizophrenia in the human frontal cortex. Nat Neurosci. 2016;19(1):40–7.

74. Wang J, et al. Genome-wide expression analysis reveals diverse effects of acute nicotine exposure on neuronal function-related genes and pathways. Front Psychiatry. 2011;2:5.

75. Census U.D.o.C.B. Current population survey, January 2007: Tobacco use supplement file: Tech. Doc. CPS-07. Washington, DC: US Department of Commerce; 2007.

76. Bhagwat SV, et al. Preferential effects of nicotine and 4-(N-methyl-N-nitrosamine)-1-(3-pyridyl)-1-butanone on mitochondrial glutathione S-transferase A4-4 induction and increased oxidative stress in the rat brain. Biochem Pharmacol. 1998;56(7):831–9.

77. Helen A, et al. Antioxidant effect of onion oil (*Allium cepa. Linn*) on the damages induced by nicotine in rats as compared to alpha-tocopherol. Toxicol Lett. 2000;116(1–2):61–8.

78. Guan ZZ, Yu WF, Nordberg A. Dual effects of nicotine on oxidative stress and neuroprotection in PC12 cells. Neurochem Int. 2003;43(3):243–9.

Asymmetric DNA methylation of CpG dyads is a feature of secondary DMRs associated with the *Dlk1*/*Gtl2* imprinting cluster in mouse

Megan Guntrum, Ekaterina Vlasova and Tamara L. Davis[*]

Abstract

Background: Differential DNA methylation plays a critical role in the regulation of imprinted genes. The differentially methylated state of the imprinting control region is inherited via the gametes at fertilization, and is stably maintained in somatic cells throughout development, influencing the expression of genes across the imprinting cluster. In contrast, DNA methylation patterns are more labile at secondary differentially methylated regions which are established at imprinted loci during post-implantation development. To investigate the nature of these more variably methylated secondary differentially methylated regions, we adopted a hairpin linker bisulfite mutagenesis approach to examine CpG dyad methylation at differentially methylated regions associated with the murine *Dlk1*/*Gtl2* imprinting cluster on both complementary strands.

Results: We observed homomethylation at greater than 90% of the methylated CpG dyads at the IG-DMR, which serves as the imprinting control element. In contrast, homomethylation was only observed at 67–78% of the methylated CpG dyads at the secondary differentially methylated regions; the remaining 22–33% of methylated CpG dyads exhibited hemimethylation.

Conclusions: We propose that this high degree of hemimethylation could explain the variability in DNA methylation patterns at secondary differentially methylated regions associated with imprinted loci. We further suggest that the presence of 5-hydroxymethylation at secondary differentially methylated regions may result in hemimethylation and methylation variability as a result of passive and/or active demethylation mechanisms.

Keywords: Genomic imprinting, DNA methylation, 5-Hydroxymethylcytosine, *Gtl2*, Gametic DMR, Somatic DMR, Epigenetics

Background

Genomic imprinting in mammals results in the parent of origin-specific monoallelic expression of a subset of genes. Achieving the appropriate balance of gene expression from the maternally and paternally contributed genomes via the establishment of parental allele-specific imprinting marks is crucial for normal growth and development. Therefore, it is critical to understand the mechanisms responsible for controlling the expression of imprinted genes. To date, approximately 150 mammalian genes have been identified as imprinted [1, 2]. Most

imprinted genes are found within clusters that contain a CpG-rich imprinting control region (ICR) that functions both to specify parental origin and to regulate imprinted expression of the genes within the cluster [3, 4]. Monoallelic expression of imprinted genes is achieved via multiple mechanisms, including epigenetic modifications such DNA methylation and histone modifications, as well as the activity of long noncoding RNAs [3, 4].

Differential DNA methylation at imprinted loci has been shown to play an important role in distinguishing the parental alleles and regulating their expression [5–9]. Differentially methylated regions (DMRs) associated with imprinted genes fall into two categories: primary and secondary DMRs. Primary, or gametic, DMRs serve

*Correspondence: tdavis@brynmawr.edu
Department of Biology, Bryn Mawr College, 101 N. Merion Avenue,
Bryn Mawr, PA 19010-2899, USA

as imprinting control regions (ICRs), functioning both to specify parental origin and as a shared regulatory element that controls the expression of genes throughout the associated imprinting cluster. Primary DMRs acquire DNA methylation on one of the two parental alleles during gametogenesis and remain differentially methylated from fertilization throughout development, thereby marking parental origin [3]. The differentially methylated state of primary DMRs can affect expression in a variety of ways. For example, primary DMRs can regulate gene expression through their differential ability to bind enhancer blocking proteins, thereby influencing the activity of an insulator [10, 11]. In other cases, primary DMRs are located at promoters, where they have been shown to directly influence the expression of long noncoding RNAs that subsequently regulate the expression of other genes in the imprinting cluster [12–15]. In contrast, secondary DMRs acquire parent of origin-specific DNA methylation after implantation [16–19]. Secondary DMRs are generally located at promoters or within gene bodies, and the acquisition of parental allele-specific DNA methylation at these sequences is dependent on differential methylation of the associated ICR, while the converse is not true [8, 9, 20]. While secondary DMRs do not function as primary imprinting marks, methylation of these regions frequently corresponds with gene silencing and may play a role in maintaining imprinted expression [17, 21–23].

The DNA methylation associated with primary DMRs is very stable, with the methylated allele displaying 90–100% methylation at the cytosines located in CpG dinucleotides throughout development [5, 19, 24–27]. DNA methylation at secondary DMRs is less consistent. For example, methylation at secondary DMRs located at the *H19* and *Gtl2* promoters average 70 and 78.9%, respectively, as compared to methylation at their respective primary DMRs, which average ~90 and 95.8% [5, 28]. We recently illustrated that the highly variable DNA methylation pattern at the secondary DMR associated with the imprinted *Dlk1* gene is asymmetric, with 35% of the methylated CpG dyads displaying hemimethylation [18]. The trend that DNA methylation is more stable at primary DMRs than at secondary DMRs associated with imprinted genes has also been observed at human imprinted loci [29].

Our current study investigates the nature of secondary DMRs associated with imprinted loci and potential causes of methylation instability, such as a failure to maintain DNA methylation and/or active demethylation catalyzed by the TET enzymes [30–34]. To test the hypothesis that variably methylated secondary DMRs display higher levels of hemimethylation than stably methylated primary DMRs, we analyzed DNA methylation at two additional DMRs associated with the *Dlk1/Gtl2* imprinting cluster: the IG-DMR, a primary DMR, and the *Gtl2*-DMR, a secondary DMR [14, 28]. We also quantified the level of 5-hydroxymethylation (5-hmC) at the IG-, *Gtl2*- and *Dlk1*-DMRs to determine whether there is a correlation between high levels of hemimethylation and high levels of 5-hydroxymethylation. Our results support the hypothesis that high levels of 5-hmC may contribute to methylation instability at secondary DMRs associated with imprinted genes.

Results

CpG dyads within the *Gtl2*-DMR display high levels of hemimethylation

To determine whether asymmetric methylation is unique to the *Dlk1*-DMR or is a feature of other secondary DMRs associated with imprinted loci, we examined CpG dyad methylation at the linked *Gtl2*-DMR. We had previously analyzed DNA methylation on the coding strand of the *Gtl2*-DMR and observed moderate variability in the methylation status, with the 5′ half of the region analyzed displaying lower average DNA methylation levels than the 3′ half [28]. We therefore assessed the DNA methylation status of cytosines located in 22 pairs of complementary CpG dinucleotides spanning this region to determine whether these CpG dyads were homomethylated versus hemimethylated (Fig. 1). All of our experiments were conducted using F_1 hybrid tissues collected from crosses between C57BL/6 (B6) and a specially derived strain containing *Mus musculus castaneus* (CAST)-derived sequences from chromosome 12 on an otherwise C57BL/6 genetic background (CAST12) [18, 28], allowing us to distinguish paternally inherited alleles from maternally inherited alleles based on sequence polymorphisms (detailed in the "Methods").

We analyzed CpG dyad methylation in DNA derived from four developmental stages: 7.5 d.p.c. embryo, 14.5 d.p.c. embryo, 5 d.p.p. liver and adult liver. The DNA methylation patterns on each parental allele were consistent throughout development and were also similar in tissues obtained from reciprocal crosses (Figs. 2, 3). Across all four developmental stages, cytosines in CpG dinucleotides were methylated 80-93% of the time on paternal alleles and 6–10% of the time on maternal alleles (Table 1). We assessed the significance of these results at each developmental stage using a Mann–Whitney U test and found that the median level of DNA methylation was significantly higher on the paternal alleles as compared to the maternal alleles in all of the tissues examined, with P values ranging from <0.0001 to 0.0147 (Table 2; Additional File 1). Furthermore, P values derived from Mann–Whitney U tests illustrate that median DNA methylation levels did not vary significantly across development on

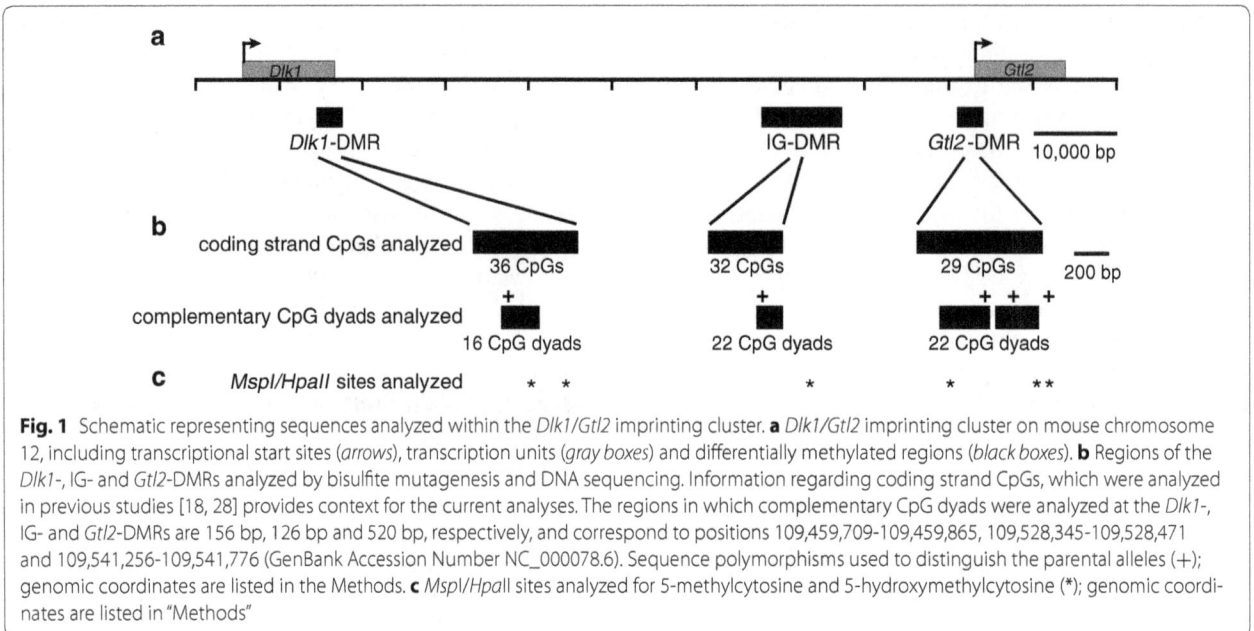

Fig. 1 Schematic representing sequences analyzed within the *Dlk1/Gtl2* imprinting cluster. **a** *Dlk1/Gtl2* imprinting cluster on mouse chromosome 12, including transcriptional start sites (*arrows*), transcription units (*gray boxes*) and differentially methylated regions (*black boxes*). **b** Regions of the *Dlk1*-, IG- and *Gtl2*-DMRs analyzed by bisulfite mutagenesis and DNA sequencing. Information regarding coding strand CpGs, which were analyzed in previous studies [18, 28] provides context for the current analyses. The regions in which complementary CpG dyads were analyzed at the *Dlk1*-, IG- and *Gtl2*-DMRs are 156 bp, 126 bp and 520 bp, respectively, and correspond to positions 109,459,709-109,459,865, 109,528,345-109,528,471 and 109,541,256-109,541,776 (GenBank Accession Number NC_000078.6). Sequence polymorphisms used to distinguish the parental alleles (+); genomic coordinates are listed in the Methods. **c** *MspI/HpaII* sites analyzed for 5-methylcytosine and 5-hydroxymethylcytosine (*); genomic coordinates are listed in "Methods"

either the paternal or the maternal allele (Additional File 1). Average DNA methylation levels did not vary substantially between the 5′ half versus the 3′ half of the analyzed region. These results confirm that the *Gtl2*-DMR is differentially methylated throughout development.

Homomethylation was observed at 68–78% of the CpG dyads containing methylated cytosine, while hemimethylation was detected at 22–32% of these CpG dyads (Table 1). The levels of homo- and hemimethylation at the *Gtl2*-DMR were similar to the overall average of 65% homomethylation and 35% hemimethylation observed at the *Dlk1*-DMR [18]. When we restricted our analysis to the same four developmental stages assessed in this study, we observed 69–74% homomethylation at methylated CpG dyads within the *Dlk1*-DMR and 26–31% hemimethylation (Table 1). We tested the homo- and hemimethylation levels at the *Gtl2*- and *Dlk1*-DMRs for statistical independence by performing Chi-squared analysis and determined that hemimethylation levels are not significantly different at these loci ($P = 0.1318$). Therefore, we conclude that hemimethylation levels are similar at two distinct secondary DMRs located within the *Dlk1/Gtl2* imprinting cluster.

CpG dyads within the IG-DMR display low levels of hemimethylation

We next assessed hemimethylation levels at the IG-DMR, which serves as the imprinting control region for the *Dlk1/Gtl2* imprinting cluster [19, 35]. We analyzed 22 CpG dyads located within the IG-DMR (Fig. 1). We had previously analyzed DNA methylation on the coding

strand of this region and had found it to lack variability, with paternally inherited alleles showing near 100% DNA methylation and maternally inherited alleles displaying less than 10% DNA methylation [28]. Consistent with our previous findings, we observed methylation at 96 and 12% of paternally versus maternally inherited CpG dinucleotides located within the IG-DMR, respectively (Fig. 4; Table 1). The median levels of DNA methylation were significantly higher on paternally derived alleles as compared to maternally derived alleles for all tissues analyzed, with P values ranging from <0.0001 to <0.01 (Table 2; Additional File 1), confirming that this region is differentially methylated throughout development. There were no significant differences in the DNA methylation profile of maternal alleles across development (Additional File 1). In contrast, while median DNA methylation levels on paternal alleles was not significantly different between the 14.5 d.p.c. embryo, 5 d.p.p. liver or adult liver samples, the distribution of DNA methylation on paternal alleles derived from 7.5 d.p.c. embryos was different from the distribution in 14.5 d.p.c. embryos ($P = 0.0021$), 5 d.p.p. liver ($P = 0.0178$) and adult liver ($P = 0.0006$). The significance of these results may be attributed to the fact that paternal alleles derived from 7.5 d.p.c. embryos contain more unmethylated cytosines than paternal alleles derived from other developmental stages (Fig. 4; Additional File 1).

Of the CpG dyads displaying cytosine methylation within the IG-DMR, 92% were homomethylated, while 8% were hemimethylated (Table 1). The frequency of hemimethylation was higher on maternally inherited

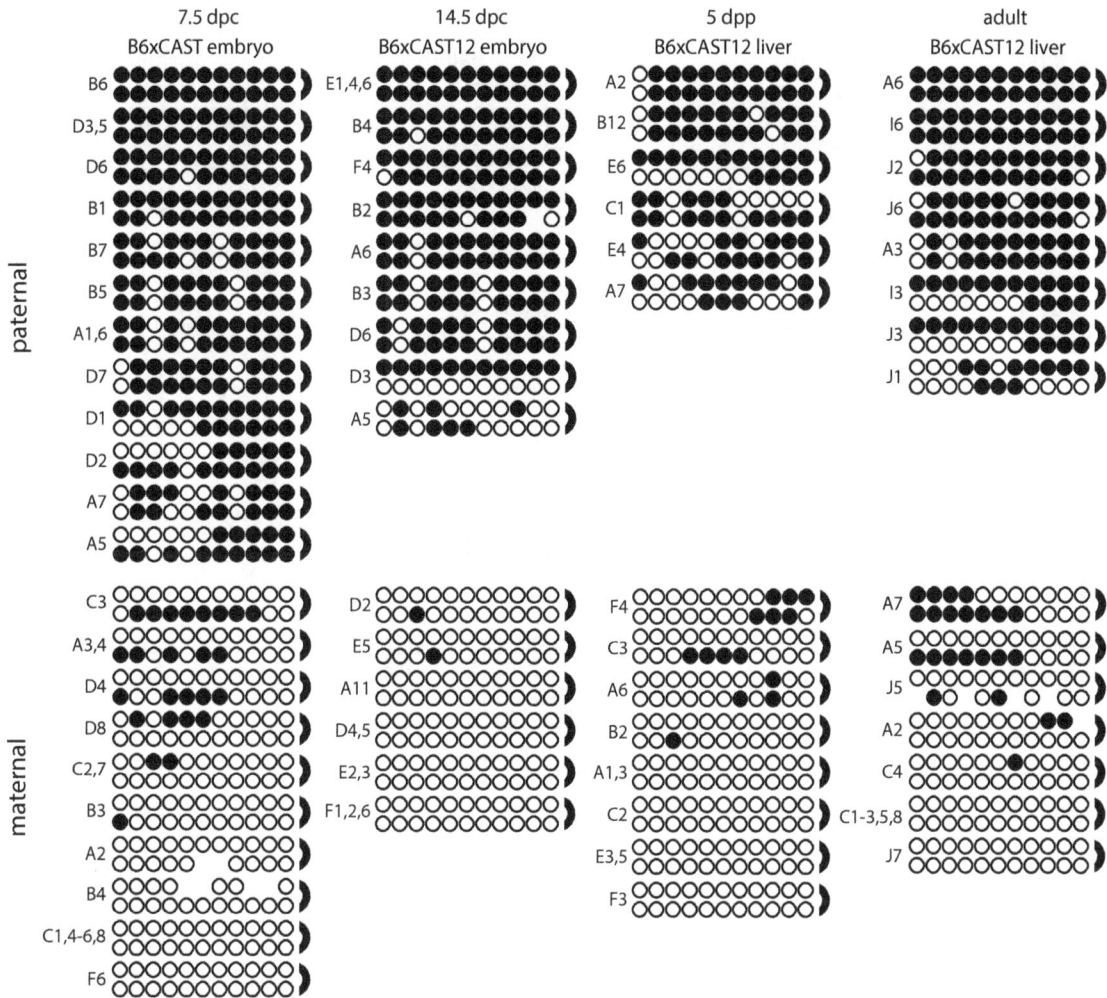

Fig. 2 DNA methylation in the 5′ portion of the *Gtl2*-DMR displays a high level of hemimethylation. Bisulfite mutagenesis and sequencing of F_1 hybrid DNA derived from 7.5 d.p.c. B6 × CAST embryos, 14.5 d.p.c. B6 × CAST12 embryos, 5 d.p.p. B6 × CAST12 liver and adult B6 × CAST12 liver. Individual *circles* in each *row* represent one of the 11 potentially methylated CpG dinucleotides analyzed, and each paired *row* of *circles* represents the complementary strands of an individual subclone; *semicircles* to the *right* indicate the location of the linker connecting the complementary strands. *Filled circles* represent methylated cytosines, *open circles* represent unmethylated cytosines, and absent *circles* represent ambiguous data. Labels to the *left* identify the PCR subclone analyzed; *letters* represent independent amplification reactions, while *numbers* represent individual subclones. Subclones derived from the same amplification that have identical sequence and methylation patterns are grouped together, as it was not possible to determine whether these amplicons were derived from the same or different template molecules

alleles, which were much more sparsely methylated than paternally inherited alleles (Table 1). On average, the level of hemimethylation we observed at the IG-DMR at each developmental stage and across development was lower than the level we observed at either the *Dlk1*-DMR or the *Gtl2*-DMR (Fig. 5). We assessed the significance of this result using a Chi-squared test of independence. The level of hemimethylation at the primary IG-DMR is significantly different than the level at either of the secondary DMRs (IG- vs. *Dlk1*-DMR, P-5.93 × 10^{-65}; IG- vs. *Gtl2*-DMR, $P = 8.76 × 10^{-57}$), supporting our hypothesis

that high levels of hemimethylation are characteristic of secondary, but not primary, DMRs associated with imprinted loci.

For all of our DNA methylation analyses, we employed a conservative approach whereby we grouped subclones that were derived from the same PCR amplification and had identical sequence and DNA methylation patterns, as it is not possible to determine whether these amplicons are derived from the same or different template molecules. As it is possible that some of the grouped subclones actually represent independent samples, we

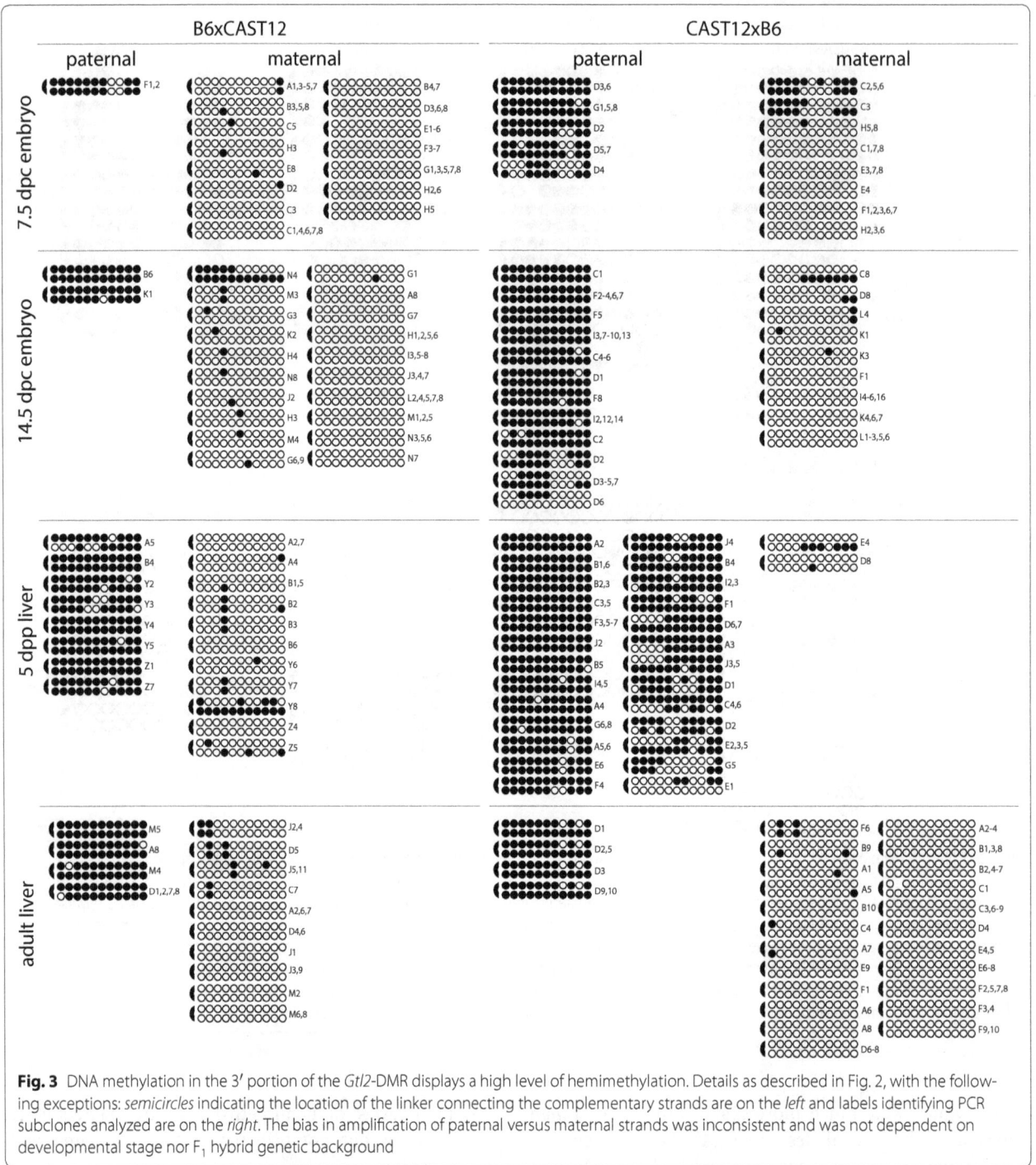

Fig. 3 DNA methylation in the 3′ portion of the *Gtl2*-DMR displays a high level of hemimethylation. Details as described in Fig. 2, with the following exceptions: *semicircles* indicating the location of the linker connecting the complementary strands are on the *left* and labels identifying PCR subclones analyzed are on the *right*. The bias in amplification of paternal versus maternal strands was inconsistent and was not dependent on developmental stage nor F₁ hybrid genetic background

performed the same analyses for ungrouped data sets in which each subclone was treated as an independent sample. Ungrouping the identical subclones resulted in greater significant differences between the parental alleles at both the *Gtl2*-DMR and the IG-DMR, confirming their differentially methylated status (Table 2; Additional File 1). While ungrouping identical subclones

reduced hemimethylation values at both the *Gtl2*- and IG-DMRs, from 31.8 to 29.8 and 8.4 to 6.1%, respectively (Additional file 3: Table S2, Additional file 4: Table S3), the difference in hemimethylation levels between the primary IG-DMR and the secondary *Gtl2*-DMR remains highly significant as assessed using a Chi-squared test of independence ($P = 2.81 \times 10^{-59}$).

Table 1 Average levels of DNA methylation on paternal and maternal alleles at the *Dlk1*-, IG- and *Gtl2*-DMRs across four developmental stages

		Dlk1-DMR		IG-DMR		*Gtl2*-DMR, 5'	*Gtl2*-DMR, 3'	
		BxC	CxB	BxC	CxB	BxC	BxC	CxB
% methylation (# methyl-ated/total)	P	42.5% (510/1199)	42% (541/1287)	96.1% (2960/3080)	95.5% (2982/3123)	79.5% (699/879)	92.7% (306/330)	85.9% (878/1034)
	M	16.6% (251/1516)	28.4% (367/1292)	11.2% (110/918)	12.2% (75/616)	9.5% (64/671)	6.2% (76/1231)	6.4% (59/923)
	total	28% (761/2715)	35.2% (908/2579)	76.8% (3070/3998)	81.8% (3057/3739)	49.2% (763/1550)	24.5% (382/1561)	47.9% (937/1957)
% homomethylation (# homomethyl-ated/total)	P	79.5% (221/278)	70.6% (223/316)	94.2% (1436/1524)	92.9% (1436/1545)	76.3% (302/396)	88.9% (144/162)	82.5% (397/481)
	M	63.4% (97/153)	65.7% (140/213)	39.7% (31/78)	70.5% (31/44)	12.2% (7/57)	35.7% (20/56)	31.1% (14/45)
	total	73.8% (318/431)	68.6% (363/529)	91.6% (1467/1602)	92.3% (1467/1589)	68.2% (309/453)	75.2% (164/218)	78.1% (411/526)
% hemimethylation (# hemimethyl-ated/total)	P	20.5% (57/278)	29.4% (93/316)	5.8% (88/1524)	7.1% (109/1545)	23.7% (94/396)	11.1% (18/162)	17.5% (84/481)
	M	36.6% (56/153)	34.3% (73/213)	60.3% (47/78)	29.5% (13/44)	87.7% (60/57)	64.3% (36/56)	68.9% (31/45)
	total	26.2% (113/431)	31.4% (166/529)	8.4% (135/1602)	7.7% (122/1589)	31.8% (144/453)	24.8% (54/218)	21.9% (115/526)

Percent methylation and number of sites analyzed on the paternal and maternal alleles in DNA derived from individual developmental stages (7.5 and 14.5 d.p.c. embryos and 5 d.p.p. and adult liver) are given in Additional file 2: Table S1. Data for the *Dlk1*-DMR were calculated from Gagne et al. [18]

Table 2 Average levels of DNA methylation on the paternal and maternal alleles are significantly different at the *Gtl2*- and IG-DMRs

	Genomic DNA sample	Grouped			Ungrouped		
		% methylation, paternal alleles	% methylation, maternal alleles	P value	% methylation, paternal alleles	% methylation, maternal alleles	P value
Gtl2-DMR, 5', BxC	7.5 d.p.c. embryo	82.2% (217/264)	11.7% (25/214)	0.0001	83.4% (257/308)	9.2% (32/346)	<0.0001
	14.5 d.p.c. embryo	82.5% (235/285)	1.5% (2/132)	0.0024	84.8% (279/329)	0.9% (2/220)	0.0001
	5 d.p.p. liver	69.7% (92/132)	8% (14/176)	0.0024	69.7% (92/132)	6.4% (14/220)	0.0014
	adult liver	78.3% (155/198)	15.4% (23/149)	0.0021	78.3% (155/198)	9.7% (23/237)	0.0004
IG-DMR, BxC	7.5 d.p.c. embryo	94.2% (912/968)	1.5% (2/132)	<0.01	94.5% (1621/1716)	1.1% (2/176)	<0.01
	14.5 d.p.c. embryo	96.9% (597/616)	11.1% (39/352)	0.0002	97.6% (816/836)	9.8% (39/396)	<0.0001
	5 d.p.p. liver	96.8% (852/880)	17.4% (45/258)	0.0003	97.6% (1159/1188)	16.8% (58/346)	<0.0001
	adult liver	97.2% (599/616)	13.6% (24/176)	<0.01	97.7% (1333/1364)	13.6% (24/176)	<0.01

Percent methylation and number of sites analyzed on the paternal and maternal alleles in DNA derived from 7.5 and 14.5 d.p.c. embryos and 5 d.p.p. and adult liver. Grouped data were derived when subclones from the same PCR with identical DNA methylation patterns and sequences were grouped as a single sample, as illustrated in Figs. 2, 4. Ungrouped data were derived when subclones from the same PCR with identical DNA methylation patterns and sequences were treated as independent samples. *P* values were calculated using a Mann–Whitney *U* test

High levels of hemimethylation correlate with high levels of 5-hydroxymethylcytosine at the *Dlk1*-DMR, but not at the *Gtl2*-DMR

We hypothesized that hemimethylation at the *Dlk1*- and *Gtl2*-DMRs could arise as a result of sequential oxidation of 5-methylcytosine (5-mC) by the TET enzymes followed by either passive depletion of methylation following DNA replication or thymine DNA glycosylase-mediated base excision repair, ultimately resulting in demethylation of that residue [30–34]. If the TET enzymes are responsible for demethylation of cytosines leading to high levels of hemimethylation at secondary

DMRs associated with imprinted loci, we would expect to see 5-hydroxymethylcytosine (5-hmC), an oxidation intermediate in this pathway, at these loci [30, 31]. We therefore assessed the relative levels of 5-mC and 5-hmC at the *Dlk1*-, *Gtl2*- and IG-DMRs in genomic DNA isolated from 9.5 and 14.5 d.p.c. embryos and from 5 d.p.p. and adult liver. To conduct this analysis, we glucosylated genomic DNA derived from each of the four developmental stages listed above, digested glucosylated and unglucosylated samples with *Msp*I, *Hpa*II or no enzyme, amplified the resulting products using qPCR and calculated percent 5-hmC based on the method previously

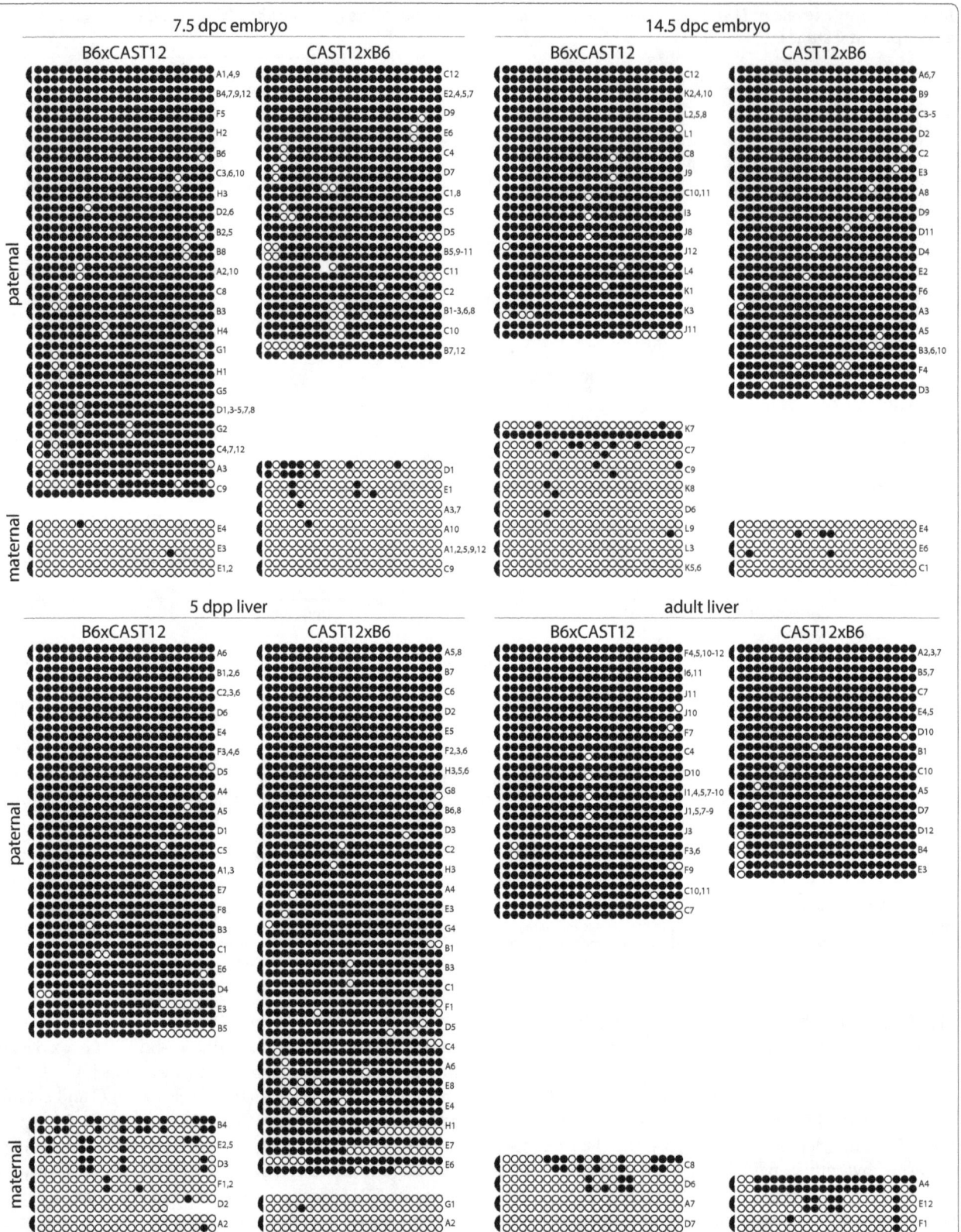

Fig. 4 DNA methylation at the IG-DMR displays low levels of hemimethylation. Details as described in Fig. 2, with the following exceptions: *individual circles* in each row represent one of the 22 potentially methylated CpG dinucleotides analyzed, *semicircles* indicating the location of the linker connecting the complementary strands are on the *left* and labels identifying PCR subclones analyzed are on the *right*

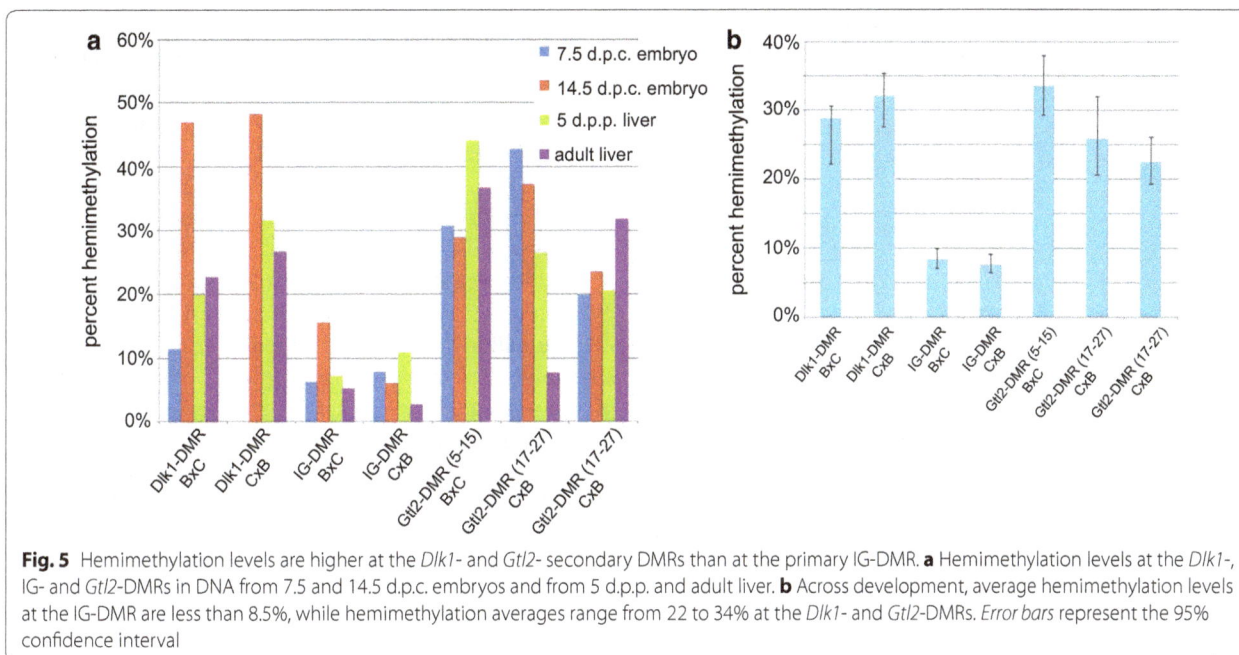

Fig. 5 Hemimethylation levels are higher at the *Dlk1*- and *Gtl2*- secondary DMRs than at the primary IG-DMR. **a** Hemimethylation levels at the *Dlk1*-, IG- and *Gtl2*-DMRs in DNA from 7.5 and 14.5 d.p.c. embryos and from 5 d.p.p. and adult liver. **b** Across development, average hemimethylation levels at the IG-DMR are less than 8.5%, while hemimethylation averages range from 22 to 34% at the *Dlk1*- and *Gtl2*-DMRs. *Error bars* represent the 95% confidence interval

Table 3 Average 5-hydroxymethylation levels at *Msp*I sites located in the *Dlk1*-, IG- and Gtl2-DMRs

	Dlk1, site A	*Dlk1*, site B	IG-DMR	*Gtl2*, site A	*Gtl2*, site B	*Gtl2*, site C
9.5 d.p.c. embryo (n)	21.5% (4)	6.6% (3)	2.9% (3)	0% (4)	1.4% (4)	0.79% (1)
14.5 d.p.c. embryo (n)	39% (3)	10% (3)	3.1% (3)	0.8% (3)	1.6% (3)	1.2% (1)
5 d.p.p. liver (n)	10.2% (8)	6.5% (8)	3.7% (3)	2.5% (4)	0.9% (6)	1.4% (1)
Adult liver (n)	8.1% (3)	1.7% (3)	8.7% (3)	1.9% (3)	0.4% (3)	2.4% (1)

n = number of biological replicates; each biological replicate was composed of three technical replicates; data for individual biological replicates are given in Additional file 5: Table S4

described by Magalhães et al. [36]. We used 9.5 d.p.c. embryos as our earliest developmental time point, rather than 7.5 d.p.c., in order to have sufficient DNA for these analyses.

At the IG-DMR, where we observed low levels of hemimethylation, we detected correspondingly low levels of 5-hmC in DNA derived from 9.5 to 14.5 d.p.c. embryos and from 5 d.p.p. liver: 3–4% of the methylation detected at the *Msp*I site located within the IG-DMR was 5-hmC (Table 3). Higher 5-hmC levels were detected in adult liver, which also had more variation between biological replicates (Additional file 5: Table S4). In contrast, our analysis of two *Msp*I sites located within the *Dlk1*-DMR detected 5-hmC levels ranging from 7 to 40% in embryos and neonatal liver, with highest levels observed in 14.5 d.p.c. embryos (Table 3). This higher level of 5-hmC at the *Dlk1*-DMR correlates with the high levels of hemimethylation detected at this locus. We therefore anticipated that we would observe similarly high levels of 5-hmC at the *Gtl2*-DMR, which also displays high hemimethylation

levels. However, the level of 5-hmC at the *Gtl2*-DMR inversely correlated with hemimethylation levels at this locus. While we detected high levels of hemimethylation at the *Gtl2*-DMR, we observed very low levels of 5-hmC at three *Msp*I sites located within this locus (Table 3).

5-hmC is absent at the *Dlk1*-, *Gtl2*- and IG-DMRs in triple TET knockout ES cells

To validate the assay used to assess 5-hmC levels in the *Dlk1*-, *Gtl2*- and IG-DMRs, we conducted an analysis of 5-hmC in DNA derived from wild-type and triple TET knockout embryonic stem (ES) cells [37]. In wild-type ES cells, we detected 5-hmC at 12, 14 and 10% of the methylated cytosines located within *Msp*I sites at the *Dlk1*-, *Gtl2*- and IG-DMRs, respectively (Table 4). In ES cells with a triple knockout of the TET genes, 5-hmC was undetectable at all three loci (Table 4). These data validate that the assay employed to detect 5-hmC and confirm that the presence of 5-hmC at these loci is dependent on functional TET enzyme activity.

Table 4 5-Hydroxymethylation at the *Dlk1-*, IG- and G*tl2*-DMRs in embryonic stem cells is dependent on the activity of TET enzymes

	WT ES cells (%)	TET-KO ES cells (%)
Dlk1-DMR, site A	12.0	0
IG-DMR	9.9	0
Gtl2-DMR	13.6	0

Discussion

Differential DNA methylation plays an important role in the regulation of imprinted genes, directly affecting the activity of ICRs as well as directly or indirectly regulating the expression of genes within imprinting clusters [34 and references therein]. Stably maintaining DNA methylation at imprinted loci is critical for normal growth and development, and aberrant DNA methylation patterns are associated not only with abnormal expression of imprinted genes, but also with multiple imprinting disorders [38, 39]. Therefore, understanding the normal methylation patterns and how they are altered are important for elucidating the regulation of imprinted genes.

Investigation of DNA methylation at imprinted loci has identified a difference in methylation stability at primary versus secondary DMRs: methylation patterns at primary DMRs appear to be very stable and consistent, while methylation at secondary DMRs is more variable in both mice and humans [5, 18, 19, 24–29]. Our previous analyses identified a correlation between variable methylation patterns and high levels of hemimethylation at CpG dyads at the secondary DMR associated with the imprinted *Dlk1* gene [18]. To test the hypothesis that hemimethylation is a normal characteristic of secondary DMRs associated with imprinted loci, we utilized a hairpin bisulfite mutagenesis approach to examine CpG dyad methylation at another secondary DMR (the *Gtl2*-DMR) as well as at a primary DMR that serves as an ICR (the IG-DMR associated with the *Dlk1/Gtl2* imprinting cluster). We found high levels of hemimethylation at the *Gtl2*-DMR, but not at the IG-DMR, supporting our hypothesis (Figs. 2, 3, 4).

It is possible that some of the hemimethylation we observed resulted from errors associated with bisulfite mutagenesis, including failed conversion of unmethylated cytosines and inappropriate conversion of methylated cytosines [40]. Either one of these errors could lead to a hemimethylated CpG dyad. It is unlikely that these errors had an appreciable effect on the hemimethylation levels we observed, as the high molarity, high temperature, short reaction time methodology we employed have been shown to result in low bisulfite conversion error

rates [40, 41]. To test this assumption, we directly calculated the failed conversion rate by examining 20,868 non-CpG cytosine locations in our bisulfite-treated samples. We identified 291 cytosines, some of which may have arisen as a result of PCR-induced error rather than failed conversion (Additional file 6: Table S5). The failed conversion rate we observed of 1.39% is similar to the error rate reported by Genereux et al. [40] for hairpin-linked molecules treated under high molarity, high temperature, 90-min reaction time mutagenesis conditions. We therefore conclude that the predicted and observed error rates are unlikely to account for an appreciable amount of the hemimethylation we observed, particularly at the secondary DMRs, where hemimethylation levels range from 22 to 35%. Furthermore, even if hemimethylation levels vary slightly due to conversion errors, it would not affect our overall conclusion that the difference in hemimethylation levels between primary and secondary DMRs is highly significant.

Based on the observation that secondary DMRs have high levels of hemimethylation, we further proposed that oxidation of 5-methylcytosine by the TET enzymes could result in hemimethylation either by impeding the activity of DNMT1, resulting in replication-dependent passive demethylation, or by further processing of the oxidized products 5-fC and 5-caC and their removal via base excision repair, a mechanism of active demethylation [30–34]. In support of this hypothesis, we found low levels of 5-hmC at the IG-DMR, which had correspondingly low levels of hemimethylation, and high levels of 5-hmC at the *Dlk1*-DMR, which had correspondingly high levels of hemimethylation. However, at the *Gtl2*-DMR, where we observed high levels of hemimethylation, we detected low levels of 5-hmC.

The fact that we observed low levels of 5-hmC at the *Gtl2*-DMR may be a consequence of our experimental approach. The earliest developmental stage at which we assessed 5-hmC was embryonic day 9.5, when methylation of the *Gtl2*-DMR is already relatively stable as compared to methylation at earlier embryonic stages such as 6.5 and 7.5 d.p.c. [28]. While we detected very little 5-hmC in DNA derived from 9.5 to 14.5 d.p.c. embryos and from neonatal and adult tissues, it is possible that the *Gtl2*-DMR contains high levels of 5-hmC earlier in development, when DNA methylation patterns at this locus are more labile. Alternatively, the absence of 5-hmC at the *Gtl2*-DMR may point to the relative importance of maintaining DNA methylation at this locus in order to appropriately silence *Gtl2* expression on the paternal allele and achieve appropriate imprinted expression patterns across the *Dlk1/Gtl2* imprinting cluster. Indeed, studies have shown that loss of methylation on the paternally inherited *Gtl2*-DMR correlates with expression of *Gtl2* from this allele [42, 43]. In contrast to the substantiated role of differential DNA methylation at the *Gtl2*-DMR, we and others have questioned whether differential methylation at the *Dlk1*-DMR, which is located in the 5th exon of *Dlk1*, plays an important regulatory role [18, 19].

To further investigate the hypothesis that 5-hmC contributes to DNA methylation instability at secondary DMRs associated with imprinted genes, we are currently examining CpG dyad methylation patterns and 5-hmC levels at additional imprinted loci. The work described herein focused on analyses of primary and secondary DMRs associated with the *Dlk1/Gtl2* imprinting cluster on mouse chromosome 12. Our current inquiries include an examination of CpG dyad methylation patterns at imprinted loci distributed across multiple genomic locations, including both paternally and maternally methylated DMRs, and our preliminary results suggest that the relationship between 5-hmC and DNA methylation variability at secondary DMRs may be generalizable (Davis Laboratory, unpublished data).

The presence of 5-hmC at secondary DMRs associated with imprinted genes suggests that the TET proteins play a role in modulating DNA methylation stability at these loci. Several studies have shown that the TET proteins play a role in epigenetic reprogramming throughout development [44 and references therein]. For example, the presence of 5-hmC at repressed promoters and in the gene bodies of expressed genes in ES cells is concomitant with the association of TET1 and TET2, respectively, with these sequences [45–48]. Recent reports have also illustrated the role of TET enzymes in epigenetic reprogramming during primordial germ cell development, including the erasure of imprinting marks [49–51], and that a double knockout of TET1 and TET2 disrupts normal methylation and expression patterns at imprinted loci [52]. Liu and colleagues further reported that some imprinted loci, such as *H19*, appear to be more sensitive to TET activity based on methylation patterns in wild-type *vs.* TET knockout (TET-KO) ES cells, suggesting that the TET proteins may have different effects at different imprinted loci [53]. Interestingly, the study conducted by Liu et al. [53] also showed that there were no substantial changes in methylation at the IG-DMR in the TET-KO ES cells as compared to wild-type ES cells, consistent with our observation that there is very little 5-hmC at the IG-DMR in embryos and neonatal liver (Table 3). Indeed, the IG-DMR has a very stable allele-specific DNA methylation pattern, in accordance with its role as an imprinting control element critical for normal growth and development [9,

54]. In contrast, the consistently higher levels of 5-hmC across we observed across development at the *Dlk1*-DMR suggests that 5-hmC may be a more stable epigenetic mark as this locus, as it is in the developing brain [data herein; 55].

We have yet to determine whether the high levels of hemimethylation at secondary DMRs plays a functional role. Arand et al. [56] illustrated a correlation between the reduction in DNA methylation levels and concomitant increase in hemimethylation levels during embryogenesis and primordial germ cell development, and suggested that hemimethylation plays a role in impairing maintenance methylation in order to keep methylation levels low. Similarly, Jin et al. [57] suggested that TET1 may act to reduce DNA methylation levels at hypomethylated CpG islands. Therefore, it would be interesting to look at allele-specific distribution of 5-hmC at imprinted loci to see whether TETs play a role in keeping unmethylated DMRs unmethylated. The assay we used for the 5-hmC studies described herein was not allele-specific, as that would have required a strain-specific polymorphism in close proximity to an *Msp*I site within the regions of analysis. The development of methods to assess 5-hmC in an allele-specific way would allow us to address this question in order to determine the significance of this epigenetic modification.

Conclusions

Secondary DMRs associated with imprinted loci have low DNA methylation fidelity as compared to primary DMRs that serve as imprinting control regions. Our current analyses illustrate that the variable DNA methylation patterns at secondary DMRs correlate with high levels of hemimethylation at CpG dyads and high levels of 5-hmC. Our work therefore supports the hypothesis that secondary DMRs have a unique epigenetic profile that distinguishes them from primary DMRs. Our studies further provide insight into the molecular mechanisms responsible for methylation instability at secondary DMRs, as oxidation of 5-mC to 5-hmC by the TET enzymes ultimately leads to a loss of DNA methylation passively due to reduced DNMT1 fidelity and/or actively via further oxidation followed by DNA repair. These results are significant as they highlight the complexities associated with the maintenance of the epigenetic profile at secondary DMRs: Differential DNA methylation is maintained at these loci despite activities that function to reduce methylation levels. Further investigation is warranted to understand how parent of origin-specific DNA methylation is established and maintained at secondary DMRs.

Methods

Mice

C57BL/6J (B6) and *Mus musculus castaneus* (CAST) mice were purchased from the Jackson Laboratory. To facilitate the isolation of F_1 hybrid mice, a strain of mice that served as the source of the *M. m. castaneus* allele (CAST12) was constructed as previously described [28]. Natural matings between B6 and CAST were used to generate F_1 hybrid males for spermatozoa collection; all other F_1 hybrid tissues used for bisulfite analyses were generated from natural matings between B6 and CAST12 mice. For all F_1 hybrid tissues, the maternal allele is located on the left.

DNA purification and bisulfite analysis

DNA was isolated from 7.5 d.p.c. embryos using the DNeasy Blood & Tissue Kit (Qiagen, Germantown, MD, cat#69504). DNA was isolated from 9.5 to 14.5 d.p.c. embryos and from 5 d.p.p. and adult liver following proteinase K digestion and a series of phenol/chloroform extractions as described previously [58]. Prior to bisulfite mutagenesis, complementary strands of DNA were covalently attached as follows: for IG-DMR analyses, 0.5 μg of genomic DNA was digested with 1 μl *Spe*I (NEB, Ipswich, MA, cat#R0133) and ligated to 1 μg of phosphorylated hairpin linker IG-DMR-HP (5′-CTAGAGC-GATGCGTTCGAGCATCGCT-3′) [59]; for *Gtl2*-DMR analyses, 0.5 μg of genomic DNA was digested with 1 μl *Ban*I (NEB, Ipswich, MA, cat#R0118) and ligated to 1 μg of phosphorylated hairpin linker Gtl2-HP-3 (5′-GTA-CAGCGATGCGTTCGAGCATCGCT-3′). 0.5 μg of hairpin-linked, ligated DNA was denatured by incubating in freshly prepared 3 M NaOH for 20 min at 42 °C, and then subjected to bisulfite mutagenesis using an EZ DNA Methylation-Direct kit (Zymo Research, Irvine, CA, cat#D5020). All mutagenized DNAs were subjected to multiple independent PCR amplifications to ensure analysis of different strands of DNA; subclones derived from independent PCR amplifications are distinguished by different letters of the alphabet. PCR contamination was ruled out via analysis of no template negative control amplifications for both the first and second rounds of PCR. Data from multiple independent tissue samples derived from the same developmental stage were combined, as we did not detect variation between biological replicates. Primer pairs used for nested amplification of the mutagenized DNA were designed to incorporate at least one SNP as well as CpG dinucleotides within the previously analyzed DMRs [28]. All base pair numbers are from GenBank Accession Number NC_000078.6.

Primers and PCR cycling conditions for the IG-DMR and for two adjacent regions within the *Gtl2*-DMR are detailed in Additional file 7: Table S6. Expected second round PCR products for the IG-DMR and the two regions of the *Gtl2*-DMR are 412 bp, 721 bp, and 695 bp, respectively. Subcloning of amplified products was achieved using a pGEM-T Easy vector (Promega Corporation, Madison, WI, cat#A1360). Sequencing reactions were outsourced to Genewiz (South Plainfield, NJ) or were performed using a Thermo Sequenase Cycle Sequencing Kit (Affymetrix, Cleveland, OH, cat#78500) and analyzed on a 4300 DNA Analyzer (LI-COR Biosciences, Lincoln, NE). Sequence polymorphisms used to distinguish B6 *vs.* CAST DNA (B/C): IG-DMR, G/A at bp#109,528,369; *Gtl2*-DMR, G/A at bp#109,541,531, A/G at bp#109,541,671, AA/GC at bp#109,541,820-109,541,821. Percent methylation was calculated based on data obtained from both complementary strands. Percent hemimethylation was calculated by dividing the number of hemimethylated CpG dyads by the number of hemimethylated plus homomethylated CpG dyads.

5-hydroxymethylation analysis

For 5-hydroxymethylation analyses, DNA was isolated from 9.5 d.p.c. embryos, 14.5 d.p.c. embryos, 5 d.p.p. liver and adult liver as described above. DNA derived from three different genetic backgrounds [C57BL/6 J, B6x(CAST or CAST12) and (CAST or CAST12)xB] was used to ensure that genetic background did not affect the outcome. In addition, DNA was isolated from wild-type and TET triple knockout embryonic stem cells [37]. 5-hydroxymethylation levels were assessed using an EpiMark 5-hmC and 5-mC Analysis Kit (NEB, Ipswich, MA, cat#E3317). Briefly, 2.5 μg genomic DNA or 2 μg ES cell DNA was glucosylated using 30 units of T4 ß-glucosyltransferase at 37 °C overnight. Glucosylated and unglucosylated control DNAs were treated with *Msp*I, *Hpa*II or no restriction endonuclease at 37 °C overnight. Following treatment with proteinase K, products were amplified via PCR and quantitative PCR (StepOnePlus, Applied Biosystems). Primers and PCR cycling conditions used are detailed in Additional file 8: Table S7. qPCR was performed in triplicate for each of three independent biological samples. Amount of 5-mC and 5-hmC in each sample was calculated according to Magalhães et al. [36]. Genomic coordinates for *Msp*I/*Hpa*II sites are: *Dlk1*-DMR-A, bp#109,459,830; *Dlk1*-DMR-B, bp#109,460,017; IG-DMR, bp#109,528,624; *Gtl2*-DMR-A, bp#109,541,314; *Gtl2*-DMR-B, bp#109,541,776; *Gtl2*-DMR-C, bp#109,541,811.

Additional files

Additional file 1. Data used for statistical analyses of DNA methylation levels at the *Gtl2*- and IG-DMRs. The numerical data used to perform Mann-Whitney U tests and the resulting P values are contained in this file. Data from the *Gtl2*-DMR 5′ region, *Gtl2*-DMR 3′ region and the IG-DMR are presented in separate sheets. Within each sheet, data from each of the developmental stages are presented in chronological order, as they are in the Results, Figures, and Tables. Each sheet presents the information for a specific locus, tissue, cross (maternal allele x paternal allele), and parental allele analyzed, as indicated in columns A-D. % methylation (column F) was calculated by dividing the number of methylated CpG sites observed in a given subclone (column E) by the total number of CpG sites analyzed within the subclone; the raw data used to make these calculations are found in Figs. 2, 3 and 4. For the *Gtl2*-DMR 3′ region and the IG-DMR, P values were calculated independently for BxC samples vs. CxB samples. In addition, P values were calculated for the combined BxC + CxB samples, as some of the BxC and CxB sample sizes were too small to accurately perform Mann-Whitney U tests. Data presented in sheets labelled "grouped" combine subclones derived from the same PCR with identical DNA methylation patterns and identical sequences as a single sample, while every subclone analyzed is treated as an independent sample in sheets labelled "ungrouped".

Additional file 2: Table S1. Percent methylation on paternal and maternal alleles at DMRs across four developmental stages.

Additional file 3: Table S2. Comparison of percent methylation at the Gtl2-DMR with like subclones grouped vs. ungrouped.

Additional file 4: Table S3. Comparison of percent methylation at the IG-DMR with like subclones grouped vs. ungrouped.

Additional file 5: Table S4. Percent 5-hydroxymethylcytosine at the Dlk1-, IG- and Gtl2-DMRs at four developmental stages.

Additional file 6: : Table S5. Quantification of bisulfite conversion failure.

Additional file 7: Table S6. Primer and PCR cycling conditions for amplification of bisulfite-mutagenized DNA.

Additional file 8: Table S7. Primers and PCR cycling conditions for 5-hmC analyses.

Abbreviations
ICR: imprinting control region; DMR: differentially methylated region; IG-DMR: intergenic DMR; d.p.c.: days post-coitum; d.p.p.: days postpartum; B6: C57BL/6; C or CAST: *Mus musculus castaneus*; C12 or CAST12: *Mus musculus castaneus* chromosome 12 on a C57BL/6 background; PCR: polymerase chain reaction; 5-mC: 5-methylcytosine; 5-hmC: 5-hydroxymethylcytosine; KO: knockout.

Authors' contributions
MG and EV carried out molecular genetic studies. TLD conceived of the study and experimental design, carried out molecular genetic studies and wrote the manuscript. All authors read and approved the final manuscript.

Acknowledgements
We thank Marisa Bartolomei for providing DNA derived from wild-type and TET triple knockout embryonic stem cells. We thank Jessica Arbon and Chris Pathmanabhan for their contributions toward this work, Joshua Shapiro for advice on statistical analyses and Marisa Bartolomei for careful reading of the manuscript and thoughtful discussion.

Competing interests
The authors declare that they have no competing interests.

Funding
This work was supported by awards from the Bryn Mawr College Faculty Research Fund and National Science Foundation Grant 1514600 to TLD. In addition, MG and KV were supported in part by the Bryn Mawr College Summer Science Research program.

References
1. Morison IM, Ramsay JP, Spencer HG. A census of mammalian imprinting. Trends Genet. 2005;21:457–65.
2. Williamson CM, Blake A, Thomas S, Beechey CV, Hancock J, Cattanach BM, Peters J. World wide web site, mouse imprinting data and references. Oxfordshire: MRC Hartwell; 2013.
3. Barlow D, Bartolomei MS. Genomic imprinting in mammals. Cold Spring Harbor Perspect. Biol. 2014. doi:10.1101/cshperspect.a018382.
4. Adalsteinsson BT, Ferguson-Smith AC. Epigenetic control of the genome: lessons from genomic imprinting. Genes. 2014;5:635–55. doi:10.3390/genes5030635.
5. Tremblay KD, Duran KL, Bartolomei MS. A 5′ 2-kilobase-pair region of the imprinted mouse H19 gene exhibits exclusive paternal methylation throughout development. Mol Cell Biol. 1997;17:4322–9.
6. Shemer R, Birger Y, Riggs AD, Razin A. Structure of the imprinted mouse Snrpn gene and establishment of its parental-specific methylation pattern. Proc Natl Acad Sci USA. 1997;94:10267–72.
7. Yoon B, Herman H, Hu B, Park YJ, Lindroth A, Bell A, West AG, Chang Y, Stablewski A, Piel JC, Loukinov DI, Lobanenkov VV, Soloway PD. Rasgrf1 imprinting is regulated by a CTCF-dependent methylation-sensitive enhancer blocker. Mol Cell Biol. 2005;25:11184–90.
8. Lopes S, Lewis A, Hajkova P, Dean W, Oswald J, Forné T, Murrell A, Constância M, Bartolomei M, Walter J, Reik W. Epigenetic modifications in an imprinting cluster are controlled by a hierarchy of DMRs suggesting long-range chromatin interactions. Hum Mol Genet. 2003;12:295–305.
9. Kagami M, O'Sullivan MJ, Green AJ, Watabe Y, Arisaka O, Masawa N, Matsuoka K, Fukami M, Matsubara K, Kato F, Ferguson-Smith AC, Ogata T. The IG-DMR and the MEG3-DMR at human chromosome 14q32.2: hierarchical interaction and distinct functional properties as imprinting control centers. PLoS Genet. 2010;6:1–13 (**PMID: 20585555**).
10. Bell AC, Felsenfeld G. Methylation of a CTCF-dependent boundary controls imprinted expression of the Igf2 gene. Nature. 2000;405:482–5.
11. Hark AT, Schoenherr CJ, Katz DJ, Ingram RS, Levorse JM, Tilghman SM. CTCF mediates methylation-sensitive enhancer-blocking activity at the H19/Igf2 locus. Nature. 2000;405:486–9.
12. Sleutels F, Zwart R, Barlow DP. The non-coding Air RNA is required for silencing autosomal imprinted genes. Nature. 2002;415:810–3.
13. Nagano T, Mitchell JA, Sanz LA, Pauler FM, Ferguson-Smith AC, Feil R, Fraser P. The Air noncoding RNA epigenetically silences transcription by targeting G9a to chromatin. Science. 2008;322:1717–20. doi:10.1126/science.1163802.
14. Latos PA, Pauler FM, Koerner MV, Şenergin HB, Hudson QJ, Stocsits RR, Allhoff W, Stricker SH, Klement RM, Warczok KE, Aumayr K, Pasierbek P, Barlow DP. Airn transcriptional overlap, but not its lncRNA products, induces imprinted Igf2r silencing. Science. 2012;338:1469–72. doi:10.1126/science.1228110.
15. Mancini-DiNardo D, Steele SJS, Levorse JM, Ingram RS, Tilghman SM. Elongation of the Kcnq1ot1 transcript is required for genomic imprinting of neighboring genes. Genes Dev. 2006;20:1268–82. doi:10.1101/gad.1416906.
16. Hanel ML, Wevrick R. Establishment and maintenance of DNA methylation patterns in mouse Ndn: implications for maintenance of imprinting in target genes of the imprinting center. Mol Cell Biol. 2001;21:2384–92.
17. Bhogal B, Arnaudo A, Dymkowski A, Best A, Davis TL. Methylation at mouse Cdkn1c is acquired during postimplantation development and functions to maintain imprinted expression. Genomics. 2004;84:961–70.

18. Gagne A, Hochman A, Qureshi M, Tong C, Arbon J, McDaniel K, Davis TL. Analysis of DNA methylation acquisition at the imprinted Dlk1 locus reveals asymmetry at CpG dyads. Epigenet Chromatin. 2014;7:9.

19. Takada S, Paulsen M, Tevendale M, Tsai C-E, Kelsey G, Cattanach BM, Ferguson-Smith AC. Epigenetic analysis of the Dlk1-Gtl2 imprinted domain on mouse chromosome 12: implications for imprinting control from comparison with Igf2-H19. Hum Mol Genet. 2002;11:77–86.

20. Beygo J, Elbracht M, de Groot K, Begemann M, Kanber D, Platzer K, Gillessen-Kaesbach G, Vierzig A, Green A, Heller R, Buiting K, Eggermann T. Novel deletions affecting the MEG2-DMR provide further evidence for a hierarchical regulation of imprinting in 14q32. Eur J Hum Genet. 2015;23:180–8. doi:10.1038/ejhg.2014.72.

21. John RM, Lefebvre L. Developmental regulation of somatic imprints. Differentiation. 2011;81:270–80.

22. Stöger R, Kubicka P, Liu C-G, Kafri T, Razin A, Cedar H, Barlow DP. Maternal-specific methylation of the imprinted mouse Igf2r locus identifies the expressed locus as carrying the imprinting signal. Cell. 1993;73:61–71.

23. Kobayashi H, Sakurai T, Sato S, Nakabayashi K, Hata K, Kono T. Imprinted DNA methylation reprogramming during early mouse embryogenesis at the Gpr1-Zdbf2 locus is linked to long cis-intergenic transcription. FEBS Lett. 2012;586:827–33.

24. Yatsuki H, Joh K, Higashimoto K, Soejima H, Arai Y, Wang Y, Hatada I, Obata Y, Morisaki H, Zhang Z, Nakagawachi T, Satoh Y, Mukai T. Domain regulation of imprinting cluster in Kip2/Lit1 subdomain on mouse chromosome 7F4/F5: large-scale DNA methylation analysis reveals that DMR-Lit1 is a putative imprinting control region. Genome Res. 2002;12:1860–70.

25. Arnaud P, Monk D, Hichins M, Gordon E, Dean W, Beechey CV, Peters J, Craigen W, Preece M, Stanier P, Moore GE, Kelsey G. Conserved methylation imprints in the human and mouse GRB10 genes with divergent allelic expression suggests differential reading of the same mark. Hum Mol Genet. 2003;12:1005–19.

26. Ono R, Shiura H, Aburatani H, Kohda T, Kaneko-Ishino T, Ishino F. Identification of a large novel imprinted gene cluster on mouse proximal chromosome 6. Genome Res. 2003;13:1696–705.

27. Coombes C, Arnaud P, Gordon E, Dean W, Coar EA, Williamson CM, Feil R, Peters J, Kelsey G. Epigenetic properties and identification of an imprint mark in the Nesp-Gnasx1 domain of the mouse Gnas imprinted locus. Mol Cell Biol. 2003;23:5475–88.

28. Nowak K, Stein G, Powell E, He LM, Naik S, Morris J, Marlow S, Davis TL. Establishment of paternal allele-specific DNA methylation at the imprinted mouse Gtl2 locus. Epigenetics. 2011;6:1012–20.

29. Woodfine K, Huddleston JE, Murrell A. Quantitative analysis of DNA methylation at all human imprinted regions reveals preservation of epigenetic stability in adult somatic tissue. Epigenet Chromatin. 2011;4:1. doi:10.1186/1756-8935-4-1.

30. Tahiliani M, Koh KP, Shen Y, Pastor WA, Bandukwala H, Brudno Y, Agarwal S, Iyer LM, Liu DR, Aravind L, Rao A. Conversion of 5-methylcytosine to 5-hydroxymethylcytosine in mammalian DNA by MLL partner TET1. Science. 2009;324:930–5.

31. Ito S, Shen L, Dai Q, Wu SC, Collins LB, Swenberg JA, He C, Zhang Y. Tet proteins can convert 5-methylcytosine to 5-formylcytosine and 5-carboxylcytosine. Science. 2011;333:1300–3.

32. He YF, Li BZ, Li Z, Liu P, Wang Y, Tang Q, Ding J, Jia Y, Chen Z, Li L, Sun Y, Li X, Dai Q, Song CX, Zhang K, He C, Xu GL. Tet-mediated formation of 5-carboxylcytosine and its excision by TDG in mammalian DNA. Science. 2011;333:1303–7.

33. Kohli RM, Zhang Y. TET enzymes, TDG and the dynamics of DNA demethylation. Nature. 2013;502:472–9. doi:10.1038/nature12750.

34. Valinluck V, Sowers LC. Endogenous cytosine damage products alter the site selectivity of human DNA maintenance methyltransferase DNMT1. Cancer Res. 2007;67:946–50.

35. Lin S-P, Coan P, Teixeira da Rocha S, Seitz H, Cavaille J, Teng P-W, Takada S, Ferguson-Smith AC. Differential regulation of imprinting in the murine embryo and placenta by the Dlk1-Dio3 imprinting control region. Development. 2007;134:417–26.

36. Magalhães HR, Leite SBP, de Paz CCP, Duarte G, Ramos ES. Placental hydroxymethylation vs methylation at the imprinting control region 2 on chromosome 11p15.5. Braz J Med Biol Res. 2013;46:916–9. doi:10.1590/1414-431X20133035.

37. Hu X, Zhang L, Mao S-Q, Li Z, Chen J, Zhang R-R, Wu H-P, Gao J, Guo F, Liu W, Xu G-F, Dai H-Q, Shi YG, Li X, Hu B, Tang F, Pei D, Xu G-L. Tet and TDG mediate DNA demethylation essential for mesenchymal-to-epithelial transition in somatic cell reprogramming. Cell Stem Cell. 2014;14:512–22. doi:10.1016/j.stem.2014.01.001.

38. Mackay DJG, Callaway JLA, Marks SM, White HE, Acerini CL, Boonen SE, Dayanikli P, Firth HV, Goodship JA, Haemers AP, Hahnemann JMD, Kordonouri O, Masoud AF, Oestergaard E, Storr J, Ellard S, Hattersley AT, Robinson DO, Temple IK. Hypomethylation of multiple imprinted loci in individuals with transient neonatal diabetes is associated with mutations in ZFP57. Nat Genet. 2008;40:949–51. doi:10.1038/ng.187.

39. Eggermann T, Perez de Nanciares G, Maher ER, Temple IK, Tümer Z, Monk D, Mackay DJ, Grønskov K, Ricci A, Linglart A, Netchine I. Imprinting disorders: a group of congenital disorders with overlapping patterns of molecular changes affecting imprinted loci. Clin Epigenet. 2015;7:123. doi:10.1186/s13148-015-0143-8.

40. Genereux DP, Johnson WC, Burden AF, Stöger R, Laird CD. Errors in the bisulfite conversion of DNA: modulating inappropriate- and failed-conversion frequencies. Nucleic Acids Res. 2008;36(22):e150. doi:10.1093/nar/gkn691.

41. Holmes EE, Jung M, Meller S, Leisse A, Sailer V, Zech J, Mengdehl M, Garbe L-A, Uhl B, Kristiansen G, Dietrich D. Performance evaluation of kits for bisulfite-conversion of DNA from tissues, cell lines, FFPE tissues, aspirates, lavages, effusions, plasma, serum and urine. PLoS ONE. 2014;9(4):e93933. doi:10.1371/journal.pone.0093933.

42. Steshina EY, Carr MS, Glick EA, Yevtodiyenk A, Appelbe OK, Schmidt JV. Loss of imprinting at the Dlk1-Gtl2 locus caused by insertional mutagenesis in the Gtl2 5′ region. BMC Genet. 2006;7:44.

43. Sekita Y, Wagatsuma H, Irie M, Kobayashi S, Kohda T, Matsuda J, Yokoyama M, Ogura A, Schuster-Gossler K, Gossler A, Ishino F, Kaneko-Ishino T. Aberrant regulation of imprinted gene expression in Gtl2^{lacZ} mice. Cytogenet Genome Res. 2006;113:223–9.

44. Li D, Guo B, Wu H, Tan L, Lu Q. TET family of dioxygenases: crucial roles and underlying mechanisms. Cytogenet Genome Res. 2015;146:171–80. doi:10.1159/000438853.

45. Pastor WA, Pape UJ, Huang Y, Henderson HR, Liste R, Ko M, McLoughlin EM, Brudno Y, Mahapatra S, Kapranov P, Tahiliani M, Daley GQ, Liu XS, Ecker JR, Milos PM, Agarwal S, Rao A. Genome-wide mapping of 5-hydroxymethylcytosine in embryonic stem cells. Nature. 2011;473:394–7. doi:10.1038/nature10102.

46. Wu H, D'Alessio AC, Ito S, Xia K, Wang Z, Cui K, Zhao K, Sun E, Zhang Y. Dual functions of Tet1 in transcriptional regulation in mouse embryonic stem cells. Nature. 2011;473:389–93. doi:10.1038/nature09934.

47. Stroud H, Feng S, Morey Kinney S, Pradhan S, Jacobsent SE. 5-hydroxymethylcytosine is associated with enhancers and gene bodies in human embryonic stem cells. Genome Biol. 2011;12(6):R54. doi:10.1186/gb-2011-12-6-r54.

48. Huang Y, Chavez L, Chang X, Wang X, Pastor WA, Kang J, Zepeda-Martinez JA, Pape UJ, Jacobsen SE, Peters B, Rao A. Distinct roles of the methylcytosine oxidases Tet1 and Tet2 in mouse embryonic stem cells. Proc Natl Acad Sci USA. 2014;111:1361–6. doi:10.1073/pnas.1322921111.

49. Yamaguchi S, Hong K, Liu R, Inoue A, Shen L, Zhang K, Zhang Y. Dynamics of 5-methylcytosine and 5-hydroxymethylcytosine during germ cell reprogramming. Cell Res. 2013;23:329–39.

50. Hackett JA, Sengupta R, Zylicz JJ, Murakami K, Lee C, Down TA, Surani MA. Germline DNA demethylation dynamics and imprint erasure through 5-hydroxymethylcytosine. Science. 2013;339:448–52.

51. Yamaguchi S, Shen L, Liu Y, Sendler D, Zhang Y. Role of Tet1 in erasure of genomic imprinting. Nature. 2013;504:460–4. doi:10.1038/nature12805.

52. Dawlaty MM, Breiling A, Le T, Raddatz G, Barrasa MI, Cheng AW, Gao Q, Powell BE, Li Z, Xu M, Faull KF, Lyko F, Jaenisch R. Combined deficiency of Tet1 and Tet2 causes epigenetic abnormalities but is compatible with postnatal development. Dev Cell. 2013;24:310–23. doi:10.1016/j.devcell.2012.12.015.

53. Liu L, Mao S-Q, Ray C, Zhang Y, Bell FT, Ng S-F, Xu G-L, Li X. Differential regulation of genomic imprinting by TET proteins in embryonic stem cells. Stem Cell Res. 2015;15:435–43. doi:10.1016/j.scr.2015.08.010.

54. Georgiades P, Watkins M, Surani MA, Ferguson-Smith AC. Parental origin-specific developmental defects in mice with uniparental disomy for chromosome 12. Development. 2000;127:4719–28.

55. Hahn MA, Qiu R, Wu X, Li AX, Zhang H, Wang J, Jui J, Jin SG, Jiang Y, Pfeifer GP, Lu Q. Dynamics of 5-hydroxymethylcytosine and chromatin marks in mammalian neurogenesis. Cell Rep. 2013;3:291–300.

56. Arand J, Wossidlo M, Lepikhow K, Peat JR, Reik W, Walte J. Selective impairment of methylation maintenance is the major cause of DNA methylation reprogramming in the early embryo. Epigenet Chromatin. 2015;8:1.

57. Jin C, Lu Y, Jelinek J, Liang S, Estecio MRH, Barton MC. Issa J-PJ. TET1 is a maintenance DNA demethylase that prevents methylation spreading in differentiated cells. Nucleic Acids Res. 2014;42:6956–71. doi:10.1093/nar/gku372.

58. Davis TL, Trasler JM, Moss SB, Yang GJ, Bartolomei MS. Acquisition of the H19 methylation imprint occurs differentially on the parental alleles during spermatogenesis. Genomics. 1999;58:18–28.

59. Laird CD, Pleasant ND, Clark AD, Sneeden JL, Hassan KMA, Manley NC, Vary JC, Morgan T, Hansen RS, Stöger R. Hairpin-bisulfite PCR: assessing epigenetic methylation patterns on complementary strands of individual DNA molecules. Proc Natl Acad Sci USA. 2004;101:204–9.

3

Genome-wide search for Zelda-like chromatin signatures identifies GAF as a pioneer factor in early fly development

Arbel Moshe and Tommy Kaplan*⦿

Abstract

Background: The protein Zelda was shown to play a key role in early Drosophila development, binding thousands of promoters and enhancers prior to maternal-to-zygotic transition (MZT), and marking them for transcriptional activation. Recently, we showed that Zelda acts through specific chromatin patterns of histone modifications to mark developmental enhancers and active promoters. Intriguingly, some Zelda sites still maintain these chromatin patterns in Drosophila embryos lacking maternal Zelda protein. This suggests that additional Zelda-like pioneer factors may act in early fly embryos.

Results: We developed a computational method to analyze and refine the chromatin landscape surrounding early Zelda peaks, using a multichannel spectral clustering. This allowed us to characterize their chromatin patterns through MZT (mitotic cycles 8–14). Specifically, we focused on H3K4me1, H3K4me3, H3K18ac, H3K27ac, and H3K27me3 and identified three different classes of chromatin signatures, matching "promoters," "enhancers" and "transiently bound" Zelda peaks. We then further scanned the genome using these chromatin patterns and identified additional loci—with no Zelda binding—that show similar chromatin patterns, resulting with hundreds of Zelda-independent putative enhancers. These regions were found to be enriched with GAGA factor (GAF, Trl) and are typically located near early developmental zygotic genes. Overall our analysis suggests that GAF, together with Zelda, plays an important role in activating the zygotic genome.

Conclusions: As we show, our computational approach offers an efficient algorithm for characterizing chromatin signatures around some loci of interest and allows a genome-wide identification of additional loci with similar chromatin patterns.

Keywords: Zelda, Maternal-to-zygotic transition, Histone modifications, Enhancers, Chromatin, Spectral clustering, Chromatin search, GAGA factor

Background

The process of transcription is vital to all living organisms and is tightly regulated by multiple mechanisms, including the packaging of DNA into chromatin. In eukaryotic cells, the DNA is wrapped around nucleosomes to form chromatin. This packaging is used differentially to control in what conditions and cell types a gene is more accessible—and active—and in which conditions it is tightly packed and silenced. This packaging of DNA was shown to be mediated by various mechanisms, including the deposition of covalent modifications (e.g., acetylation, methylation, phosphorylation or ubiquitylation) at different residues of the core histone proteins that are assembled into a nucleosome [1]. These histone modifications influence various processes along the DNA. For example, H3K4me1 and H3K27ac (namely, mono-methylation of Lysine 4 or acetylation of Lysine 27 in histone H3) are found at nucleosomes carrying active regulatory regions, while H3K27me3 is known to be a repressive mark [2–4].

*Correspondence: tommy@cs.huji.ac.il
School of Computer Science and Engineering, The Hebrew University of Jerusalem, Jerusalem 91904, Israel

The packaging of DNA into chromatin, including nucleosome positioning and their histone modifications, are fundamental to the proper activity of regulatory regions, by controlling which regions of the genome are accessible for protein binding and which are not [5–7]. Of specific interest are promoter regions, located near the transcription start site (TSS) of genes; and enhancers, that regulate gene expression from afar, often up to 1 Mb away from their target genes. In addition, chromatin marks and structural proteins allow distal enhancers to fold in 3D into close spatial proximity to their target genes [8–10].

Comparison of tissue- and condition-specific chromatin data highlights the tight regulation of gene expression by chromatin, determining which genomic regions are active and which are not, and as a consequence which genes will be transcribed. This raises the question of causality. Who regulates packaging? Or how does the genome get packed initially in the proper architecture, e.g., to drive early developmental expression?

In practically all animals, early developmental stages begin with maternal proteins and RNA that control the first hours in the fertilized egg. At this stage, these proteins control the first wave of zygotic expression and direct the first mitotic divisions [11]. After that, the embryo undergoes a process called maternal-to-zygotic transition (MZT), in which the zygotic genome is activated and takes control of mRNA and protein production. Finally, maternal mRNA and proteins are degraded.

In the fruit fly *Drosophila melanogaster*, embryonic development is characterized by a series of 13 rapid replication cycles, occurring during the first 2 h after fertilization. The division of cells slows at the 14th mitotic cycle, and zygotic transcription initiates. This marks the end of the *D. melanogaster* maternal-to-zygotic transition. This process is crucial for the normal development of the embryo and is tightly controlled, in both time and space, by the gradual activation of a cascade of transcription factors [12, 13]. This required multiple molecular mechanisms that include chromatin, nucleosomes, DNA accessibility, steric hindrance between DNA-binding proteins and more.

Previous research showed that many of the early transcribed genes in Drosophila embryos contain a specific DNA motif of 7 bp, CAGGTAG, that occurs within their regulatory regions, including both promoters and enhancers [14–16]. This motif was later identified to be the binding site of the zinc-finger transcription factor Zelda (vielfaltig, vfl) [17]. Following studies, by us and others, showed that Zelda is present in the embryonic nucleus as early as mitotic cycle 2 and binds thousands of genomic loci, including the promoters and enhancers of thousands of early developmental genes (Fig. 1) [12, 16–18].

Computational and experimental studies, by us and others, suggested that Zelda acts as a pioneer factor, binding mostly inaccessible DNA regions and making the chromatin accessible for other transcription factors to bind, thus marking thousands of genes and regulatory regions for activation which drives the first transcriptional program of the developing embryo [9, 16, 19].

Indeed, experimental studies showed that early Zelda binding is a predictor for open chromatin and transcription factor binding at mitotic cycle 14 [13, 15–17, 20]. The entire set of molecular mechanisms by which Zelda functions to access the genome and mark it for activation is yet to be discovered, as is the role of additional proteins in this crucial stage in embryonic gene expression.

Several studies mapped the chromatin landscape in *D. melanogaster*, most of which in cells or during very broad temporal windows [21–24]. Of particular interest are few studies, in which early and manually staged Drosophila embryos were used to portray Zelda binding locations and multiple histone marks (including H3K27ac, H3K4me1, H3K4me3, H327me3, and H3K36me3) at several time points throughout MZT and early embryonic development [5, 16]. As was shown, early Zelda binding often results in open chromatin and characteristic histone marks, including H3K27ac, H3K4me1, and H3K27me3 peaks [17, 25].

Intriguingly, the causal role of Zelda in proper establishment of chromatin domains and normal gene expression was also examined. Embryos lacking zygotic Zelda expression as well as embryos with mutations in specific Zelda binding sites (near early genes) showed developmental abnormalities [14, 17]. Embryos lacking maternal Zelda expression (zld^{M-}) showed reduced accessibility for many, but not all, distal enhancers regulating early fly development, while maintaining near-normal promoter accessibility [5, 20].

The latter results suggest that perhaps, in addition to Zelda, there might be another protein that marks early developmental genes for activation. To answer this question, we revisited the chromatin patterns surrounding Zelda binding sites and developed a computational statistical model to characterize histone modifications and their dynamics. Several previous methods faced a similar situation, where a set of genomic loci are given, and unsupervised machine learning techniques should be used to re-orient them [26] (e.g., for asymmetric chromatin signatures, often found at transcription start sites), re-align them using profile alignment dynamic programming algorithms [27], and cluster them into several distinct classes. At a second stage, we used our computational model to scan the genome and identify additional genomic loci that show similar chromatin patterns. Previous algorithms (e.g., RFECS [28]) used machine learning to train a set of binary classification trees (random forest) and classify the

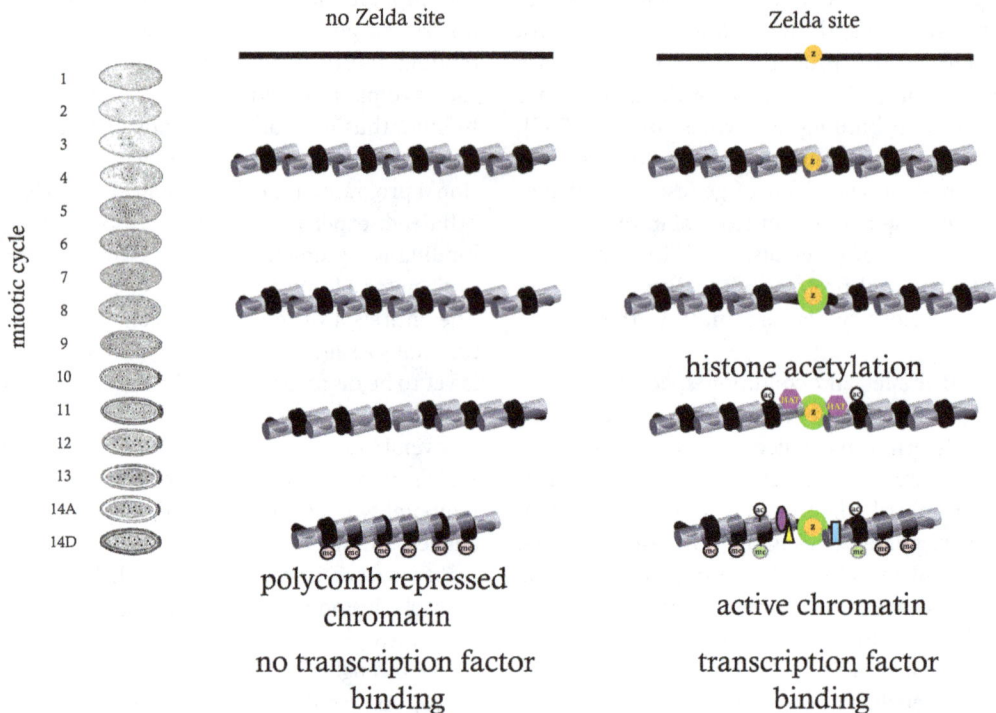

Fig. 1 The role of Zelda during zygotic genome activation. Zelda binds to regulatory regions in pre-MZT embryos as early as mitotic cycle 2. This leads to histone acetylation and nucleosome remodeling around ZLD binding sites, which facilitates binding by other transcription factors and deposition of active histone marks. At the same time, Zelda binding prevents deposition of H3K27me3 marks and formation of repressive chromatin structure. Adapted from Li et al. [5]

chromatin data around each putative enhancer locus (± 1 kb window). Instead, we applied a regularized correlation-based approach, where each genomic locus is compared to the (multidimensional) chromatin signature of each cluster. This allowed us to maintain the overall typical shape of each histone modification, while allowing for different ChIP signal intensities.

As we show, our approach identified about 2000 genomic regions with Zelda-like chromatin signatures and dynamics, with no Zelda binding. As we show, these regions are enriched for the protein GAF, whose early binding could establish open chromatin and activation of regulatory regions.

Results

To identify additional Zelda-like regulators that act as pioneer factors during early Drosophila development, we begin by characterizing the chromatin landscape induced around early Zelda sites.

Zelda peaks vary in their chromatin signatures

We have previously shown that early Zelda peaks, identified via ZLD ChIP-seq in hand-sorted fly embryos from mitotic cycle 8, are associated with open chromatin regions and transcription factors binding later on, toward the end of the maternal-to-zygotic transition (mitotic cycle 14) (Fig. 1) [12, 13, 16].

However, a close examination of some strong early Zelda sites (Fig. 2) suggests that there is more than one typical chromatin signature. Specifically, shown are two Zelda peaks, one at an intron of the *schnurri* gene (*shn*, Fig. 2a) and one at the promoter of *bitesize* (btsz, Fig. 2b). The former shows a ZLD peak surrounded by H3K4me1-marked nucleosomes and flanked by H3K27 tri-methylation. The latter shows the promoter Zelda peak is surrounded by H3K4me1 nucleosomes, as well as H3K4me3 on one side of the ZLD peak. Intriguingly, as shown in Fig. 2c, some early Zelda peaks seem to show no chromatin pattern whatsoever (by cycle 14) and are practically indistinguishable from their surroundings.

We therefore hypothesized that different Zelda peaks could have different histone modification patterns, which may not be obscured when only considering the average, common pattern, near Zelda peaks.

Chromatin-based re-orientation of early Zelda peaks

As a first step toward a more descriptive and accurate characterization of chromatin near Zelda sites, we first wanted to check which histone modifications are symmetric around Zelda peaks and which are not [26]. For this, we

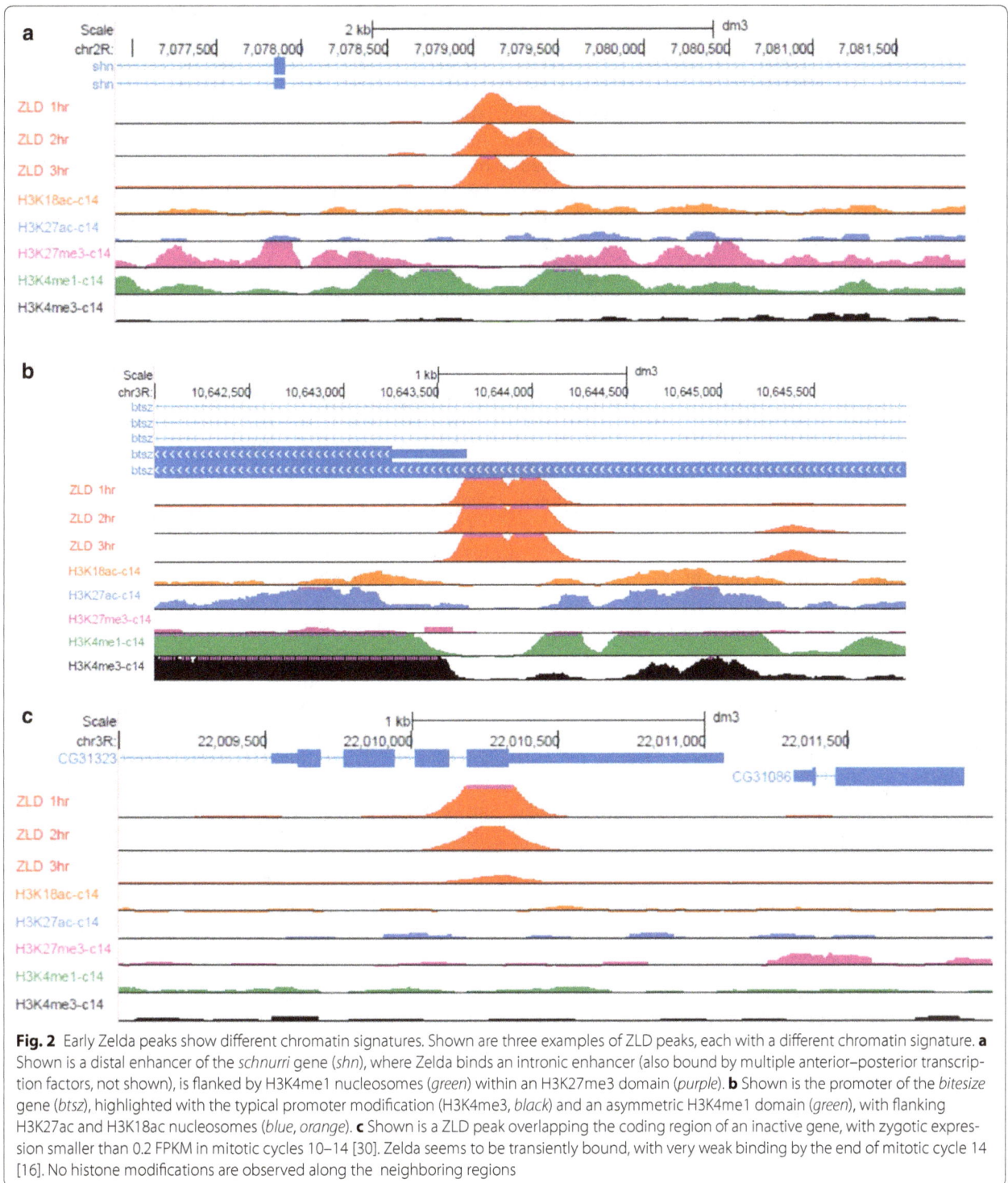

Fig. 2 Early Zelda peaks show different chromatin signatures. Shown are three examples of ZLD peaks, each with a different chromatin signature. **a** Shown is a distal enhancer of the *schnurri* gene (*shn*), where Zelda binds an intronic enhancer (also bound by multiple anterior–posterior transcription factors, not shown), is flanked by H3K4me1 nucleosomes (*green*) within an H3K27me3 domain (*purple*). **b** Shown is the promoter of the *bitesize* gene (*btsz*), highlighted with the typical promoter modification (H3K4me3, *black*) and an asymmetric H3K4me1 domain (*green*), with flanking H3K27ac and H3K18ac nucleosomes (*blue, orange*). **c** Shown is a ZLD peak overlapping the coding region of an inactive gene, with zygotic expression smaller than 0.2 FPKM in mitotic cycles 10–14 [30]. Zelda seems to be transiently bound, with very weak binding by the end of mitotic cycle 14 [16]. No histone modifications are observed along the neighboring regions

implemented an expectation maximization (EM)-like iterative orientation method that automatically infers, based on histone modification data, the orientation of each Zelda peak, while iteratively improving the typical chromatin "signature" surrounding Zelda peaks [26, 29].

Our algorithm begins with a random orientation of each Zelda site. We then calculate the average ChIP-seq signal over all Zelda peaks in a 10 Kb region (Fig. 3a). Then, every iteration, we enumerate over each Zelda peak and compare the similarity of its surrounding 10 Kb for the average signal in each of the two orientations (+ or −). This is done while considering all chromatin marks together. Regions with higher similarity for the inverted signatures are then flipped. Finally, we calculate the updated average chromatin signatures, and a new iteration begins. Figure 3b shows the final chromatin signatures, after the algorithm converges (typically, in less than 10 iterations).

We decided to choose a fully unsupervised chromatin-based re-orientation, rather than relying on gene directionality, as we believe that at least some of the modifications are "one-sided," and therefore would contribute to inferring the orientation of most Zelda peaks. Indeed, while most of the modifications maintain a symmetrical formation, both H3K4me3 (promoter) and H3K36me3 (gene body) show strong asymmetric bias, consistent with the orientation of the underlying gene (Fig. 3b). Also notable is the input IP signature, which is almost entirely flat (except for a tiny bump near Zelda), suggesting that the asymmetric signatures are not due to the additional degree of freedom per Zelda site.

Chromatin-based clustering of Zelda peaks

To test whether there are different types of Zelda peaks, each with a different combination of histone modifications patterns, we turned to develop a computational model that will allow us to characterize the chromatin landscape and temporal dynamics at each early Zelda peak. Such an approach would not only shed light on the different functional roles of Zelda peaks, but would also allow us to scan the genome using a refined model, and identify additional loci, independent of ZLD binding, that undergo similar chromatin dynamics during MZT.

As shown in Figs. 2 and 3, it is not enough to only consider the maximal signal of each histone modification, as the modifications we look at are not characterized by a single peak, but rather show a unique spatial signature, which is often asymmetric. We therefore want a specialized method that considers both the pattern of each histone modification and the overall combination of different modifications and their heights.

For this, we decided to examine a 10-Kb window surrounding each Zelda peak and apply a spectral clustering algorithm that will consider all histone modifications simultaneously. We begin by describing a distance function between the chromatin signature of two loci that would capture both the "landscape shape" of each modification, its overall magnitude, and the combinatorial nature of multiple modifications.

An advantage of spectral clustering is the fact that the algorithm relies on the adjacency (or connectivity) between objects within each cluster, rather than their spatial shape.

Our initial chromatin ChIP-seq data consisted of nine histone modifications (H3K4ac, H3K4me1, H3K4me3, H3K9ac, H3K18ac, H3K27ac, H3K27me3, H3K36me3, and H4K5ac) in four time points throughput MZT (mitotic cycles 8, 11, 13, and 14) [5]. We therefore selected the top 2000 early (cycle 8) Zelda peaks—prior to MZT and the activation of most zygotic genes, and before the establishment of chromatin marks [5, 30]—and for each considered a surrounding window of 10 Kb for five regulatory histone modifications (H3K4me1, H3K4me3, H3K27ac, H3K27me3, and H3K18ac), as measured in mitotic cycles 13 and 14. H3K36me3 was often depleted from Zelda sites and localized mainly in

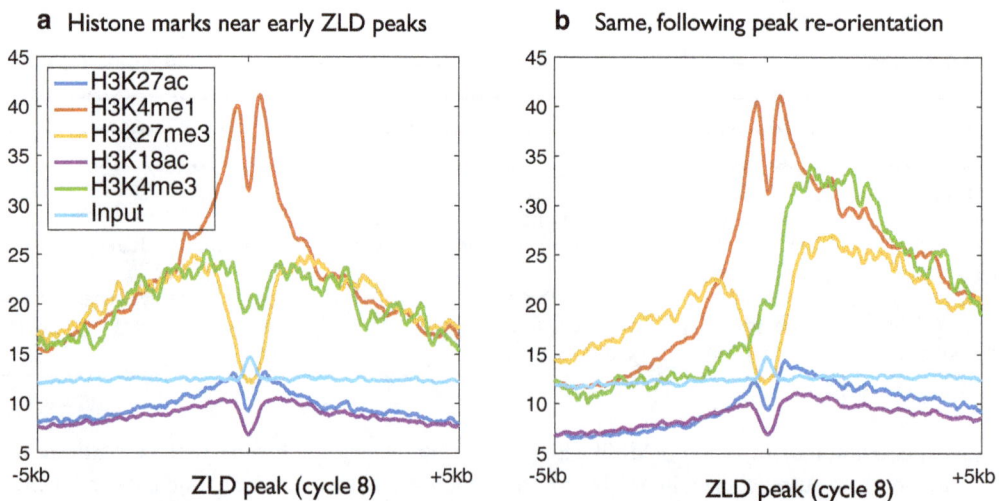

Fig. 3 Chromatin marks surrounding Zelda peaks prior to and following peak re-orientation. **a** Average chromatin marks at mitotic cycle 14, from Li et al. [5], centered on 2000 early Zelda peaks (cycle 8). Clearly visible are relative enrichment near Zelda site for the enhancer mark H3K4me1 (*orange*) and depletion of the repressive H3K27me3 (*yellow*). **b** Same, following re-orientation of Zelda peaks based on all five modifications

---

---

gene bodies, and we decided not to use it for the clustering. The idea here was not to obtain a definitive metric for chromatin, but rather to identify the regulation-specific chromatin landscapes near Zelda peaks.

This resulted in representing each genomic locus i by ten vectors $\mathbf{x}_i^m$: one for each of the five histone modifications at each of the two time points (marked by m). We have down-sampled the chromatin data to 10 bp resolution, resulting in vectors of length 1000. To calculate the distance $\mathbf{dist}_{i,j}^m$ between the two vectors representing the landscape of the modification m surrounding two loci i and j, we first used the root mean squared deviation (RMSD, see Eq. 1, "Methods") between the two vectors $\mathbf{x}_i^m$ and $\mathbf{x}_j^m$ and then converted these into weights (or adjacency scores), using a Gaussian kernel with a modification-specific parameter σ_m (Eq. 2, "Methods"). This parameter sets the standard deviation parameter of the Gaussian kernel, thus normalizing different histone modifications and assigning each a similar importance. We have arbitrarily set σ_m for each modification/time point to equal to the 10th percentile in the distribution of pairwise distances for that specific modification (see "Methods"). Finally, we summed each of the ten adjacency matrices (Eq. 3) and applied standard (symmetrically normalized) spectral clustering, followed by k-means [31]. The value for k (i.e., number of clusters) was chosen using the eigengap heuristic ("Methods"). Overall, the clustering process resulted with three main clusters, each with a unique combinatorial chromatin signature (Fig. 4 and Additional file 1: Figure S1) . Similar results with different bin sizes (20 bp, 50 bp, 100 bp) resulted with almost identical orientation or cluster assignment (<1% change).

The first cluster (Fig. 4, left) is composed of 620 early Zelda peaks (of the initial 2000 peaks analyzed) and shows enrichment for H3K4me1 (orange), on either side of early Zelda peaks, and H3K4me3 mark (green), enriched only downstream of Zelda peaks. While these marks are already observed in mitotic cycle 13 (bottom row), they dramatically strengthen by cycle 14 (top) [5].

The second cluster (Fig. 4, center, 660 Zelda peaks) shows narrow symmetric H3K4me1 peaks (orange) on either sides of Zelda, flanked by wider H3K27me3 domains (yellow). Intriguingly, while this pattern resembles the chromatin signature of "poised enhancers," described by Rada-Iglesias et al. [3] in human embryonic stem cells (hESCs), a closer examination suggests that most of these genomic regions are actually active enhancer regions (see below). As we noted previously [5], H3K27me3 is deposited almost exclusively by mitotic cycle 14, following MZT.

Finally, the third cluster (Fig. 4, right, 720 Zelda peaks) showed almost no enriched chromatin marks, suggesting

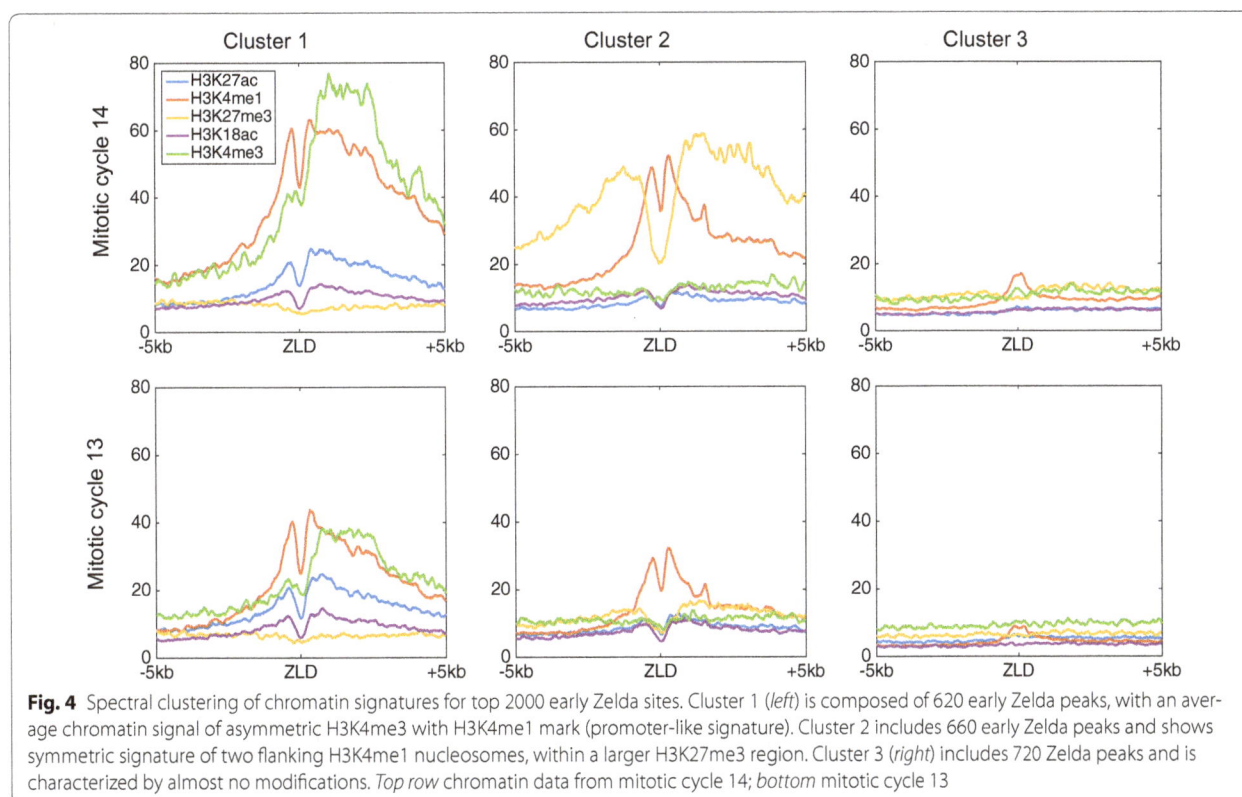

Fig. 4 Spectral clustering of chromatin signatures for top 2000 early Zelda sites. Cluster 1 (left) is composed of 620 early Zelda peaks, with an average chromatin signal of asymmetric H3K4me3 with H3K4me1 mark (promoter-like signature). Cluster 2 includes 660 early Zelda peaks and shows symmetric signature of two flanking H3K4me1 nucleosomes, within a larger H3K27me3 region. Cluster 3 (right) includes 720 Zelda peaks and is characterized by almost no modifications. Top row chromatin data from mitotic cycle 14; bottom mitotic cycle 13

that these are transient Zelda binding sites that are bound early and then vacated with no apparent biological effect [5].

Functional characterization of three Zelda clusters

As we showed, spectral clustering of early Zelda peaks resulted in three clusters, each displaying a unique combination of histone modifications. We next turned to examine whether these clusters demonstrate different functional parameters.

First, we wanted to test whether the detected differences in chromatin packaging affect transcription factor binding. For this, we analyzed genome-wide binding data for 17 early development transcription factors, as measured after the completion of the maternal-to-zygotic transition at mitotic cycle 14 [13]. For each of the 2000 early Zelda peaks that were clustered, we calculated the

number of unique anterior–posterior (A–P) and dorsal–ventral (D–V) transcription factors that are bound at the same locus. As shown in Fig. 5a, early Zelda-bound regions in cluster 1 show strong accumulation of transcription factors by cycle 14 (>4 TFs on average), with an even stronger signal for cluster 2 Zelda sites (~6 TFs on average), compared to ~2 TFs for cluster 3 regions.

This raises the hypothesis that while most early Zelda sites become accessible during MZT and facilitate the binding of regulatory proteins, perhaps the sites captured by cluster 3 do not.

For this, we turned to examine the dynamics of Zelda binding itself, as measured by ChIP-seq in three time points throughout MZT [16]. As shown in Fig. 5b, a similar number of early Zelda peaks in cluster 1 exhibit increased vs. decreased ChIP signal during MZT (37% increasing, 40% decreasing). Cluster 2 Zelda peaks are

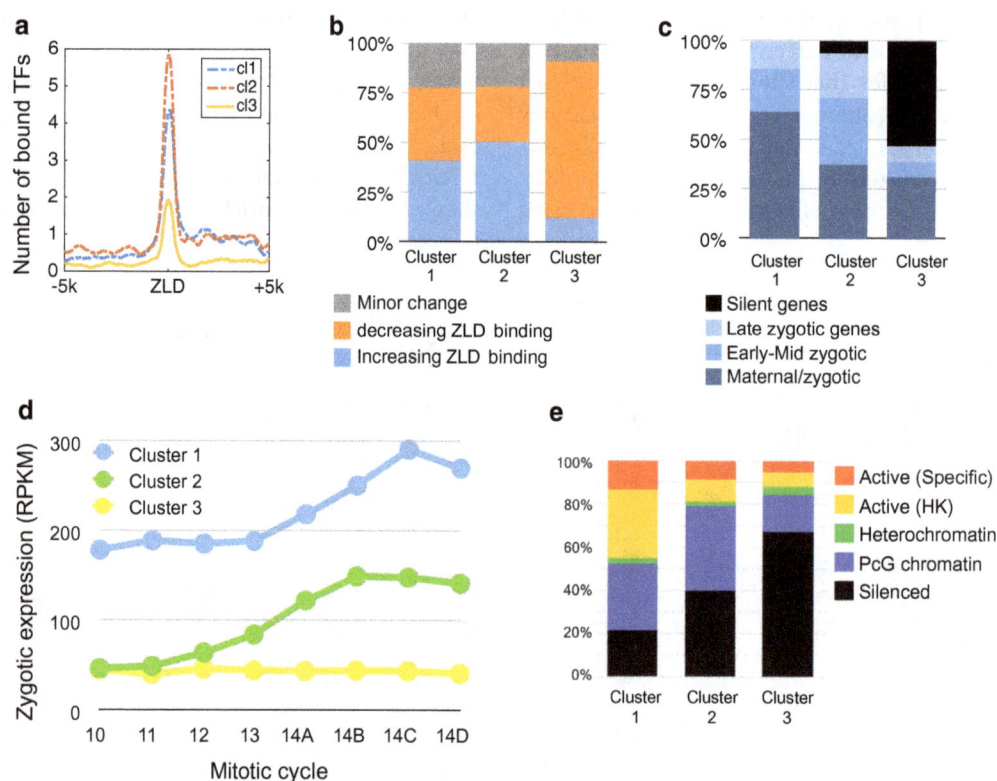

Fig. 5 Biological characteristics of the three clusters. **a** By examining the binding of 17 anterior–posterior and dorsal–ventral transcription factor (TFs) in early fly development, we calculated the average number of TFs bound for each locus with each cluster [13]. Early Zelda peaks in clusters 1 and 2 seem to be bound (by the end of mitotic cycle 14) by >4 and ~6 factors, respectively, compared to <2 factors bound near early Zelda peaks in cluster 3. **b** Analysis of temporal dynamics in Zelda binding [16]. In clusters 1 (*left*) and 2 (*center*), about 37% of the ZLD peaks are increasing, 40% decreasing, and ~20% show minor changes in peak height. In contrast, 79% of cluster 3 regions (*right*) show reduced ZLD binding. **c** By associating each Zelda site locus to the nearest TSS, we show that cluster 1 peaks are mostly associated with maternally deposited genes that are also transcribed in early developmental stages ("maternal/zygotic," e.g., house-keeping genes). Cluster 2 is more associated with zygotically transcribed genes, while cluster 3 is mostly not expressed [30]. **d** Average expression levels [30] for gene associated with clustered Zelda peaks, at eight time points throughout MZT. **e** Different chromatin patterns as annotated into five chromatin types using HMM with 53 chromatin proteins in Drosophila Kc167 cells [21]. Types include tissue-specific active genes (*red*), "house-keeping"-like active genes (*yellow*), and heterochromatin (*green*), polycomb-related (*blue*) or silenced genomic regions (*black*)

mostly increasing (50%, compared to 28% peaks with weaker ZLD ChIP-seq signal from mitotic cycle 8–14). Conversely, the vast majority (79%) of Zelda peaks in cluster 3 show a decrease in Zelda binding, consistently with the observed reduction in transcription factor binding. As shown in Additional file 2: Figure S2, cluster 3 Zelda peaks are initially showing a lower average ChIP-seq signal for Zelda binding, and indeed most peaks show a consistent decrease throughout the MZT.

To test the functional effect of these three groups in regulating gene expression, we associated every Zelda peak with the nearest gene. Indeed, we observed differences in the distance distribution to the nearest TSS. While over 38% of cluster 1 peaks were directly located at gene promoters (hence the asymmetric H3K4me3 signature shown in Fig. 4, left), only 25% of the peaks from cluster 2, and 11% of cluster 3 Zelda peaks were located at promoters (Additional file 3: Figure S3). Opposite trends were observed for distal "intergenic" peaks, consisting of 3% of cluster 1 peaks, compared to 23 and 25% of clusters 2 and 3, respectively. Considering the chromatin signature (H3K4me1 and H3K27me3) and the high number of TF bound at these loci, it is not unlikely that most of the distal regions (in cluster 2) act as regulatory enhancers.

Following Harrison et al. [16], where we used single-embryo RNA-seq data from mitotic cycles 10–14 [30] to classify early developmental genes into temporal groups, we have now compared these annotations to the genes associated with each of the peaks in the three clusters. As shown in Fig. 5c, peaks in cluster 1 were exclusively associated with genes expressed both maternally and in the zygote (maternal/zygotic) or early developmental zygotic genes. Zelda peaks from cluster 2 (i.e., the putative enhancer-like group) were even more enriched with early, strictly zygotic genes. Cluster 3 was dramatically enriched with silenced genes or house-keeping genes (expressed both maternally and in the zygote). As Fig. 5d and Additional file 4: Figure S4A show, genes associated with cluster 1 peaks are already expressed (strongly) by mitotic cycle 10 [30], while cluster 2 Zelda peaks are mostly associated with early developmental genes activated during MZT.

Finally, we compared the three clusters to HMM-based annotations of the fly genome into five chromatin types, based on the binding patterns of 53 chromatin proteins in Drosophila Kc167 cells [21]. While these results are based on a cultured cell-line—compared to in vivo ChIP-seq or gene expression data from embryos—they do originate from embryonic tissue (developmental stages 13–15). These results demonstrate how early Zelda binding shapes the chromatin to form the basis for later stages of embryonic development. As shown in Fig. 5e, the three classes show district patterns, with cluster 1 mostly

enriched for the active genes (either "house-keeping"-like genes in yellow, or more tissue-specific genes, red), as well as genomic regions that are packed in polycomb (blue) or repressed (black) in Kc167 cells.

Functional GO annotation analysis for the genes associated with each Zelda cluster also supports our separation into three classes, where cluster 1 peaks are associated with terms such as anatomical morphogenesis and developmental genes ($p < 8e-28$ and $p < 3.3e-22$, respectively), while cluster 2 peaks are more associated with transcription factors ($p < 5.6e-45$) and patterning ($p < 1.1e-40$; Additional file 4: Figure S4).

De novo chromatin-based identification of Zelda-like loci

We next turned to use the three chromatin signatures obtained using the spectral clustering algorithm and identify additional loci showing similar chromatin patterns. While Zelda plays a major role in shaping the chromatin landscape and accessibility during the MZT, marking them for activation and enabling the binding of regulatory proteins, many early Drosophila genes are still expressed in Zelda maternal mutants embryos [5, 20]. This suggests that there might be some other mechanisms (besides Zelda) that may also lead to similar chromatin signatures of active and regulatory chromatin (Additional file 5: Figure S5, Additional file 6: Figure S6).

For this, we cross-correlated the genome with the chromatin patterns of the promoter-like (cluster 1) and enhancer-like (cluster 2) signatures (Fig. 4). We used the same set of histone modification marks as used for clustering (H3K27ac, H3K27me3, H3K4me1, H3K4me3, and H3K18ac) at cycle 14, with the same 10 Kb windowing. Initially, we scanned the genome with each of the histone modification patterns separately and computed the average correlation for all modifications. This naive analysis failed to obtain satisfying results, as the relative strength of different histone modification (ChIP-seq signal) was not preserved. As a result, many loci showed high correlation all query patterns, but their combined chromatin signature was skewed (e.g., strong H3K27me3 flanks, with very weak H3K4me1 signal). Next, we tried to search for all modifications simultaneously by concatenating the histone modifications into one long vector and then cross-correlating against the genome. These results were still rather disappointing, since correlation by itself does not account for the overall magnitude of the histone modification pattern, but only the overall shape.

As a trade-off between shape and strength consideration, we added a regularization term that combined the correlation coefficient obtained for each genomic locus (Eq. 7) by the empirical likelihood (or the relative frequency) of the height of ChIP-seq signal, among the bona fide Zelda peaks. This reflects the empirical prior probability of obtaining a peak of such height among positive samples.

We then scanned the *D. melanogaster* genome at both orientations and extracted genomic loci with high local score. As shown in Fig. 6, our de novo chromatin-based identification of Zelda-like loci retrieved the majority of bona fide Zelda peaks in clusters 1 and 2. When considering the top 2,500 hits identified by the promoter-like (cluster 1) signature, we identified over 75% of the original Zelda peaks in that cluster, with the addition of ~400 (15%) regions with similar chromatin patterns. Similarly, by scanning the genome with the enhancer-like (cluster 2) pattern, we managed to locate more than 80% of original early Zelda peaks, with an addition of ~200 (7%) loci not associated with Zelda peaks. When examining in vivo binding of A–P and D–V transcription factors (as before), we observed an average of 5.2 bound factors among the novel enhancer-like regions (compared to ~6 TFs found at the top Zelda-bound regions and <2 factors bound at random regions).

Motif analysis of discovered putative regulatory sites

To further examine the functional role of the newly discovered enhancer-like regions, we ran a de novo motif analysis on the top 1600 enhancer-like peaks, using HOMER [32]. Our motif analysis, based on 100 bp long sequences centered on the mid-points of "hit" 10 Kb windows, identifying three major motifs de novo (Fig. 6a). Motif analysis with 250 bp or 500 bp long sequences yielded similar results. The most significant motif identified among Zelda-like regions is CAGGTAG ($p < 1e-40$), the known recognition element of Zelda [14–17]. This enrichment is not surprising, considering the enrichment of Zelda sites within our set of chromatin-based enhancer-like loci. The second motif identified by HOMER [32], with a p value of 1e−18, is TTATGA and is similar to the known recognition site of two *Drosophila* proteins, Caudal (Cad) and Abdominal-B (Abd-B). The latter shows very low expression levels, and only toward the end of cycle 14 [30], while the former is a key transcription factor that is expressed in a gradient toward the posterior end of early Drosophila embryos and acts as a morphogen for abdomen/tail formation. Motif 3, with a p value of 1e−17, is a poly(GA)-rich motif and is similar to the known recognition site of the protein GAGA factor (GAF, also referred to as Trithorax-like, or Trl).

To determine whether Caudal or GAGA factor (GAF) may act as pioneer factors independently of Zelda, we examined ChIP-seq and ChIP-chip data for Caudal and GAF and calculated their overlap with Zelda peaks [13,

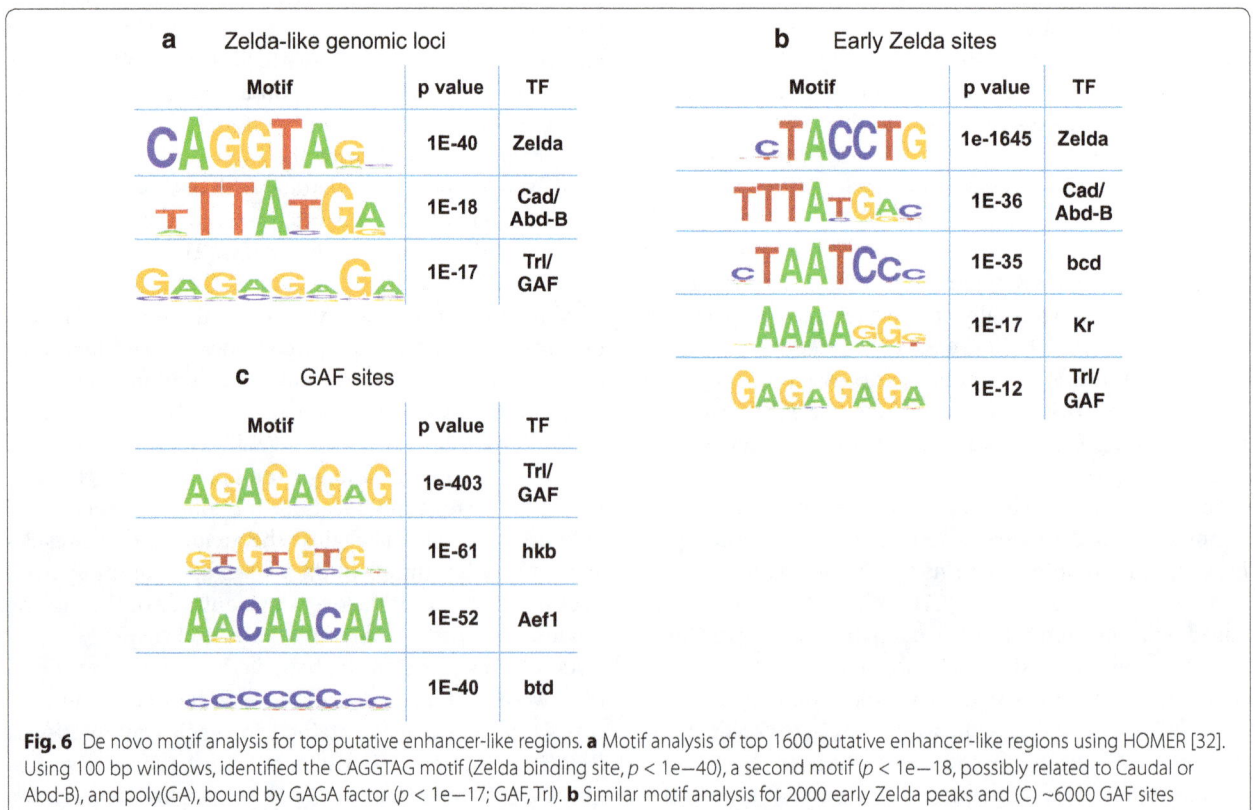

Fig. 6 De novo motif analysis for top putative enhancer-like regions. **a** Motif analysis of top 1600 putative enhancer-like regions using HOMER [32]. Using 100 bp windows, identified the CAGGTAG motif (Zelda binding site, $p < 1e-40$), a second motif ($p < 1e-18$, possibly related to Caudal or Abd-B), and poly(GA), bound by GAGA factor ($p < 1e-17$; GAF, Trl). **b** Similar motif analysis for 2000 early Zelda peaks and (C) ~6000 GAF sites

16, 24]. Overall, almost all Caudal peaks (94%) seem to overlap with Zelda binding, compared to only half (56%) of GAF peaks. We therefore continued to study the role of GAF as a pioneer factor.

For comparison, similar motif analysis (Fig. 6b) for the initial set of 2000 early Zelda peaks identified CAGGTAG ($p <$ 1e−1645), Caudal ($p <$ 1e−36), bicoid ($p <$ 1e−35), Kruppel ($p <$ 1e−17) and GAF (Trl, $p <$ 1e−12, barely above the significant threshold following correction for multiple hypothesis testing). Similar analysis for 5927 GAF binding sites (Fig. 6c) identifies the Trl/GAF motif (p value $<$1e−403), as well as the recognition motifs for hkb ($p <$ 1e−61), Aef1 ($p <$ 1e−52), btd ($p <$ 1e−52).

GAGA factor (GAF) acts as a Zelda-like pioneer factor

As our motif analysis showed, the GAGA motif (to which GAF binds) is enriched among the putative enhancer-like loci, with about half of GAF sites showing no Zelda binding. To test how prevalent GAF binding is at those enhancer-like loci, we also analyzed GAF ChIP data from hours 0–8 of *D. melanogaster* development [24]. Indeed, the average ChIP strength for the predicted enhancer-like loci is more than threefold higher, compared to flanking regions (77 vs. 22). It should be noted that similarly to the motif analysis where only ~15% of the enhancer-like regions contained a GAGA motif, analysis of ChIP data identified GAF binding in about 20% of these regions (compared to over 75% for Zelda binding).

Finally, we turned to directly test the ability of GAF to act as a pioneer factor with or without Zelda. Recently, we measured DNA accessibility via formaldehyde-assisted isolation of regulatory elements (FAIRE) experiments [20]. The data consist of FAIRE throughout 2–3 h of development in wild-type embryos, as well as embryos maternally depleted for Zelda (zld^{M-}).

To test the hypothesis that GAF and Zelda could both act as a pioneer factors, we decided to analyze ZLD and GAF binding to every putative enhancer-like region and to compare these data with their overall DNA accessibility data (as measured by FAIRE) in WT and in zld^{M-} mutants.

For ZLD, we expected to see a positive correlation between ZLD binding and DNA accessibility (i.e., regions with strong ZLD binding are more likely to be accessible) in WT, but not in zld^{M-} mutants. In addition, we expect to see that enhancers with strong GAF binding and weak Zelda binding are less affected by the deletion of *zld*, namely to show strong accessibility in both WT and zld^{M-} mutants FAIRE data.

Due to the relatively low number of data points (2500 enhancer-like regions divided into 36 groups according to their ZLD and GAF ChIP signals), we only show the

average FAIRE accessibility for each group. As Fig. 7a shows for the WT embryos, an increase in both ZLD binding (moving upwards) and GAF binding (moving to the right), results in increased DNA accessibility (darker shade of blue). Conversely, the accessibility matrix for zld^{M-} mutants is less sensitive to ZLD levels, especially on the left hand side, where GAF levels are very low, yet it is almost unaffected by the deletion of zld among the strong GAF sites (Fig. 7b, right hand side). These data suggest that GAGA factor may act as a pioneer factor and contribute to the accessibility and chromatin landscape of DNA regions, much like Zelda (yet for fewer genomic loci), independently of Zelda.

Discussion

In this work, we developed a computational method aimed to analyze the chromatin landscape surrounding some regions of interest (here, binding of the protein Zelda in early embryonic development of the fruit fly *D. melanogaster*), refined it by clustering into sub-groups of unique characteristics, and then scanned the entire genome for novel regions showing similar chromatin characteristics.

As we have shown, much of our method's accuracy stems from the initial re-orientation and clustering procedure, which together allow us to derive a specialized mixture model rather than averaging away the chromatin signal.

We have devised methodologies to simultaneously analyze multiple genomic tracks (i.e., different histones modifications) at different time points, while maintaining the spatial signature of each modification. Such models are more compelling than naive combinatorial models that only consider the heights of the various modifications but ignore the overall shape. As we have shown, many of the observed signals come in unique shapes, including a dip at the ZLD site with two flanking nucleosomes marked with one modification, all within some domain of different modification.

We have analyzed early Zelda peaks throughout the Drosophila genome and identified three typical chromatin signatures they are associated with, including (1) active promoters, (2) active enhancer regions, and (3) genomic regions that were transiently bound by Zelda. We then characterized each of the classes and scanned the genome for additional regions matching the enhancer-like chromatin signature. As we have shown, many of the retrieved regions were shown to be bound by either Zelda or GAF, suggesting both act as pioneer factors by making these genomic regions accessible and depositing the appropriate histone modifications, thus marking them for activation during the maternal-to-zygotic transition. The entire role of GAF in MZT as

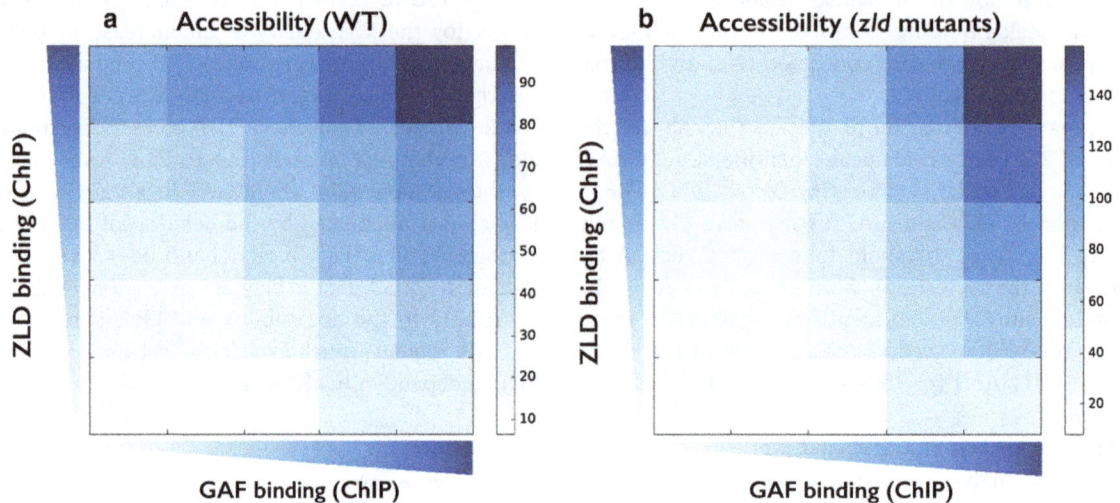

Fig. 7 DNA accessibility in WT and zld mutants. **a** Shown is the average FAIRE signal (DNA accessibility) in WT embryos for 2500 enhancer-like loci that are binned based on their ZLD (*Y*-axis) and GAF (*X*-axis) ChIP binding. **b** Same as A, using FAIRE data from maternal zld mutants [20]. Comparison of the two matrices shows that genomic regions with no GAF binding (in WT) are mostly affected by deletion of zld (left hand side of both matrices), while regions bound by GAF (right hand side) still show high DNA accessibility, suggesting that GAF could ac as a pioneer factor in early fly development, together or independently of Zelda

well as the possible role for additional factors is yet to be studied.

While these questions are key to understanding the processes that initiate and shape embryonic gene expression in the fruit fly *D. melanogaster*, the method presented here is more general and would allow—given a query set of input loci—the identification of similar elements in any genome. In that sense, we believe that our method could be of great interest to any genome-wide research project.

Conclusions

As more genome-wide chromatin data accumulate in multiple conditions, computational methods emerge as a possible means to characterize key loci and identify novel genomic regions with similar chromatin signatures. Our work addresses these tasks by re-orienting and clustering the chromatin landscape surrounding regions of interest and then scanning the entire genome for novel regions showing similar characteristics. Using pre-MZT regions bound by the pioneer factor Zelda, we identify thousands of additional Zelda-like enhancer regions and mark the protein GAF (Vfl) as an additional factor in regulation of early developmental genes. These results are significant as they untangle the complexity in chromatin signatures near Zelda sites, and in linking GAF to chromatin accessibility and gene regulation, as well as developing a novel end-to-end method for studying genomic data.

Methods

Multichannel chromatin representation of ZLD sites

To characterize the chromatin signature around Zelda sites, we considered the top 2000 Zelda peaks in mitotic cycle 8 [16] and looked at five histone modifications (H3K27ac, H3K27me3, H3K4me1, H3K4me3, and H3K18ac) at two time points (mitotic cycles 13 and 14) [5]. For each peak, we considered a window of 10 Kb (at 10 bp resolution) and constructed a vector of length 1000 for each modification. This resulted in a 10×1000 matrix for each Zelda site.

Chromatin-based re-orientation of Zelda peaks

These 2000 Zelda peaks (namely a 3D matrix of $2000 \times 10 \times 1000$) was subjected to an Expectation Maximization (EM)-like iterative orientation algorithm [26, 29]. For every peak, we randomly sampled an orientation ($+/-$) and calculate the average ChIP-seq signal. Then, for 20 iterations (or until convergence), we assigned each peak with the orientation minimizing the overall distance to the average chromatin signature (as in the Max–Max variation of the algorithm).

Peak annotation and GO term analysis

Peaks were associated with nearest genes and annotated with HOMER [32], using "annotatePeaks.pl" with default parameters (dm3). GO term annotations are based on "annotatePeaks.pl" with the "-go" option.

Chromatin-based distance and adjacency functions

To define a distance function between two loci x_i and x_j, we begin by focusing on one histone modification m:

$$\text{dist}_{i,j}^m = \sqrt{\frac{\sum \left(\vec{x}_i^m - \vec{x}_j^m\right)^2}{n}} \quad (1)$$

We then use a standard Gaussian kernel to translate distances into weights in the spectral clustering adjacency matrix

$$w_{i,j}^m = e^{-(\text{dist}_{i,j}^m)^2/2\sigma_m^2} \quad (2)$$

where σ_m is a modification-specific parameter (set below). We then sum the adjacency values over all 10 modifications to obtain one value:

$$w_{i,j} = \sum_{m \in \text{mods}} w_{i,j}^m \quad (3)$$

Spectral clustering

Given $N < 10 \times 1000 >$ matrices $\{x_i\}$ that correspond to the histone modification data for each peak, we defined the graph Laplacian:

$$L = D - W \quad (4)$$

where W is the adjacency matrix (as defined above, Eq. 4), and D is the diagonal degree matrix, defined as:

$$D_{i,j} = \begin{cases} 0 & i \neq j \\ \sum_k w_{i,k} & i = j \end{cases} \quad (5)$$

Following Ng, Jordan, and Weiss [31], we transform L into the symmetric normalized Laplacian

$$L^{\text{sym}} = D^{-1/2}LD^{-1/2} \quad (6)$$

thus normalizing the rows and columns of the graph Laplacian. We then select the first k eigenvectors of L^{sym}, project the N data points onto this k-dimensional subspace, and normalize the projected data (rows of length k) to L^2 norm of 1 (i.e., project each data point to the k-dimensional sphere). Finally, we apply a standard k-means clustering algorithm (where k here equals the number of eigenvectors k used for the spectral projection).

To choose k, we applied a standard eigengap (or spectral gap) heuristic, namely calculating the difference between successive eigenvalues (λ_k and λ_{k+1}) and choosing the first to surpass some threshold.

Empirically normalized distance/adjacency function

In Eq. 2 we described the Gaussian kernel used to translate average Euclidean distances into adjacency weights. As each genomic track (histone modification, time) presents different absolute values, it is crucial to normalize

each Gaussian specifically using its own scaling parameter σ_m. For this, we calculated the empirical distribution of (unnormalized) pairwise distances for each modification m (Additional file 7: Figure S7, purple line), and set σ_m to be the 10th percentile (red vertical line, Additional file 7: Figure S7). This way, we ensured that for each modification exactly 10% of the pairwise distances (purple regions) will have adjacency values larger than $\exp(-\frac{1}{2}) = 0.606$ in the adjacency matrix. This independently normalizes each modification to its own distribution of pairwise distances, thus setting similar "weight" to each modification, allowing for the summation of different adjacency matrices into one (Eq. 3).

Genome-wide scanning of similar chromatin loci

Following the spectral clustering of Zelda peak into three classes, we wished to identify novel enhancer regions with similar chromatin signature to that of Zelda cluster 2 peaks. For this, we scanned the genome in 10 Kb windows (stride of 100 bp) and cross-correlated the chromatin signature at each window to the average histone modification pattern of cluster 2 (enhancer-like regions). We scored each locus i by first calculating the Pearson correlation between its surrounding x_i^m (where m denotes the histone modification/time point) versus the chromatin signature for cluster 2 peaks (denoted by cl_2^m).

This approach, as opposed to just comparing the maximal peak height within each window, gives spatial resolution and identifies the unique shapes we observe for the different modifications (e.g., two symmetric narrow peaks, flanking domains, asymmetrical peak).

To incorporate prior knowledge and capture the overall complex shape of enhancer-like peaks, namely the typical combination of heights for each modification/time point, we also included a "prior" term P:

$$\text{Score}_i^m = \text{Corr}(\text{cl}_2^m, x_i^m) \cdot P_m(\text{height}_i^m) \quad (7)$$

where height_i^m denotes the average ChIP-seq signal modification m within its 10 Kb surrounding locus i, and $P_m(\text{height}_i^m)$ denotes the empirical probability function (or relative frequency) of such mean ChIP-seq signal for modification m among bona fide Zelda peaks, as calculated using the chromatin signatures around the initial set of 2000 early Zelda peaks within cluster 2. This penalizes loci with similar shape (hence, high Pearson correlation coefficients) but in different ChIP magnitude.

Finally, we calculated the overall similarity score for each locus, by summing over the modification-specific scores:

$$\text{Score}_i = \sum_m \text{Score}_i^m \quad (8)$$

and identified the top local 2500 maxima genome wide.

Additional files

Additional file 1: Figure S1. Chromatin signatures around clustered early Zelda peaks. Heatmap and average ChIP-seq signal for five histone modifications at two time points (mitotic cycle 13, left) and 14 (right) around top 2,000 early Zelda peaks. Peaks were divided into three clusters, including cluster 1 (blue line; top heatmaps), cluster 2 (orange line; center heatmaps), and cluster 3 (yellow line, bottom heatmaps).

Additional file 2: Figure S2. Average ChIP-seq binding of Zelda. (A) Average ZLD ChIP-seq signal at three time points, including mitotic cycle 8 (1 h after fertilization), mitotic cycle 11 (2 h), and mitotic cycle 14 (3 h) for clustered Zelda peaks. (B) Same, shown as metaplot (top) or heatmaps (top heatmap: cluster 1; middle: cluster 2; bottom: cluster 3).

Additional file 3: Figure S3. Functional annotation of Zelda peaks. Early 2,000 Zelda peaks (in three clusters) were annotated using HOMER into several classes including promoter/TSS, Exonic, Intronic, and Intergenic loci.

Additional file 4: Figure S4. Gene expression and GO term annotations. (A) Transcription levels of gene associated with early 2,000 Zelda peaks (in three clusters), along eight time points throughout the maternal-to-zygotic transition from mitotic cycle 10–14D [30]. (B–D) GO term enrichments for gene associated with cluster 1 Zelda peaks (B); cluster 2 Zelda peaks (C); and cluster 3 Zelda peaks (D).

Additional file 5: Figure S5. Analysis of chromHMM states. Chromatin data were analyzed by chromHMM [33] by first binarizing the chromatin data (default parameters) and then segmenting the genome into seven chromatin classes. (A) Shown are the average number of A–P and D–V transcription factors bound for each state. (B) chromHMM regions were associated with genes, and the average expression levels along MZT is shown as in Fig. 5d.

Additional file 6: Figure S6. Chromatin signatures and functional annotations of GAF peaks. 5,927 GAF peaks from in vivo GAF binding in *Drosophila melanogaster* embryos (hours 0–8 of development) [24] were analyzed, similarly to our analysis of 2,000 early Zelda peaks. Peaks were re-oriented and clustered into three clusters. Also shown are ZLD in vivo binding data, similarly to Additional file 2: Figure S2. (B) Annotation of GAF peaks, in clusters, shows enrichment of promoters/TSS (41–60% of peaks), and intronic (22–35%) peaks.

Additional file 7: Figure S7. Distance–weight functions for various histone marks. Plotted are the empirical pairwise distance distribution (purple line) of various histone modifications at mitotic cycles 13 (bottom) and 14 (top), over pairs of early Zelda peaks $<x_i, x_j>$. Vertical red lines correspond to the 10th percentile in each distance distribution. This value is assigned as σ_m (horizontal red line). The blue line shows the matching Gaussian kernel function (based on each σ_m), used to transform pairwise distances (X-axis) to weights (Y-axis) when building each Laplacian matrix of spectral clustering.

Abbreviation
MZT: maternal-to-zygotic transition.

Authors' contributions
AM and TK designed the study and collected data. AM analyzed the data. AM and TK wrote the manuscript. Both authors read and approved the final manuscript.

Acknowledgements
Not applicable.

Competing interests
The authors declare that they have no competing interests.

Funding
This work has been supported by the Israeli Centers of Excellence (I-CORE) for Gene Regulation in Complex Human Disease (Grant No. 41/11) and for Chromatin and RNA in Gene Regulation (Grant No. 1796/12).

References
1. Kouzarides T. Chromatin modifications and their function. Cell. 2007;128(4):693–705.
2. Liu CL, Kaplan T, Kim M, Buratowski S, Schreiber SL, Friedman N, Rando OJ. Single-nucleosome mapping of histone modifications in S. cerevisiae. PLoS Biol. 2005;3(10):e328.
3. Rada-Iglesias A, Bajpai R, Swigut T, Brugmann SA, Flynn RA, Wysocka J. A unique chromatin signature uncovers early developmental enhancers in humans. Nature. 2011;470(7333):279–83.
4. Heintzman ND, Stuart RK, Hon G, Fu Y, Ching CW, Hawkins RD, Barrera LO, Van Calcar S, Qu C, Ching KA, et al. Distinct and predictive chromatin signatures of transcriptional promoters and enhancers in the human genome. Nat Genet. 2007;39(3):311–8.
5. Li X-Y, Harrison MM, Villalta JE, Kaplan T, Eisen MB. Establishment of regions of genomic activity during the Drosophila maternal to zygotic transition. eLife. 2014;3:e03737.
6. Kaplan T, Li X-Y, Sabo PJ, Thomas S, Stamatoyannopoulos JA, Biggin MD, Eisen MB. Quantitative models of the mechanisms that control genome-wide patterns of transcription factor binding during early Drosophila development. PLoS Genet. 2011;7(2):e1001290.
7. Thomas S, Li X-Y, Sabo PJ, Sandstrom R, Thurman RE, Canfield TK, Giste E, Fisher W, Hammonds A, Celniker SE, et al. Dynamic reprogramming of chromatin accessibility during Drosophila embryo development. Genome Biol. 2011;12(5):R43.
8. Schübeler D. Enhancing genome annotation with chromatin. Nat Genet. 2007;39(3):284–5.
9. Zaret KS, Carroll JS. Pioneer transcription factors: establishing competence for gene expression. Genes Dev. 2011;25(21):2227–41.
10. Smallwood A, Ren B. Genome organization and long-range regulation of gene expression by enhancers. Curr Opin Cell Biol. 2013;25(3):387–94.
11. Tadros W, Lipshitz HD. The maternal-to-zygotic transition: a play in two acts. Development. 2009;136(18):3033–42.
12. Li X-Y, Macarthur S, Bourgon R, Nix D, Pollard DA, Iyer VN, Hechmer A, Simirenko L, Stapleton M, Luengo Hendriks CL, et al. Transcription factors bind thousands of active and inactive regions in the Drosophila blastoderm. PLoS Biol. 2008;6(2):e27.
13. Macarthur S, Li X-Y, Li J, Brown JB, Chu HC, Zeng L, Grondona BP, Hechmer A, Simirenko L, Keränen SVE, et al. Developmental roles of 21 Drosophila transcription factors are determined by quantitative differences in binding to an overlapping set of thousands of genomic regions. Genome Biol. 2009;10(7):R80.
14. ten Bosch JR, Benavides JA, Cline TW. The TAGteam DNA motif controls the timing of Drosophila pre-blastoderm transcription. Development. 2006;133(10):1967–77.

15. Bradley RK, Li X-Y, Trapnell C, Davidson S, Pachter L, Chu HC, Tonkin LA, Biggin MD, Eisen MB. Binding site turnover produces pervasive quantitative changes in transcription factor binding between closely related Drosophila species. PLoS Biol. 2010;8(3):e1000343.

16. Harrison MM, Li X-Y, Kaplan T, Botchan MR, Eisen MB. Zelda binding in the early *Drosophila melanogaster* embryo marks regions subsequently activated at the maternal-to-zygotic transition. PLoS Genet. 2011;7(10):e1002266.

17. Liang H-L, Nien C-Y, Liu H-Y, Metzstein MM, Kirov N, Rushlow C. The zinc-finger protein Zelda is a key activator of the early zygotic genome in Drosophila. Nature. 2008;456(7220):400–3.

18. Nien C-Y, Liang H-L, Butcher S, Sun Y, Fu S, Gocha T, Kirov N, Manak JR, Rushlow C. Temporal coordination of gene networks by Zelda in the early Drosophila embryo. PLoS Genet. 2011;7(10):e1002339.

19. Satija R, Bradley RK. The TAGteam motif facilitates binding of 21 sequence-specific transcription factors in the Drosophila embryo. Genome Res. 2012;22(4):656–65.

20. Schulz KN, Bondra ER, Moshe A, Villalta JE, Lieb JD, Kaplan T, McKay DJ, Harrison MM. Zelda is differentially required for chromatin accessibility, transcription factor binding, and gene expression in the early Drosophila embryo. Genome Res. 2015;25(11):1715–26.

21. Filion GJ, van Bemmel JG, Braunschweig U, Talhout W, Kind J, Ward LD, Brugman W, de Castro IJ, Kerkhoven RM, Bussemaker HJ, et al. Systematic protein location mapping reveals five principal chromatin types in Drosophila cells. Cell. 2010;143(2):212–24.

22. modENCODE Consortium, Roy S, Ernst J, Kharchenko PV, Kheradpour P, Nègre N, Eaton ML, Landolin JM, Bristow CA, Ma L, et al. Identification of functional elements and regulatory circuits by Drosophila modENCODE. Science. 2010;330(6012):1787–97.

23. Kharchenko PV, Alekseyenko AA, Schwartz YB, Minoda A, Riddle NC, Ernst J, Sabo PJ, Larschan E, Gorchakov AA, Gu T, et al. Comprehensive analysis of the chromatin landscape in *Drosophila melanogaster*. Nature. 2011;471(7339):480–5.

24. Nègre N, Brown CD, Ma L, Bristow CA, Miller SW, Wagner U, Kheradpour P, Eaton ML, Loriaux P, Sealfon R, et al. A cis-regulatory map of the Drosophila genome. Nature. 2011;471(7339):527–31.

25. Sun Y, Nien C-Y, Chen K, Liu H-Y, Johnston J, Zeitlinger J, Rushlow C. Zelda overcomes the high intrinsic nucleosome barrier at enhancers during Drosophila zygotic genome activation. Genome Res. 2015;25(11):1703–14.

26. Kundaje A, Kyriazopoulou-Panagiotopoulou S, Libbrecht M, Smith CL, Raha D, Winters EE, Johnson SM, Snyder M, Batzoglou S, Sidow A. Ubiquitous heterogeneity and asymmetry of the chromatin environment at regulatory elements. Genome Res. 2012;22(9):1735–47.

27. Wang J, Lunyak VV, Jordan IK. Chromatin signature discovery via histone modification profile alignments. Nucleic Acids Res. 2012;40(21):10642–56.

28. Rajagopal N, Xie W, Li Y, Wagner U, Wang W, Stamatoyannopoulos J, Ernst J, Kellis M, Ren B. RFECS: a random-forest based algorithm for enhancer identification from chromatin state. PLoS Comput Biol. 2013;9(3):e1002968.

29. Dempster A, Laird N, Rubin D. Maximum likelihood from incomplete data via the EM algorithm. J Roy Stat Soc Ser B (Methodol). 1977;39:1–38.

30. Lott SE, Villalta JE, Schroth GP, Luo S, Tonkin LA, Eisen MB. Noncanonical compensation of zygotic X transcription in early *Drosophila melanogaster* development revealed through single-embryo RNA-seq. PLoS Biol. 2011;9(2):e1000590.

31. Ng AY, Jordan MI, Weiss Y. On spectral clustering: analysis and an algorithm. NIPS. 2001;14(2):849–56.

32. Heinz S, Benner C, Spann N, Bertolino E, Lin YC, Laslo P, Cheng JX, Murre C, Singh H, Glass CK. Simple combinations of lineage-determining transcription factors prime cis-regulatory elements required for macrophage and B cell identities. Mol Cell. 2010;38(4):576–89.

33. Ernst J, Kellis M. ChromHMM: automating chromatin-state discovery and characterization. Nat Methods. 2012;9(3):215–6.

4

Vitamin C induces specific demethylation of H3K9me2 in mouse embryonic stem cells via Kdm3a/b

Kevin T. Ebata[1†], Kathryn Mesh[1†], Shichong Liu[2], Misha Bilenky[3], Alexander Fekete[4], Michael G. Acker[4], Martin Hirst[3,5], Benjamin A. Garcia[2] and Miguel Ramalho-Santos[1*]

Abstract

Background: Histone methylation patterns regulate gene expression and are highly dynamic during development. The erasure of histone methylation is carried out by histone demethylase enzymes. We had previously shown that vitamin C enhances the activity of Tet enzymes in embryonic stem (ES) cells, leading to DNA demethylation and activation of germline genes.

Results: We report here that vitamin C induces a remarkably specific demethylation of histone H3 lysine 9 dimethylation (H3K9me2) in naïve ES cells. Vitamin C treatment reduces global levels of H3K9me2, but not other histone methylation marks analyzed, as measured by western blot, immunofluorescence and mass spectrometry. Vitamin C leads to widespread loss of H3K9me2 at large chromosomal domains as well as gene promoters and repeat elements. Vitamin C-induced loss of H3K9me2 occurs rapidly within 24 h and is reversible. Importantly, we found that the histone demethylases Kdm3a and Kdm3b are required for vitamin C-induced demethylation of H3K9me2. Moreover, we show that vitamin C-induced Kdm3a/b-mediated H3K9me2 demethylation and Tet-mediated DNA demethylation are independent processes at specific loci. Lastly, we document Kdm3a/b are partially required for the upregulation of germline genes by vitamin C.

Conclusions: These results reveal a specific role for vitamin C in histone demethylation in ES cells and document that DNA methylation and H3K9me2 cooperate to silence germline genes in pluripotent cells.

Keywords: Vitamin C, Histone methylation, Histone lysine demethylase, Epigenetics, Embryonic stem cells

Background

Epigenetic information encoded in chromatin is crucial for cell identity and differentiation [1]. One major layer of chromatin-level regulation is histone methylation. In particular, histone lysine residues can be modified by mono-methylation (me1), dimethylation (me2), or tri-methylation (me3). Histone methyltransferases deposit methyl groups, which can then be recognized by reader proteins that modulate gene expression [2]. Depending on the reader proteins recruited to specific methylated histone residues, the corresponding genes may be repressed or activated. For example, H3K4me3 is associated with the recruitment of factors that promote gene activation, whereas H3K9me2/3 is associated with recruitment of repressive factors. In turn, histone methyl groups are removed by histone demethylases, most of which belong to the family of Fe(II)- and 2-oxoglutarate-dependent dioxygenases [2, 3]. The balance of the activity of histone methylases and demethylases determines the overall histone methylation patterns within a cell. Shifts in this balance underlie the extensive remodeling of histone methylation patterns that occurs during development.

The essential nutrient vitamin C has historically been described as a co-factor of collagen prolyl hydroxylases,

*Correspondence: mrsantos@ucsf.edu
†Kevin T. Ebata and Kathryn Mesh contributed equally to this work
[1] Eli and Edythe Broad Center of Regeneration Medicine and Stem Cell Research, University of California, San Francisco, San Francisco, CA, USA
Full list of author information is available at the end of the article

the prototypical members of the family of Fe(II)- and 2-oxoglutarate-dependent dioxygenases, by recycling Fe(III) to Fe(II) [4]. The more recent discoveries that enzymes involved in DNA and histone demethylation are also Fe(II)- and 2-oxoglutarate-dependent dioxygenases [5] raises the interesting possibility that epigenetic programs may be modulated by diet [6]. In support of this notion, we and others reported that vitamin C enhances the activity of Tet enzymes, which are also Fe(II)- and 2-oxoglutarate-dependent dioxygenases, leading to widespread DNA demethylation and a blastocyst-like state in Embryonic Stem (ES) cells [7, 8].

Several studies have explored DNA methylation and histone mark patterning in ES cells during the transition from formative or primed (hypermethylated) to naïve (hypomethylated) state [9–12]. This transition can be modeled in vitro by switching ES cells maintained in serum containing media to a serum-free media supplemented with GSK3β and ERK1/2 inhibitors (2i) [9, 10]. In the presence of 2i, ES cells show a gradual global decrease in DNA methylation and concurrent reduction of H3K9me2 while maintaining H3K9me3 [11, 12]. Additionally, the distribution of H3K27me3 is globally redistributed with reduced levels at promoters in 2i [9] and accumulation at transposons after conversion to 2i + vitamin C [12]. Together, these studies identified a global reprogramming in DNA and histone methylation patterns during the primed to naïve state. However, the specific impact of vitamin C on histone methylation patterns and the potential interplay with DNA methylation and gene expression in ES cells maintained in the naïve (2i) state has not been investigated.

We report here that vitamin C leads to a remarkably specific and global reduction of histone H3 lysine 9 dimethylation (H3K9me2) in naïve mouse ES cells. The histone demethylases Kdm3a and Kdm3b are required for vitamin C-induced H3K9me2 demethylation, in a manner that is independent of Tet-mediated DNA demethylation. DNA methylation and H3K9me2 cooperated to repress germline gene expression in ES cells. These results highlight that histone methylation patterns are not indiscriminately sensitive to vitamin C. Moreover, our findings uncover a specific role for vitamin C in the epigenetic program of pluripotent cells, with implications for the regulation of germline development.

Results

Vitamin C induces a specific loss of H3K9me2 in ES cells

We used mass spectrometry to perform an unbiased quantitative analysis of H3 N-terminal post-translational modifications (PTMs) in naïve mouse ES cells with or without vitamin C treatment for 72 h. Analysis of the percentages of unmodified H3K4, H3K9, H3K27, or H3K36 revealed that only unmodified H3K9 increased

significantly in vitamin C-treated ES cells (Fig. 1a). Interestingly, there is a concomitant reduction in H3K9me2, but not H3K9me3, upon vitamin C treatment (Fig. 1b, c). H3K9me2 is reduced about 2.6-fold in vitamin C-treated ES cells, whereas there are minimal to no changes in other H3 PTM analyzed (Fig. 1c). Western blot of H3K9 PTMs confirmed a global reduction of H3K9me2 following vitamin C treatment, while H3K9me3, H3K9me1, and H3K9ac did not show significant changes (Fig. 1d). In agreement with the mass spectrometry data, we did not observe changes by Western blot in other H3 PTMs following vitamin C treatment (Additional file 1: Figure S1A). The lack of changes in H3K9me1 suggests that vitamin C promotes demethylation of both methyl groups of H3K9me2, leading to the increase in unmodified H3K9. Alternatively, vitamin C may promote demethylation of H3K9me2 to H3K9me1 and of H3K9me1 to H3K9me0 on short time scales such that overall H3K9me1 levels are not changed. Immunofluorescence for H3K9 PTMs showed an overall decrease in H3K9me2 staining intensity with vitamin C treatment as well as a change from diffuse nuclear staining to a punctate staining pattern that correspond to Dapi-dense heterochromatin (Fig. 1e; Additional file 1: Figure S1B). Residual H3K9me2 at condensed heterochromatin may be insensitive to vitamin C because it is less accessible to histone demethylases, and/or because it is a transient step on the way to H3K9me3.

The histone methyltransferase G9a and its binding partner G9a-like protein (GLP) can generate H3K9me2, but not H3K9me3 [13]. We therefore compared H3K9me2 levels in vitamin C-treated ES cells to $G9a^{-/-}$ and $GLP^{-/-}$ ES cells. These mutant ES cells have almost undetectable levels of H3K9me2, a much more profound loss of this mark than that induced by vitamin C treatment (Additional file 2: Figure S2A). Consistent with a previous study [14], H3K9me2 in $G9a^{-/-}$ and $GLP^{-/-}$ ES cells is present in a punctate pattern similar to vitamin C treatment (Additional file 2: Figure S2B), and this minimal residual H3K9me2 in $G9a^{-/-}$ and $GLP^{-/-}$ ES cells is further reduced by vitamin C (Additional file 2: Figure S2A). Overall, these data suggest that vitamin C induces partial loss of H3K9me2 generated by G9a/GLP.

H3K9me2 is reduced at large chromosomal domains and gene promoters in ES cells treated with vitamin C

We next examined the effect of vitamin C treatment on the genome-wide distribution of H3K9me2 at a finer resolution using chromatin immunoprecipitation (ChIP) followed by next-generation sequencing (ChIP-seq). A rescaling factor was applied to correct for greater sequencing depth in the H3K9me2 data from vitamin C-treated ES cells (Additional file 3: Figure S3, see "Methods" section). The overall distribution of H3K9me2 is not

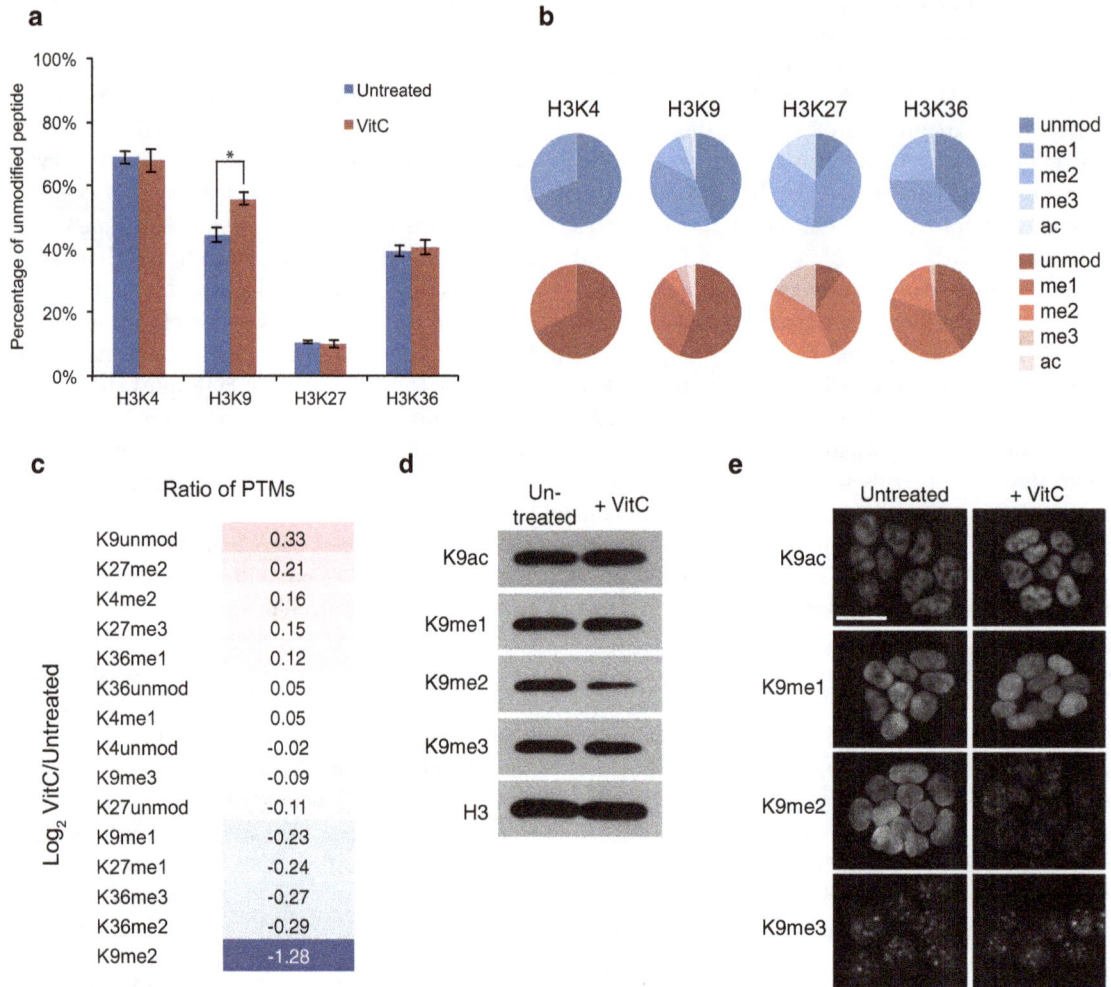

Fig. 1 Vitamin C leads to a global reduction in H3K9me2 in ES cells. **a** Percentage of peptide with unmodified K4, K9, K27 or K36 residues on histone H3 from histone mass spectrometry performed on biological triplicates of untreated and vitamin C-treated ES cells. The percentage of unmodified H3K9 is significantly increased with vitamin C treatment. *Asterisk* represents $P < 0.05$ by t test. **b** Pie charts displaying the prevalence of various PTMs on histone H3 K4, K9, K27 and K36 residues in untreated (*blue*) and vitamin C-treated (*red*) ES cells. **c** Log$_2$ ratio of indicated H3 PTMs in vitamin C-treated versus untreated ES cells. H3K9me2 shows the largest fold change and is decreased following vitamin C treatment. Unmodified H3K9 shows the largest increase. **d** Western blot for H3K9 PTMs in untreated and vitamin C-treated ES cells. **e** Immunofluorescence for H3K9 PTMs in untreated and vitamin C-treated ES cells. *Scale bar* represents 20 μm

affected with vitamin C treatment (Fig. 2a), but the level of H3K9me2 signal is reduced genome-wide (Fig. 2b), consistent with the western blot, Immunofluorescence and mass spectrometry data. Importantly, we found that gene promoters display a generalized reduction in H3K9me2 signal in vitamin C-treated ES cells (Fig. 2c).

To validate the ChIP-seq data, ChIP-qPCR for H3K9me2 was performed at promoters of germline genes that are de-repressed in ES cells following vitamin C treatment [7]. There is a two to threefold decrease in H3K9me2 in vitamin C-treated ES cells compared to untreated ES cells across all genes analyzed (Fig. 2d). A similar pattern is observed at repeat elements known to

be marked by H3K9me2 (Additional file 4: Figure S4). Specificity of the ChIP was confirmed by immunoprecipitation with anti-IgG antibody as well as a primer set for the promoter of *Gapdh*, which is devoid of H3K9me2 (Fig. 2d). Taken together, these results indicate that vitamin C induces a widespread loss of H3K9me2 in ES cells at large chromosomal regions, repeat elements and promoters of repressed germline genes.

Kdm3a and Kdm3b mediate vitamin C-induced loss of H3K9me2

We next assessed the kinetics and reversibility of H3K9me2 loss upon vitamin C treatment. H3K9me2 is

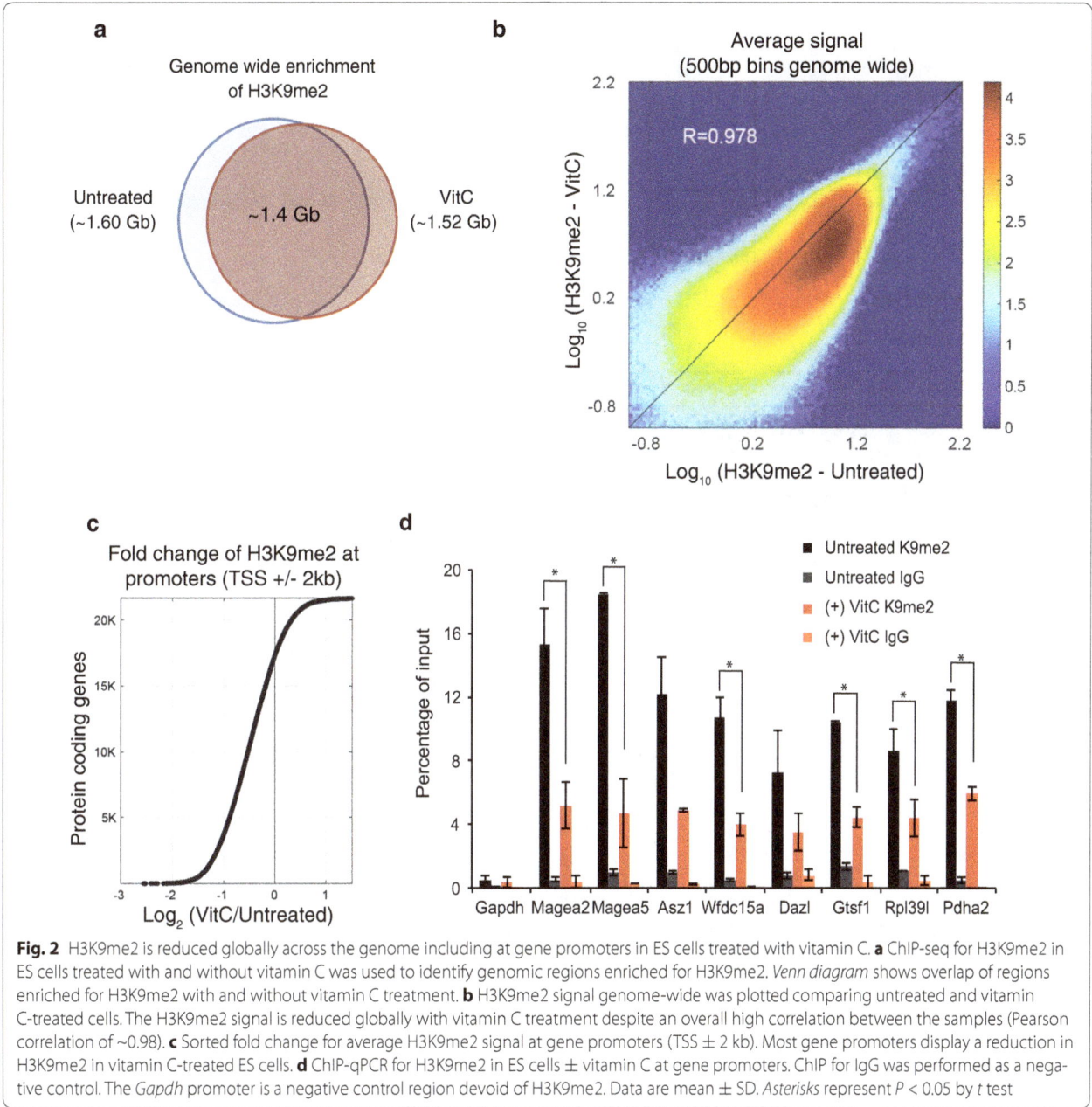

Fig. 2 H3K9me2 is reduced globally across the genome including at gene promoters in ES cells treated with vitamin C. **a** ChIP-seq for H3K9me2 in ES cells treated with and without vitamin C was used to identify genomic regions enriched for H3K9me2. *Venn diagram* shows overlap of regions enriched for H3K9me2 with and without vitamin C treatment. **b** H3K9me2 signal genome-wide was plotted comparing untreated and vitamin C-treated cells. The H3K9me2 signal is reduced globally with vitamin C treatment despite an overall high correlation between the samples (Pearson correlation of ~0.98). **c** Sorted fold change for average H3K9me2 signal at gene promoters (TSS ± 2 kb). Most gene promoters display a reduction in H3K9me2 in vitamin C-treated ES cells. **d** ChIP-qPCR for H3K9me2 in ES cells ± vitamin C at gene promoters. ChIP for IgG was performed as a negative control. The *Gapdh* promoter is a negative control region devoid of H3K9me2. Data are mean ± SD. *Asterisks* represent $P < 0.05$ by t test

reduced after 24 h of vitamin C treatment, with no further reduction by 72 h (Fig. 3a). The loss of H3K9me2 is reversible and returns to baseline levels 72 h after vitamin C withdrawal (Fig. 3a). We therefore explored whether vitamin C may lead to reversible loss of H3K9me2 by shifting the balance of methylation/demethylation toward demethylation. Enzymes of the Kdm3 family are responsible for demethylating H3K9me1 and H3K9me2 [15]. There are three Kdm3 family members: *Kdm3a*, *Kdm3b*, and *Kdm3c* (also known as *Jmjd1a*, *Jmjd1b*, and *Jmjd1c*, respectively), although Kdm3c may

not have catalytic activity [16]. Importantly, there is no change in the expression of Kdm3 family members following vitamin C treatment (Additional file 5: Figure S5A). We tested which Kdm3 family members might be involved in vitamin C-induced H3K9me2 demethylation, using RNAi. Knockdown efficiency and specificity was confirmed by qRT-PCR (Additional file 5: Figure S5B). Knockdown of Kdm3a or Kdm3b, but not Kdm3c, attenuated vitamin C-induced demethylation of H3K9me2 (Fig. 3b). Double knockdown of Kdm3a and Kdm3b completely abolished vitamin C-induced demethylation of

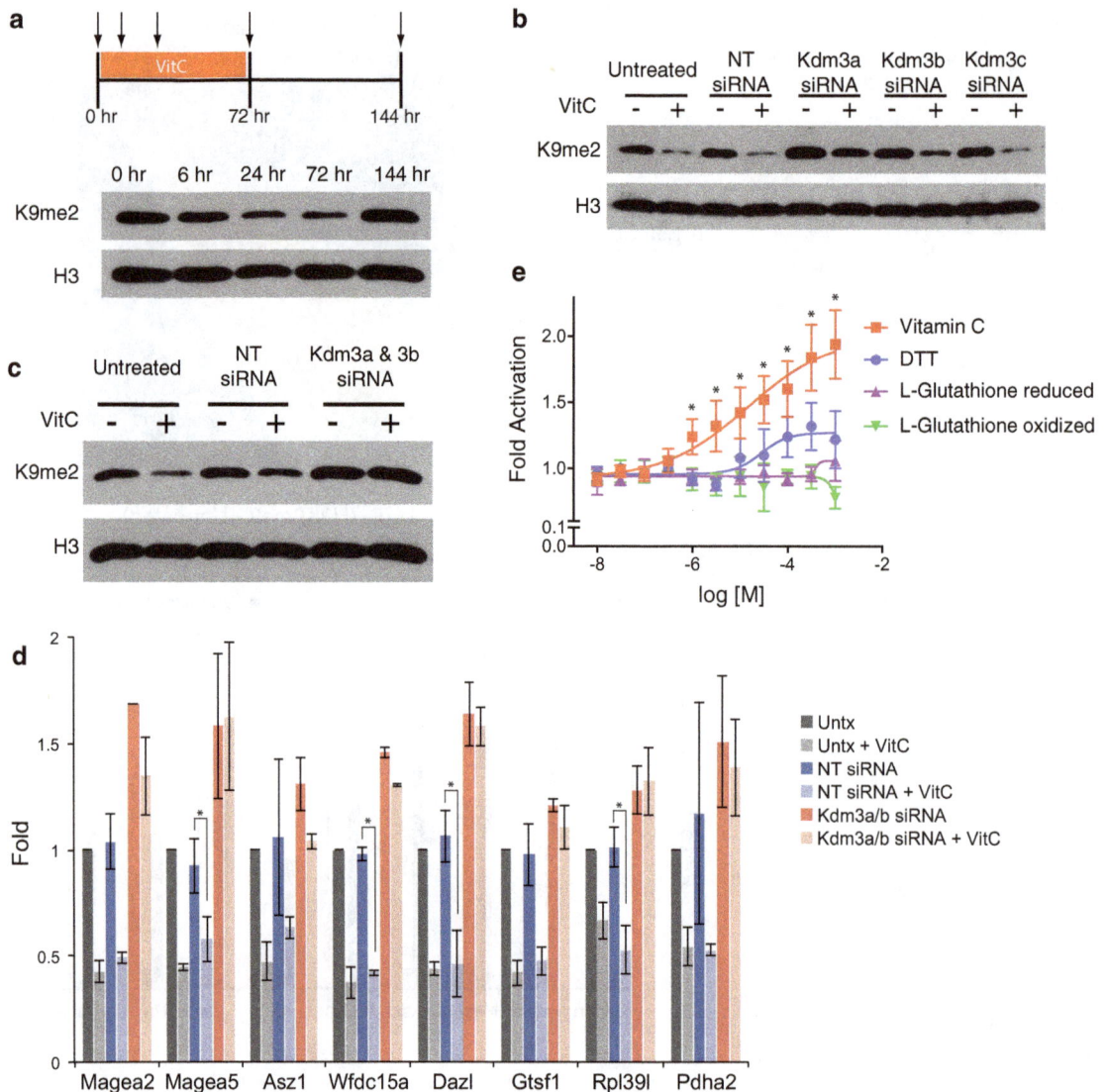

Fig. 3 Kdm3a and Kdm3b mediate vitamin C-induced loss of H3K9me2. **a** Kinetics and reversibility of H3K9me2 loss following vitamin C treatment. H3K9me2 levels were measured by western blot at 6, 24, and 72 h following vitamin C treatment. Vitamin C was then removed, and H3K9me2 levels were measured 72 h later (144 h). **b** Western blot for H3K9me2 in ES cells ± vitamin C with siRNA knockdown of Kdm3 family enzymes or a non-targeting control. **c** Western blot for H3K9me2 in ES cells ± vitamin C with double knockdown of Kdm3a/3b or non-targeting control. **d** ChIP-qPCR for H3K9me2 at gene promoters in ES cells ± vitamin C with double knockdown of Kdm3a/3b or non-targeting control. Data are presented as fold change relative to the untreated control. Data are mean ± SD. *Asterisks* represent $P < 0.05$ by t test. **e** In vitro activity of recombinant KDM3A toward demethylation of a synthetic H3K9me1 peptide, in the presence of vitamin C, DTT or glutathione (see "Methods" section for details). Data are mean ± SD. *Asterisks* are $P < 0.05$ by t test for vitamin C compared to the oxidized form of glutathione

H3K9me2 (Fig. 3c). ChIP-qPCR for H3K9me2 confirmed that Kdm3a/3b are required for vitamin C-induced loss of H3K9me2 (Fig. 3d). Moreover, we found that vitamin C enhances the in vitro activity of recombinant human KDM3A in demethylating a peptide at H3K9 (Fig. 3e; Additional file 6: Figure S6). These results indicate that vitamin C acts via Kdm3a/b enzymes to promote demethylation of H3K9me2.

Interaction between vitamin C-induced changes in 5-methylcytosine/5-hydroxymethylcytosine and vitamin C-induced reduction in H3K9me2

Vitamin C has previously been shown to enhance Tet activity in ES cells leading to an increase in 5-hydroxymethylcytosine (5-hmC) and DNA demethylation at gene promoters [7, 8]. There is extensive crosstalk between DNA methylation and histone methylation [17]. DNA demethylation could conceivably impact chromatin accessibility and affect the ability of H3K9me2 demethylases to act. We therefore sought to determine the interaction between vitamin C-induced changes in 5-hmC/5-mC and vitamin C-induced reduction in H3K9me2. Tet1/2 double knockout (Tet DKO) ES cells, which do not increase 5-hmC or demethylate promoters in response to vitamin C, still display a global loss of H3K9me2 following vitamin C treatment (Fig. 4a). Thus, vitamin C-induced loss of H3K9me2 can occur independently of vitamin C-induced Tet-mediated deposition of hydroxymethylation. Additionally, global 5-hmC is increased following vitamin C treatment in $G9a^{-/-}$ and $GLP^{-/-}$ ES cells to levels comparable to those in parental wild-type ES cells (Fig. 4b). Finally, we focused on germline genes that lose H3K9me2 (Fig. 2d), lose DNA

methylation and are induced [7] in response to vitamin C. Kdm3a/b double knockdown ES cells have a similar extent of vitamin C-induced DNA demethylation as non-targeting siRNA-transfected ES cells; however, they appear to have slightly elevated baseline DNA methylation levels (Fig. 4c). These data indicate that vitamin C-induced DNA demethylation can occur independently of H3K9me2 demethylation at germline genes sensitive to vitamin C levels.

Kdm3a/3b are partially required for vitamin C-induced upregulation of germline genes in ES cells

DNA methylation and H3K9me2 are both considered repressive marks that inhibit gene expression when present at gene promoters. We previously showed that vitamin C treatment in ES cells leads to upregulation of

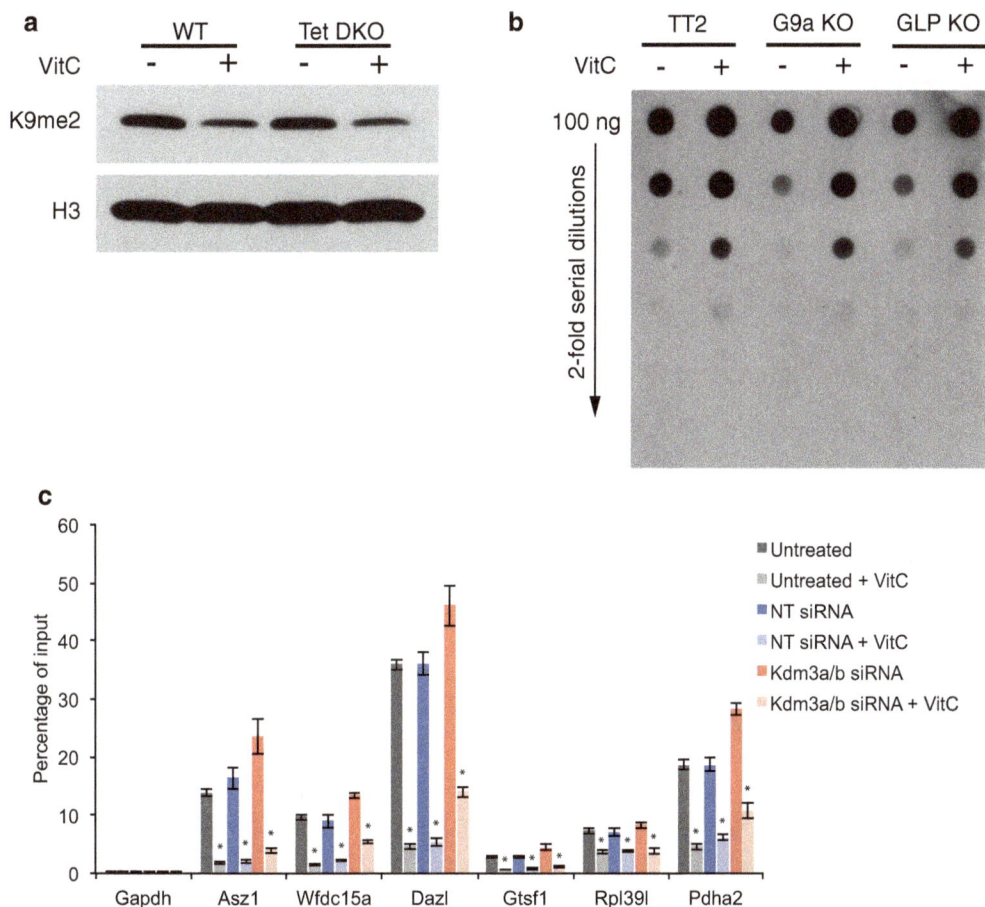

Fig. 4 Interaction between vitamin C-induced increase in 5-hydroxymethylation and vitamin C-induced loss of H3K9me2. **a** Western blot for H3K9me2 in wild-type and Tet1/2 double knockout (Tet DKO) ES cells ± vitamin C. Vitamin C-induced loss of H3K9me2 still occurs in Tet DKO ES cells that do not undergo vitamin C-induced increase in 5-hmC. **b** Dot blot for 5-hydroxymethylcytosine (5-hmC) in wild-type parental TT2, G9a knockout, and GLP knockout ES cells ± vitamin C. Vitamin C-induced increase in 5-hmC still occurs in G9a and GLP knockout ES cells depleted of H3K9me2. **c** meDIP-qPCR on ES cells ± vitamin C that were untreated, treated with a non-targeting (NT) siRNA, or treated with siRNAs against Kdm3a and Kdm3b. Vitamin C-induced DNA demethylation occurs in the absence of Kdm3a/b and loss of H3K9me2. Data are mean ± SD. *Asterisks* represent $P < 0.05$ by t test

germline genes and that this upregulation is partially Tet-dependent [7]. Our time-course experiments indicate that loss of H3K9me2 occurs after 24 h (Fig. 3a), whereas DNA demethylation occurs more slowly and takes 72 h [7]. We sought to determine whether loss of H3K9me2 contributes to the activation of germline genes by vitamin C. First, we analyzed the kinetics of germline gene induction following vitamin C treatment, between 12 and 72 h, and found that vitamin C-induced gene induction occurs progressively over this time period (Fig. 5a). Next, we found that double knockdown of Kdm3a/b attenuates vitamin C-induced germline gene induction (Fig. 5b). Overall, these data suggest that DNA methylation and H3K9me2 are partially redundant mechanisms of repression of germline genes.

Discussion

We report here that vitamin C leads to the specific and widespread loss of H3K9me2 in ES cells, via a mechanism that requires the Kdm3a/3b histone demethylases and is independent of Tet-mediated DNA demethylation. Vitamin C has also been shown to increase the efficiency of reprogramming of somatic cells to the induced pluripotent stem cell state, and this effect is likely mediated by both Tet enzymes and histone demethylases [18, 19].

The remarkable stability of other histone marks in this study raises the possibility that not all histone demethylases are equally dependent on vitamin C. Although our analysis of global histone marks by Western and mass spectrometry showed a specific loss of H3K9me2, it does not exclude the possibility that vitamin C enhances the activity of other histone demethylases at a local level.

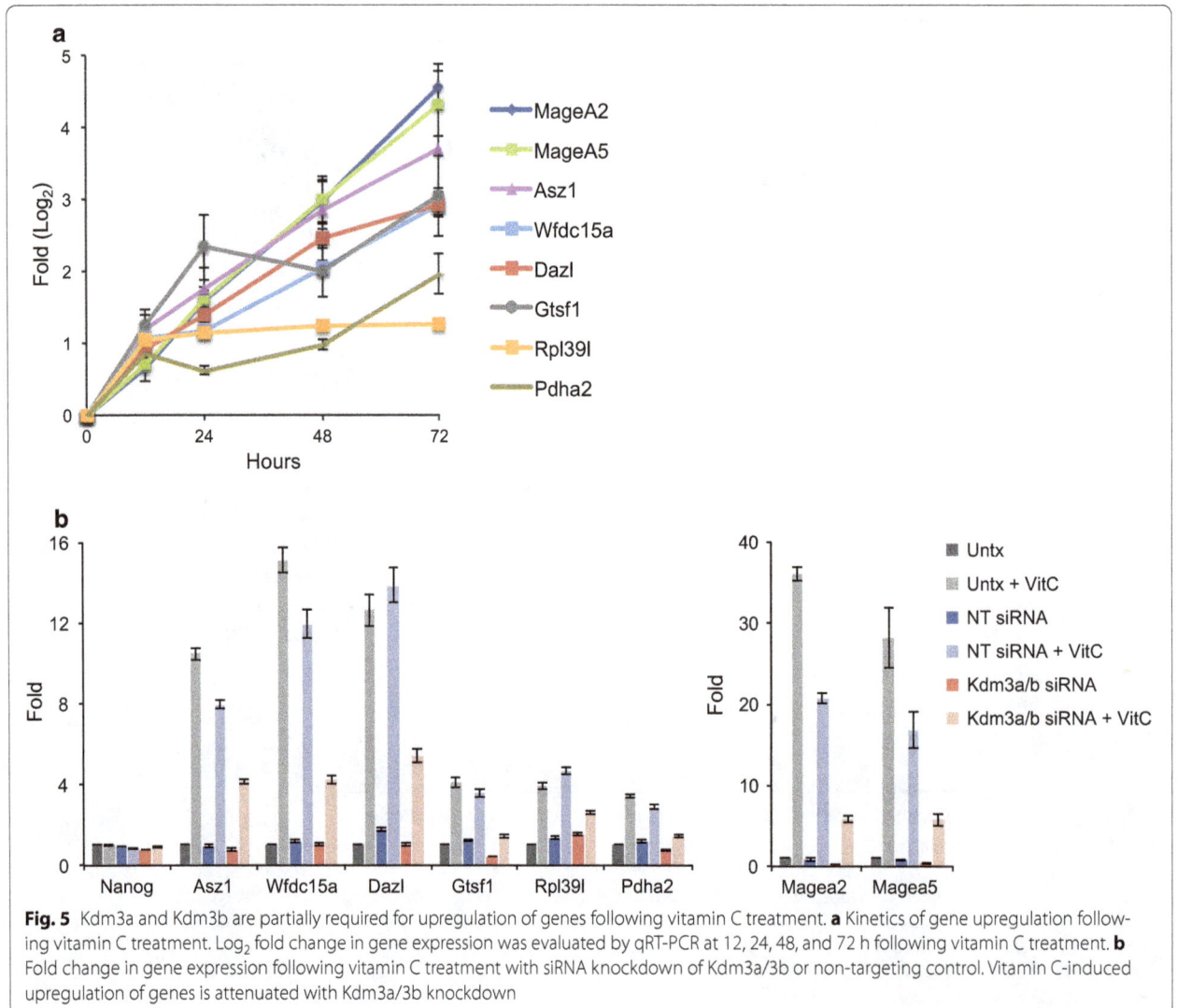

Fig. 5 Kdm3a and Kdm3b are partially required for upregulation of genes following vitamin C treatment. **a** Kinetics of gene upregulation following vitamin C treatment. Log$_2$ fold change in gene expression was evaluated by qRT-PCR at 12, 24, 48, and 72 h following vitamin C treatment. **b** Fold change in gene expression following vitamin C treatment with siRNA knockdown of Kdm3a/3b or non-targeting control. Vitamin C-induced upregulation of genes is attenuated with Kdm3a/3b knockdown

Interestingly, in contrast to the loss of H3K9me2 we observed in ES cells, vitamin C was shown to promote loss of H3K36me2/3 via KDM2A/2B in mouse embryonic fibroblasts. These results suggest that different cell types, varying in many factors including expression levels of different histone demethylases, may affect the extent to which histone methylation patterns are sensitive by vitamin C [20]. It will be important to study the effect of vitamin C on histone methylation patterns in other cell types and investigate a potential structural basis for the differential sensitivity of histone demethylases to vitamin C.

Given the genome-wide loss of H3K9me2 (this study) and DNA methylation [7] induced by vitamin C, it is interesting that the transcriptome of ES cells is largely unchanged, with the exception of a subset of germline-associated genes [7]. It is possible that other genetic programs remain silenced by other repressive mechanisms, such as H3K27me3 [21], which we show here is largely insensitive to vitamin C, and/or the lack of transcriptional activators. We found that vitamin C-induced loss of H3K9me2 is independent of Tet-mediated DNA demethylation at specific germline genes and that both H3K9me2 and DNA methylation contribute to their repression in ES cells. Our results indicate that pluripotent cells have the ability to express germline genes, but that these are repressed by H3K9me2 and DNA methylation. Interestingly, both H3K9me2 and DNA methylation are detected at high levels in pluripotent epiblast cells in vivo and are extensively erased during germline (but not soma) development [22–25]. Moreover, methods to generate germline-like cells from pluripotent stem cells in vitro involve culture in the presence of knockout serum replacement (KSR), which contains vitamin C [26, 27]. It will be of interest to assess whether vitamin C regulates H3K9me2, DNA methylation or gene expression during germline development, in vivo or in vitro. In addition, our results raise the possibility that vitamin C availability may regulate other processes where H3K9me2 has been shown to play a role, such as in hematopoietic lineage commitment [28], brain development [29] or adult metabolism [30].

Conclusions

In this study, we investigated the role of vitamin C in regulating histone methylation in naïve ES cells. We performed comprehensive profiling of histone methylation by mass spectrometry following vitamin C treatment and found a specific reduction in H3K9me2. The distribution pattern of H3K9me2 was not changed by vitamin C, instead levels were reduced throughout the genome, including at gene promoters. Mechanistic studies showed that vitamin C-induced loss of H3K9me2 is via Kdm3a/b

enzymes. Vitamin C reduces both H3K9me2 and DNA methylation independently at germline gene promoters, allowing expression of these genes in ES cells.

Methods
ES cell lines and cell culture
$Oct4$-GiP, TT2, $G9a^{-/-}$, $GLP^{-/-}$, V6.5, and $Tet1^{-/-}$ $Tet2^{-/-}$ ES cells were used in this study. ES cells were cultured on tissue culture plates coated with 0.1% gelatin in 2i medium, which consists of N2B27 base medium [31] supplemented with the MEK inhibitor, PD0325901 (1 µM, Stemgent), the GSK3β inhibitor, CHIR99021 (3 µM, Selleck Chemicals), and with ESGRO leukemia inhibitory factor (LIF) at 1000 U/ml (Millipore). Vitamin C (L-ascorbic acid 2-phosphate, Sigma, A8960) was added on day 1 after seeding at 100 µg/ml. Medium was replaced daily.

Immunofluorescence
Cultured cells were fixed in 4% paraformaldehyde for 15 min, washed 3× with PBS, and blocked with blocking solution (PBS + 0.5% Tween20 + 5% FBS) for 1 h at room temperature (RT). Primary antibodies were diluted in blocking solution and incubated with cells overnight at 4 °C. Primary antibodies included H3K9ac (1:1000, Active Motif, 39917), H3K9me1 (1:2000, Abcam, ab9045), H3K9me2 (1:1000, Abcam, ab1220), H3K9me3 (1:1000, Diagenode, pAb-056-050). Cells were then washed for 5 min 3× in PBS and incubated with secondary antibodies in blocking solution for 2 h at RT. Secondary antibodies included 594-conjugated donkey anti-mouse (1:1000, Life Technologies, A21203) and 594-conjugated donkey anti-rabbit (1:1000, Life Technologies, A21207). DAPI (1 µg/ml) was added with secondary antibodies. Cells were washed for 5 min 3× in PBS prior to imaging.

Histone extraction for quantitative mass spectrometry and Western blotting
Core histones were extracted from cultured ES cells using the Histone Purification Mini Kit (Active Motif, 40026). Following extraction, core histones were precipitated in 4% perchloric acid overnight at 4 °C, washed, and resuspended in water.

Western blot
Isolated histones were resolved on a 4–20% gradient TGX gel and transferred to PVDF membranes. Membranes were blocked in blocking solution (Li-Cor Odyssey Buffer diluted 1:1 with PBS) for 1 h at RT. Primary antibodies were diluted in blocking solution and incubated for 3 h at RT. Membranes were washed for 10 min 3× in TBST (TBS + 0.5% Tween) and incubated with secondary antibodies diluted in blocking solution for 2 h

at RT. Membranes were washed for 10 min 3× in TBST and visualized with Pierce ECL Plus.

Quantitative mass spectrometry (qMS)

Approximately 20 µg of extracted histones was resuspended in 30 µl of 100 mM ammonium bicarbonate, at pH 8.0. Chemical propionylation derivatization, digestion and desalting of histones followed by analysis by LC–MS and MS/MS were performed as described previously [32, 33]. In brief, purified peptides were loaded onto 75-µm-ID fused-silica capillary columns packed with 12 cm of C18 reversed-phase resin (Reprosil-pur 120 C18, aq-3 µm particles, Fisher Scientific). Peptides were separated using EASY-nLC nano-HPLC (Thermo Scientific, Odense, Denmark) and introduced into a hybrid linear quadrupole ion trap–Orbitrap mass spectrometer (ThermoElectron) and resolved with a gradient from 0 to 35% solvent B ($A = 0.1\%$ formic acid; $B = 95\%$ MeCN, 0.1% formic acid) over 30 min and from 34 to 100% solvent B in 20 min at a flow-rate of 250 nL/min. The Orbitrap was operated in data-dependent mode essentially as previously described [33]. Relative abundances of peptide species were calculated by chromatographic peak integration of full MS scans using EpiProfile [34]. Where necessary, peptide and PTM identity were verified by manual inspection of MS/MS spectra.

Chromatin immunoprecipitation

Isolated ES cells were resuspended in PBS and crosslinked with 1% formaldehyde for 5 min at RT. Crosslinking was quenched with 125 mM glycine. Crosslinked material was sonicated on a Covaris sonicator for 12 min at duty 5%, intensity 3, and bursts 200. ChIP was performed using the Diagenode LowCell# Kit with a mouse anti-H3K9me2 antibody (Abcam, ab1220). A mouse anti-IgG antibody (Abcam, ab18413) was used as control. Isolated DNA was used for qPCR analysis with the KAPA SYBR Fast ABI Prism qPCR kit on an Applied BioSystems 7900HT Sequence Detection System. Primer sequences are listed in Additional file 7: Table S1. For ChIP-seq isolated DNA was resuspended in an elution buffer (10 mM Tris–HCl pH 7.6, 200 mM NaCl, 5 mM EDTA, 0.5% SDS) with 10 µl RNAse and reverse-cross-linked at 65 °C for 4 h. Proteinase K was added and samples were incubated at 55 °C for 1 h. DNA was purified with QIAquick and quantified with a Qubit Fluorometer (Invitrogen). ChIP-seq libraries were constructed from immunoprecipitated DNA as described [35]. Libraries were sequenced using paired-end 100nt sequencing V3 chemistry on an HiSeq 2000 sequencer following the manufacturer's protocols (Illumina, Hayward, CA.) using multiplex custom index adapters added during library construction to distinguish pooled samples.

Dot blot analysis

Dot blot analysis was performed as previously described [7]. Briefly, isolated DNA was denatured and then serially diluted twofold. DNA samples were spotted on a nitrocellulose membrane using a Bio-Dot apparatus (Bio-Rad) and immunoblotted with rabbit anti-5-hydroxymethylcytosine polyclonal antibody (Active Motif, 1:5000) in Odyssey:PBS (Li-Cor). The membrane was washed and then incubated with HRP-conjugated goat anti-rabbit IgG (Abcam, 1:10,000) secondary antibody in Odyssey:PBS. The membrane was visualized by chemiluminescence with GE ECL Plus.

5mC DNA immunoprecipitation

DNA immunoprecipitation was performed using the Diagenode MagMeDIP as previously described [7]. Briefly, DNA was sonicated into short fragments with a Diagenode Bioruptor. Sonicated DNA (1 µg) was immunoprecipitated with 1 µg of mouse anti-5-methylcytosine monoclonal antibody (Active Motif, 1 µg/µl). Isolation of immunoprecipitated DNA was performed according to the kit instructions, and qPCR was performed in combination with the KAPA SYBR Fast ABI Prism qPCR kit on an Applied BioSystems 7900HT Sequence Detection System. Primer sequences are listed in Additional file 7: Table S1.

qRT-PCR

Total RNA was isolated from cultured cells using Qiagen RNeasy with on-column DNase I treatment. cDNA was generated from 1 µg of RNA using random hexamers to prime the reaction. Quantitative RT-PCR was performed with the KAPA SYBR Fast ABI Prism qPCR kit on an Applied BioSystems 7900HT sequence detection system. Primer sequences are listed in Additional file 7: Table S1. The relative amount of each gene was normalized using two housekeeping genes (*L7* and *Ubb*), unless otherwise indicated.

siRNA transfection

Oct4-GiP ES cells were transfected in suspension with Dharmacon siGENOME SMARTpool siRNAs (4 siRNAs per gene) against *Kdm3a*, *Kdm3b*, or *Kdm3c* at 50 nM and seeded into tissue culture plates (Day 0). Cells were re-transfected on day 2 followed by treatment with vitamin C for 24 or 72 h. For double knockdown of Kdm3a and Kdm3b, cells were transfected with 50 nM of both Kdm3a and Kdm3b siRNAs for a total final concentration of 100 nM. Dharmacon siGENOME non-targeting siRNA (NT2) was used as a control. Dharmafect reagent was used for transfections according to manufacturer's instructions.

KDM3a biochemical assay

The KDM3A assay was performed in 384-well black flat-bottom plates (Corning, Cat# 3654). Enzyme and substrate solutions were prepared in assay buffer consisting of 5 mM HEPES pH 7.0, 15 mM HEPES pH 7.5, 25 mM NaCl, 1 μM (or 1 mM) Alpha-ketoglutarate, 3.75 μM $FeSO_4$, 0.014% Tween 20, 0.03% BSA. Dose–response concentrations of vitamin C (MP Biochemicals, 100769), DTT (Sigma, D0632) or glutathione (Affymetrix, 16315; Chem-Impex, 00158) were as indicated in the graphs. The KDM3A enzyme was produced in house [16]. H3K9me1 substrate was purchased from Perkin Elmer. Plates were covered and incubated for 30 min at room temperature. The reaction is stopped with detection buffer consisting of 100 mM HEPES pH 7.0, 800 mM KF and 0.2% BSA containing the detection reagents Eu-antiH3K9me0 (purchased from Perkin Elmer) and SA:APC (i.e., XL-665, purchased from Cisbio). The final concentrations of the various reagents were: KDM3A 2 nM, H3K9me1 500 nM, Eu-antiH3K9me0 0.75 nM, SA:APC 1.67 μg/ml. Plates were read in a Perkin Elmer Envision Plate Reader (ex/em 340/615 and 340/665).

Bioinformatics

ChIP-seq raw sequences were examined for quality, sample swap and reagent contamination using custom in house scripts. Sequence reads were aligned to mm10 using BWA 0.5.7 and default parameters and assessed for overall quality using Findpeaks 3.1 [36]. For rescaling, we calculated average ChIP-seq coverage in 500 bp genomic bins. We considered ~150 bins with the highest signal in untreated data samples and calculated distributions of the fold change between vitamin C-treated and untreated data. These bins typically correspond to alignment artifacts and have extreme coverage not due to enrichment. They appear in both IP and control data sets with a very similar coverage profile, as validated by visual inspection of the majority of those locations in the UCSC genome browser. The alignment artifacts have a universal nature (mostly due to incomplete knowledge of genome) with the expectation that for similar cell types and the same read length and DNA fragment length distributions, the difference in the coverage for these 'outliers' will be a reflection of the differences in the sequencing depth. We observed that, as expected, for the input sample, the fold change for the total genomic coverage between vitamin C-treated and untreated data is in agreement with median fold change measured from 'outliers' bins. For the H3K9me2 signal, these twofold change values are very different, and for the 'outliers' the fold change is larger than a value of 1, suggesting that vitamin C-treated data had to be rescaled with a factor ~0.57 (Additional file 3: Figure S3). We verified that the value for the scaling factor depends only weakly on the exact threshold for choice of the outliers within a reasonable range. Using the rescaled data, the coverage profiles for DNA fragments corresponding to properly paired aligned reads were calculated and used to determine the average coverage in 500 bp bins genome wide (5,459,336 bins) and in the promoter regions (TSS ± 2 Kb) of the coding genes (21,958 genes, Ensembl v81).

Additional files

Additional file 1: Figure S1. Evaluation of changes in H3 PTMs following vitamin C treatment. A) Western blot for several H3 PTMs in ES cells ± vitamin C. B) Immunofluorescence for H3K9me2 and corresponding DAPI staining in untreated and vitamin C-treated ES cells. Merged images show H3K9me2 in green and DAPI staining in red. H3K9me2 immunofluorescence is also shown in Fig. 1e. Scale bar represents 20 μm.

Additional file 2: Figure S2. Analysis of H3K9me2 in G9a and GLP knockout ES cells treated with vitamin C. A) Western blot for H3K9me2 in wild-type parental TT2, G9a knockout, and GLP knockout ES cells ± vitamin C. B) Immunofluorescence for H3K9me2 in GiP ES cells ± vitamin C and untreated wild-type TT2, G9a knockout, and GLP knockout ES cells. GiP ES cells treated with vitamin C show a H3K9me2 staining pattern that is similar to G9a and GLP knockout ES cells. Scale bar represents 20 μm.

Additional file 3: Figure S3. Differences in sequencing between untreated and vitamin C-treated H3K9me2 ChIP-seq samples. Average coverage in 500 bp bins was calculated for H3K9me2 ChIP-seq and DNA input samples. The top 150 bins with the highest signal in untreated samples were used to calculate distributions of the fold change between vitamin C-treated and untreated data. These bins with high signal typically correspond to alignment artifacts (see "Methods" section). As expected for the Input sample, the fold change for the total genomic coverage between vitamin C-treated and untreated data is in agreement with the median fold change measured for the outlier top 150 bins. For the H3K9me2 ChIP-seq samples, the difference between the twofold changes suggests that the vitamin C sequencing data has to be rescaled with a factor of ~0.57.

Additional file 4: Figure S4. Analysis of H3K9me2 at repetitive elements in ES cells treated with vitamin C. ChIP-qPCR for H3K9me2 in ES cells ± vitamin C at the repetitive element families indicated. ChIP for IgG was performed as a negative control. Data are mean ± SD. Asterisks represent $P < 0.05$ by t test.

Additional file 5: Figure S5. Effect of vitamin C treatment and siRNA knockdown on the expression of Kdm enzymes. A) Expression of Kdm family enzymes as a percentage of housekeeping gene in ES cells ± vitamin C. B) Gene expression levels of Kdm3a, Kdm3b, and Kdm3c following siRNA knockdown. Data are presented as fold change relative to the untreated control. A non-targeting (NT) siRNA was also used as a control. Each siRNA was applied in the presence or absence of vitamin C to show that vitamin C treatment does not affect knockdown efficiency.

Additional file 6: Figure S6. Effect of α-ketoglutarate on recombinant KDM3A activity with vitamin C, DTT and glutathione. In vitro activity of recombinant KDM3A toward demethylation of a synthetic H3K9me1 peptide, in the presence of vitamin C, DTT or glutathione, at 1 μM α-KG (see "Methods" section for details). At this lower concentration of α-KG, both vitamin C and DTT can enhance activity of KDM3A, but the effect of DTT saturates, whereas vitamin C does not. Data are mean ± SD. Asterisks represent $P < 0.05$ by t test for vitamin C compared to the oxidized form of Glutahione.

Additional file 7: Table S1. Primer list.

Author contributions

M.R.-S. directed the study. K.T.E. and K.M. designed and performed experiments. K.T.E., K.M. and M.R.-S. analyzed and interpreted the data. M.B. and M.H. developed and performed ChIP sequencing data processing. S.L. and B.A.G. performed histone methylation quantitative mass spectrometry and analysis. A.F. and M.A. performed in vitro histone demethylation assays. K.T.E., K.M., and M.R.-S. wrote the manuscript with input from all the authors. All authors read and approved the final manuscript.

Author details

[1] Eli and Edythe Broad Center of Regeneration Medicine and Stem Cell Research, University of California, San Francisco, San Francisco, CA, USA. [2] Epigenetics Program, Department of Biochemistry and Biophysics, Perelman School of Medicine, University of Pennsylvania, Philadelphia, PA, USA. [3] Canada's Michael Smith Genome Sciences Centre, BC Cancer Agency, Vancouver, BC, Canada. [4] Novartis Institutes for Biomedical Research, Cambridge, MA, USA. [5] Department of Microbiology and Immunology, Centre for High-Throughput Biology, University of British Columbia, Vancouver, BC, Canada.

Acknowledgements

We thank Matthew C. Lorincz and Yoichi Shinkai for TT2, $G9a^{-/-}$, $GLP^{-/-}$ and M.M. Dawlaty and R. Jaenisch for the V6.5, $Tet1^{-/-}$ $Tet2^{-/-}$ ES cells.

Competing interests

The authors declare that they have no competing interests.

Funding

This work was supported by NIH Grant R01GM110174 to B.A.G., Grants CCSRI 22R22583 and TFRI NI 22R22545 to M.H., and by NIH Grants R01OD012204 and R01GM113014 to M.R.-S.

References

1. Mohn F, Weber M, Rebhan M, Roloff TC, Richter J, Stadler MB, Bibel M, Schübeler D. Lineage-specific polycomb targets and de novo DNA methylation define restriction and potential of neuronal progenitors. Mol Cell. 2008;30:755–66.
2. Black JC, Van Rechem C, Whetstine JR. Histone lysine methylation dynamics: establishment, regulation, and biological impact. Mol Cell. 2012;48:491–507.
3. Klose RJ, Kallin EM, Zhang Y. JmjC-domain-containing proteins and histone demethylation. Nat Rev Genet. 2006;7:715–27.
4. Du J, Cullen JJ, Buettner GR. Ascorbic acid: chemistry, biology and the treatment of cancer. Biochim Biophys Acta. 2012;1826:443–57.
5. Horton JR, Upadhyay AK, Qi HH, Zhang X, Shi Y, Cheng X. Enzymatic and structural insights for substrate specificity of a family of jumonji histone lysine demethylases. Nat Struct Mol Biol. 2009;17:38–43.
6. Monfort A, Wutz A. Breathing-in epigenetic change with vitamin C. EMBO Rep. 2013;14:337–46.
7. Blaschke K, Ebata KT, Karimi MM, Zepeda-Martínez JA, Goyal P, Mahapatra S, Tam A, Laird DJ, Hirst M, Rao A, Lorincz MC, Ramalho-Santos M. Vitamin C induces Tet-dependent DNA demethylation and a blastocyst-like state in ES cells. Nature. 2013;500:222–6.
8. Yin R, Mao S-Q, Zhao B, Chong Z, Yang Y, Zhao C, Zhang D, Huang H, Gao J, Li Z, Jiao Y, Li C, Liu S, Wu D, Gu W, Yang Y-G, Xu G-L, Wang H. Ascorbic

acid enhances tet-mediated 5-methylcytosine oxidation and promotes DNA demethylation in mammals. J Am Chem Soc. 2013;135:10396–403.
9. Marks H, Kalkan T, Menafra R, Denissov S, Jones K, Hofemeister H, Nichols J, Kranz A, Francis Stewart A, Smith A, Stunnenberg HG. The transcriptional and epigenomic foundations of ground state pluripotency. Cell. 2012;149:590–604.
10. Leitch HG, McEwen KR, Turp A, Encheva V, Carroll T, Grabole N, Mansfield W, Nashun B, Knezovich JG, Smith A, Surani MA, Hajkova P. Naive pluripotency is associated with global DNA hypomethylation. Nat Struct Mol Biol. 2013;20:311–6.
11. von Meyenn F, Iurlaro M, Habibi E, Liu NQ, Salehzadeh-Yazdi A, Santos F, Petrini E, Milagre I, Yu M, Xie Z, Kroeze LI, Nesterova TB, Jansen JH, Xie H, He C, Reik W, Stunnenberg HG. Impairment of DNA methylation maintenance is the main cause of global demethylation in naive embryonic stem cells. Mol Cell. 2016;62:848–61.
12. Walter M, Teissandier A, Pérez-Palacios R, Bourc'his D. An epigenetic switch ensures transposon repression upon dynamic loss of DNA methylation in embryonic stem cells. Elife. 2016;5:e11418.
13. Tachibana M, Ueda J, Fukuda M, Takeda N, Ohta T, Iwanari H, Sakihama T, Kodama T, Hamakubo T, Shinkai Y. Histone methyltransferases G9a and GLP form heteromeric complexes and are both crucial for methylation of euchromatin at H3-K9. Genes Dev. 2005;19:815–26.
14. Tachibana M, Sugimoto K, Nozaki M, Ueda J, Ohta T, Ohki M, Fukuda M, Takeda N, Niida H, Kato H, Shinkai Y. G9a histone methyltransferase plays a dominant role in euchromatic histone H3 lysine 9 methylation and is essential for early embryogenesis. Genes Dev. 2002;16:1779–91.
15. Yamane K, Toumazou C, Tsukada Y-I, Erdjument-Bromage H, Tempst P, Wong J, Zhang Y. JHDM2A, a JmjC-containing H3K9 demethylase, facilitates transcription activation by androgen receptor. Cell. 2006;125:483–95.
16. Brauchle M, Yao Z, Arora R, Thigale S, Clay I, Inverardi B, Fletcher J, Taslimi P, Acker MG, Gerrits B, Voshol J, Bauer A, Schübeler D, Bouwmeester T, Ruffner H. Protein complex interactor analysis and differential activity of KDM3 subfamily members towards H3K9 methylation. PLoS ONE. 2013;8:e60549.
17. Cedar H, Bergman Y. Linking DNA methylation and histone modification: patterns and paradigms. Nat Rev Genet. 2009;10:295–304.
18. Chen J, Liu H, Liu J, Qi J, Wei B, Yang J, Liang H, Chen Y, Chen J, Wu Y, Guo L, Zhu J, Zhao X, Peng T, Zhang Y, Chen S, Li X, Li D, Wang T, Pei D. H3K9 methylation is a barrier during somatic cell reprogramming into iPSCs. Nat Genet. 2013;45:34–42.
19. Hu X, Zhang L, Mao S-Q, Li Z, Chen J, Zhang R-R, Wu H-P, Gao J, Guo F, Liu W, Xu G-F, Dai H-Q, Shi YG, Li X, Hu B, Tang F, Pei D, Xu G-L. Tet and TDG mediate DNA demethylation essential for mesenchymal-to-epithelial transition in somatic cell reprogramming. Cell Stem Cell. 2014;14:512–22.
20. Wang T, Chen K, Zeng X, Yang J, Wu Y, Shi X, Qin B, Zeng L, Esteban MA, Pan G, Pei D. The histone demethylases Jhdm1a/1b enhance somatic cell reprogramming in a vitamin-C-dependent manner. Cell Stem Cell. 2011;9:575–87.
21. Boyer LA, Plath K, Zeitlinger J, Brambrink T, Medeiros LA, Lee TI, Levine SS, Wernig M, Tajonar A, Ray MK, Bell GW, Otte AP, Vidal M, Gifford DK, Young RA, Jaenisch R. Polycomb complexes repress developmental regulators in murine embryonic stem cells. Nat Cell Biol. 2006;441:349–53.
22. Hajkova P, Erhardt S, Lane N, Haaf T, El-Maarri O, Reik W, Walter J, Surani MA. Epigenetic reprogramming in mouse primordial germ cells. Mech Dev. 2002;117:15–23.
23. Seki Y, Hayashi K, Itoh K, Mizugaki M, Saitou M, Matsui Y. Extensive and orderly reprogramming of genome-wide chromatin modifications associated with specification and early development of germ cells in mice. Dev Biol. 2005;278:440–58.
24. Seki Y, Yamaji M, Yabuta Y, Sano M, Shigeta M, Matsui Y, Saga Y, Tachibana M, Shinkai Y, Saitou M. Cellular dynamics associated with the genome-wide epigenetic reprogramming in migrating primordial germ cells in mice. Development. 2007;134:2627–38.
25. Seisenberger S, Peat JR, Reik W. Conceptual links between DNA methylation reprogramming in the early embryo and primordial germ cells. Curr Opin Cell Biol. 2013;25:281–8.
26. Hayashi K, Ohta H, Kurimoto K, Aramaki S, Saitou M. Reconstitution of the mouse germ cell specification pathway in culture by pluripotent stem cells. Cell. 2011;146:519–32.

27. Irie N, Weinberger L, Tang WWC, Kobayashi T, Viukov S, Manor YS, Dietmann S, Hanna JH, Surani MA. SOX17 is a critical specifier of human primordial germ cell fate. Cell. 2015;160:253–68.
28. Chen X, Skutt-Kakaria K, Davison J, Ou Y-L, Choi E, Malik P, Loeb K, Wood B, Georges G, Torok-Storb B, Paddison PJ. G9a/GLP-dependent histone H3K9me2 patterning during human hematopoietic stem cell lineage commitment. Genes Dev. 2012;26:2499–511.
29. Tsukada Y-I, Ishitani T, Nakayama KI. KDM7 is a dual demethylase for histone H3 Lys 9 and Lys 27 and functions in brain development. Genes Dev. 2010;24:432–7.
30. Inagaki T, Tachibana M, Magoori K, Kudo H, Tanaka T, Okamura M, Naito M, Kodama T, Shinkai Y, Sakai J. Obesity and metabolic syndrome in histone demethylase JHDM2a-deficient mice. Genes Cells. 2009;14:991–1001.
31. Ying Q-L, Wray J, Nichols J, Batlle-Morera L, Doble B, Woodgett J, Cohen P, Smith A. The ground state of embryonic stem cell self-renewal. Nature. 2008;453:519–23.
32. Lin S, Garcia BA. Examining histone posttranslational modification patterns by high-resolution mass spectrometry. Meth Enzymol. 2012;512:3–28.
33. Sridharan R, Gonzales-Cope M, Chronis C, Bonora G, McKee R, Huang C, Patel S, Lopez D, Mishra N, Pellegrini M, Carey M, Garcia BA, Plath K. Proteomic and genomic approaches reveal critical functions of H3K9 methylation and heterochromatin protein-1γ in reprogramming to pluripotency. Nat Cell Biol. 2013;15:872–82.
34. Yuan Z-F, Lin S, Molden RC, Cao X-J, Bhanu NV, Wang X, Sidoli S, Liu S, Garcia BA. Epiprofile quantifies histone peptides with modifications by extracting retention time and intensity in high-resolution mass spectra. Mol Cell Proteomics. 2015;14:1696–707.
35. Gascard P, Bilenky M, Sigaroudinia M, Zhao J, Li L, Carles A, Delaney A, Tam A, Kamoh B, Cho S, Griffith M, Chu A, Robertson G, Cheung D, Li I, Heravi-Moussavi A, Moksa M, Mingay M, Hussainkhel A, Davis B, Nagarajan RP, Hong C, Echipare L, O'Geen H, Hangauer MJ, Cheng JB, Neel D, Hu D, McManus MT, Moore R, et al. Epigenetic and transcriptional determinants of the human breast. Nat Commun. 2015;6:6351.
36. Fejes AP, Robertson G, Bilenky M, Varhol R, Bainbridge M, Jones SJM. Find-Peaks 3.1: a tool for identifying areas of enrichment from massively parallel short-read sequencing technology. Bioinformatics. 2008;24:1729–30.

Regions of common inter-individual DNA methylation differences in human monocytes: genetic basis and potential function

Christopher Schröder[1†], Elsa Leitão[2†], Stefan Wallner[3], Gerd Schmitz[3], Ludger Klein-Hitpass[4], Anupam Sinha[5], Karl-Heinz Jöckel[6], Stefanie Heilmann-Heimbach[7,8], Per Hoffmann[7,8,9,10], Markus M. Nöthen[7,8], Michael Steffens[11], Peter Ebert[12,13], Sven Rahmann[1] and Bernhard Horsthemke[2*] ●

Abstract

Background: There is increasing evidence for inter-individual methylation differences at CpG dinucleotides in the human genome, but the regional extent and function of these differences have not yet been studied in detail. For identifying regions of common methylation differences, we used whole genome bisulfite sequencing data of monocytes from five donors and a novel bioinformatic strategy.

Results: We identified 157 differentially methylated regions (DMRs) with four or more CpGs, almost none of which has been described before. The DMRs fall into different chromatin states, where methylation is inversely correlated with active, but not repressive histone marks. However, methylation is not correlated with the expression of associated genes. High-resolution single nucleotide polymorphism (SNP) genotyping of the five donors revealed evidence for a role of *cis*-acting genetic variation in establishing methylation patterns. To validate this finding in a larger cohort, we performed genome-wide association studies (GWAS) using SNP genotypes and 450k array methylation data from blood samples of 1128 individuals. Only 30/157 (19%) DMRs include at least one 450k CpG, which shows that these arrays miss a large proportion of DNA methylation variation. In most cases, the GWAS peak overlapped the CpG position, and these regions are enriched for CREB group, NF-1, Sp100 and CTCF binding motifs. In two cases, there was tentative evidence for a *trans*-effect by KRAB zinc finger proteins.

Conclusions: Allele-specific DNA methylation occurs in discrete chromosomal regions and is driven by genetic variation in *cis* and *trans*, but in general has little effect on gene expression.

Keywords: DNA methylation, Haplotype, Genome-wide association study, Differentially methylated regions, Inter-individual variability, Allele-specific methylation, Whole genome bisulfite sequencing, SNP genotyping, Methylation array

Background

Allele-specific DNA methylation occurs at distinct regions of the mammalian genome: (1) at imprinted loci as a result of genomic imprinting in the germline, (2) at gene promoters on the silent X chromosome in females as a result of X inactivation during early embryogenesis and (3) at non-imprinted autosomal loci as a consequence of genetic variation in *cis* (haplotype-dependent allele-specific methylation, hap-ASM [1–5]). In contrast to genomic imprinting and X inactivation, which always result in methylation of one allele in each cell of an individual, hap-ASM can be present on both alleles, on just one allele or on none of the alleles, dependent on the individual's genotype. In practice, however, most often DNA methylation levels other than 100, 50 or 0% are observed. This is because of extensive cell-to-cell

*Correspondence: bernhard.horsthemke@uni-due.de
†Christopher Schröder and Elsa Leitão contributed equally to this work
² Institute of Human Genetics, University of Duisburg-Essen, University Hospital Essen, Hufelandstraße 55, 45147 Essen, Germany
Full list of author information is available at the end of the article

heterogeneity (epigenetic mosaicism) as well as tissue heterogeneity. Epigenetic mosaicism means that even in a pure, isogenic cell population, cells differ from each other with respect to DNA methylation at a given locus, probably because the two alleles of a single nucleotide polymorphism (SNP) do not always dictate or prevent methylation of their genomic environment, but only increase or decrease the possibility that methylation occurs. This probability may even vary across tissues and may also be affected by environmental factors. As a consequence of this, a given genotype can be associated with multiple epigenotypes.

It has been suggested that hap-ASM may contribute to phenotypic variation, although there is no direct evidence for this to date. Indirect evidence comes from methylation quantitative trait loci (mQTL) studies, expression quantitative trait loci (eQTL) studies and genome-wide association studies (GWASs) [2, 6, 7]. These investigations have shown correlations between DNA sequence, methylation levels, gene expression levels and phenotypic traits, but it remains to be determined whether hap-ASM mediates the effect of DNA sequence variation on gene expression levels and phenotypic traits (active role), whether it stabilizes gene expression levels that have been brought about by SNP-sensitive transcription factors (passive role), or whether it occurs on certain haplotypes without having a function (no role).

Most mQTL studies were performed with methylation sensitive microarrays such as the Illumina 450k array. Owing to the low probe density of this array (it assays only 450,000 CpGs (1.6%) out of 28,000,000 CpGs), only single CpG sites or a combination of CpGs scattered over regions with poorly defined borders have been studied. Most hap-ASM studies used bisulfite sequencing (Sanger sequencing of subcloned bisulfite PCR products), which provides base-pair resolution, but were targeted at a few candidate regions only. A vast improvement in the field are techniques which enrich for all genomic regions known to impact gene regulation (hybrid capture kits; see for example [2]) or for all highly methylated regions (antibody-based approaches; see for example [8]). An unbiased survey of all differentially methylated regions, however, requires whole genome bisulfite sequencing (WGBS). WGBS has recently become the gold standard of genome-wide methylation analysis, but owing to the high costs involved, most often only a small number of samples are studied with this technique. However, one advantage of using allele-specific analysis compared to the QTL approaches is the smaller sample size requirements [9]. Nevertheless, sophisticated bioinformatic tools are necessary to reliably detect differentially methylated regions (DMRs) in a limited number of datasets, and candidate DMRs have to be validated by array-based techniques and/or targeted approaches.

Hap-ASM can seriously confound comparative methylome analyses in humans, if the samples are from different individuals. We have recently observed that DNA methylation differences between individuals may be larger than between distinct cell types [10], which has prompted us to identify and characterize inter-individual differences in DNA methylation in a more systematic way. For identifying regions showing common allele-specific DNA methylation, we have searched for blocks of co-varying CpGs (COMETs; [11]) that occur in only two states/epialleles (mainly methylated or mainly unmethylated) in each cell. At such a DMR, any individual has one of three epigenotypes: methylated/methylated, methylated/unmethylated, or unmethylated/unmethylated. To reduce the number of possible confounders in the DMR discovery phase, we restricted our analysis to a single cell type (monocytes) and used cells isolated by the same procedure (elutriation) from donors of the same sex (males). Based on epigenomic datasets generated by the same laboratory and bioinformatics pipeline according to standards set by the International Human Epigenetic Consortium (IHEC), this approach has enabled us to identify a significant number of high-confident DMRs and to link them to chromatin states and genetic variation.

Results

Identification of differentially methylated regions (DMRs)

Our previous DNA methylation analysis [10] had been performed in human monocytes and macrophages from two male donors (Hm03 and Hm05; for a cluster analysis see Additional file 1). For identifying inter-individual DNA methylation differences in human monocytes in a systematic way (see Fig. 1 for an overview of our study), we included three additional WGBS datasets produced by our laboratory (M55900 and Hm01 [12] as well as Hm02 (this study; for quality parameters of the methylomes see Additional file 2). In addition, we downloaded five publicly available IHEC WGBS datasets on human monocytes from other male donors, two from the BLUEPRINT consortium and three from the Canadian Epigenetics, Environment and Health Research Consortium (CEEHRC). A principle component analysis (PCA; Additional file 3), however, revealed that the data are very heterogeneous: While our five methylomes fall right on top of each other, the other methylomes are very different from each other and from our methylomes. The differences are probably due to the use of different cell purification methods, WGBS library preparation protocols, sequencing chemistries and bioinformatics pipelines. Therefore, we proceeded only with our five methylomes.

For identifying regions of common inter-individual DNA methylation differences, we devised a novel

Exploratory cohort — 5 individuals

Genomic data — Monocytes WGBS; 2.5 million SNP genotypes

DMRs discovery — 2 synthetic methylomes; 22% DMRs discovered; 157 DMRs (min 4 CpGs, min diff 0.8); 2 DMRs overlap with Do et al.

DMRs characterization and correlation with nearby SNPs — DMR environment; Histone marks; Gene expression; Correlation with SNPs; 29 DMRs overlap with 19 CGIs; 5 clusters according to histone marks; Associated with 240 genes; 50% DMRs correlate (>0.9) to the genotype of a nearby SNP; DNA methylation inversely correlated to active histone marks; Enrichment/depletion in chromatin states; Correlation between methylation and nearby SNP genotype validated by targeted Bis-seq

Large validation cohort — 1128 individuals

Genomic data — Whole-blood and monocyte 450k arrays (Reinius et al.); Whole-blood 450k arrays; ~500.000 SNP genotypes

High correlation of monocyte and whole-blood methylation levels (50 CpGs); Methylation levels for 50 CpGs; Methylation level distributions; 30/157 DMRs have a 450k CpG (50 CpGs); 50 GWAS

Genome wide association studies — 48/50 correlation peaks near CpG position; 30 lead SNPs; + 471 SNP in high LD with lead SNPs

Causative SNPs characterization — 14/30 DMRs: SNPs within or <100bp from the DMR (24 SNPs); Mainly intergenic or intronic SNPs; 23% known protein binding events in ±57kb from the DMRs occur within the DMRs or <100bp; TFs binding; Motifs changed

Major findings — Majority of DMRs have lower methylation than flanking regions; Correlation between changes in chromatin state and methylation; No correlation between methylation and gene expression; Allele-specific methylation proved on the read level for 10 DMRs (12 SNPs); DNA methylation depends on genetic variation in *cis*; Enrichment in: CREB group NF-1 Sp100 CTCF

Fig. 1 Overview of the study

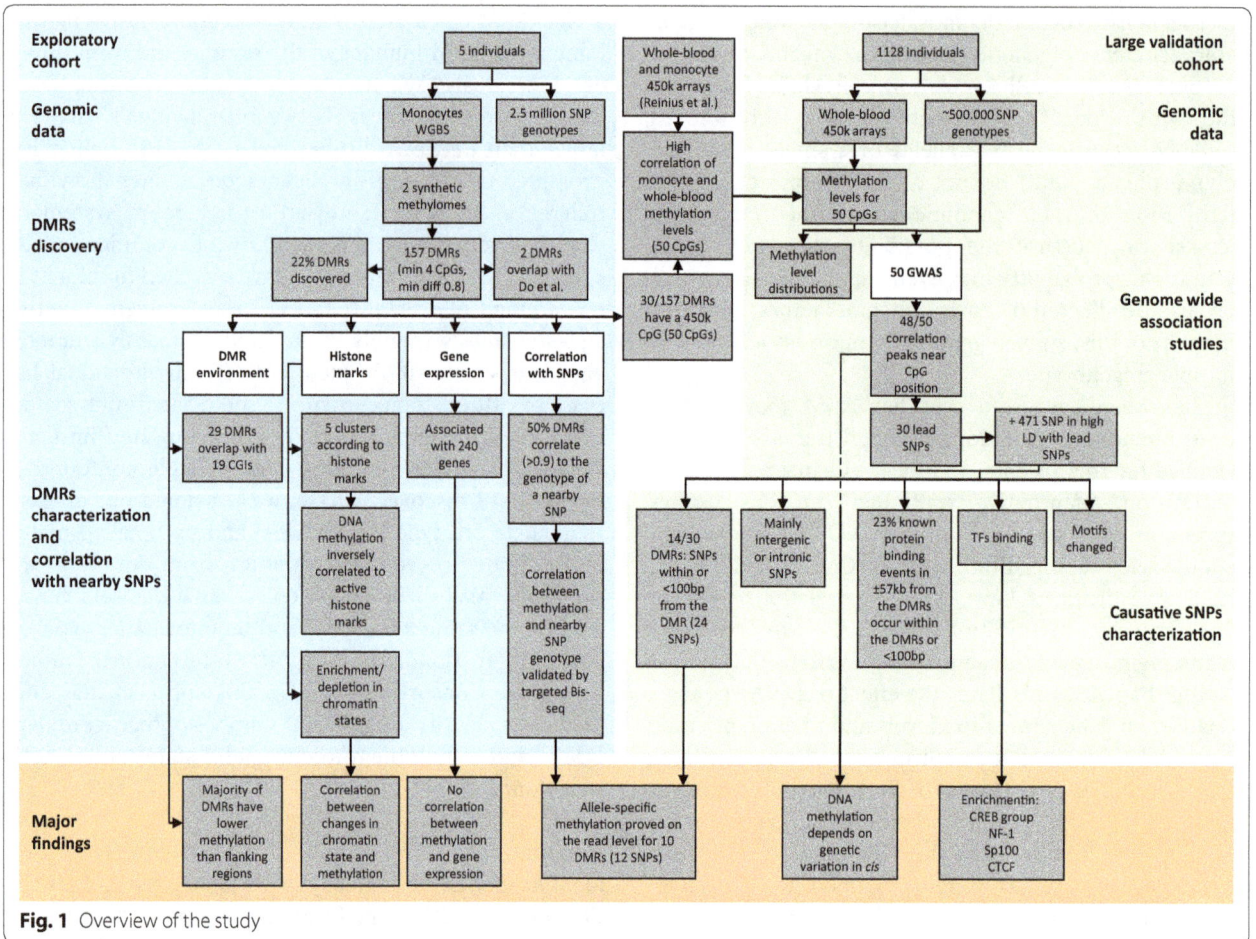

bioinformatic strategy: We created two synthetic methylomes, one with the highest methylation value of each CpG in the five samples and one with the lowest methylation value (Fig. 2a). We then used a modified version of Bsmooth [13] to detect differentially methylated regions (DMRs) between the two synthetic methylomes (see "Methods" section). Defining a DMR as a region of at least 4 CpGs with a methylation level difference of at least 0.8, we identified 157 DMRs ($p < 0.001$; Additional file 4). The threshold of 0.8 implies that a DMR is homozygously methylated in at least one individual and homozygously unmethylated in at least another individual, i.e., methylation differences in this region are very common.

The DMRs cover 1165 CpGs, have a size range of 9 to 1495 bp and encompass 4 to 44 CpGs (Additional files 4 and 5). The region with the highest number of CpGs is shown in Fig. 2b. Five DMRs have previously been reported by others: Two regions (DMR87 and DMR134) overlap previously designated hap-ASM DMRs [2], two DMRs contain a previously reported SNP mQTL (DMR25—rs6760544 [5], and DMR104—rs11158727 [6]),

and one DMR contains a previously reported ASM–SNP (DMR24—rs1530562 [14]). The majority of the DMRs are either intergenic (79/157) or intronic (57/157), while 13/157 span over an exon–intron boundary and 7/157 are located within an exon (Additional file 4). In comparison with randomly chosen regions, intergenic DMRs are highly overrepresented, whereas intragenic and exonic DMRs are highly underrepresented (Additional file 6).

Since non-imprinted autosomal CpG islands (CGIs) are typically unmethylated, we find it surprising that 29/157 DMRs overlap a CpG island (CGI), although in general CGI-DMRs are also underrepresented (Additional file 6). Most of these CGI-DMRs (19/29) are intragenic (either intron ($n = 8$), exon ($n = 5$) or intron–exon boundary ($n = 6$)). In 24 of these CGI-DMRs, all CpGs are within a CGI, in 4 cases there is a partial overlap with at least 50% of the CpGs belonging to a CGI, and in one case, the CGI is within the DMR. In some cases, closely linked DMRs affect the same CGI, probably because the DMR detection algorithm separated a large DMR into two or more DMRs. In total, 19 CGIs overlap a DMR, 11 of which are

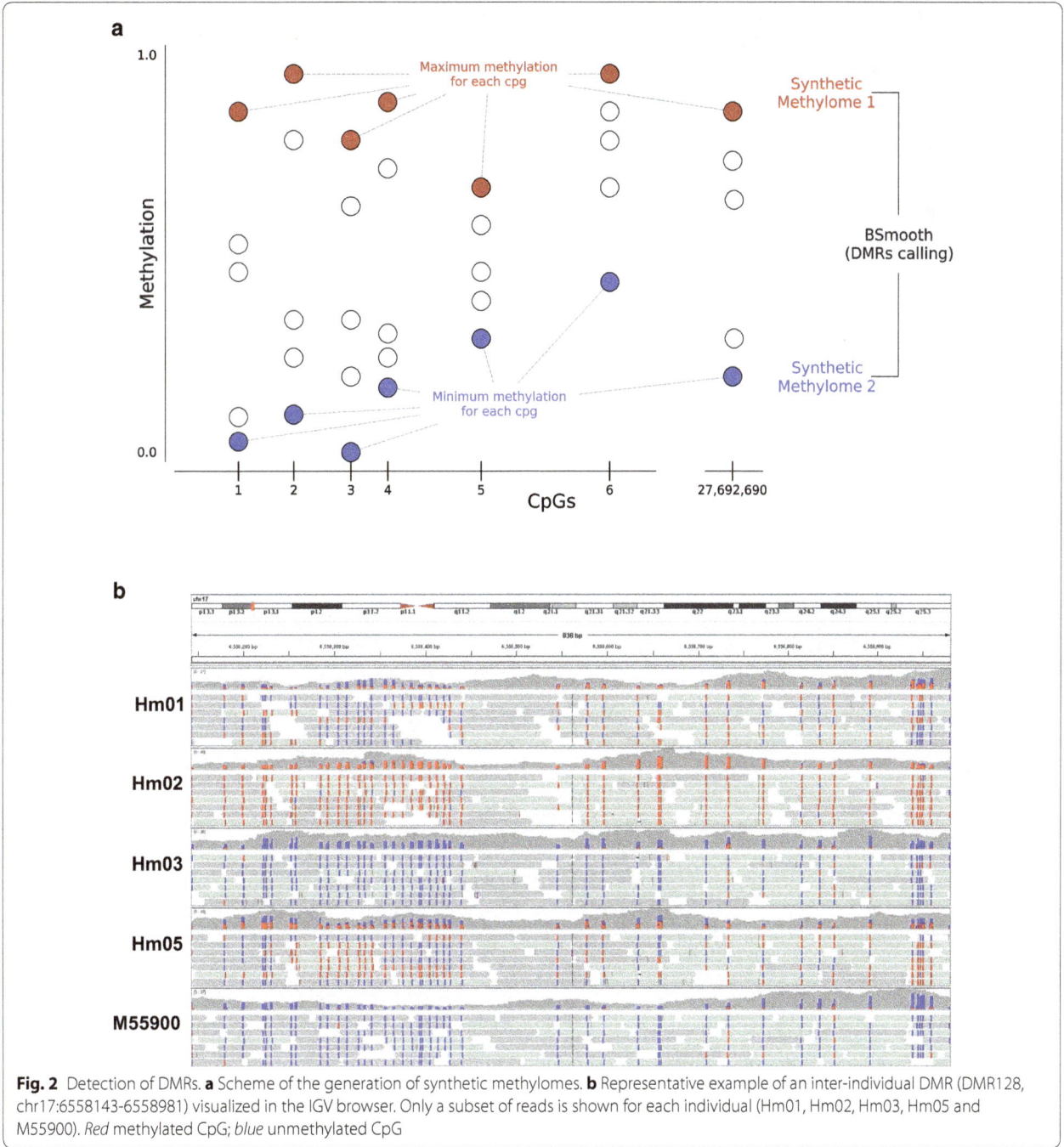

Fig. 2 Detection of DMRs. **a** Scheme of the generation of synthetic methylomes. **b** Representative example of an inter-individual DMR (DMR128, chr17:6558143-6558981) visualized in the IGV browser. Only a subset of reads is shown for each individual (Hm01, Hm02, Hm03, Hm05 and M55900). *Red* methylated CpG; *blue* unmethylated CpG

intragenic. Some of the CGIs are orphan CGIs, i.e., they are not associated with an annotated transcription start site [15].

We analyzed five samples, and our stringent settings allowed us to detect regions with common methylation differences. We expect that increasing the sample number could lead to the discovery of additional DMRs with rarer epialleles. Thus, we asked how many DMRs with a minor epigenetic allele frequency >0.05 might be present in the human population. Assuming that DNA methylation is allele-specific in these regions (for validation see below) and that the Hardy–Weinberg equilibrium applies in this situation, we estimate that there are 692 such DMRs. Of these, we have detected 23%.

Genomic environment of the DMRs

Since WGBS provides DNA methylation levels of all CpGs, we could investigate whether certain haplotypes (see below) act to decrease DNA methylation in a highly methylated domain or act to increase DNA methylation in a lowly methylated domain. For this, we compared the mean methylation level of the DMRs in the five donors to that of their flanking regions. In order to avoid border effects, we ignored the next three CpGs on each side of the DMRs and analyzed the following 10 CpGs. We observed that the mean methylation level of the DMRs is 0.49, which is close to the expected methylation level if on average methylated and unmethylated alleles occur at a similar frequency in the five donors. In contrast, both upstream and downstream flanking regions have a much higher level of methylation (average 0.72), which is highly significant ($p = 4.86 \times 10^{-20}$ and $p = 1.71 \times 10^{-21}$, respectively, Wilcoxon rank-sum test) (Additional file 4). Indeed, the vast majority of DMRs (107) have mean methylation levels lower than the two adjacent regions, while only 7 have higher methylation than both surrounding sequences (Fig. 3). In 40 DMRs, the methylation is intermediate from that of both flanking regions. For the remaining 3 DMRs, data are lacking for one of the flanks.

Chromatin states of the DMRs

For investigating the functional significance of the DMRs, we looked at six histone modifications (H3K4me1, H3K4me3, H3K27ac, H3K36me3, H3K27me3 and H3K9me3), which had been determined in the same

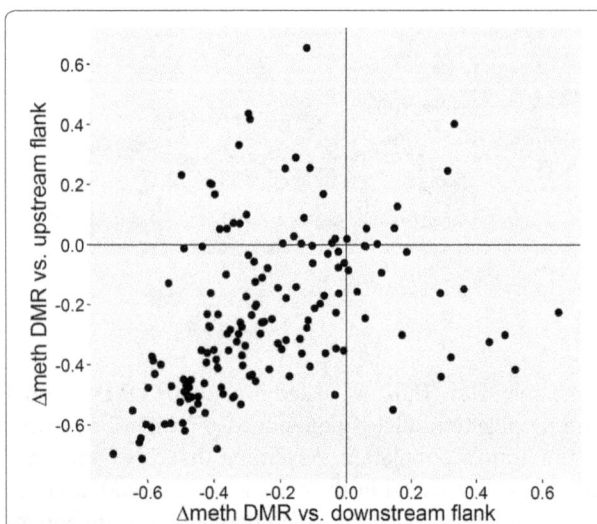

Fig. 3 Differences in DNA methylation of the DMRs and the upstream and downstream flanking regions. Most of the DMRs are flanked by regions with higher methylation levels (quadrant at *lower left*)

monocyte samples from donors Hm03 and Hm05 [10]. Using the k-means algorithm (with $k = 5$ classes) to cluster 2 kb sequences centered on the DMRs according to the ChIP signal across all six histone marks, we found that the DMRs have different histone modifications patterns (Fig. 4a). Clusters 1 and 4 are enriched for H3K27ac (albeit weakly in cluster 4) and H3K4me1, cluster 1 also for H3K4me3. These marks are indicative of active enhancers and promoters. Cluster 2 is strongly enriched for the repressive mark H3K27me3 and weakly enriched for the repressive mark H3K9me3. Cluster 3 is weakly enriched for H3K27ac and strongly enriched for H3K36me3, suggesting that these DMRs are transcribed elements. Cluster 5 is weakly enriched for the repressive mark H3K27me3. In summary, approximately 50% of the DMRs (84/157 in Hm03 and 82/157 in Hm05) carry strong or weakly repressive histone marks (clusters 2 and 5). The same is true for the subset of the 29 CGI-DMRs: 11/29 CGI-DMRs belong to cluster 2 in both donors, and 5/29 CGI-DMRs belong to cluster 2 in one of the two donors (Additional file 4). Most of the DMRs belonging to histone cluster 1 are intragenic (14/22 Hm03, 14/19 Hm05). On the other hand, most of the DMRs belonging to histone cluster 5 are intergenic (35/56 Hm03, 37/52 Hm05).

Independent clustering was performed for Hm03 and Hm05, since the two donors differ in the DNA methylation values of the DMRs. When we looked at the correlation between differences in DNA methylation levels and differences in histone modification levels between the two donors, we found that DNA methylation was inversely correlated with the active histone marks (linear regression), although the differences in histone modifications were small, but it was not correlated with the repressive histone marks (Fig. 4b and Additional file 7). In summary, this analysis suggests that the DMRs have different chromatin states and are more correlated to active than to repressive histone marks.

Based on the combination of the different histone marks in Hm03 and Hm05 monocytes, we segmented the genome into 18 chromatin states with the help of ChromHMM [16] and investigated whether certain chromatin states are over- or underrepresented (Additional file 8). We found that in both datasets 1_TssA, 5_Tx and 17_ReprPCWk were underrepresented and that 16_ReprPC and 2_TssFlnk or 4_TssFlnkD were overrepresented. Next, we investigated whether DMRs having (1) the same chromatin states in Hm03 and Hm05, or (2) different states in both donors, have similar distributions of absolute methylation differences between the two donors. In fact, the distribution is significantly different: In the DMRs that have different chromatin states in the two donors, methylation differences are higher compared to

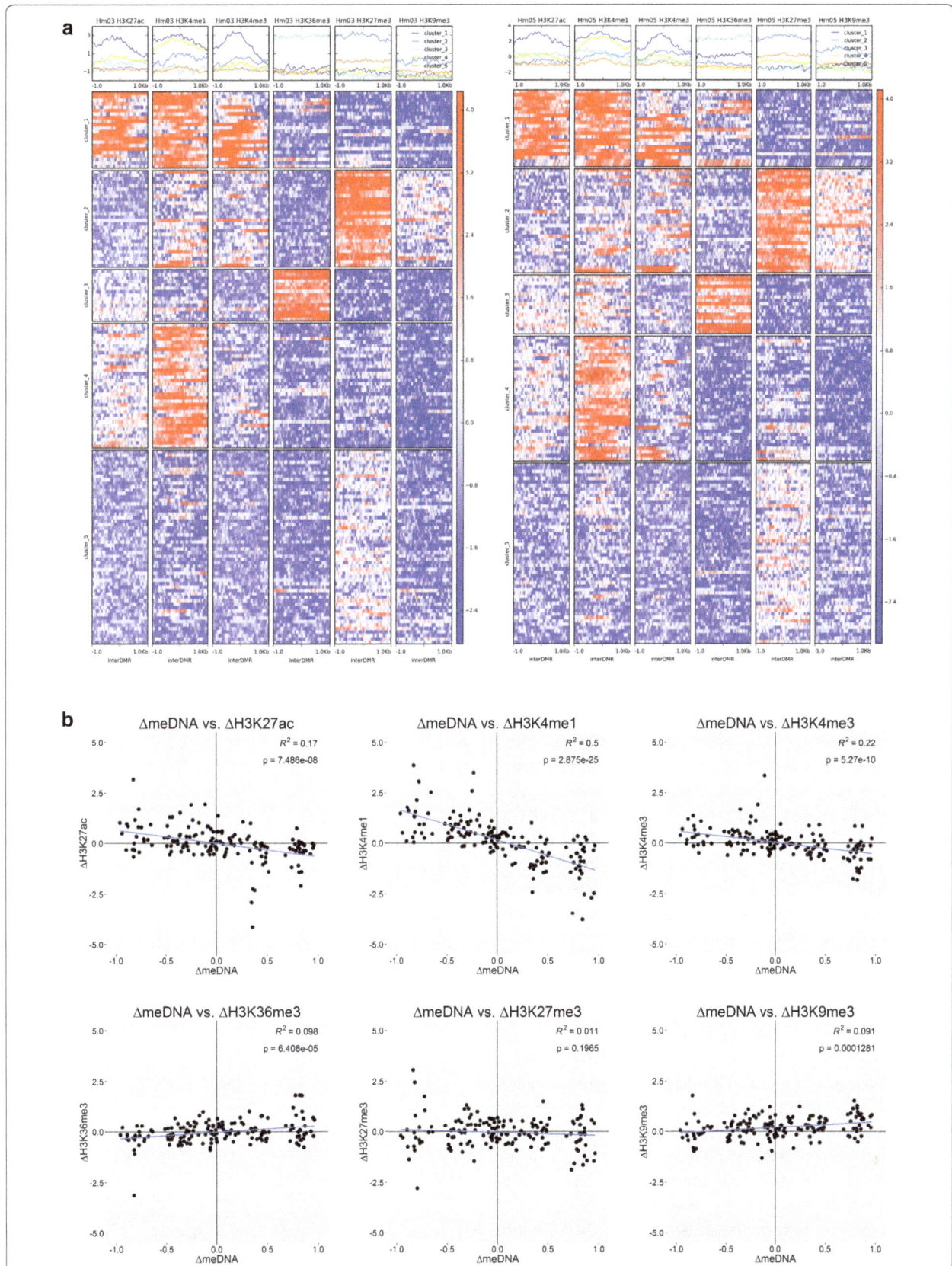

(See figure on previous page.)
Fig. 4 Histone modifications of 2 kb regions centered on the 157 inter-individual DMRs. **a** Heatmaps of histone modification signals for Hm03 (*left*) and Hm05 (*right*). Heatmaps show log2 ratio ChIP signal over input for six different histone modifications. **b** Scatter plots showing difference in histone modification signals between Hm05 and Hm03 as a function of methylation differences between the two donors. Active histone marks are inversely correlated with DNA methylation (linear regression)

the others ($p = 2.33 \times 10^{-5}$, Wilcoxon rank-sum test). As shown by the violin plots in Additional file 9, there are many DMRs with the same chromatin state and the same level of DNA methylation in the two donors (methylation difference <0.1), but there are very few DMRs with different chromatin states and the same DNA methylation. The relative abundance of DMRs with different chromatin states and methylation differences around 0.4 may be explained by homozygosity for a state in one donor and heterozygosity in the other donor, while differences around 0.8 may occur in DMRs where the two donors are homozygous for opposite states. In summary, these findings show that there is a correlation between DNA methylation and active chromatin states.

Location of the DMRs and putative target genes
The analysis of the 157 DMRs with Genomic Regions Enrichment of Annotations Tool (GREAT), which identifies *cis*-regulatory elements and their target genes, showed that 155/157 DMRs are associated with at least one gene and that in the majority of cases these are far away (see Additional files 4 and 10). In total, 240 different genes were identified. There was no significant enrichment of GO terms. The expression levels of these genes in donors Hm03 and Hm05 were not different from those genes that are not associated with a DMR (17,544; $p = 0.45$, Wilcoxon rank-sum test) (Additional file 11). As shown in Fig. 5a, there is no correlation between the differences in gene expression levels and the differences in methylation levels in these donors. The same is also true for the subset of genes that are associated with a DMR belonging to the histone modification clusters 1, 3 or 4 (active and transcribed DMRs), the subset of genes that are associated with a DMR which has a different chromatin state in donors Hm03 and Hm05, or the subset of genes that harbor a DMR (data not shown).

Since DNA methylation might affect alternative transcript initiation or splicing without changing total mRNA levels [17], we further investigated the genes harboring a DMR. As shown in Fig. 5b, there was no significant correlation between differences in methylation levels of the 77 intragenic DMRs and differences in transcript isoform expression of the host genes ($r^2 = 0.006$, $p = 0.085$).

Due to the fact that the gene list used by GREAT does not include all long non-coding RNA genes, we queried the database for annotated human lncRNAs (LNCipedia)

to identify lncRNA genes overlapping the DMRs. We found nine such genes (Additional file 4), and the two genes that are expressed in monocytes have equal RNA levels in both donors.

Correlation of DMR methylation levels with nearby SNPs
Next, we asked whether DNA polymorphisms within the DMRs or close by could be responsible for the

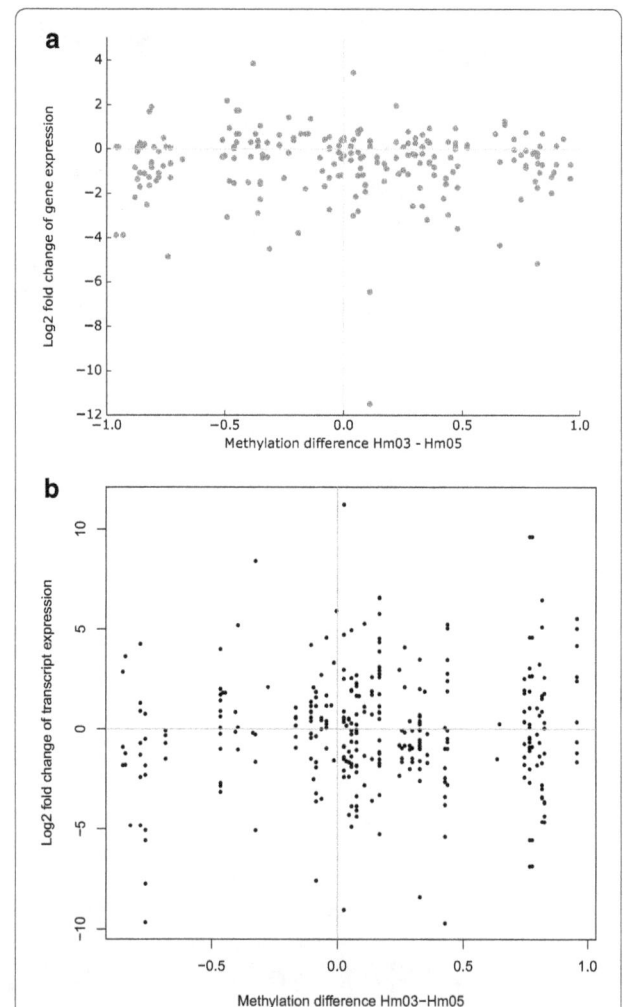

Fig. 5 Correlation between DNA methylation and gene expression. **a** Scatter plot of the differences in gene expression levels of the putative target genes identified by GREAT and the differences in DMR methylation in donors Hm03 and Hm05. **b** Scatter plot of the differences in transcript isoform levels of genes harboring a DMR and the differences in methylation of the 77 intragenic DMRs

inter-individual differences in DNA methylation. We genotyped the five donors for 2.5 million SNPs and found that 82/157 (52%) DMRs have a methylation level that is highly correlated (score > 0.9; see Methods) with the genotype of at least one nearby SNP (±6 kb from the center of the DMR; Additional file 12; for details see Methods). In 21/157 DMRs, that SNP is located within the corresponding DMR, and in 18/157 it is located <200 bp from the corresponding DMR border.

Validation of selected DMRs

We selected seven DMRs for validation that matched each of the following criteria: (1) SNP correlation score >0.9, (2) at least one CpG present on 450k arrays and (3) a methylation level of 33–67% in at least one of the five donors. Validation was performed by targeted deep bisulfite sequencing of four monocyte samples used for WGBS (Hm01, Hm02, Hm03 and Hm05) as well as two additional samples (Hm06 and Hm10), whom we genotyped for the 13 SNPs highly correlated with those DMRs (Additional file 12). For 6/7 DMRs, we observed a correlation between the DMR methylation levels and the genotype of at least one of the correlating SNPs (Additional file 13). In these cases, the homozygotes showed either the highest or lowest DMR methylation level, depending on the SNP allele, while the heterozygotes presented intermediate levels of methylation. For DMR12, which was no longer correlated with SNP rs692963 when two additional individuals where analyzed, it is possible that a correlating SNP lies >6 kb from the center of the DMR (as shown below, it is indeed). Regarding DMR128, in which the correlating SNP (rs9911968) is located within the DMR, we further analyzed heterozygotes for this SNP and calculated the methylation levels for reads containing the A or the G allele. We observed a significant difference in the methylation levels depending on the SNP allele present in the read, with the vast majority of the A allele containing reads being methylated, while the reads containing the G allele were unmethylated (Fig. 6a and Additional file 14). This demonstrates that DMR128 is subject to allele-specific methylation.

To verify that the same is true for other DMRs, we selected 11 regions that matched the criteria (1) and (3) above and the criterion that the correlating SNP locates within the DMR or in close vicinity (<200 bp from the corresponding DMR border). We performed targeted deep bisulfite sequencing of samples heterozygous for the corresponding correlating SNPs. For 8/11 DMRs, there are statistically significant differences between the two alleles ($p < 0.05$, two-tailed paired Student's t test), proving on the read level that allele-specific methylation occurs in these DMRs (Fig. 6a and Additional file 14), and validating the methodology we used to discover

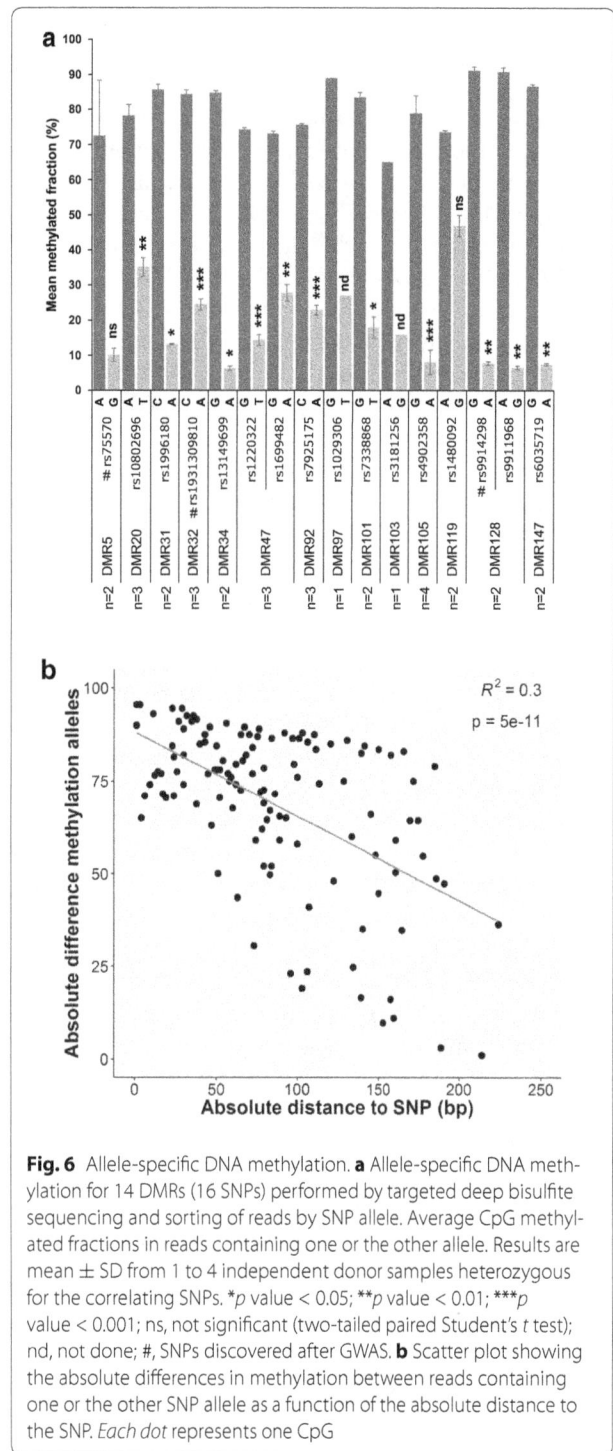

Fig. 6 Allele-specific DNA methylation. **a** Allele-specific DNA methylation for 14 DMRs (16 SNPs) performed by targeted deep bisulfite sequencing and sorting of reads by SNP allele. Average CpG methylated fractions in reads containing one or the other allele. Results are mean ± SD from 1 to 4 independent donor samples heterozygous for the correlating SNPs. *p value < 0.05; **p value < 0.01; ***p value < 0.001; ns, not significant (two-tailed paired Student's t test); nd, not done; #, SNPs discovered after GWAS. **b** Scatter plot showing the absolute differences in methylation between reads containing one or the other SNP allele as a function of the absolute distance to the SNP. *Each dot* represents one CpG

the DMRs. For 2/3 DMRs that fail to reach statistical significance (DMR97 and DMR103), we had only one sample. We also observed that the difference in methylation between the methylated and the unmethylated allele diminishes with the distance to the correlated SNP

(Fig. 6b), reassuring the relevance of these SNPs genotype on the methylation levels.

Genome-wide association studies (GWAS)

To validate the association between SNP genotypes and DNA methylation states in an independent and larger cohort, we investigated SNP and DNA methylation data of 1128 probands from the Heinz-Nixdorf Recall Study [18, 19]. In this cohort, DNA methylation levels had been determined in blood DNA with the help of Illumina 450k microarrays. Only 30/157 (19%), DMRs include one or more Illumina 450k CpGs (total: 50 CpGs), which shows that these arrays miss a large proportion of DNA methylation variation. In at least 29/50 cases, the distribution of the methylation levels showed three distinct peaks, suggesting that there are two epialleles (high and low methylation) (Fig. 7a and Additional file 15).

First, we checked whether monocyte and whole blood methylation levels were correlated. For this, we analyzed previously published Illumina 450k microarray data generated for whole blood and CD14+ monocytes samples from six healthy male donors [20]. The comparison of monocyte and whole blood methylation levels for the 50 CpGs revealed a high correlation (>0.92) in all individuals (Additional file 16).

For each of the 50 CpGs, we performed a GWAS with ~500,000 SNP, in which CpG methylation was treated as a quantitative trait. In 47/50 cases, there was a single correlation peak, which overlapped the CpG position (Fig. 7b, c; Additional files 17 and 18). For the CpG in DMR94 on chromosome 12, there was a correlation peak at the CpG position ($p = 1.59 \times 10^{-15}$) and at a locus on chromosome 19 ($p = 1.40 \times 10^{-41}$). For the CpG in DMR53 on chromosome 6, there was a single correlation peak on chromosome 7 ($p = 6.01 \times 10^{-11}$). We did not find any evidence for misannotation or cross-hybridisation of the array probes in these cases, but noted that the two GWAS peaks located on different chromosomes than the DMRs overlapped with genes coding for KRAB zinc finger transcription factors (*ZNF573* and *ZNF92*, for the DMR94 and DMR53 GWAS peaks, respectively). Since a p value threshold of 5×10^{-8} has become a standard for genome-wide significance in GWAS, the extremely low p values at the *ZNF573* and *ZNF92* loci point to a *trans*-acting effect.

For each GWAS, the SNP at the CpG locus with lowest p value (in most cases $p < 10^{-2000}$) was designated as lead-SNP (total $n = 30$). Using this genome-wide approach, we were able to confidently detect correlating SNPs outside the 12-kb window, e.g., in DMR12, for which the

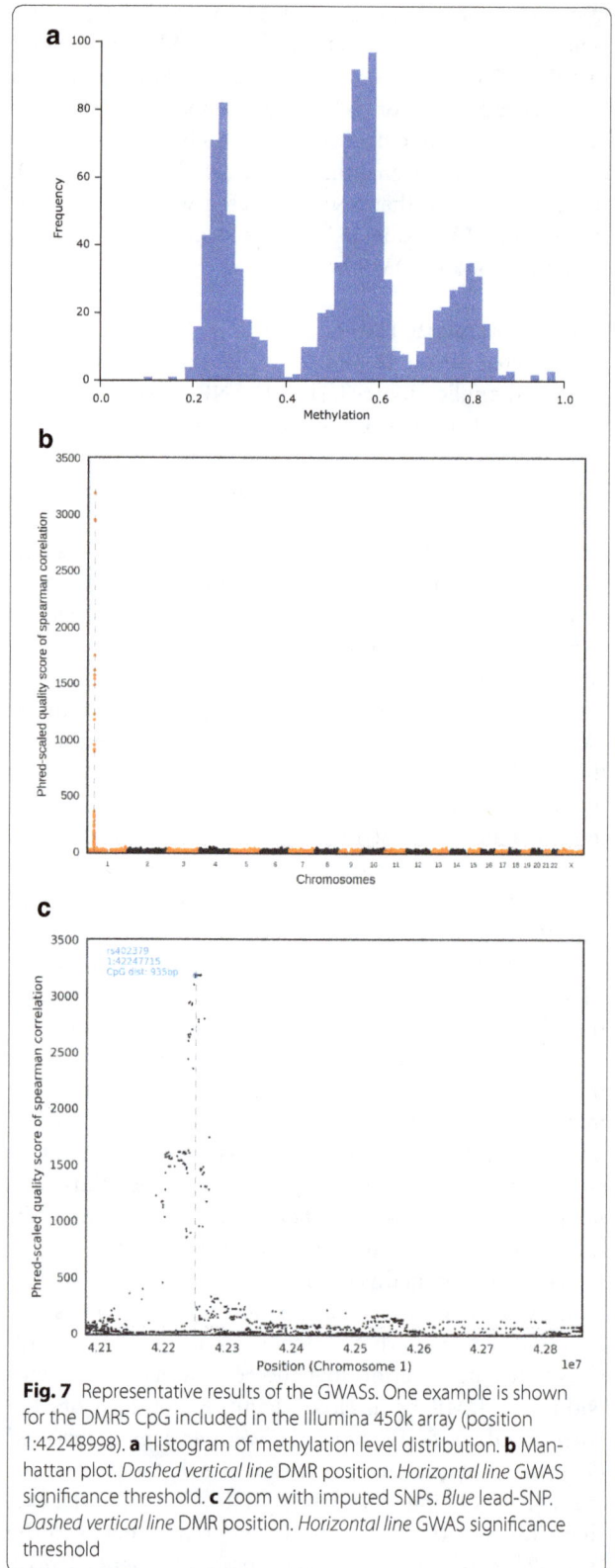

Fig. 7 Representative results of the GWASs. One example is shown for the DMR5 CpG included in the Illumina 450k array (position 1:42248998). **a** Histogram of methylation level distribution. **b** Manhattan plot. *Dashed vertical line* DMR position. *Horizontal line* GWAS significance threshold. **c** Zoom with imputed SNPs. *Blue* lead-SNP. *Dashed vertical line* DMR position. *Horizontal line* GWAS significance threshold

putatively correlated SNP detected in a 12-kb window failed validation in six individuals, the lead-SNP is actually located 40 kb from the DMR (Additional file 18). We used HaploReg [21] to retrieve SNPs in high linkage disequilibrium (LD, $r^2 > 0.8$) to the lead-SNPs (total $n = 471$), which are located within the corresponding DMR or up to ~116 kb from it, and mainly in intronic or intergenic regions (Additional file 19). For three of the DMRs validated previously, we confirmed the occurrence of ASM at the read level at SNPs in high LD with the lead-SNPs (Fig. 6a). These findings validate and extend the exploratory study described above.

Analysis of transcription factor-binding sites in and around the DMRs

Next, we analyzed whether the lead-SNPs and the highly correlated SNPs (total $n = 501$) might affect transcription factor-binding sites. Exploration of SNP annotation data from HaploReg database revealed that 23% of known protein binding events (Encode ChIPseq data) occur within the DMRs or <100 bp away (Additional file 20). The remaining events occur over a region up to 57 kb away from the DMR. The top five proteins found to bind within the DMRs or in close vicinity (<100 bp) are CTCF, CMYC, CEBPB, RAD21 and SMC3 (Additional file 19).

TRANSFAC analysis showed that the SNP regions are enriched for CREB group, NF-1, Sp100 and CTCF binding motifs (Additional file 21), and further analysis of their HaploReg annotations revealed that most of the SNPs are likely to alter regulatory motifs (Additional file 19).

Discussion

The use of whole genome bisulfite sequencing in human monocytes as part of full IHEC epigenomes and a novel bioinformatic approach has enabled us to identify and characterize regions of common inter-individual differences in DNA methylation at base-pair resolution, where allele-specific methylation is mainly caused by *cis*-acting genetic variation in transcription factor-binding sites. In two cases, we have obtained tentative evidence for *trans*-acting genetic variation in KRAB zinc finger genes. High-resolution WGBS has also allowed us to determine the methylation level of the genomic environment of these regions: Most of them are flanked by highly methylated DNA, which shows that certain haplotypes act to decrease DNA methylation in a highly methylated domain. Almost none of the DMRs described here has been identified before, and unexpectedly some overlap with a CpG island (CGI). Overall, however, gene promoters are underrepresented among the DMRs. Differences in DNA methylation are correlated with differences in active histone modifications and chromatin states, but in general not with differences in expression levels of the putative target genes, suggesting that other regulatory mechanisms are preponderant over DNA methylation in maintaining expression levels.

It is not possible to determine the exact number of DMRs, because this number obviously depends on the criteria used for DMR detection. Using WGBS data from five individuals and stringent thresholds, we have detected 157 inter-individual DMRs in monocytes and estimate that there are 692 regions with a minor epiallele frequency >0.05 in the human population. Interestingly, Do et al. arrive at similar figure ($n = 792$), but there is hardly any overlap between their DMRs and ours (only two). Most probably, the DMRs identified by us are not targeted by the Agilent SureSelect Methyl-seq capture kit used by Do et al. [2], which queries only 3.7/28 million CpGs (13.2%). Since this kit focuses on regions where methylation is known to impact gene regulation, our DMRs appear to lie in regions of unknown function (see also below). On the other hand, we may have missed the DMRs identified by Do et al., because they did not pass our stringent criteria for significance and/or different tissues were used.

By defining a DMR as a region with a methylation difference between the two synthetic methylomes of at least 0.8, we have identified regions of common methylation differences which are characterized by allele-specific DNA methylation in the majority of cells. Regions which show allele-specific DNA methylation in only a fraction of cells would not have passed the 0.8 threshold. This is one reason why the number of hap-ASM regions identified in this way is much smaller than the number of loci typically identified in mQTL studies (see for example [7]). In mQTL studies, where the genotype of each tested SNP is correlated with the methylation level of each tested CpG, CpGs often have normalized methylation values on a continuous scale between 0.0 and 1.0. At the level of individual cells, however, DNA methylation (unlike mRNA levels, for example) cannot be a quantitative trait, because a given CpG in a DNA molecule can only be methylated or unmethylated, allowing for three discrete epigenotypes per cell only. Therefore, mQTLs studies primarily measure the proportion of cells with one or two methylated alleles or—in other words—the probability that a CpG becomes methylated, rather than allelic methylation differences *per se*. Mean methylation values across a region with several CpGs could be a quantitative trait, if the methylation of the individual CpGs were poorly correlated. In our validation studies based on deep bisulfite sequencing, we have not observed such heterogeneous patterns.

We note, similar to Do et al. [2], that most of the hap-ASM regions are not covered by the Illumina 450k array.

In our case, the array missed 80% of the DMRs. For almost all of the tested DMRs that did not have a CpG on the 450k array, allele-specific methylation was proven to occur on the read level at SNPs within or in the close vicinity to the DMR. Thirty of the 157 DMRs could be studied in a large cohort on 450k methylation arrays. In many of these cases, we saw a trimodal distribution of the methylation values, reflecting the three possible epigenotypes and indicating low epigenetic mosaicism. In the other cases, the CpG tested may not be representative for the DMR, because it is close to the border of the DMR, for example.

In our GWAS studies, we found that the methylation levels were significantly correlated with the genotype of nearby SNPs, often with p values $<10^{-2000}$, which also validates these DMRs. In two cases (DMR53 and DMR94), we obtained tentative evidence for the existence of *trans*-acting loci. Interestingly, in both cases the GWAS peak was over KRAB zinc finger transcription factors genes, namely *ZNF92* and *ZNF573*. Unfortunately, the two proteins are poorly characterized, but KRAB zinc fingers are known to interact—among other proteins—with TRIM28, which plays a role in maintaining DNA methylation [22–24]. This finding certainly requires validation in another cohort as well as more detailed molecular studies.

For investigating the possible role of the DMRs, we made use of the histone modification and gene expression data that we have on the Hm03 and Hm05 monocytes as part of the full IHEC epigenomes. We find that the DMRs lie in regions with different chromatin states including active and repressive chromatin and that they are enriched for regions flanking transcription start sites (TssFlnk), but depleted for strong transcription (Tx), active transcription start sites (TssA) and weakly repressed sites (ReprPCwk). With regard to active chromatin, we note that McClay et al., Do et al. and Cheung et al. also observed that mQTLs are enriched for TssFlnk regions [1, 2, 7]. The correlated SNP regions are bound by CTCF, CMYC, CEBPB, RAD21, SMC3 and other transcription factors. Enrichment of CTCF- and RAD21-binding sites in TssFlnkD regions has also been observed by the Epigenome Roadmap Consortium [25]. CTCF, RAD21 and SMC3 play an important role in chromatin architecture [26]. The relevance of CTCF binding for hap-ASM in other DMRs has previously been reported [2, 14, 27]. Together, these data suggest that SNPs cause hap-ASM through affecting the binding of transcription factors to the DNA, most likely mediated through chromatin looping in the case of SNPs located far away from the DMR. However, which SNP and which transcription factor affect DNA methylation is difficult to pinpoint, since genetic variants are often in linkage disequilibrium

and may have either a direct influence on transcription factor binding by disrupting the recognition motif, or indirect by affecting cooperative and collaborative transcription factor binding, or altering the chromatin state or conformation affecting the stability of interactions between transcription factors and with DNA [28].

There is a significant inverse correlation between DNA methylation and active histone marks, although the differences in histone modifications are small, but no correlation with repressive histones marks. It is not possible to decide whether the SNPs cause differences in certain histone modifications that favor or hinder DNA methylation, or whether the SNPs cause differences in DNA methylation that affect the recruitment of histone modifying enzymes. Based on what is known about the interplay between DNA and histone modifications [29, 30], we tend to believe that the first scenario is true.

Surprisingly, differences in DNA methylation and histone modifications do not appear to affect gene expression levels. It could be argued that these differences poise the genes for expression in response to external stimuli, which we did not test, but then we would expect that the genes were related to monocyte and macrophage function, which we did not find in our gene ontology analysis. Another possibility is that the methylation differences affect other gene features such as alternative transcript initiation or splicing, which may be true for the subset of intragenic DMRs. Overall, however, there was no correlation between the levels of transcript isoforms and DNA methylation, although it remains possible that in a few exceptional cases the methylation level does affects a transcript isoform. This needs to be investigated further by a series of detailed molecular studies. Still another possibility is that GREAT did not identify the real target genes, which may encode long non-coding RNAs expressed at low levels, especially in large introns and intergenic regions. However, only very few DMRs overlap a lncRNA gene, and most of these are not expressed in monocytes. In summary, we conclude that the majority of the DMRs do not seem to have a strong gene regulatory function under the tested conditions. While this hypothesis may not be welcomed by everybody in the epigenetic field, it is in line with other observations. Gibbs et al. found that only 4.8% of significant mQTLs were also an eQTL [6], and McClay et al. suggested that many mQTLs, especially those located in repressive chromatin, lack functional consequence [7]. Do et al. have also recently hypothesized that only a minority of hap-ASM DMRs are likely to have important effects on gene expression by being located in crucial regulatory regions [9]. However, although some of our DMRs are located in regions with signatures of active enhancers and promoters, they do not seem to affect the expression of target genes. Overall, these results question a major role of hap-ASM in phenotypic variation.

Unexpectedly, a significant fraction of the DMRs overlapping a CGI (see below) carry the repressive histone mark H3K27me3, irrespective of whether they are methylated or not, and there is no correlation with the expression levels of the putative target genes. In imprinted DNA methylation, silent X-associated DNA methylation and cell-type-specific DNA methylation (for the latter see for example [10]), specific DNA sequences are subject to stable transcriptional silencing even in the presence of all of the factors required for their expression [31]. In contrast, a significant proportion of hap-ASM appears to occur in regions where certain haplotypes fail to keep them methylation-free in the presence of the DNA methylation machinery, without affecting gene expression levels.

Since CGIs are almost exclusively unmethylated in all tissue types, regardless of state of expression [31], the observation that 29/157 (~20%) of our DMRs overlap a CGI was unexpected. Assuming that there are 692 such DMRs (see "Results" section) and that a similar fraction of the undetected DMRs overlaps with a CGI, we estimate that ~100/~30,000 CGIs might be affected by hap-ASM in human monocytes. Since hap-ASM shows considerable tissue heterogeneity [2], which substantiates the notion that transcription factors are instrumental in setting up hap-ASM patterns, more than 100 CGIs may be affected. Although the total number of such CGIs is probably small, we find it surprising that hap-ASM affects CGIs at all. It remains to be determined what makes certain CGIs susceptible to hap-ASM. It is probably a combination of transcription factor-binding sites (or a lack thereof) that—on certain haplotypes—fails to protect a CGI against the invasion of methylation from the surrounding region. This is probably true also for other hap-ASM regions. The finding that most of these regions are flanked by highly methylated DNA on both sides suggests that in general DNA-binding factors prevent DNA methylation. In only few cases, the flanking DNA is lowly methylated, and here DNA-binding factors may attract DNA methylation.

Conclusions

We have identified novel regions of common inter-individual DNA methylation differences in human monocytes. Our study supports and extends the observation that allelic DNA methylation differences can be caused by genetic variation *in cis*. Interestingly, DNA methylation at some loci may also be affected by genetic variation *in trans*, namely at KRAB zinc finger genes. In general, hap-ASM, especially hap-ASM in repressive chromatin domains, appears to have little functional consequences.

Methods
Monocytes isolation
Primary human monocytes were isolated from healthy normolipidemic volunteers (Hm02, Hm06 and Hm10) by leukapheresis and counterflow elutriation as described previously [32].

DNA extraction from monocytes and tissues
DNA was isolated from monocytes using QIAamp columns (Qiagen, Germany) and quantified with a Nanodrop 100 spectrophotometer (Peqlab, Germany).

Whole genome bisulfite sequencing and analysis
Generation of whole genome bisulfite sequencing data from monocytes obtained from donor Hm02 was performed as described previously [10, 12].

Detecting DMRs
We used the WGBS datasets from Hm02 and four additional donors (Hm01, Hm03, Hm05 and M55900; see Data retrieval and deposition) to generate two synthetic methylomes, one with the highest methylation level of each CpG in the five samples and one with the lowest methylation level. We modified BSmooth [13] to identify differentially methylated regions with a minimum difference of 0.8 between the two synthetic methylomes. BSmooth is designed to compare a group of multiple cases against a group of multiple controls. Because we have no class labels, our data consist of two single synthetic methylomes (min, max) and therefore of case and control groups of one sample each. The main formula of the BSmooth algorithm

$$t(c) = \frac{\Delta(c)}{\left[\sigma(c)\sqrt{\frac{1}{n_2} + \frac{1}{n_2}}\right]}$$

calculates a signal-to-noise statistic $t(c)$ for each CpG c with $\Delta(c)$ referring to the mean methylation differences of both groups.

In our case, we reduced this formula to $t(c) := (\max(c) - \min(c))$ for each CpG c. The terms $\max(c)$ and $\min(c)$ simulate the process of creating synthetic methylomes by selecting the maximum and minimum methylation level of a CpG over all samples. DMRs are formed by consecutive groups of CpGs with $t(c) > v$ or $t(c) < -v$ with a threshold $v > 0$. We use $v = 0.5$ as parameter for the DMR calling.

A DMR's border may differ in shape, and DMR calling algorithms often cannot identify them exactly. In contrast to BSmooth, we calculate a DMR's methylation level by building a weighted average methylation level

$$\mu_i(d) = \frac{1}{|C(d)|} \sum_{c \in C(d)} \sigma(c) m_i(c)$$

for DMR d, its set of CpGs $C(d)$, the methylation levels $m_i(c)$ and standard deviation over all samples $\sigma(c)$ of CpG $c \in C(d)$ in sample i. We call $\mu_i(d)$ the core methylation of d in sample i. The core methylation is less influenced by the inaccurate DMRs borders. We only keep DMRs with high (core) mean methylation differences ≥ 0.8 and sufficiently long DMRs consisting of 4 CpGs or more.

To test the DMRs for statistical significance, we calculated an empirical p value by simulating 1000 sets of five samples according to the null model that there is no methylation difference as follows:

Let $n_{s,c}$ be the coverage and $m_{s,c}$ the methylation count for observed sample s at CpG c. First, we calculate the average methylation

$$p_c = \frac{\sum m_{s,c}}{\sum n_{s,c}}$$

value for each CpG c.

For each of the five observed samples s, we simulate a corresponding null sample o.

We set the coverage of CpG c in sample o to $n_{o,c} = n_{s,c}$.

The methylation count $M_{o,c}$ for each c is randomly chosen with binomial probability

$$P\big(M_{o,c} = m\big) = \binom{n_{o,c}}{m} \cdot p_c^m \cdot (1 - p_c)^{n_{o,c}-m}$$

Therefore, the coverage of each CpG in the null sample is equal to the coverage in the corresponding observed sample, while differences in methylation are only caused by finite sampling size. Per definition, there exist no DMRs for the null samples, and every detected DMR is a false positive. We applied our algorithm for DMR detection to the null samples. We repeated the process 1000 times.

The algorithm did not detect any DMRs. This leads to an empirical p value <0.001 for each DMR.

Calculation of the DMR detection rate

We assume that a single SNP is responsible for the methylation of a DMR and that the probability of being a causative SNP is independent of its allele frequency. We further assume that the epigenotypes follow the Hardy–Weinberg equilibrium with $P(AA) = p^2$, $P(aa) = q^2$ and $P(Aa \text{ or } aA) = 2pq$ with some frequencies p and q such that $p + q = 1$. Our approach is only able to detect DMRs where at least one out of n samples is fully methylated and at least one sample is unmethylated.

The probability of obtaining such an event can be derived from an urn model with three different types of balls. Two types with probabilities p^2 and $q^2 = (1 - p)^2$,

respectively, represent the two different homozygous SNP states, and the third type with probability $2pq$ represents the heterozygous SNP state. For n samples, the probability to draw at least one ball of each of the first two types is $P(p, q) = 1 - [(1 - p^2)^n + (1 - q^2)^n - (2pq)^n]$, where we have applied the inclusion–exclusion principle to the complementary event.

We used the known allele frequencies of all SNPs with a minor allele frequency >0.05 contained in dbSNP [33] to estimate the fraction of detectable DMRs. For $n = 5$, we estimate that we can detect 23% of DMRs with a minor epigenetic allele frequency >0.05.

SNP genotyping

For donors Hm01, Hm02, Hm03, Hm05 and M55900, 2.5 million SNPs were genotyped using Illumina's Omni2.5Exome Bead Array. For donors Hm06 and Hm10, SNP genotypes were inferred from the targeted bisulfite sequencing data (see below), or by Sanger sequencing regions amplified with primers listed in Additional file 22.

DMR SNP correlation score calculation

The mean methylation level of a sample in a region with allele-specific methylation is expected to be either close to 0.0, close to 1.0 or about 0.5. Due to inaccurate DMR borders, finite sequencing coverage and noise, measured values may differ from this expectation. We assume three possible classes "full-methylated," "half-methylated" and "unmethylated" for this epigenotype.

In order to compare these epigenotypes with SNP genotypes, we have to classify the methylation level of each sample for each DMR. To avoid fixed thresholds for class assignment, we calculate the posterior probabilities of mean DMR methylation level to fall into each of the classes. We consider the empirical distribution (histogram) of $157 \times 5 = 768$ core methylation levels $\mu_i(d)$ of each sample i and DMR d, which contains data from all three classes. This empirical distribution can be decomposed into a three-component mixture of beta distributions. A beta distribution is a continuous probability distribution on the unit interval $[0, 1]$ that is frequently used to model data that naturally takes values between 0 and 1 [34] such as methylation levels. Each component beta distributions have two parameters α and β that determine the shape of the beta distribution. We used the betamix software [35] to robustly fit a three-component beta mixture model to the observed histogram.

Let α_i and β_i be the beta distribution parameters and π_k the mixture coefficient of component $k \in \{0, 1, 2\}$ after fitting. For a single sample, a DMR with a core methylation level μ and one SNP genotype $g \in \{0, 1, 2\}$, the posterior probability of g given μ is given by

$$L(g,\mu) = \frac{\pi_g b_{\alpha_g,\beta_g}(\mu)}{\sum_k \pi_k b_{\alpha_k,\beta_k}(\mu)}.$$

To calculate a score based on multiple samples, we extend the formula. Let $g_i(s) \in \{0, 1, 2\}$ be the genotype and $\mu_i(d)$ be the core methylation level (see "Detecting DMRs" section) for DMR d, sample i and SNP s. The joint posterior probability

$$\text{score}(s,d) = \prod_i L\big(g_i(s), \mu_i(d)\big)$$

is given by the product of the single posterior probabilities over all samples. We use this posterior probability as a score to assess whether DMR d and SNP s are co-varying. The scores are separately calculated for each DMR and each SNP within a range of ±6 kb of the DMR's location. For $n = 5$ samples, we used 0.9 as a threshold to call an SNP correlated with a DMR.

GWAS analysis
The SNP array data were produced with three different SNP array types: Omni1_Quad_v1 (334 probands), OmniExpress_12v1.0 (627 probands) and OmniExpress_12v1.1 (170 probands). The data were normalized and CpG methylation levels extracted using RnBeads v1.2.2 [36].

We filtered each array separately by removing SNPs that failed the Hardy–Weinberg test at a significance threshold of 0.001, having a minor allele frequency less than 0.01 or a missing rate greater than 0.1 using plink v1.07 [37]. The arrays were merged by plink and the data again filtered by plink using the previously described parameters. This merged data served as genotypes for the GWASs.

The Spearman correlations and p value calculation between methylation levels and SNP genotypes were performed using NumPy v1.11.0 and SciPy v0.14.0 [38].

For the imputation, a region was chosen that includes all SNPs with a p value $<5 \times 10^{-8}$ but to a maximum of ±1 Mb of the CpGs position. The arrays were then converted to ped-format using gtool v0.7.5 [39] and separately imputed using impute2 [40] for the determined regions with the phase 3 data of the 1000 genomes project [41]. The imputed data were reconverted to bed-format again using gtool and merged under the previously given filter parameters by plink.

DMR validation by targeted deep bisulfite sequencing
Bisulfite-converted DNA was obtained using 500 ng of monocytes DNA (Hm01, Hm02, Hm03, Hm05, Hm06 and Hm10) and the EZ DNA Methylation-Gold Kit (Zymo Research) according to the manufacturer's instructions. Locus-specific bisulfite amplicon libraries were amplified by PCR employing bisulfite tagged primers (Additional file 22) designed using the MethPrimer [42] and BiSearch [43, 44] tools and HotStarTaq Master Mix (Qiagen). Sample-specific barcode sequences (MID, multiplex identifiers) and universal linker tags (454 adaptor sequences) were added by performing a second PCR. Samples were prepared and sequenced on a Roche/454 GS Junior system (Roche Diagnostics) with special filter settings applied to increase the yield of reads [45]. Automated CpG methylation analysis was performed using the Amplikyzer software [46] with minimum bisulfite conversion rate set to 95%, leading to an average of 2450 reads per sample (minimum 187).

Histone modification ChIPseq heatmaps
Heatmaps visualizing the ChIP log2-ratio between signal and input across six histone modifications in two biological replicates were generated using deepTools [47] as previously described [10], except that we plotted data from 2-kb regions centered on the middle of 157 DMRs and clustered them using $k = 5$ clusters in the k-means algorithm. Independent clustering was performed for Hm03 and Hm05, since the two donors differ in the DMRs DNA methylation values.

Chromatin segmentation by chromatin states
Chromatin segmentation of samples Hm03 and Hm05 was performed with ChromHMM [16]. We estimated the p value for over- and underrepresented chromatin states by simulating 1 million datasets, consisting of 151 regions each and equal size distribution compared to our DMRs. Each of the simulated regions was selected from non-repetitive regions covering at least 4 CpGs. For each of our 157 DMRs and each region of the simulated datasets, the overlap between its coordinates and the chromatin states of each sample (Hm03, Hm05) was calculated. We then compared the count of overlapping chromatin states of a sample for our DMRs and each random set. The empirical p value for overrepresentation for state x is the fraction of random sets that have a higher count for x than the DMRs. The empirical p value for underrepresentation for state x is the fraction of sets that have a lower count for x than the DMRs. We partitioned the absolute methylation differences between Hm03 and Hm05 into (1) a set for the DMRs with different chromatin states in the two donors and (2) another set for the DMRs with the same states. We consider a state as different in the two donors, if the intersection of the DMR overlapping chromatin states is empty. The two sets of methylation differences were then compared by applying the Wilcoxon rank-sum test to determine if the methylation differences are independent from a chromatin state difference.

GREAT analysis

The bioinformatic tool GREAT was used to predict DMR functions by analyzing the annotations of nearby genes [48], under species assembly GRCh37 with whole genome background and choosing the "Basal plus extension" association rule setting with default parameters of 5.0 kb upstream, 1.0 kb downstream and up to 1000.0 kb distal.

Distribution of gene expression levels

To obtain the gene expression rates for each gene, we summed the transcript per million (tpm) values as calculated by kallisto with default parameters [49]. Since GREAT uses only the extremely high-confidence genes prediction subset of the UCSC Known Genes, we reduced the kallisto gene list to this subset ($n = 17,784$). We then partitioned the mean expression rates of Hm03 and Hm05 for each gene into two groups: genes that are associated with a DMR as identified by GREAT ($n = 240$) and genes that are not associated (17,544). We compared the expression rates of these two groups by applying a Wilcoxon rank-sum test to test for differences.

Identification of transcription factor-binding motifs

We used the TRANSFAC database (professional version, release 2015.3, [50]) to determine, if certain motifs were enriched in the SNP regions (501 regions: SNP $\pm$ 100 bp). Example regions as provided by TRANSFAC served as background, and the parameters were set to default.

Differential transcript expression

We ran Tophat 2.0.11 [51], with Bowtie 2.2.1 [52] and NCBI build 37.1 using the following parameters: – library-type fr-firststrand and –b2-very-sensitive setting, to generate the mapping files from total-RNA of Hm03 and Hm05 samples. Subsequently, StringTie [53] with NCBI build 37.1 was run in -e -b -G mode to generate files for analysis with Ballgown [54]. Differential transcript expression analysis was performed using Ballgown, and an FDR cutoff of 0.05 was chosen to extract the differentially expressed transcripts.

Long non-coding RNAs

LNCipedia 4.0 [55, 56] (GRCh37/hg19) was used to extract high-confidence lncRNA regions. Bedtools was subsequently run in "intersect" mode to get the overlap of the differentially methylated regions with the lncRNA regions.

Data retrieval

The full epigenome data from Hm03 and Hm05 monocytes (Study Accession ID: EGAS00001001595, Dataset Accession ID: EGAD00001002201) as well as the methylome data from M55900 (ENA PRJEB5800) and Hm01 (EGAS00001000719) have previously been produced by our group [10, 12]. The BLUEPRINT and CEEHRC WGBS datasets on human monocytes from other male donors were retrieved from the IHEC Data Portal (http://epigenomesportal.ca/ihec/grid.html; [57]). In addition, 450k array data of monocytes and whole blood DNA obtained from six individuals were downloaded from the gene expression omnibus (GSE35069) [20].

Additional files

Additional file 1. Cluster analysis of Hm03 and Hm05 monocytes and macrophages of the 1000 most variable CpGs. CpG SNPs were excluded from the analysis. The difference between donors is greater than between cell types.

Additional file 2. Quality parameters of WGBS datasets.

Additional file 3. Principal component analysis (PCA) of ten monocyte methylomes from males generated by three IHEC consortia: DEEP (red, our datasets), BLUEPRINT (green) and CEEHRC (blue).

Additional file 4. Annotated list of DMRs including environment and GREAT target genes.

Additional file 5. Histogram of DMR sizes.

Additional file 6. Enrichment and depletion of DMRs for gene features.

Additional file 7. Correlation between differences in DNA methylation and histone modifications. Scatter plots showing, for each of the six histone marks, the difference in histone signals at the DMRs between Hm05 and Hm03 as a function of methylation differences between the two donors.

Additional file 8. Under- and overrepresentation of chromatin states.

Additional file 9. Methylation differences vs. changes in chromatin state. Distribution of DNA methylation differences between donors Hm03 and Hm05 in DMRs that have the same (left) or a different (right) chromatin state in both donors as determined by ChromHMM.

Additional file 10. DMRs target genes identified by GREAT. Number of DMR target genes (a) and their distance from the DMR (b).

Additional file 11. Expression levels of DMR related genes (240) vs. genes not associated with a DMR (17,544). *tpm* transcripts per million.

Additional file 12. List of DMRs with correlated SNPs in 12 kb window. SNPs within the same haplotype block are separated by a comma. Different haplotype blocks are separated by a slash.

Additional file 13. Validation of SNP correlations in seven DMRs using monocytes from six independent donor samples. Graphs showing relationship between the methylation levels as quantified by targeted deep bisulfite sequencing and the genotype of nearby SNPs. Hm01, Hm02, Hm03, Hm05, Hm06 and Hm10: donors.

Additional file 14. Amplikyzer comparative methylation plots. Plots show CpG methylation averages for 14 DMRs after sorting reads by allele of the correlated SNPs (16 SNPs). Each plot shows data from 1 to 4 independent donor samples heterozygous for the correlating SNPs. The two alternative alleles are defined with respect to the forward strand. SNPs rs1996180, rs13130981 and rs7925175 are A/C SNPs, but the C is converted to a T after bisulfite conversion. Asterisks mark CpGs that are outside the DMR borders.

Additional file 15. Histograms of 450k methylation levels in the 1128 probands.

Additional file 16. Scatter plots of monocyte vs. whole blood correlation of DNA methylation. Plots show the correlation between monocyte and whole blood methylation levels in six healthy male individuals for the 50 CpGs that are included in the Illumina 450k array. Analysis performed with Illumina 450k array data previously published [20].

Additional file 17. Manhattan plots of GWASs. *Dashed vertical line* DMR position. *Horizontal line* GWAS significance threshold.

Additional file 18. Zoom-ins with imputed SNPs. *Blue* lead-SNP. *Dashed vertical line* DMR position. *Horizontal line* GWAS significance threshold.

Additional file 19. HaploReg annotations of the 30 lead-SNP and SNPs in the corresponding haplotype blocks. Haplotype block: SNPs in high linkage disequilibrium, $r^2 > 0.8$. SNP positions were converted to hg19 coordinates.

Additional file 20. Distance of SNPs with known binding proteins to the corresponding DMR border. Number of known proteins binding to lead-SNPs or to SNPs in high LD with the lead-SNPs vs. their distance to the corresponding DMR border. Data from Encode ChIPseq obtained via the HaploReg database.

Additional file 21. TRANSFAC motif enrichment in 501 SNP regions (SNP ± 100 bp). Yes and No denote the relative number of sites for the selected matrix in the DMRs as compared to the background dataset.

Additional file 22. Primer sequences and PCR conditions for targeted bisulfite sequencing and genotyping 13 SNPs.

Abbreviations
ASM: allele-specific methylation; CGI: CpG island; ChIP: chromatin immunoprecipitation; DMR: differentially methylated region; eQTL: expression quantitative trait loci; GO: gene ontology; GREAT: Genomic Regions Enrichment of Annotations Tool; GWAS: genome-wide association study; H3K27ac: histone H3 lysine 27 acetylation; H3K27me3: histone H3 lysine 27 tri-methylation; H3K36me3: histone H3 lysine 36 tri-methylation; H3K4me1: histone H3 lysine 4 mono-methylation; H3K4me3: histone H3 lysine 4 tri-methylation; H3K9me3: histone H3 lysine 9 tri-methylation; IHEC: International Human Epigenome Consortium; LD: linkage disequilibrium; mQTL: methylation quantitative trait loci; SNP: single nucleotide polymorphism; WGBS: whole genome bisulfite sequencing.

Authors' contributions
BH conceived and supervised the study and wrote the first draft. CS performed DMR detection, GWAS and correlation analyses, was involved in the bioinformatic and statistical analyses and wrote the first draft. EL performed the targeted methylation analyses, was involved in the bioinformatic and statistical analyses and wrote the first draft. SW provided samples. GS provided samples. LKH performed next generation sequencing. AS performed RNA analysis. KHJ provided samples. SHH performed the microarray analyses. PH performed the microarray analyses. MMN performed the microarray analyses. MS was involved in the bioinformatic and statistical analyses. PE was involved in the bioinformatic and statistical analyses. SR supervised the bioinformatic and statistical analyses. All authors read and approved the final manuscript.

Author details
[1] Genome Informatics, Institute of Human Genetics, University of Duisburg-Essen, University Hospital Essen, Essen, Germany. [2] Institute of Human Genetics, University of Duisburg-Essen, University Hospital Essen, Hufelandstraße 55, 45147 Essen, Germany. [3] Institute for Clinical Chemistry and Laboratory Medicine, University Hospital Regensburg, Regensburg, Germany. [4] Institute of Cell Biology, University Hospital Essen, Essen, Germany. [5] Institute of Clinical Molecular Biology, Kiel University, University Hospital, Kiel, Germany. [6] Institute of Medical Informatics, Biometry and Epidemiology, University Hospital Essen, Essen, Germany. [7] Institute of Human Genetics, School of Medicine, University Hospital of Bonn, University of Bonn, Bonn, Germany. [8] Department of Genomics, Life and Brain Center, University of Bonn, Bonn, Germany. [9] Institute of Medical Genetics and Pathology, University Hospital Basel, Basel, Switzerland. [10] Human Genomics Research Group, Department of Biomedicine, University of Basel, Basel, Switzerland. [11] Research Division, Federal Institute for Drugs and Medical Devices (BfArM), Bonn, Germany. [12] Max Planck Institute for Informatics, Saarland Informatics Campus, Saarbrücken, Germany. [13] Saarbrücken Graduate School of Computer Science, Saarland Informatics Campus, Saarbrücken, Germany.

Acknowledgements
We thank Thomas Wienker, Michael Zeschnigk and Thomas Manke for helpful discussions, Claudia Haak, Sabine Kaya and Claudia Mertel for expert technical assistance, and Giedion Zipprich for uploading the data to EGA. This study makes use of data generated by the Blueprint Consortium. A full list of the investigators who contributed to the generation of the data is available from www.blueprint-epigenome.eu. Funding for the project was provided by the European Union's Seventh Framework Programme (FP7/2007–2013) under grant agreement no 282510 BLUEPRINT. This research used data shared by the McGill Epigenomics Mapping Centre.

Competing interests
The authors declare that there is no competing interests.

Funding
This research was funded by the Federal Ministry of Education and Research under the Project Number 01KU1216 (Deutsches Epigenom Programm, DEEP).

References
1. Cheung WA, Shao X, Morin A, Siroux V, Kwan T, Ge B, Aissi D, Chen L, Vasquez L, Allum F, et al. Functional variation in allelic methylomes underscores a strong genetic contribution and reveals novel epigenetic alterations in the human epigenome. Genome Biol. 2017;18:50.
2. Do C, Lang CF, Lin J, Darbary H, Krupska I, Gaba A, Petukhova L, Vonsattel JP, Gallagher MP, Goland RS, et al. Mechanisms and disease associations of haplotype-dependent allele-specific DNA methylation. Am J Hum Genet. 2016;98:934–55.
3. Hellman A, Chess A. Extensive sequence-influenced DNA methylation polymorphism in the human genome. Epigenet Chromatin. 2010;3:11.
4. Kerkel K, Spadola A, Yuan E, Kosek J, Jiang L, Hod E, Li K, Murty VV, Schupf N, Vilain E, et al. Genomic surveys by methylation-sensitive SNP analysis identify sequence-dependent allele-specific DNA methylation. Nat Genet. 2008;40:904–8.
5. Schalkwyk LC, Meaburn EL, Smith R, Dempster EL, Jeffries AR, Davies MN, Plomin R, Mill J. Allelic skewing of DNA methylation is widespread across the genome. Am J Hum Genet. 2010;86:196–212.
6. Gibbs JR, van der Brug MP, Hernandez DG, Traynor BJ, Nalls MA, Lai SL, Arepalli S, Dillman A, Rafferty IP, Troncoso J, et al. Abundant quantitative trait loci exist for DNA methylation and gene expression in human brain. PLoS Genet. 2010;6:e1000952.

7. McClay JL, Shabalin AA, Dozmorov MG, Adkins DE, Kumar G, Nerella S, Clark SL, Bergen SE, Swedish Schizophrenia C, Hultman CM, et al. High density methylation QTL analysis in human blood via next-generation sequencing of the methylated genomic DNA fraction. Genome Biol. 2015;16:291.

8. Illingworth RS, Gruenewald-Schneider U, De Sousa D, Webb S, Merusi C, Kerr AR, James KD, Smith C, Walker R, Andrews R, Bird AP. Inter-individual variability contrasts with regional homogeneity in the human brain DNA methylome. Nucleic Acids Res. 2015;43:732–44.

9. Do C, Shearer A, Suzuki M, Terry MB, Gelernter J, Greally JM, Tycko B. Genetic-epigenetic interactions in cis: a major focus in the post-GWAS era. Genome Biol. 2017;18:120.

10. Wallner S, Schroder C, Leitao E, Berulava T, Haak C, Beisser D, Rahmann S, Richter AS, Manke T, Bonisch U, et al. Epigenetic dynamics of monocyte-to-macrophage differentiation. Epigenet Chromatin. 2016;9:33.

11. Libertini E, Heath SC, Hamoudi RA, Gut M, Ziller MJ, Czyz A, Ruotti V, Stunnenberg HG, Frontini M, Ouwehand WH, et al. Information recovery from low coverage whole-genome bisulfite sequencing. Nat Commun. 2016;7:11306.

12. Rademacher K, Schroder C, Kanber D, Klein-Hitpass L, Wallner S, Zeschnigk M, Horsthemke B. Evolutionary origin and methylation status of human intronic CpG islands that are not present in mouse. Genome Biol Evol. 2014;6:1579–88.

13. Hansen KD, Langmead B, Irizarry RA. BSmooth: from whole genome bisulfite sequencing reads to differentially methylated regions. Genome Biol. 2012;13:R83.

14. Paliwal A, Temkin AM, Kerkel K, Yale A, Yotova I, Drost N, Lax S, Nhan-Chang CL, Powell C, Borczuk A, et al. Comparative anatomy of chromosomal domains with imprinted and non-imprinted allele-specific DNA methylation. PLoS Genet. 2013;9:e1003622.

15. Illingworth RS, Gruenewald-Schneider U, Webb S, Kerr AR, James KD, Turner DJ, Smith C, Harrison DJ, Andrews R, Bird AP. Orphan CpG islands identify numerous conserved promoters in the mammalian genome. PLoS Genet. 2010;6:e1001134.

16. Ernst J, Kellis M. ChromHMM: automating chromatin-state discovery and characterization. Nat Methods. 2012;9:215–6.

17. Lev Maor G, Yearim A, Ast G. The alternative role of DNA methylation in splicing regulation. Trends Genet. 2015;31:274–80.

18. Erbel R, Mohlenkamp S, Moebus S, Schmermund A, Lehmann N, Stang A, Dragano N, Gronemeyer D, Seibel R, Kalsch H, et al. Coronary risk stratification, discrimination, and reclassification improvement based on quantification of subclinical coronary atherosclerosis: the Heinz Nixdorf Recall study. J Am Coll Cardiol. 2010;56:1397–406.

19. Schmermund A, Mohlenkamp S, Stang A, Gronemeyer D, Seibel R, Hirche H, Mann K, Siffert W, Lauterbach K, Siegrist J, et al. Assessment of clinically silent atherosclerotic disease and established and novel risk factors for predicting myocardial infarction and cardiac death in healthy middle-aged subjects: rationale and design of the Heinz Nixdorf RECALL Study. Risk factors, evaluation of coronary calcium and lifestyle. Am Heart J. 2002;144:212–8.

20. Reinius LE, Acevedo N, Joerink M, Pershagen G, Dahlen SE, Greco D, Soderhall C, Scheynius A, Kere J. Differential DNA methylation in purified human blood cells: implications for cell lineage and studies on disease susceptibility. PLoS ONE. 2012;7:e41361.

21. Ward LD, Kellis M. HaploReg v4: systematic mining of putative causal variants, cell types, regulators and target genes for human complex traits and disease. Nucleic Acids Res. 2016;44:D877–81.

22. Li XJ, Ito M, Zhou F, Youngson N, Zuo XP, Leder P, Ferguson-Smith AC. A maternal-zygotic effect gene, Zfp57, Maintains both maternal and paternal imprints. Dev Cell. 2008;15:547–57.

23. Messerschmidt DM, de Vries W, Ito M, Solter D, Ferguson-Smith A, Knowles BB. Trim28 is required for epigenetic stability during mouse oocyte to embryo transition. Science. 2012;335:1499–502.

24. Quenneville S, Verde G, Corsinotti A, Kapopoulou A, Jakobsson J, Offner S, Baglivo I, Pedone PV, Grimaldi G, Riccio A, Trono D. In embryonic stem cells, ZFP57/KAP1 recognize a methylated hexanucleotide to affect chromatin and DNA methylation of imprinting control regions. Mol Cell. 2011;44:361–72.

25. Roadmap Epigenomics C, Kundaje A, Meuleman W, Ernst J, Bilenky M, Yen A, Heravi-Moussavi A, Kheradpour P, Zhang Z, Wang J, et al. Integrative analysis of 111 reference human epigenomes. Nature. 2015;518:317–30.

26. Merkenschlager M, Nora EP. CTCF and cohesin in genome folding and transcriptional gene regulation. Annu Rev Genomics Hum Genet. 2016;17:17–43.

27. Kaplow IM, MacIsaac JL, Mah SM, McEwen LM, Kobor MS, Fraser HB. A pooling-based approach to mapping genetic variants associated with DNA methylation. Genome Res. 2015;25:907–17.

28. Deplancke B, Alpern D, Gardeux V. The genetics of transcription factor DNA binding variation. Cell. 2016;166:538–54.

29. Cedar H, Bergman Y. Linking DNA methylation and histone modification: patterns and paradigms. Nat Rev Genet. 2009;10:295–304.

30. Rose NR, Klose RJ. Understanding the relationship between DNA methylation and histone lysine methylation. Biochim Biophys Acta Gene Regul Mech. 2014;1839:1362–72.

31. Bestor TH, Edwards JR, Boulard M. Notes on the role of dynamic DNA methylation in mammalian development. Proc Natl Acad Sci USA. 2015;112:6796–9.

32. Ecker J, Langmann T, Moehle C, Schmitz G. Isomer specific effects of conjugated linoleic acid on macrophage ABCG1 transcription by a SREBP-1c dependent mechanism. Biochem Biophys Res Commun. 2007;352:805–11.

33. Sherry ST, Ward MH, Kholodov M, Baker J, Phan L, Smigielski EM, Sirotkin K. dbSNP: the NCBI database of genetic variation. Nucleic Acids Res 2001;29:308–11.

34. Ji Y, Wu C, Liu P, Wang J, Coombes KR. Applications of beta-mixture models in bioinformatics. Bioinformatics. 2005;21:2118–22.

35. Schröder C, Rahmann S. A Hybrid parameter estimation algorithm for beta mixtures and applications to methylation state classification. In: Frith M, Storm Pedersen CN, editors. Algorithms in bioinformatics: 16th international workshop, WABI 2016, Aarhus, August 22–24, 2016 Proceedings. Cham: Springer; 2016. p. 307–19.

36. Assenov Y, Muller F, Lutsik P, Walter J, Lengauer T, Bock C. Comprehensive analysis of DNA methylation data with RnBeads. Nat Methods. 2014;11:1138–40.

37. Purcell S, Neale B, Todd-Brown K, Thomas L, Ferreira MA, Bender D, Maller J, Sklar P, de Bakker PI, Daly MJ, Sham PC. PLINK: a tool set for whole-genome association and population-based linkage analyses. Am J Hum Genet. 2007;81:559–75.

38. van der Walt S, Colbert SC, Varoquaux G. The NumPy array: a structure for efficient numerical computation. Comput Sci Eng. 2011;13:22–30.

39. Freeman C, Marchini J. GTOOL. Wellcome trust centre for human genetics. Oxford: University of Oxford; 2007.

40. Howie BN, Donnelly P, Marchini J. A flexible and accurate genotype imputation method for the next generation of genome-wide association studies. PLoS Genet. 2009;5:e1000529.

41. Genomes Project C, Auton A, Brooks LD, Durbin RM, Garrison EP, Kang HM, Korbel JO, Marchini JL, McCarthy S, McVean GA, Abecasis GR. A global reference for human genetic variation. Nature. 2015;526:68–74.

42. Li LC, Dahiya R. MethPrimer: designing primers for methylation PCRs. Bioinformatics. 2002;18:1427–31.

43. Aranyi T, Varadi A, Simon I, Tusnady GE. The BiSearch web server. BMC Bioinform. 2006;7:431.

44. Tusnady GE, Simon I, Varadi A, Aranyi T. BiSearch: primer-design and search tool for PCR on bisulfite-treated genomes. Nucleic Acids Res. 2005;33:e9.

45. Beygo J, Ammerpohl O, Gritzan D, Heitmann M, Rademacher K, Richter J, Caliebe A, Siebert R, Horsthemke B, Buiting K. Deep bisulfite sequencing of aberrantly methylated loci in a patient with multiple methylation defects. PLoS ONE. 2013;8:e76953.

46. Rahmann S, Beygo J, Kanber D, Martin M, Horsthemke B, Buiting K. Amplikyzer: automated methylation analysis of amplicons from bisulfite flowgram sequencing. PeerJ PrePrints. 2013;1:e122v2.

47. Ramirez F, Dundar F, Diehl S, Gruning BA, Manke T. deepTools: a flexible platform for exploring deep-sequencing data. Nucleic Acids Res. 2014;42:W187–91.

48. McLean CY, Bristor D, Hiller M, Clarke SL, Schaar BT, Lowe CB, Wenger AM, Bejerano G. GREAT improves functional interpretation of cis-regulatory regions. Nat Biotechnol. 2010;28:495–501.

49. Bray NL, Pimentel H, Melsted P, Pachter L. Near-optimal probabilistic RNA-seq quantification. Nat Biotechnol. 2016;34:525–7.

50. Matys V, Kel-Margoulis OV, Fricke E, Liebich I, Land S, Barre-Dirrie A, Reuter I, Chekmenev D, Krull M, Hornischer K, et al. TRANSFAC and its module

TRANSCompel: transcriptional gene regulation in eukaryotes. Nucleic Acids Res. 2006;34:D108–10.

51. Kim D, Pertea G, Trapnell C, Pimentel H, Kelley R, Salzberg SL. TopHat2: accurate alignment of transcriptomes in the presence of insertions, deletions and gene fusions. Genome Biol. 2013;14:R36.

52. Langmead B, Salzberg SL. Fast gapped-read alignment with Bowtie 2. Nat Methods. 2012;9:357–9.

53. Pertea M, Pertea GM, Antonescu CM, Chang TC, Mendell JT, Salzberg SL. StringTie enables improved reconstruction of a transcriptome from RNA-seq reads. Nat Biotechnol. 2015;33:290–5.

54. Pertea M, Kim D, Pertea GM, Leek JT, Salzberg SL. Transcript-level expression analysis of RNA-seq experiments with HISAT, StringTie and Ballgown. Nat Protoc. 2016;11:1650–67.

55. Volders PJ, Helsens K, Wang X, Menten B, Martens L, Gevaert K, Vandesompele J, Mestdagh P. LNCipedia: a database for annotated human lncRNA transcript sequences and structures. Nucleic Acids Res. 2013;41:D246–51.

56. Volders PJ, Verheggen K, Menschaert G, Vandepoele K, Martens L, Vandesompele J, Mestdagh P. An update on LNCipedia: a database for annotated human lncRNA sequences. Nucleic Acids Res. 2015;43:4363–4.

57. Bujold D, Morais DA, Gauthier C, Cote C, Caron M, Kwan T, Chen KC, Laperle J, Markovits AN, Pastinen T, et al. The international human epigenome consortium data portal. Cell Syst. 2016;3(496–499):e492.

Additional sex combs interacts with enhancer of zeste and trithorax and modulates levels of trimethylation on histone H3K4 and H3K27 during transcription of *hsp70*

Taosui Li[1], Jacob W. Hodgson[1], Svetlana Petruk[2], Alexander Mazo[2] and Hugh W. Brock[1]*

Abstract

Background: Maintenance of cell fate determination requires the Polycomb group for repression; the trithorax group for gene activation; and the enhancer of trithorax and Polycomb (ETP) group for both repression and activation. *Additional sex combs* (*Asx*) is a genetically identified ETP for the *Hox* loci, but the molecular basis of its dual function is unclear.

Results: We show that in vitro, Asx binds directly to the SET domains of the histone methyltransferases (HMT) enhancer of zeste [E(z)] (H3K27me3) and Trx (H3K4me3) through a bipartite interaction site separated by 846 amino acid residues. In *Drosophila* S2 cell nuclei, Asx interacts with E(z) and Trx in vivo. *Drosophila Asx* is required for repression of heat-shock gene *hsp70* and is recruited downstream of the *hsp70* promoter. Changes in the levels of H3K4me3 and H3K27me3 downstream of the *hsp70* promoter in *Asx* mutants relative to wild type show that *Asx* regulates H3K4 and H3K27 trimethylation.

Conclusions: We propose that during transcription Asx modulates the ratio of H3K4me3 to H3K27me3 by selectively recruiting the antagonistic HMTs, E(z) and Trx or other nucleosome-modifying enzymes to *hsp70*.

Keywords: *Additional sex combs*, SET domain, Trithorax, Enhancer of zeste, *hsp70* transcriptional elongation, Histone trimethylation

Background

Polycomb group (PcG) and trithorax group (trxG) proteins maintain gene repression and activation, respectively, during metazoan development [1–3]. In *Drosophila melanogaster*, *Asx* was originally identified as a PcG mutant because of prominent posterior transformations caused by derepression of *Hox* genes [4–6]. Subsequently, it was observed that embryos mutant for *Asx* exhibit both anterior and posterior transformations, because *Hox* genes are improperly activated and derepressed, respectively [6–8]. Consistent with this model, *Asx* mutants enhance the homeotic transformation of trxG [8] and PcG [9, 10] mutations. Genes with these characteristics have been termed enhancers of trithorax and Polycomb (ETP) [11, 12]. Genetic analysis suggests that Asx is required for both trxG and PcG function.

Various enzymatic activities are associated with trxG and PcG proteins, including trimethylation of histone H3 lysine 4 (H3K4) and H3K27 [13, 14]. Thus, one model to explain the ETP function of Asx is that it interacts directly with E(z) and Trx to regulate H3K4 and H3K27 methylation. An alternative model is that Asx affects trimethylation of H3K4 and H3K27 indirectly by

*Correspondence: hugh.brock@ubc.ca
[1] Department of Zoology, Life Sciences Institute, University of British Columbia, 2350 Health Science Mall, Vancouver, BC V6T 1Z4, Canada
Full list of author information is available at the end of the article

regulating histone demethylases or acetyltransferases. In either model, Asx should be required to regulate levels of H3K4 and H3K27 methylation in vivo. To our knowledge, neither of these models has been tested on *Asx* or its mammalian homologs, perhaps because of difficulty of identifying a single locus at which both PcG and trxG proteins act at the same time in the same cell.

The *hsp70* gene is well characterized. Before heat-shock induction, the *hsp70* promoter region is maintained in a nucleosome-free conformation by the GAGA factor [15], with a paused Pol II located approximately 25 nucleotides downstream of the transcription starting site [16, 17]. The paused Pol II is phosphorylated at serine 5 (Ser-5) but not Ser-2 of the C-terminal domain (CTD) [18], showing that transcriptional elongation has not begun. In Drosophila, these events occur 2–4 h after egg deposition. Heat stress leads to recruitment of heat-shock factor (HSF) [19], positive transcription elongation factor b (P-TEFb), mediator and various elongation factors including Spt5, Spt6 and facilitates chromatin transcription (FACT) complex that contains Spt16 for synthesis of full-length transcripts [20–22]. P-TEFb contains Cdk9 that is required for Pol II CTD Ser-2 phosphorylation and transcription elongation [18].

Any temporal analysis of the heat-shock response in *Drosophila* later in development than the first 4 h of embryogenesis will have three phases: (1) an early phase corresponding to the switch from a promoter-paused state to elongation; (2) an intermediate phase that combines transcriptional initiation, promoter clearance and elongation; and (3) a phase in which transcription of heat-shock genes is terminated. Recruitment of the trithorax (Trx) protein complex, TAC1, is required to maintain high levels of transcriptional elongation and of H3K4 trimethylation at the *hsp70* promoter region [23]. The PcG gene *pleiohomeotic* (*pho*) is required to repress *hsp70* transcription after heat shock during termination phase [24]. Maternal deposition of *Asx* mRNA or Asx protein prevents analysis of transcriptional initiation or promoter clearance at *hsp70* in *Asx* mutants in early embryos (first 4 h of embryogenesis). In later embryos, when initiation, clearance and elongation are occurring simultaneously, it is difficult to distinguish these phases of transcription in chromatin immunoprecipitation experiments.

Here, we show that Asx interacts directly in vitro and associates in vivo with E(z) and Trx, suggesting a recruitment mechanism for modulation of trimethylation of H3K4 and H3K27 at *hsp70*. We also show that *hsp70* is an excellent target to investigate the molecular basis of Asx function as an ETP after 10 min of heat-shock induction. We show that at the *hsp70* locus, *Asx* represses *hsp70* transcription because *Asx* mutants induce induction of

the heat-shock response, but unlike the PcG gene *pho*, it is not required in the termination phase. Asx is recruited downstream of the promoter following heat stress induction, but during the first 10 min of heat-shock induction, *Asx* repression of *hsp70* is independent of changes in levels of H3K4me3 and H3K27me3. Subsequently, *Asx* modulates levels of H3K4me3 and H3K27me3, notably at transition from elongation to termination during the heat-shock cycle. This, however, does not exclude the modulation of trimethylation by recruitment of other histone methyltransferases.

Methods

Fly culture, transgenic lines, embryo imaging and cell culture

Flies were maintained at 22 °C on standard cornmeal-sucrose medium containing Tegosept as a mold inhibitor. The Asx^3 allele was maintained over a *CyO twist-GAL4, UAS-eGFP* balancer chromosome. The Asx^3 mutant is a null mutation with 1.3-kb deletion in the middle of coding region that produces a truncated protein product of approximately 800 N-terminal amino acids [4, 25] (Hodgson, unpublished).

Antibodies

The following antibodies were employed: sheep polyclonal anti-Asx (aa 75–95) IgG [25]; rabbit polyclonal anti-trimethyl histone H3K4 antibody (Active Motif, Cat.# 39159; 1:1000 dilution); rabbit polyclonal anti-trimethyl histone H3K27 (Millipore, Cat.# 07-449; 1:100 dilution); rabbit polyclonal IgG antibody (Abcam, Cat.# ab27478; 1:200 dilution) used as a negative control; rabbit polyclonal anti-E(z) (Santa Cruz, Cat.# sc-98265); rat polyclonal anti-Trx antisera and purified rabbit anti-Trx IgG (Mazo Lab). Rabbit anti-Asx antibody was raised against *Drosophila* Asx (aa 200–356) (Additional file 1: Text S1). The specificity of rabbit anti-Asx antibody was tested (Additional file 2: Fig. S1).

Construction of Asx full length and deletion mutants for cell-free expression

The DNA fragment corresponding to the full-length Asx (1669 amino acid residues) was subcloned from pBS(KS+)–Asx(1–1669) as an *Nde*I–*Kpn*I fragment into the *Nde*I–*Sma*I sites of the vector pTβSTOP (kindly provided by Robert Tjian) downstream of the T7 promoter and β globin leader sequence to generate pTβSTOP-A(1–1669). An Asx COOH terminal deletion mutant A(1–1200) was generated by replacing the wild-type 1.59-kb *Sph*I–*Kpn*I sequence in pBS(KS+)–Asx with a 0.495-kb PCR fragment produced using the following primer pairs: forward: 5′-ccggattccttgg GCA AGA CAT TAC CAG TGG CT-3′ and reverse:

5'-ccggagtggtacc TCA CAT ATT ACT GTT GTG-3'. The pBS(KS+)–Asx(1–1200) was subsequently digested with KpnI, end-repaired with T4 DNA polymerase and digested with NdeI. The truncated 3.6-kb Asx fragment was subcloned into the NdeI–SmaI site of pTβSTOP to generate pTβSTOP-A(1–1200). Four additional COOH terminal deletion fragments as well as four NH2 terminal deletion fragments of Asx were amplified from the pTβSTOP-A(1–1669) by PCR (Additional file 3: Table S1), subcloned into the NdeI–SmaI/EcoRV site of pTβSTOP and transformed into DH5α cells (Thermo Fisher). All plasmid constructs were expressed in the TNT-coupled T7 transcription/translation system (Promega) using rabbit reticulocyte lysate (IVT–RRL) for GST pull-down assays.

Construction of GST fusions of AsxETSI-2 and the SET domains of E(z) and Trx

The SET domain of E(z) (aa residues 626–740) was amplified by PCR using the primer pair shown in Additional file 3: Table S1 and subcloned into the EcoRI–XhoI sites of pGEX-6P1 to generate pGEX-6P1-E(z)SET. The SET domain of Trx (aa 3608-3759) (kindly provided by Michael Kyba) was subcloned into the EcoRI–XhoI sites of pGEX-6P1 to generate pGEX-6P1-TrxSET. The E(z)/Trx SET domain interaction site 2 of Asx (AsxETSI-2), residues 1200–150 l, was amplified using primer pair in Additional file 3: Table S1 and subcloned into the EcoRI–XhoI sites of pGEX-6P-1 to generate pGEX-6P1-AsxETSI-2. The three GST fusion constructs were each transformed into the E. coli Rosetta 2(DE3) strain (Novagen) for expression.

Expression and purification of GST-TrxSET and GST-AsxETSI-2 fusion proteins

Overnight cultures of 10 ml of cells transformed with either pGEX-6P-1, pGEX-6P1-TrxSET or pGEX-6P1-AsxETSI-2 were diluted into 240 ml Luria–Bertani (LB)/100 μg ml^{-1} Amp media and induced at an A_{600} of 0.8 units with 1 mM isopropyl-β-D-thio-galactoside (IPTG)/100 μM $ZnSO_4$ for 14 h at 23 °C. The cells were centrifuged, washed with PBS, lysed in 20 ml buffer TEEZMG—0.5 M KCl, pH 7.9, supplemented with protease inhibitors and treated with 4 mg/ml lysozyme (Sigma) for 30 min at 4 °C. Each lysate was sonicated, diluted twofold with buffer TEEZMG—0.5 M KCl/protease inhibitors and centrifuged at 14,000 rpm for 20 min at 4 °C. Extracts of GST, GST-TrxSET or GST-AsxETSI-2 were recovered in the supernatant as soluble fractions named S1.

Twenty ml of each S1 fraction was rotated with 0.5 ml of GSH-agarose pre-equilibrated with buffer TEMZG—0.3 M KCl, pH 7.9, for 3 h at 4 °C. The protein-bound resin was washed three times with buffer PBSMG—0.3 M NaCl, pH 7.2, three times with buffer PBSMG—0.8 M NaCl, pH 7.2, three times with buffer TEMZG—0.5 M KCl, pH 7.9, and once with buffer TEMZ—0.1 M KCl, pH 9.0, at 4 °C. Proteins were eluted from a column with 6 ml of buffer Elut-TEMZ—0.1 M KCl, pH 9.0/10 mM reduced glutathione–NaOH and collected in 0.5 ml fractions. Peak fractions were pooled and dialyzed into buffer Dyl-TEEMZG—0.1 M KCl, pH 7.9, for 18 h at 4 °C and stored at −80 °C. Buffers used in these experiments are described in Additional file 4: Text S2.

Expression and purification of GST-E(z)SET fusion protein

A 10-ml overnight culture of pGEX-6P1-E(z)SET was diluted into 240 ml LB/100 μg/ml Amp media and induced at an A_{600} of 0.8 units with 1 mM IPTG/100 μM $ZnSO_4$ for 3 h at 23 °C. The cells were washed in PBS, lysed in 20 ml buffer TEEZG—0.5 M KCl, pH 7.9, supplemented with protease inhibitors and treated with 4 mg/ml lysozyme (Sigma) for 30 min at 4 °C. The lysate was sonicated, diluted twofold with buffer TEEZG—0.5 M KCl, pH 7.9/protease inhibitors and centrifuged at 14,000 rpm for 20 min at 4 °C.

The pellet was resuspended in 15 ml buffer TZS—0.3 M NaCl, pH 7.9 using a Dounce homogenizer and mixed on a nutator for 2 h at 4 °C to solubilize GST-E(z)SET. The homogenate was centrifuged at 14,000 rpm for 20 min at 4 °C, and the supernatant (PI-Ext) was diluted fivefold with buffer TZD—0.3 M NaCl, pH 7.9, to reduce the sarkosyl concentration to 2% in the presence of Triton X-100 and CHAPS [26]. Fifty ml of diluted P1-Ext was mixed with 0.5 ml of GSH-agarose and pre-equilibrated with buffer TZXSC—0.3 M NaCl, pH 7.9, for 3 h at 4 °C on a rotator. The protein-bound resin was washed three times with buffer TZGXSC—0.3 M KCl, pH 7.9, three times with buffer TZGXSC—0.6 M KCl, pH 7.9, and two times with buffer TZXSGC—0.1 M NaCl, pH 9.0, at 4 °C. Proteins were eluted from a column with 6 ml of buffer Elut-TZGXSC—0.1 M NaCl, pH 9.0/10 mM reduced glutathione–NaOH, pH 9.0, and collected in 0.5 ml fractions. Peak fractions were pooled and dialyzed into 1 l buffer Dyl-TZXS—0.1 M NaCl/15% glycerol for 18 h at 4 °C and stored at −80 °C.

GST pull-down assay of ^{35}S-Asx interaction with SET domains of Trx and E(z)

For each reaction, 60 μl of packed GSH-agarose equilibrated with immobilization buffer IM-A pH 7.9 was resuspended in 400 μl buffer IM-A and mixed with 6 μg of purified GST for 120 min at 4 °C on a nutator. The resin was pelleted at 6000 rpm for 2 min, washed two times and resuspended with 300 μl buffer IM-A on ice. For each fragment tested, the immobilized GST

(GST-agarose) was split into a 250-µl aliquot to pre-clear the Asx in vitro translation mixture and a 50-µl aliquot for a control GST pull-down assay (30).

Asx protein fragments were produced by in vitro translation in rabbit reticulocyte lysate (RRL; Promega). Briefly, 1 µg supercoiled pTβSTOP plasmids containing Asx DNA fragments 1.8 kb to 5 kb long (Additional file 3: Table S1) were denatured at 80 °C, chilled on ice, expressed and labeled with ^{35}S using the 25-µl reaction TNT T7-coupled transcription/translation rabbit reticulocyte lysate kit (Promega). The lysates were subsequently adjusted to 5 mM Mg acetate and treated with DNase I and RNase A/RNase T1 (Fermentas) for 15 min at 25 °C, and the Roche protease inhibitor cocktail was added. The lysate was pre-cleared by mixing with 100 µl of buffer 2× PDB-P5 and 25 µl of GST-agarose for 30 min at 4 °C. To assay interactions, 65 µl of pre-cleared ^{35}S-Asx-containing lysate was mixed with either immobilized 1 µg GST, GST-E(z)SET or GST-TrxSET for 2 h on a nutator at 4 °C. Lysate-bound agarose beads were washed once with 300 µl buffer 1× PDB-P5, three times with buffer WB—0.6 M NaCl, pH 7.9, and once with buffer WB—0.1 M NaCl, pH 7.9. The protein-bound agarose pellet was mixed with 15 µl 2× SDS sample buffer, resolved by SDS–PAGE and analyzed by autoradiography.

GST pull-down western assays of embryo nuclear extract
Asx fragments indentified as interaction sites for SET domains of either Trx (TSI), E(z) (ESI) or both (ETSI) were subcloned into pGEX-6P-1, expressed and purified from Rosetta 2(DE3) cells as described above for GST-AsxETSI-2. Detailed methods are described in Additional file 5: Text S3.

Co-immunoprecipitation western assays of embryo nuclear extract
For co-immunoprecipitation experiments, 800 micrograms of nuclear extract Bio-Rex 70 fraction was diluted into 300 µl of 1× IP buffer containing 5% polyethylene glycol 10,000 in place of Ficoll and mixed with 1:100 dilution of purified sheep anti-Asx IgG at 4 °C for 3 h [25]. Immune complexes were precipitated with protein G-Sepharose at 25 °C for 15 min and washed six times with buffer TELG containing 0.22 and 0.26 M KCl and two times with buffer TELG containing 0.05 M KCl. The samples were subsequently resolved on SDS–polyacrylamide gels and transferred onto nitrocellulose membranes for western analysis as described above. Blots were probed with 1:3000 dilution of purified rabbit anti-Trx IgG or 1:200 dilution of rabbit anti-E(z) IgG (Santa Cruz Biotech, cat# sc-98265).

Demonstrating protein–protein interaction in situ by PLA
Drosophila S2 cells were cultured at room temperature in chamber slides, heat shocked at 37 °C for 15 min and allowed to recover 60 or 180 min at room temperature. Cells were fixed with 2% formaldehyde in culture medium for 20 min, washed with PBS, blocked and incubated overnight at 4 °C with either sheep anti-Asx and rat anti-Trx or sheep anti-Asx and rabbit anti-E(z). Proximity ligation assays (PLAs) were performed as described [27, 28] with several modifications. Secondary antibodies (Jackson Immuno Research) were conjugated to 5′-amino-modified MTPX oligonucleotides in Additional file 6: Table S2 using the Thunder-Link oligo conjugation systems (Innova Biosciences 420-0300) and stored at 4 °C. Circularization PLA 5′-phosphorylated oligonucleotides as well as detector PLA oligonucleotides were synthesized (Additional file 6: Table S2). For each reaction, 40 µl secondary antibodies with conjugated oligonucleotides were incubated on a shaker for 1 h in a humidity chamber at 37 °C. Three circularization PLA oligos were annealed to two corresponding PLA probes (Additional file 6: Table S2) and ligated with T4 DNA ligase (Thermo) for 30 min at 37 °C. A closed circle forms if proteins are in close proximity [29]. Rolling-circle amplification by phi28 polymerase (Thermo) was carried out in the presence of fluorescent-labeled detector oligonucleotides (Additional file 6: Table S2) for 100 min in the dark at 37 °C.

Immunostaining of salivary gland polytene chromosomes
Preparation and immunostaining of chromosomes have been described [23]. For immunostaining, sheep anti-Asx IgG and FITC- or Texas-Red-conjugated secondary antibodies (Jackson Immunoresearch, PA) were used at dilutions of 1:100 [25]. Images of labeled chromosomes were acquired with a Zeiss microscope equipped with a digital camera, and processed using Adobe Photoshop.

Embryo collection
Flies were acclimated to the laying chamber for 2 days before 10–14-h AEL embryos from about 300 flies were collected at 22 °C on 2% agar (supplemented with 1% sucrose/3.5% ethanol/1.5% apple cider vinegar). Embryos on laying plates were immediately washed onto a nylon sieve to remove excess yeast, dechorionated with 50% bleach and washed twice with 120 mM NaCl; 0.02% Triton X-100, followed by two washes in 1x PBS; 0.05% Triton X-100. *Asx³* homozygous mutant embryos were identified by the absence of *GFP* expression under a wide-field GFP fluorescence microscope. The wild-type strain Oregon R was used as a control.

Heat-shock induction and recovery in embryos

Wild-type or homozygous Asx^3 mutant embryos were collected in 200 µl of 1× PBT into plastic microcentrifuge tubes that were incubated in a 37 °C water bath for 5, 10 or 15 min. To study recovery, embryos heat shocked for 15 min were transferred onto a small piece of moist filter paper placed in a moist chamber and incubated for up to 180 min at room temperature, and transferred into a new tube with 200 µl of 1× PBT for further analysis.

Determining *hsp70* mRNA level in embryos

RNA preparation and first-strand cDNA synthesis were done as previously described [30]. A dilution series of *Drosophila* genomic DNA was used to generate standard curves for *hsp70* and *Ahcy89E*. The relative mRNA levels of genes were measured with comparison to the standard curve. All quantitative PCRs (qPCRs) were performed on the Step-One Plus Real-Time PCR system (ABI). The *hsp70* expression level in different samples was normalized to the expression of *Ahcy89E*, whose expression does not change during heat shock.

Chromatin immunoprecipitation (ChIP)

ChIP was carried out essentially as described [31] except as follows. Exactly 200 embryos (in 200 µl buffer) were sonicated for five pulses of 10 s at 30% power at room temperature with an ultrasonic processor (CPX 130 PB, Cole Parmer), followed by 50 s on ice to yield 500-bp fragments. Samples were mixed with 200 µl of 6 M urea and incubated for 10 min on ice, and insoluble material was removed by centrifugation at 12,000×g for 10 min at 4 °C. The supernatant was divided into 4 × 100 µl aliquots, and ChIP experiments were performed with H3K4, H3K27 and IgG antibodies. After the final washing step, 100 µl of 10% Chelex 100 resin (BioRad) was added to protein G beads with vortexing, and samples were incubated at 95 °C for 10 min. Samples were deproteinized with proteinase K (Sigma) at 55 °C, incubated for a further 10 min at 95 °C and centrifuged to recover the beads [32]. Approximately 3% of the immunoprecipitated material was assayed by qPCR using primers specific to sequences at the *hsp70* promoter downstream, *bxd* PRE and *Ubx* promoter (Additional file 7: Text S4).

Results

Asx contains a bipartite site for interaction with SET domains of both E(z) and Trx

Genetic analysis suggests that Asx is required for both trxG and PcG function. However, no genetic experiment can show that Asx has a direct effect on the histone methyltransferases (HMT) responsible for trimethylation of H3K4 and H3K27. Therefore, we looked for evidence of direct association of Asx with Trx, a key HMT for

H3K4, and E(z), the HMT for H3K27. Alignment of the protein sequences of Trx and E(z) revealed a SET domain catalytic site [33] with a 35% amino acid identity between both proteins (Fig. 1a). To determine whether association between Asx and both HMTs can occur through the SET domains of Trx and E(z), we developed a GST pull-down autoradiography (GST pull-down) assay [34]. The SET domains of Trx and E(z) were fused to GST, expressed in *E. coli* Rosetta 2 (DE3) cells, purified by GSH-agarose affinity chromatography and immobilized on GSH-agarose (Fig. 1b). A preparation of ^{35}S-Met-labeled Asx produced using coupled in vitro transcription/translation in rabbit reticulocyte lysate (IVT–RRL) (Promega) was mixed with immobilized GST-TrxSET, GST-E(z) SET and GST in control experiments. Full-length ^{35}S-Asx interacted specifically with both GST-E(z)SET and GST-TrxSET (Fig. 1c) and exhibited a difference in the salt sensitivity of association with GST-TrxSET relative to GST-E(z). Higher salt concentrations increased the association of Asx with GST-E(z)SET but decreased the association with GST-TrxSET, which may reflect ionic or hydrophobic effects on interaction [35].

To determine the strength of association of Asx with E(z) or Trx in nuclear extracts, given that in vitro, the Asx-Trx(SET) interaction is about 5× weaker than the Asx-E(z)(SET) interaction, the Bio-Rex 70 fractions were further analyzed by anti-Asx co-immunoprecipitation coupled with α-Trx or α-E(z) western blotting (Fig. 2). E(z) and Trx were resolved into two distinct salt fractions of 0.1 and 0.85 M, respectively, by Bio-Rex 70 (Fig. 2b). Asx was co-eluted with Trx in the BR70-0.85 fraction and weakly co-immunoprecipitated at 0.26 M NaCl (Fig. 2c). By contrast, there was no detectable co-elution nor co-immunoprecipitation of Asx with E(z) in either the BR70-0.1 or BR70-0.85 fractions (Fig. 2d, e). This suggests that in nuclear extracts, the interaction between Asx and E(z) is weakly ionic or transient, which may be readily disrupted on the Bio-Rex 70 resin by a salt gradient. Alternatively, the amount of Asx in association with E(z) is below the limits of detection of immunoprecipitation. A third untested explanation is that interactions between Asx and Trx complexes are more stable than interactions of the individual proteins, and vice versa for Asx and E(z).

The interaction sites of GST-E(z)SET and GST-Trx-SET on full-length Asx were mapped by determining the association of ^{35}S-Asx COOH terminal or NH$_2$ terminal deletion fragments (Fig. 3a) with GST-E(z)SET and GST-TrxSET using the GST pull-down assay (Figs. 3b, 4). Two interaction sites were identified on Asx for both GST-E(z)SET and GST-TrxSET (Fig. 3b): (1) at NH$_2$ terminal residues 1–354 termed E(z)/Trx SET domain interaction site 1 (AsxETSI-1) and (2) at COOH terminal residues 1200–1501 termed (AsxETSI-2) (Fig. 4). In addition, a

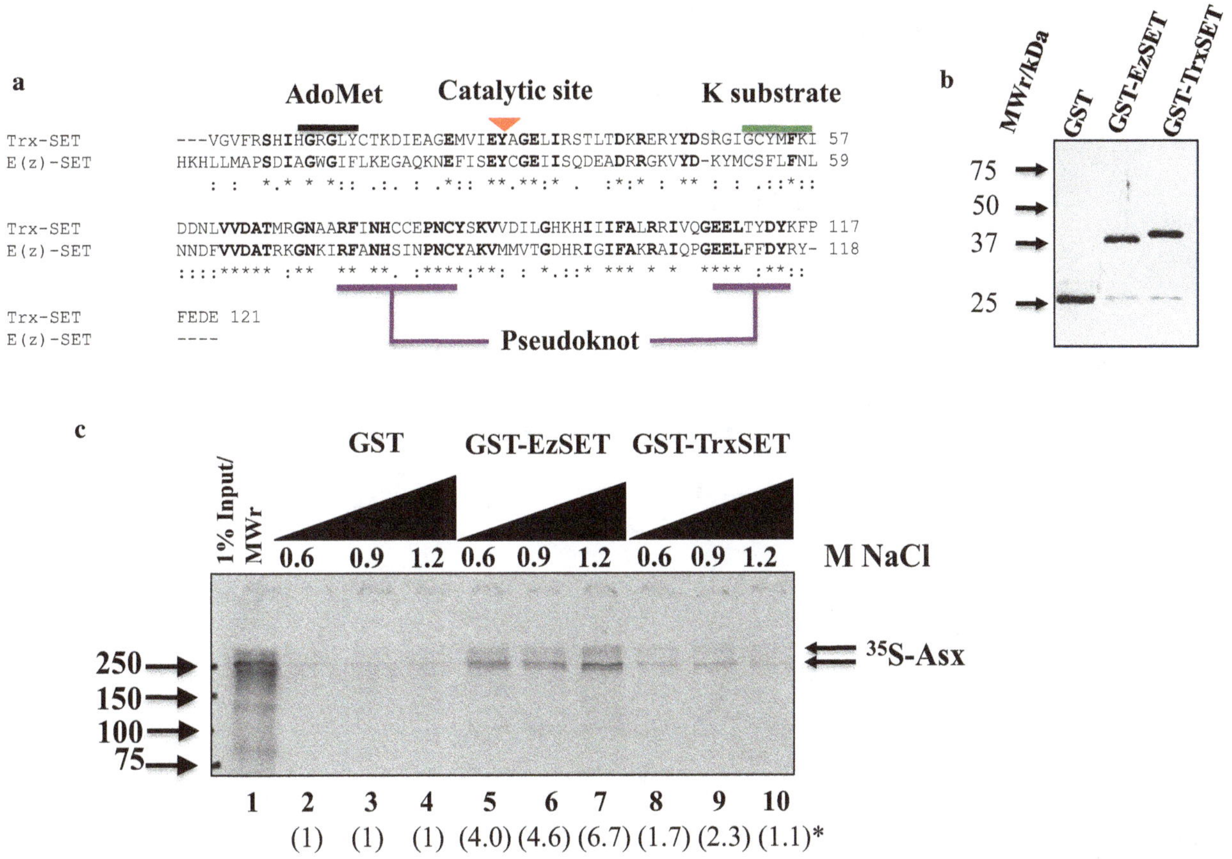

Fig. 1 Full-length Asx interacts directly with SET domains of trxG activator, Trx and PcG repressor, E(z) in a GST pull-down assay. **a** Clustal Omega sequence alignment of SET domains of Drosophila Trx and E(z), showing 35% sequence identity. **b** SDS–PAGE (13%) analysis of affinity purified E. coli-expressed GST fusions of SET domains of E(z) and Trx. **c** GST pull-down analysis of ^{35}S-methionine-labeled rabbit reticulocyte lysate-in vitro translated (RRL-IVT) Asx with GST fusions of SET domains of E(z) and Trx. The stringency of the NaCl washes indicates stronger interaction between Asx and GST-E(z)SET than GST-TrxSET. Proteins were analyzed on 13% SDS–polyacrylamide reducing gels. ()* Estimated fold binding of Asx with GST fusions relative to GST (determined by densitometry) is shown beneath each lane

weak but specific COOH interaction site at terminal residues 1501–1669 was mapped for Trx (Fig. 3b), termed Trx SET domain interaction site (AsxTSI) (Fig. 4). These results suggest that Asx contains a bipartite E(z)/Trx SET domain interaction (ETSI-1 and ETSI-2) site that has low sequence identity (14.97%) and is separated by 846 amino acids (Additional file 8: Fig. S2). It is interesting that the Asx NH$_2$ terminal residues 1–1200 which contain ETSI-1 did not show any significant binding with GST-E(z)SET and GST-TrxSET. In together with other pull-down results, it is possible that the Asx 610–1200 fragment which contains HR2–HR6 domain plays a negative role on ETSI-1 and GST-E(z)SET/GST-TrxSET interaction.

Alignment of the amino acid sequences of Asx^3 mutants and wild type [25] indicated a deletion of the COOH terminal ETSI-2 and TSI in Asx^3 mutants (Fig. 5a). To test for interaction between E(z) and Asx-ETSI-2, we developed a GST pull-down western assay using purified GST-ETSI-2 (Fig. 5b) and chromatography

fractionated embryo nuclear extracts as a source of E(z) (Fig. 5c). Similar experiments were not performed with AsxETSI-1 because of the instability of purified GST-ETSI-1. The E(z)-enriched BR70-0.1 M chromatography fraction (Fig. 5c) was mixed with immobilized GST-ETSI-2 and resolved by SDS–PAGE coupled with western blotting using anti-E(z) antibody. Two bands were specifically detected on the blot corresponding to E(z) and a breakdown fragment (Fig. 5d) consistent with the interaction of GST-E(z) with AsxETSI-2. It is possible that E(z) was fragmented due to the sub-optimized pull-down condition. These results indicate that Asx ETSI-2 (which is deleted in Asx^3) can associate with E(z) in vivo.

Asx associates with E(z) and Trx after induction and recovery from heat shock in vivo

To further investigate whether Asx weakly or transiently associates with Trx and E(z) in vivo, we performed proximity ligation assay (PLA) in situ following heat shock on

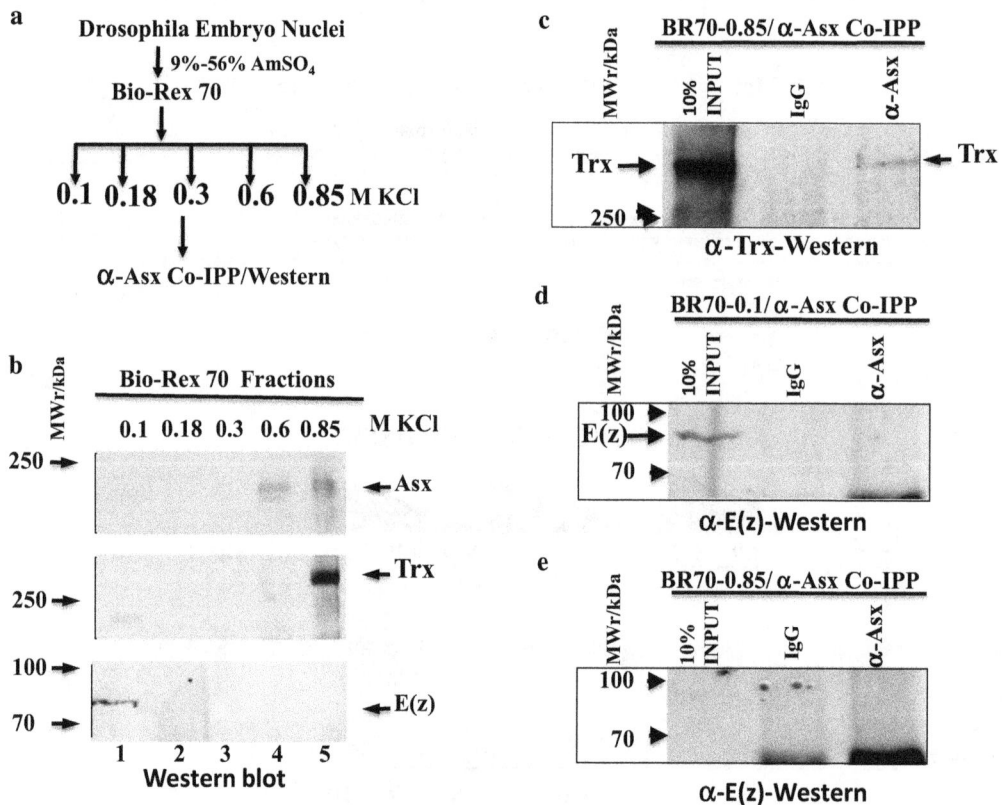

Fig. 2 Asx co-fractionates and co-immunoprecipitates with Trx in embryo nuclear extracts. **a** Schematic of Bio-Rex-70 fractionation of embryo nuclear extracts and coupled co-immunoprecipitation–western blot analysis. **b** Western blot analysis of Bio-Rex 70 fractions with anti-Asx (upper panel), anti-Trx (middle panel) and anti-E(z) (lower panel). **c** Coupled anti-Asx immunoprecipitation and anti-Trx western blot analysis of the BR70-0.85 fraction, showing Asx association with Trx. **d** Coupled anti-Asx immunoprecipitation and anti-E(z) western blot analysis of the BR70-0.1 fraction, showing lack of Asx association with E(z) in the enriched E(z) fraction. **e** Coupled anti-Asx immunoprecipitation and anti-E(z) western blot analysis of the BR70-0.85 fraction, showing lack of Asx association with E(z) in the enriched Trx fraction

Drosophila S2 cells. This assay allows very sensitive (compared to immunoprecipitation) detection of protein–protein interactions either in solution or in situ at single-cell resolution [36]. Association between Asx with Trx was only observed after heat-shock induction, and the level of association becomes significant after 60–180 min of recovery period (Fig. 6a). The association between Asx with E(z) was observed before and after heat-shock induction (Fig. 6b). These results suggest Asx associates with E(z) and Trx in vivo during heat-shock recovery. Taken together, these three protein assays indicate that Asx associates both in vitro and in vivo with E(z) and Trx.

Asx binds to *hsp70* promoter downstream region upon heat-shock induction and is required for *hsp70* repression during induction and recovery

The foregoing results suggest a mechanism whereby Asx modulates the ratio of H3K4 and H3K27 trimethylation by associating with the antagonistic HMTs during heat shock and recovery. If so, then mutations in *Asx*

should affect levels of trimethylation at histone H3K4 and H3K27 during transcription. To determine whether Asx was recruited to heat-shock loci on chromatin, polytene chromosomes were prepared from salivary glands subjected to 20 min of heat shock at 37 °C, and stained with antibodies to Asx. These were compared to preparations from glands that were not heat shocked. Asx was recruited to *hsp70* region (at 87AC) and other heat-shock loci following 20 min of heat shock (Fig. 7a). In contrast, in control polytene chromosomes that were not heat shocked, Asx was recruited to region 89E (Fig. 7a) which includes *Ultrabithorax* (*Ubx*) thus serving as a positive control for the polytene staining [25].

Pol II recruitment and initiation of *hsp70* transcription to generate the paused *hsp70* promoter occur early in embryogenesis (2–4 h after egg deposition) when maternally deposited Asx protein or *Asx* mRNA is still present. To allow enough time for maternal levels of *Asx* mRNA or protein to drop, and thus allow us to detect the embryonic effect of *Asx* mutations, heat shock was induced in

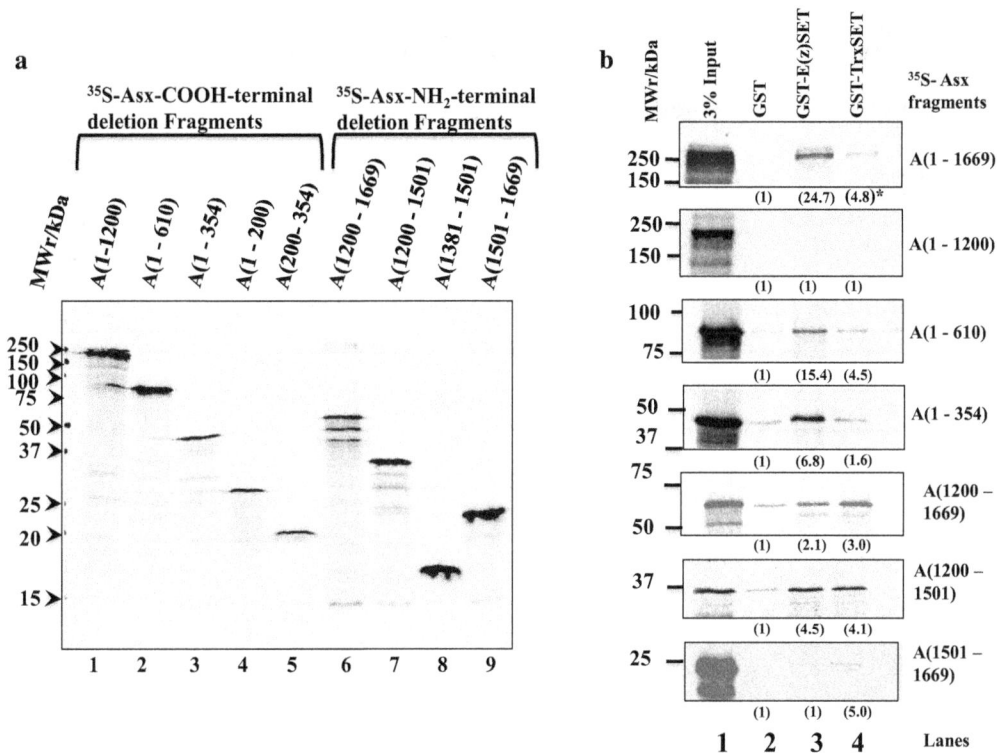

Fig. 3 Asx contains two shared interaction sites for GST-TrxSET and GST-E(z)SET. **a** SDS–PAGE analysis of expression of ^{35}S-labeled RRL-IVT deletion fragments of Asx (3% of the input for GST pull-down assay in Fig. 2b) and exposed for an hour at −70 °C (lanes 1–9). **b** Identification of two shared interaction sites for GST-E(z)SET and GST-TrxSET (1–354 and 1200–1500) and one unique, weak site for GST-TrxSET (1501–1669) on Asx. GST pull-down analysis of the Asx fragments in A) with GST-E(z)SET (lane 3) and GST-TrxSET (lane 4) compared to GST controls (lane 1). ()* Estimated fold binding of Asx with GST fusions relative to GST (determined by densitometry) is shown beneath each lane. Unbound fragments, as shown for A(1–1200), are listed in Fig. 3b

14–16-h embryos. Using this protocol, we are unable to assay the role of *Asx* in establishment of transcriptional initiation and promoter clearance of *hsp70* in early embryos. Our studies in later embryos do not attempt to distinguish the role of *Asx* in elongation between roles at the promoter in re-initiation and promoter clearance, although we biased the results toward elongation by selection of a primer downstream of the promoter (+218 to +392 bp) previously identified as a site of recruitment for Trx upon heat shock [23] for the ChIP experiments below.

The *hsp70* gene is induced in all cells of the embryo in response to thermal stress, allowing us to use whole embryos for this study [37]. To determine whether Asx binds downstream of the *hsp70* promoter downstream region, we performed ChIP and quantitative PCR (qPCR) using the anti-Asx antibody (Fig. 7b, c). In control embryos that were not subject to thermal stress (0 min of heat shock), Asx did not significantly bind at the *hsp70* promoter downstream region. After 15 min of heat shock, the level of Asx binding was threefold higher than with no heat shock. At 30-min recovery after 15-min heat

shock, Asx binding to the *hsp70* gene increased sevenfold relative to no heat shock. The level of Asx binding at *hsp70* gradually decreased from 60-min recovery to 180-min recovery (Fig. 7b). As a positive control for Asx binding in our assay conditions, we tested Asx binding to selected regions within *Ubx* promoter and *bithoraxoid* (*bxd*) Polycomb response element (PRE) in *Drosophila* embryos with the our Asx antibody in our ChIP assay [10]. The results were consistent with previous observations that Asx binds chromatin at these locations without heat-shock induction (Additional file 9: Fig. S3). These results confirm that the *hsp70* locus is a direct binding target of Asx upon heat shock, with peak binding occurring between 15-min induction/30-min recovery and 15-min induction/120-min recovery.

To investigate the effect of *Asx* on transcription during induction and recovery from heat shock, we compared the steady-state *hsp70* mRNA levels during heat shock and recovery of homozygous Asx^3 null mutants to wild-type (Oregon R) embryos (see "Methods" section). The mRNA level difference between the homozygous Asx^3 mutant and wild type was at least twofold during

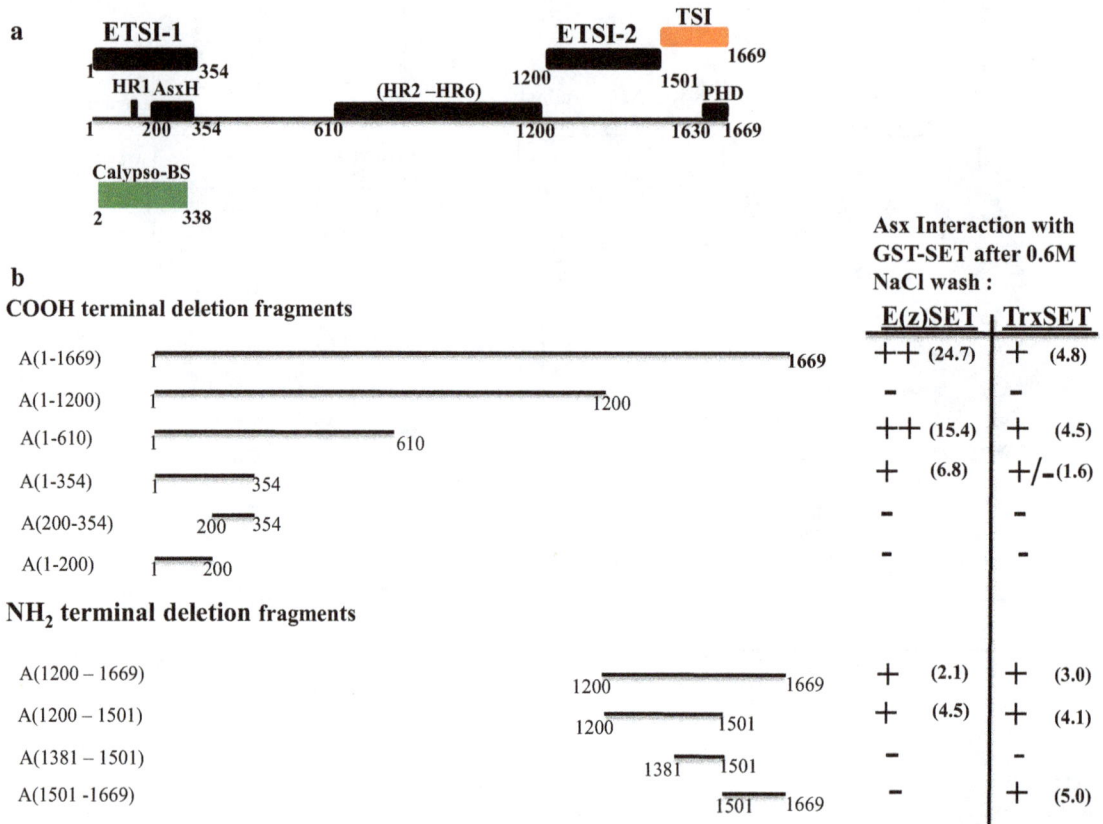

Fig. 4 Map of Asx summarizing interaction sites of E(z) and Trx SET domains indentified using a GST pull-down assay. **a** Linear map of Asx showing a bipartite recognition site for both E(z)SET and TrxSET domains. The NH2 terminal recognition site (residues 1–354) termed E(z)-Trx SET domain interaction site 1 (ETSI-1) and the COOH terminal site (residues 1200–1501) termed ETSI-2 are marked above in blue. The single weak recognition site for TrxSET domain (residues 1501–1669), termed TSI, is marked in orange. The ETSI-1 site overlaps both the conserved AsxH domain (residues 200–354) and the Calypso binding site (marked below in green). **b** Summary of the GST pull-down analysis of the interaction between ^{35}S-labeled Asx deletion fragments with GST fusion proteins of E(z) SET domain or Trx SET domain after a 0.6 M NaCl wash. Non-binding Asx fragments A(1–1200), A(1–200), A(200–354) and A(1381–1501) are indicated by (−). Scouring of the interactions of the Asx deletion fragments with GST-E(z) SET or GST-Trx SET is based on the densitometry of band intensity relative to GST [(++; > tenfold);, +; (twofold–tenfold)]

heat-shock induction and recovery, and at 90-min recovery, the difference reached a maximum level at 2.7-fold (Fig. 8a). The steady-state *Ahcy89E* mRNA levels during heat shock and recovery were stable and the mRNA level was comparable between *Asx³* mutant and wild-type embryos, indicating that *Ahcy89E* is suitable for normalizing the *hsp70* mRNA levels (Fig. 8b). Together, these data show that *Asx* represses the *hsp70* locus during heat-shock induction and recovery. After 180-min recovery, the *hsp70* mRNA level in both wild type and homozygous *Asx³* mutant decreased to the level before heat-shock induction, showing that *Asx* is not required for terminating transcription after heat-shock induction. In addition, peak transcription observed between 15-min induction/30-min recovery and 15-min induction/90-min recovery correlates with Asx binding within this period (Figs. 7b, 8a).

Asx transiently reduces H3K4 trimethylation at *hsp70* during heat-shock induction

Asx represses *hsp70* during heat shock, so in *Asx³* mutant embryos, we expected to observe increased levels of H3K4me3 downstream of the *hsp70* promoter after heat shock compared to wild type. As shown in Fig. 9a, in ChIP-qPCR experiments, the level of H3K4 trimethylation in wild type and *Asx³* mutant did not differ in the first 10 min of heat-shock induction. However, at 15-min induction and 15-min/30-min recovery (Fig. 9a), the level of H3K4 trimethylation in *Asx³* mutants was 1.8-fold higher than in wild-type (significant at $P < 0.05$), suggesting that in wild-type embryos, *Asx* reduces the level of histone H3K4me3 at these time points. Interestingly, in the later recovery phase, *Asx³* mutants showed less H3K4me3 than wild-type embryos after 120 min of recovery from 15-min heat-shock induction (significant

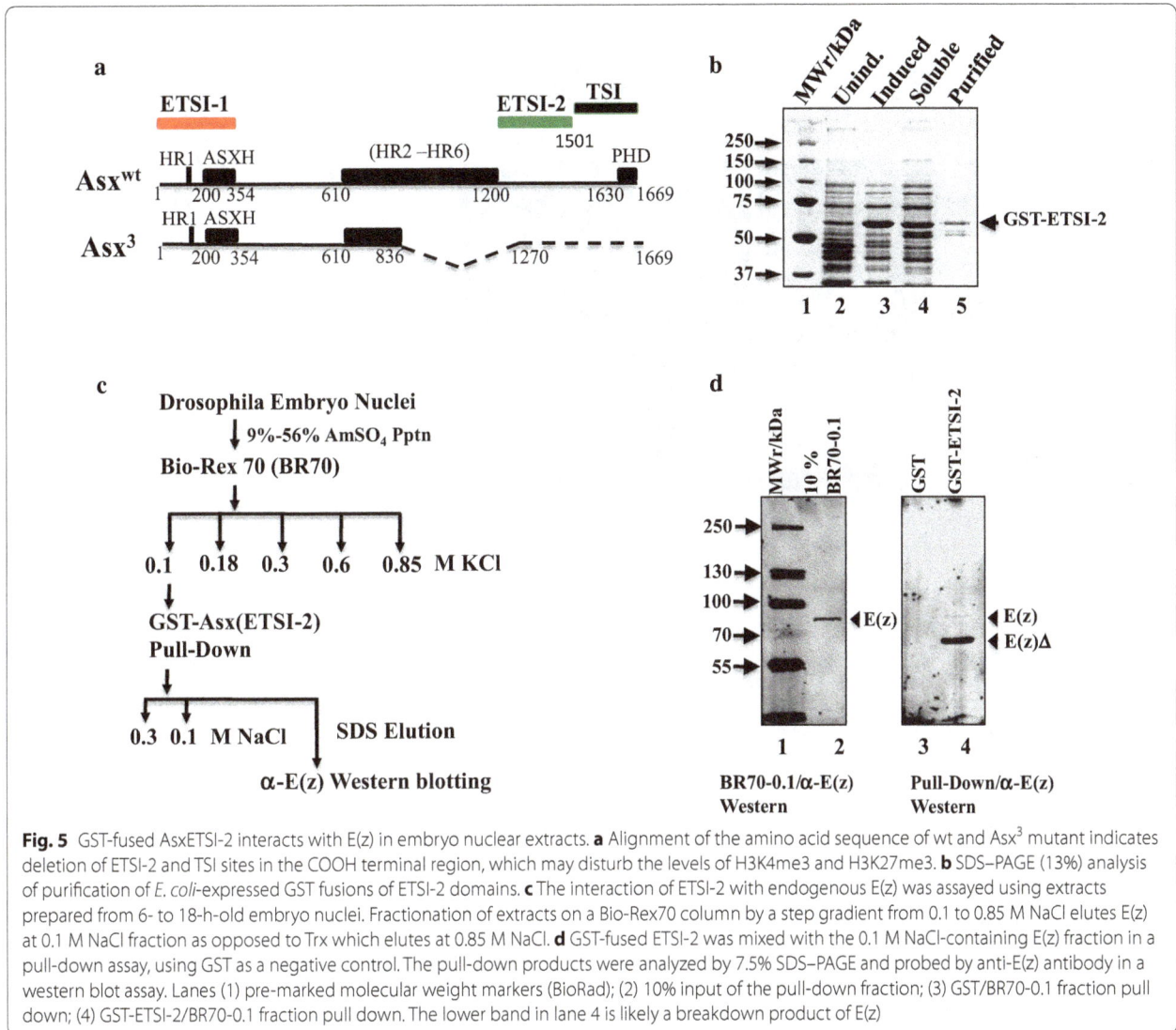

Fig. 5 GST-fused AsxETSI-2 interacts with E(z) in embryo nuclear extracts. **a** Alignment of the amino acid sequence of wt and Asx³ mutant indicates deletion of ETSI-2 and TSI sites in the COOH terminal region, which may disturb the levels of H3K4me3 and H3K27me3. **b** SDS–PAGE (13%) analysis of purification of *E. coli*-expressed GST fusions of ETSI-2 domains. **c** The interaction of ETSI-2 with endogenous E(z) was assayed using extracts prepared from 6- to 18-h-old embryo nuclei. Fractionation of extracts on a Bio-Rex70 column by a step gradient from 0.1 to 0.85 M NaCl elutes E(z) at 0.1 M NaCl fraction as opposed to Trx which elutes at 0.85 M NaCl. **d** GST-fused ETSI-2 was mixed with the 0.1 M NaCl-containing E(z) fraction in a pull-down assay, using GST as a negative control. The pull-down products were analyzed by 7.5% SDS–PAGE and probed by anti-E(z) antibody in a western blot assay. Lanes (1) pre-marked molecular weight markers (BioRad); (2) 10% input of the pull-down fraction; (3) GST/BR70-0.1 fraction pull down; (4) GST-ETSI-2/BR70-0.1 fraction pull down. The lower band in lane 4 is likely a breakdown product of E(z)

at $P < 0.05$), consistent with the rapid drop of *hsp70* mRNA level during 120-min recovery after heat-shock induction. At this time, Asx could transiently modulate the rate of transcription by governing the rate of transcript decay. We suggest that Asx directly or indirectly regulates H3K4me3 to an appropriate level during heat shock and recovery.

Asx regulates the ratio of histone H3K4 to H3K27 trimethylation at *hsp70* during heat-shock induction and recovery

If Asx represses *hsp70* by acting as a PcG protein, then we expect to observe reduced H3K27me3 levels in *Asx³* mutants compared to wild type during heat-shock induction and recovery. We performed ChIP-qPCR experiments with anti-H3K27me3 antibody on wild-type and

Asx³ mutant embryos during heat shock and recovery. The level of H3K27me3 in wild type and *Asx³* mutant did not differ markedly at 10 min of heat-shock induction. At subsequent time points from 15-min heat-shock induction to 120-min recovery after heat shock except the 30-min recovery, the level of H3K27me3 in *Asx³* mutants was 1.5-fold lower than in wild type (Fig. 9b).

Thus, the data in Fig. 9a, b show that mutation of *Asx* alters the levels of H3K4me3 and H3K27me3 in the expected way at one target gene during the same regulatory event. Interestingly, the significant increase in the ratio of H3K4me3/H3K27me3 from 15-min induction to 15-min induction/30-min recovery from heat shock in *Asx³* compared to wild type (Fig. 9c) occurs close to the peak of transcription (Fig. 8a) that coincides with the transition from promoter clearance to elongation. As

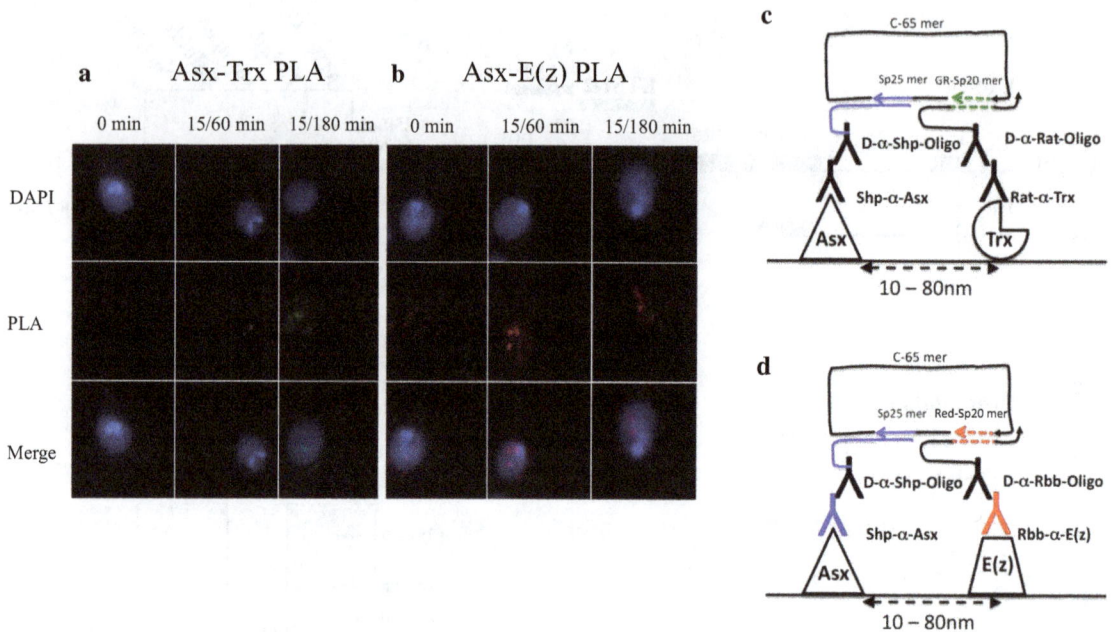

Fig. 6 Asx interacts with E(z) and Trx in *Drosophila* S2 cells in vivo. **a** Proximity ligation assay (PLA) between Asx and Trx with *Drosophila* S2 cells before and after heat-shock induction. Nuclei were labeled with DAPI in blue and PLA signals in green for Asx-Trx and in red for Asx-E(z). **b** PLA between Asx and E(z) with S2 cells before and after heat-shock induction. **c, d** Schematic diagram of PLA reaction. Oligonucleotides and antibodies used in PLA are listed in "Methods" section. A closed circle between connector and splints forms if proteins are in close proximity. Rolling-circle amplification (RCA) was carried out in the presence of fluorescent-labeled oligonucleotides. Signals would be detected only when proteins were in distance between 10 and 80 nm

positive and negative controls for the levels of H3K4me3 and H3K27me3 in our assay conditions, we compared the highest and lowest levels of recovery of H3K4me3 and H3K27me3 observed at *hsp70* in ChIP experiments to levels obtained with a fragment from within the *bxd* PRE as a positive control, and a region upstream of the *DRP12* gene as a negative control, in *Drosophila* embryos (Additional file 10: Fig. S4). The results show that significantly more H3K4me3 is recovered at the *hsp70* locus compared to *bxd*, consistent with *hsp70* being actively transcribed in all cells, whereas *bxd* is expressed only in a subset of cells. During the recovery phase from heat shock, when H3K27me3 levels are highest, they are equivalent to those observed at *bxd*. Even at the lowest levels of H3K27me3 observed at *hsp70* (15 min of heat shock, 120 min of recovery), the levels observed are significantly higher than the negative or IgG controls. Thus, we are confident that the changes we observe in H3K4me3 and K3K27me3 we observe at *hsp70* are biologically important.

Discussion

Our experiments on the role of *Asx* in regulation of the relative levels of H3K4me3 to H3K27me3 suggest that Asx recruits Trx and E(z) at different temporal stages via

interaction with their SET domains (Figs. 9, 6). The mapping data shown in Figs. 1 and 2 confirm the hypothesis that Asx interacts directly with E(z) and Trx in vitro. The mapping data also show that there are two non-overlapping Asx domains that interact with E(z) and Trx SET domains, termed ETSI-1 and ETSI-2 (Fig. 4). Sequence comparison of ETSI-1 and 2 reveals 15% sequence identity (Additional file 8: Fig. S2), suggesting that either there is conservation of 2D structure with clustered amino acid sequence conservation, or that ETSI-1 and 2 domains have a unique interaction mechanism. The third Asx interaction site, TSI (Figs. 3, 4), is a distinct Trx SET domain interaction site, which overlaps the conserved C-terminal atypical PHD motif, consistent with the suggestion that the conserved atypical PHD motif lacks structural features which restricts its binding to the N-terminal tail of H3 [38].

The identification of Asx interaction sites for the SET domains of E(z) and Trx (ETSI-1/2 and TSI) (Fig. 4) is consistent with Asx association with equivalent levels of E(z) and Trx in vivo at 60 to 180 min of recovery (Fig. 6). These findings support the hypothesis that *Asx* directly regulates H3K4me3 and H3K27me3 levels downstream of the *hsp70* promoter (Fig. 9). However, the mapping of 2 SET domain interaction sites suggests that it is unlikely

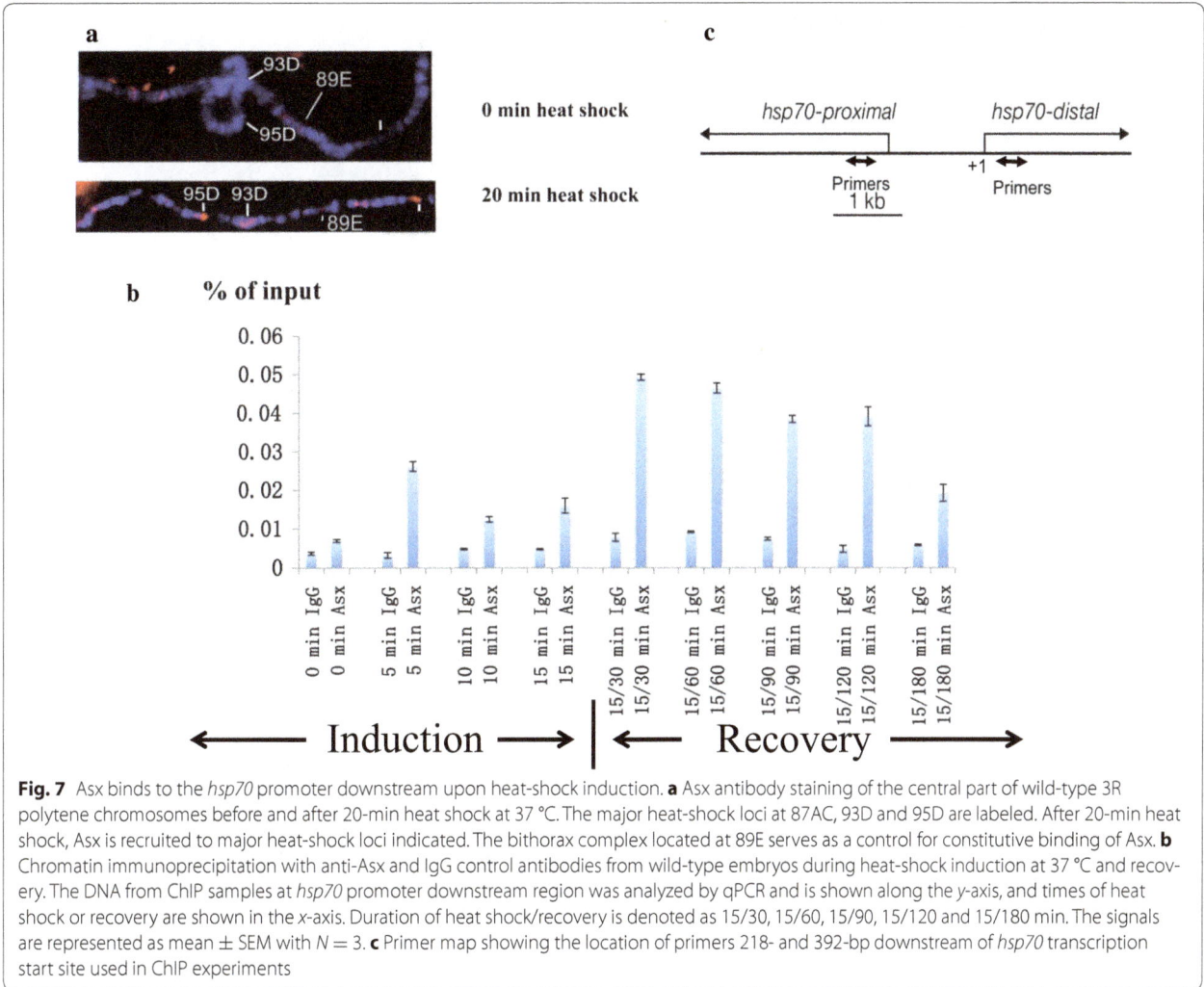

Fig. 7 Asx binds to the *hsp70* promoter downstream upon heat-shock induction. **a** Asx antibody staining of the central part of wild-type 3R polytene chromosomes before and after 20-min heat shock at 37 °C. The major heat-shock loci at 87AC, 93D and 95D are labeled. After 20-min heat shock, Asx is recruited to major heat-shock loci indicated. The bithorax complex located at 89E serves as a control for constitutive binding of Asx. **b** Chromatin immunoprecipitation with anti-Asx and IgG control antibodies from wild-type embryos during heat-shock induction at 37 °C and recovery. The DNA from ChIP samples at *hsp70* promoter downstream region was analyzed by qPCR and is shown along the y-axis, and times of heat shock or recovery are shown in the x-axis. Duration of heat shock/recovery is denoted as 15/30, 15/60, 15/90, 15/120 and 15/180 min. The signals are represented as mean ± SEM with N = 3. **c** Primer map showing the location of primers 218- and 392-bp downstream of *hsp70* transcription start site used in ChIP experiments

that Trx and E(z) compete to bind Asx. This conclusion is supported by the data in Fig. 2 showing that E(z) and Trx do not co-fractionate after chromatography of nuclear extracts on Bio-Rex 70. These results suggest that Asx association with Trx or E(z) in vivo may be influenced by the different temporal stages of heat shock and recovery. These experiments do not rule out interactions of Asx with domains outside the SET domains of Trx and E(z), but this possibility was not tested.

Comparison of existing structure function analysis of Asx homologs in *Drosophila* and mammals with our new data allows us to speculate that Asx acts as an ETP by interacting with multiple nucleosome-modifying enzymes to modulate histone modifications. Asx and its mammalian homolog ASXL1 interact with Calypso/BAP1 to form the PR-DUB complex, which deubiquitinates histone H2AK118Ub/H2AK119Ub at the *Ubx* PRE and promoter [10]. The ASXH domain is conserved between all mammalian ASXL and *Drosophila* Asx and

directly binds the deubiquitinase BAP1 [10, 39]. Interestingly, in mammals the ASXL1 DEUBAD domain (amino acid 238–290) located within the ASXH (amino acid 249–368) domain is required to activate the BAP1 enzymatic function on deubiquinating H2AK119Ub [40, 41]. Unlike *Asx*, mutation of *calypso* did not have a significant effect on the relative *hsp70* mRNA levels in *calypso²* mutant and wild-type embryos at selected time points of heat shock or recovery (Additional file 11: Fig. S5). Thus, *calypso* is not required for the activity of Asx at *hsp70* during heat-shock induction and recovery.

Both our results on Asx and recent studies on ASXL1 suggest that ETSI-1 site alone is insufficient to maintain the normal H3K27me3 level at target promoters [42]. We therefore suggest that both the ETSI-1 and ETSI-2 regions are required to recruit and/or retain the function of E(z)/EZH2 at target promoters. Consistent with this view, ASXL1 co-immunoprecipitates with the PRC2 components EZH2, SUZ12 and EED [43]. The C-terminal

a   **Relative mRNA level**

b

    Relative mRNA level

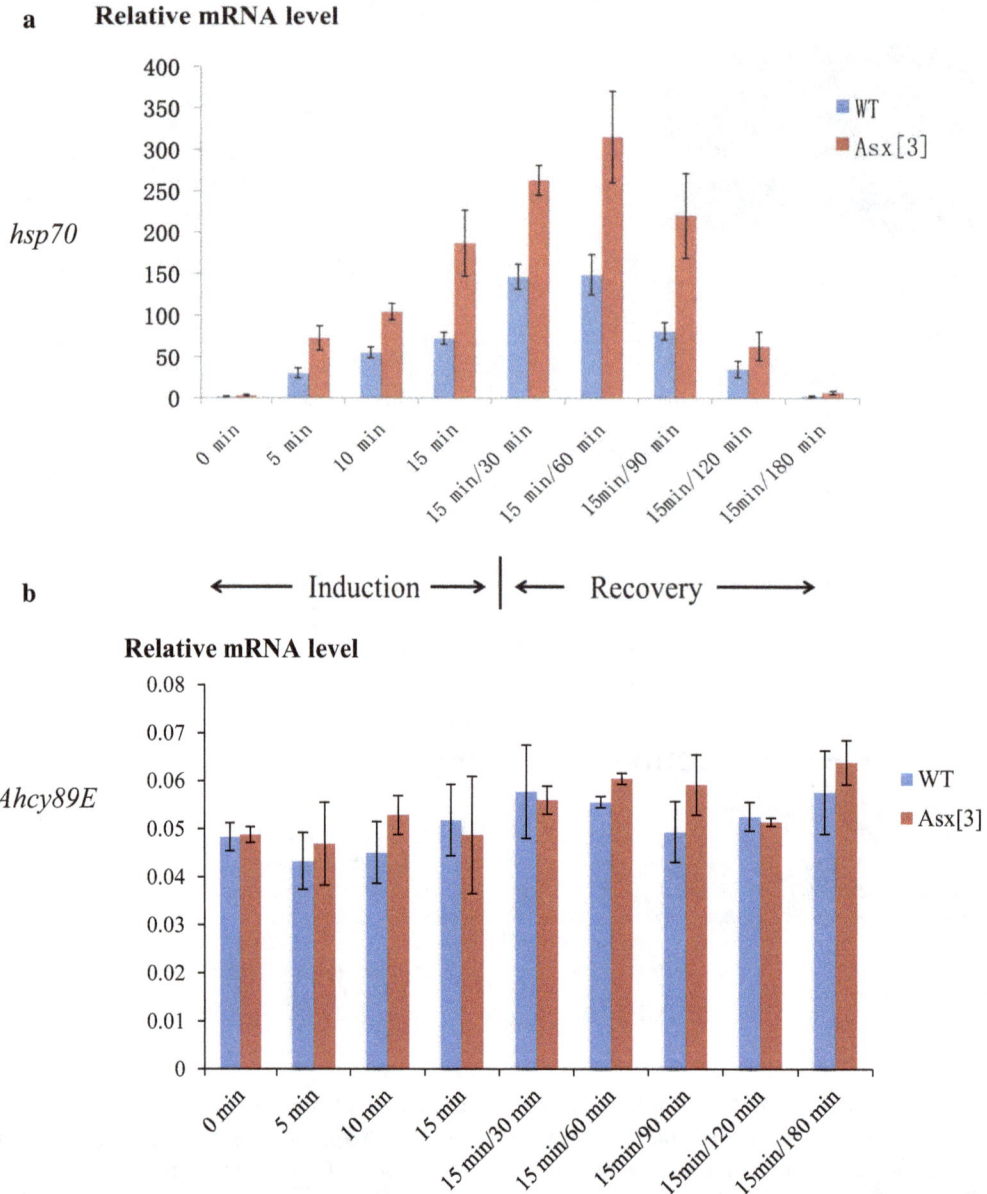

Fig. 8 *Asx* is required for *hsp70* repression during heat-shock induction and recovery. **a** The relative mRNA levels in wild-type and *Asx³* homozygous null mutant embryos were measured by RT-qPCR. *Asx* null mutations are embryonic lethal at a late embryonic stage [4], so homozygous 12–15-h *Asx³* mutants were collected using absence of expression of the GFP-marked balancer chromosome as a criterion. The duration of heat shock/recovery is as described in Fig. 7. The *x*-axis shows the heat-shock induction times at 37 °C and heat-shock recovery times after 15 min of 37 °C heat-shock induction. The *y*-axis indicates the *hsp70* mRNA level normalized to the control gene *Ahcy89E* mRNA level. The signals are represented as mean ± SEM with *N* = 3. **b** The relative mRNA levels of the control gene *Ahcy98E* are compared between *Asx³* mutant and wild-type embryos throughout the heat-shock induction and recovery. The *x*-axis shows the heat-shock induction times at 37 °C, and heat-shock recovery times after 15 min of 37 °C heat-shock induction. The *y*-axis shows the relative mRNA level of the control gene *Ahcy89E* mRNA, as mean ± SEM with *N* = 3

region of ASXL1 aligns to the Asx ETSI-2 with 15% sequence identity (Additional file 8: Fig. S2B), suggesting conservation of function. It is possible that ETSI-1 and ETSI-2 interact independently or cooperatively. In the future, it will be interesting to determine whether each of these domains simultaneously recruits Trx and E(z),

whether binding of Trx or E(z) occurs preferentially to ETSI-1 or 2 and whether histone demethylases or histone acetyl transferases also associate with these domains.

One major finding in this study is that *Asx* is required to repress the transcription of *hsp70* during heat stress and recovery. In the first 10 min of heat stress, when

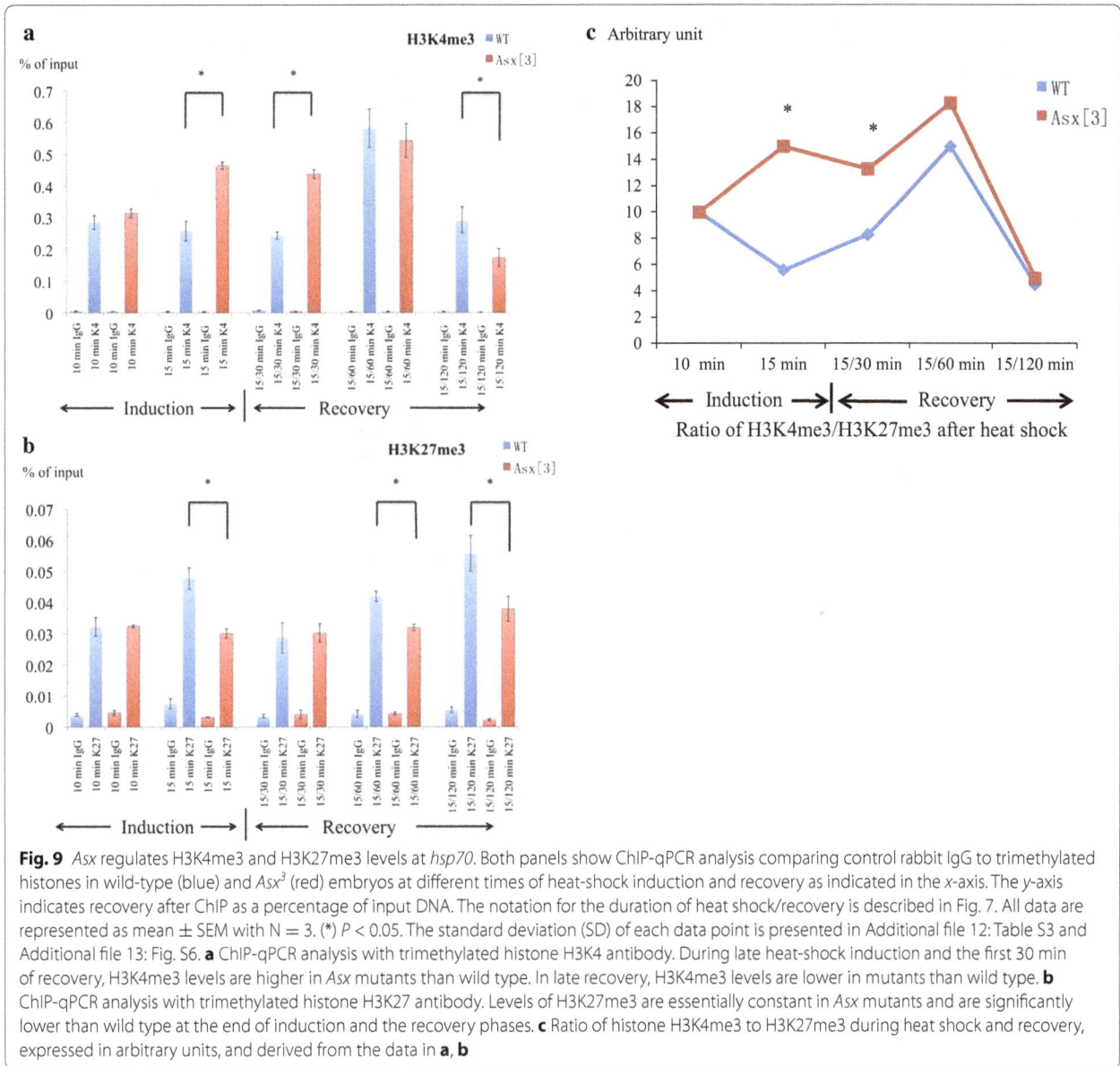

Fig. 9 *Asx* regulates H3K4me3 and H3K27me3 levels at *hsp70*. Both panels show ChIP-qPCR analysis comparing control rabbit IgG to trimethylated histones in wild-type (blue) and *Asx*[3] (red) embryos at different times of heat-shock induction and recovery as indicated in the x-axis. The y-axis indicates recovery after ChIP as a percentage of input DNA. The notation for the duration of heat shock/recovery is described in Fig. 7. All data are represented as mean ± SEM with N = 3. (*) $P < 0.05$. The standard deviation (SD) of each data point is presented in Additional file 12: Table S3 and Additional file 13: Fig. S6. **a** ChIP-qPCR analysis with trimethylated histone H3K4 antibody. During late heat-shock induction and the first 30 min of recovery, H3K4me3 levels are higher in *Asx* mutants than wild type. In late recovery, H3K4me3 levels are lower in mutants than wild type. **b** ChIP-qPCR analysis with trimethylated histone H3K27 antibody. Levels of H3K27me3 are essentially constant in *Asx* mutants and are significantly lower than wild type at the end of induction and the recovery phases. **c** Ratio of histone H3K4me3 to H3K27me3 during heat shock and recovery, expressed in arbitrary units, and derived from the data in **a**, **b**

transcriptional elongation should predominate over re-initiation, *Asx* mutants exhibit higher levels of *hsp70* transcription compared to controls (Fig. 8), so *Asx* acts as a governor to prevent rapid increase in the rate of *hsp70* transcription. Interestingly, we do not detect any significant change in H3K4me3 levels during this time (Fig. 9), implying that the *Asx* effect in the first 10 min is Trx independent [23]. Given previous observations that Asx is not recruited to polytene chromosomes in *trx* mutants and vice versa [44] and that *trx* mutations abolish induction of heat shock [23], one would predict no heat-shock response in *Asx* mutants. As noted in "Background," we would not detect a Trx-dependent effect in our experiments immediately after heat-shock induction because

recruitment of Trx and Pol II and initiation of transcription occur many hours earlier in the presence of maternal Asx. We suggest that changes in H3K4me3 levels detected after 15 min of heat stress reflect re-initiation of *hsp70* transcription.

Asx may have a previously unreported role in transcriptional elongation of *hsp70* that is independent of the catalytic activity of Trx or other H3K4 HMT because *Asx* mutants have higher transcription compared to wild type in the first 10 min of heat stress that cannot be attributed to changes in levels of H3K4me3. *Asx* may regulate factors required for *hsp70* transcriptional elongation including Mediator, elongation factors (such as Spt5, Spt6 and FACT) or the H3K36 histone methyltransferase activity

[18, 20, 22, 45]. Alternatively, *Asx* may participate in Pol II pausing and retention or release during the 0–10-min heat-shock phase before a Trx-dependent step [46].

From 15 min of heat-shock induction up to 60 min of recovery, *Asx* is required to govern the rate of *hsp70* transcription and changes the ratio of H3K4me3 to H3K27me3. As shown in Fig. 9c, the ratio of H3K4me3/H3K27me3 levels is consistently higher in *Asx* mutants compared to wild type during this period suggesting that the relative levels of H3K4 and H3K27 trimethylation regulate *hsp70* transcription. Alternatively, the ratio of H3K4me3/H3K27me3 may reflect consequences of other events regulated by *Asx* that lead to changes in the ratio.

From 90 to 180 min after recovery from heat shock, transcription of *hsp70* falls in both control and *Asx³* mutant embryos, showing that *Asx* is not required for repression of *hsp70* transcription at this phase (Fig. 8a). By contrast, the PcG gene *pho* is required for this repression because the level of *hsp70* mRNA in *pho¹* mutant larvae was significantly higher than in wild-type embryos after 30-min induction/180-min or 300-min recovery, but no significant difference was observed after 30-min induction/60-min recovery after heat shock [24]. Therefore, the timing of Asx regulation of *hsp70* transcription does not significantly overlap with the requirement for *pho*.

Asx was identified first as a PcG gene [4], and subsequent experiments in *Drosophila* supported this conclusion [5, 6]. To our knowledge, our observations provide the first example where *Asx* mutations cause simultaneous changes in both H3K4me3 and H3K27me3 at the same locus at the same time. Protein nulls such as *Asx²²P⁴* have no change in global H3K27 trimethylation and show very slight reduction in global H3K4 trimethylation [10]. The increased sensitivity of gene-specific analysis or particular features of *hsp70* regulation may allow us to detect clear effects of *Asx* mutations on relative levels of H3K4me3 and H3K27me3 that are not detected in bulk chromatin. Our data do not rule out alternative models could account for the regulation of the H3K4me3 to H3K27me3 levels: (1) Asx acts indirectly via CBP-mediated H3K27 acetylation to block methylation [47, 48]; (2) a role for two other H3K4me3 HMTs, SET1 and Trr, cannot be excluded since both contain SET domains with 50% amino acid sequence identity to the SET domain of Trx (NCBI protein sequences, see web refence below); (3) Asx affects demethylation of H3K27me3 by Jarid2 by recruiting E(z), and its HMT PRC2 subunits Su(z)12, Esc and Jarid2 [49].

Conclusions

The major finding in this study is that Asx interacts directly with E(z) and Trx in vitro and in vivo during heat-shock recovery. *Asx* is required to repress transcription of the *hsp70* locus during heat stress and the first 60 min of recovery by regulating the relative H3K4 and H3K27 trimethylation levels at the promoter. These results are consistent with genetic identification of *Asx* as an ETP required for both trxG and PcG function.

Additional files

Additional file 1: Text S1. Rabbit anti-Asx antibody.

Additional file 2: Fig. S1. Validation of Asx antibody. Western blots with *Drosophila* wild-type and *Df(2R)trix* mutant embryo extract showing the rabbit anti-Asx antibody (aa. 200–356) generated for this study binds specifically to Asx. *Df(2R)trix* mutant contains deletion of entire *Asx*. The binding level was significantly reduced in *Df(2R)trix* mutant embryo extract compared to wild-type embryo extract.

Additional file 3: Table S1. PCR primer pairs for construction of E(z) SET and Asx expression vectors.

Additional file 4: Text S2. Buffers.

Additional file 5: Text S3. GST pull-down western assays of embryo nuclear extract.

Additional file 6: Table S2. Oligonucleotides for *in situ* PLA in *Drosophila* S2 cells.

Additional file 7: Text S4. Primers for *hsp70*, *Ahcy89E*, *bxd* PRE and *Ubx* promoters.

Additional file 8: Fig. S2. Alignment of amino acid sequences of AsxETSI. (**A**) Clustal Omega alignment of AsxETSI-1 and AsxETSI-2 showing 14.97 % sequence identity. (**B**) Clustal Omega alignment of AsxETSI-2 and ASXL1 (943–1307) showing 15.1% sequence identity.

Additional file 9: Fig. S3. Asx binds to *Ultrabithorax* (*Ubx*) promoter and *bithoraxoid* (*bxd*) PRE region. (**A**) Primer map showing the location of primers used in ChIP experiments. L2, L7 and L8 primers are located within the *bxd* PRE region, 12.5-kb upstream of the *Ubx* promoter. U2 and U3 primers are located downstream of the *Ubx* promoter. (**B, C**) ChIP-qPCR analysis of anti-Asx and control rabbit IgG antibodies from wild-type embryos. The DNA recovered from ChIP samples was analyzed by qPCR and is shown along the y-axis. The signals are represented as mean ± SEM with N = 3.

Additional file 10: Fig. S4. H3K4me3 and H3K27me3 levels at *bithoraxoid* (*bxd*) Polycomb response element (PRE) and DPR12 genes compared to highest and lowest levels observed at *hsp70*. ChIP-qPCR analysis of H3K4me3 and H3K27me3 and control rabbit IgG antibodies from wild-type embryos. The DNA recovered from ChIP samples was analyzed by qPCR, and percent recovery is shown along the y-axis. The data for *hsp70* are taken from Fig 9. The signals are represented as mean ± SEM with N = 3. The *bxd* PRE primers are located between BX-C 218839 and 218959. C1 is located at +39kb to the DPR12 gene.

Additional file 11: Fig. S5. *calypso* is not required for *hsp70* repression during heat-shock induction and recovery. The relative mRNA levels in wild-type and *calypso²* homozygous null mutant embryos were measured by RT-qPCR. The x-axis shows the heat-shock induction times at 37 °C, and heat-shock recovery times after 15 min of 37 °C heat-shock induction. The y-axis indicates the *hsp70* mRNA level normalized to the control gene *Ahcy89E* mRNA level. The signals are represented as mean ± SEM with N = 3.

Additional file 12: Table S3. Standard deviation (SD) table for ChIP experiments.

Additional file 13: Fig. S6. Asx regulates H3K4me3 and H3K27me3 levels at *hsp70*. Both panels show ChIP-qPCR analysis comparing control rabbit IgG to trimethylated histones in wild-type (blue) and *Asx³* (red) embryos at different times of heat-shock induction and recovery as indicated in the x-axis. The y-axis indicates recovery after ChIP as a percentage of input DNA. The notation for the duration of heat shock/recovery is described in Fig. 7. All data are represented as mean ± SD. There was minimal difference when compared with the error bars using SEM as the error source in Fig. 9.

Abbreviations

PcG: Polycomb group; TrxG: trithorax group; ETP: enhancer of trithorax and Polycomb; Asx: additional sex combs; HMT: histone methyl transferase; E(z): enhancer of zeste; Trx: trithorax; hsp70: 70-kDa heat-shock proteins; SET: Su(var)3–9, enhancer of zeste and trithorax; CTD: C-terminal domain; HSF: heat-shock factor; ChIP: chromatin immunoprecipitation; ETSI: E(z)/Trx SET domain interaction site; TSI: Trx SET domain interaction site; PLA: proximity ligation assay; Ahcy89E: adenosylhomocysteinase 89E; PR-DUB: Polycomb repressive deubiquitinase.

Authors' contributions

HWB and AM conceived the study; TL, JWH and SP performed experiments. All authors discussed the results, analyzed the data and commented on drafts of the manuscript. TL, JWH and HWB wrote the manuscript. All authors read and approved the final manuscript.

Author details

[1] Department of Zoology, Life Sciences Institute, University of British Columbia, 2350 Health Science Mall, Vancouver, BC V6T 1Z4, Canada. [2] Department of Biochemistry and Molecular Biology, Thomas Jefferson University, Philadelphia, PA 19107, USA.

Acknowledgements

We thank Bloomington Stock Center for fly stocks. We acknowledge the assistance of Dr. Sheryl T. Smith and Maya Kupczyk during the initial phase of this project. We thank Drs. Robert Tjian and Michael Kyba for plasmid vectors.

Competing interests

The authors declare that they have no competing interests.

Funding

SP was supported by a training program from the National Cancer Institute (NCI) (CA009678). This study was supported by grants from the Natural Sciences and Engineering Research Council and the Canadian Institutes of Health Research to HWB. The funders had no role in study design, collection, analysis or interpretation of data, in the writing or in the decision to submit.

References

1. Paro R. Imprinting a determined state into the chromatin of Drosophila. Trends Genet. 1990;6(12):416–21.
2. Kennison JA. Transcriptional activation of Drosophila homeotic genes from distant regulatory elements. Trends Genet. 1993;9(3):75–9.
3. Brock HW, Fisher CL. Maintenance of gene expression patterns. Dev Dyn. 2005;232(3):633–55.
4. Jurgens G. A group of genes controlling the spatial expression of the bithorax complex in Drosophila. Nature. 1985;316(6024):153–5.
5. Simon J, Peifer M, Bender W, O'Connor M. Regulatory elements of the bithorax complex that control expression along the anterior-posterior axis. EMBO J. 1990;9(12):3945–56.
6. Sinclair DA, Campbell RB, Nicholls F, Slade E, Brock HW. Genetic analysis of the additional sex combs locus of Drosophila melanogaster. Genetics. 1992;130(4):817–25.
7. Duncan I. The bithorax complex. Annu Rev Genet. 1987;21:285–319.
8. Milne TA, Sinclair DA, Brock HW. The Additional sex combs gene of Drosophila is required for activation and repression of homeotic loci, and interacts specifically with Polycomb and super sex combs. Mol Gen Genet. 1999;261(4–5):753–61.
9. Campbell RB, Sinclair DA, Couling M, Brock HW. Genetic interactions and dosage effects of Polycomb group genes of Drosophila. Mol Gen Genet. 1995;246(3):291–300.
10. Scheuermann JC, de Ayala Alonso AG, Oktaba K, Ly-Hartig N, McGinty RK, Fraterman S, Wilm M, Muir TW, Muller J. Histone H2A deubiquitinase activity of the Polycomb repressive complex PR-DUB. Nature. 2010;465(7295):243–7.
11. Gildea JJ, Lopez R, Shearn A. A screen for new trithorax group genes identified little imaginal discs, the Drosophila melanogaster homologue of human retinoblastoma binding protein 2. Genetics. 2000;156(2):645–63.
12. Brock HW, van Lohuizen M. The Polycomb group-no longer an exclusive club? Curr Opin Genet Dev. 2001;11(2):175–81.
13. Muller J, Verrijzer P. Biochemical mechanisms of gene regulation by Polycomb group protein complexes. Curr Opin Genet Dev. 2009;19(2):150–8.
14. Schuettengruber B, Martinez AM, Iovino N, Cavalli G. Trithorax group proteins: switching genes on and keeping them active. Nat Rev Mol Cell Biol. 2011;12(12):799–814.
15. Tsukiyama T, Becker PB, Wu C. ATP-dependent nucleosome disruption at a heat-shock promoter mediated by binding of GAGA transcription factor. Nature. 1994;367(6463):525–32.
16. Rasmussen EB, Lis JT. In vivo transcriptional pausing and cap formation on three Drosophila heat shock genes. Proc Natl Acad Sci USA. 1993;90(17):7923–7.
17. Rougvie AE, Lis JT. The RNA polymerase II molecule at the 5′ end of the uninduced hsp70 gene of D. Melanogaster is transcriptionally engaged. Cell. 1988;54(6):795–804.
18. Boehm AK, Saunders A, Werner J, Lis JT. Transcription factor and polymerase recruitment, modification, and movement on dhsp70 in vivo in the minutes following heat shock. Mol Cell Biol. 2003;23(21):7628–37.
19. Westwood JT, Wu C. Activation of Drosophila heat shock factor: conformational change associated with a monomer-to-trimer transition. Mol Cell Biol. 1993;13(6):3481–6.
20. Andrulis ED, Guzman E, Doring P, Werner J, Lis JT. High-resolution localization of Drosophila Spt5 and Spt6 at heat shock genes in vivo: roles in promoter proximal pausing and transcription elongation. Genes Dev. 2000;14(20):2635–49.
21. Kaplan CD, Morris JR, Wu C, Winston F. Spt5 and spt6 are associated with active transcription and have characteristics of general elongation factors in D. Melanogaster. Genes Dev. 2000;14(20):2623–34.
22. Saunders A, Werner J, Andrulis ED, Nakayama T, Hirose S, Reinberg D, Lis JT. Tracking FACT and the RNA polymerase II elongation complex through chromatin in vivo. Science. 2003;301(5636):1094–6.
23. Smith ST, Petruk S, Sedkov Y, Cho E, Tillib S, Canaani E, Mazo A. Modulation of heat shock gene expression by the TAC1 chromatin-modifying complex. Nat Cell Biol. 2004;6(2):162–7.
24. Beisel C, Buness A, Roustan-Espinosa IM, Koch B, Schmitt S, Haas SA, Hild M, Katsuyama T, Paro R. Comparing active and repressed expression states of genes controlled by the Polycomb/Trithorax group proteins. Proc Natl Acad Sci USA. 2007;104(42):16615–20.
25. Sinclair DA, Milne TA, Hodgson JW, Shellard J, Salinas CA, Kyba M, Randazzo F, Brock HW. The Additional sex combs gene of Drosophila encodes a chromatin protein that binds to shared and unique Polycomb group sites on polytene chromosomes. Development. 1998;125(7):1207–16.
26. Tao H, Liu W, Simmons BN, Harris HK, Cox TC, Massiah MA. Purifying natively folded proteins from inclusion bodies using sarkosyl, Triton X-100, and CHAPS. Biotechniques. 2010;48(1):61–4.
27. Leuchowius KJ, Clausson CM, Grannas K, Erbilgin Y, Botling J, Zieba A, Landegren U, Soderberg O. Parallel visualization of multiple protein complexes in individual cells in tumor tissue. Mol Cell Proteom. 2013;12(6):1563–71.
28. Soderberg O, Leuchowius KJ, Gullberg M, Jarvius M, Weibrecht I, Larsson LG, Landegren U. Characterizing proteins and their interactions in cells and tissues using the in situ proximity ligation assay. Methods. 2008;45(3):227–32.
29. Klasener K, Maity PC, Hobeika E, Yang J, Reth M. B cell activation involves nanoscale receptor reorganizations and inside-out signaling by Syk. Elife. 2014;3:e02069.

30. Beck SA, Falconer E, Catching A, Hodgson JW, Brock HW. Cell cycle defects in polyhomeotic mutants are caused by abrogation of the DNA damage checkpoint. Dev Biol. 2010;339(2):320–8.

31. Petruk S, Sedkov Y, Riley KM, Hodgson J, Schweisguth F, Hirose S, Jaynes JB, Brock HW, Mazo A. Transcription of bxd noncoding RNAs promoted by trithorax represses Ubx in cis by transcriptional interference. Cell. 2006;127(6):1209–21.

32. Nelson JD, Denisenko O, Bomsztyk K. Protocol for the fast chromatin immunoprecipitation (ChIP) method. Nat Protoc. 2006;1(1):179–85.

33. Dillon SC, Zhang X, Trievel RC, Cheng X. The SET-domain protein superfamily: protein lysine methyltransferases. Genome Biol. 2005;6(8):227.

34. Frangioni JV, Neel BG. Solubilization and purification of enzymatically active glutathione S-transferase (pGEX) fusion proteins. Anal Biochem. 1993;210(1):179–87.

35. Kauzmann W. Some factors in the interpretation of protein denaturation. Adv Protein Chem. 1959;14:1–63.

36. Greenwood C, Ruff D, Kirvell S, Johnson G, Dhillon HS, Bustin SA. Proximity assays for sensitive quantification of proteins. Biomol Detect Quantif. 2015;4:10–6.

37. Lakhotia SC, Prasanth KV. Tissue- and development-specific induction and turnover of hsp70 transcripts from loci 87A and 87C after heat shock and during recovery in *Drosophila melanogaster*. J Exp Biol. 2002;205(Pt 3):345–58.

38. Aravind L, Iyer LM. The HARE-HTH and associated domains: novel modules in the coordination of epigenetic DNA and protein modifications. Cell Cycle. 2012;11(1):119–31.

39. Fisher CL, Berger J, Randazzo F, Brock HW. A human homolog of Additional sex combs, ADDITIONAL SEX COMBS-LIKE 1, maps to chromosome 20q11. Gene. 2003;306:115–26.

40. Sahtoe DD, van Dijk WJ, Ekkebus R, Ovaa H, Sixma TK. BAP1/ASXL1 recruitment and activation for H2A deubiquitination. Nat Commun. 2016;7:10292.

41. Katoh M. Functional and cancer genomics of ASXL family members. Br J Cancer. 2013;109:299.

42. Balasubramani A, Larjo A, Bassein JA, Chang X, Hastie RB, Togher SM, Lahdesmaki H, Rao A. Cancer-associated ASXL1 mutations may act as gain-of-function mutations of the ASXL1-BAP1 complex. Nat Commun. 2015;6:7307.

43. Abdel-Wahab O, Adli M, LaFave LM, Gao J, Hricik T, Shih AH, Pandey S, Patel JP, Chung YR, Koche R, Perna F, Zhao X, Taylor JE, Park CY, Carroll M, Melnick A, Nimer SD, Jaffe JD, Aifantis I, Bernstein BE, Levine RL. ASXL1 mutations promote myeloid transformation through loss of PRC2-mediated gene repression. Cancer Cell. 2012;22(2):180–93.

44. Petruk S, Smith ST, Sedkov Y, Mazo A. Association of trxG and PcG proteins with the bxd maintenance element depends on transcriptional activity. Development. 2008;135(14):2383–90.

45. Park JM, Werner J, Kim JM, Lis JT, Kim YJ. Mediator, not holoenzyme, is directly recruited to the heat shock promoter by HSF upon heat shock. Mol Cell. 2001;8(1):9–19.

46. Samarakkody A, Abbas A, Scheidegger A, Warns J, Nnoli O, Jokinen B, Zarns K, Kubat B, Dhasarathy A, Nechaev S. RNA polymerase II pausing can be retained or acquired during activation of genes involved in the epithelial to mesenchymal transition. Nucleic Acids Res. 2015;43(8):3938–49.

47. Tie F, Banerjee R, Stratton CA, Prasad-Sinha J, Stepanik V, Zlobin A, Diaz MO, Scacheri PC, Harte PJ. CBP-mediated acetylation of histone H3 lysine 27 antagonizes Drosophila Polycomb silencing. Development. 2009;136(18):3131–41.

48. Petruk S, Sedkov Y, Smith S, Tillib S, Kraevski V, Nakamura T, Canaani E, Croce CM, Mazo A. Trithorax and dCBP acting in a complex to maintain expression of a homeotic gene. Science. 2001;294(5545):1331–4.

49. Herz HM, Mohan M, Garrett AS, Miller C, Casto D, Zhang Y, Seidel C, Haug JS, Florens L, Washburn MP, Yamaguchi M, Shiekhattar R, Shilatifard A. Polycomb repressive complex 2-dependent and -independent functions of Jarid2 in transcriptional regulation in Drosophila. Mol Cell Biol. 2012;32(9):1683–93.

Web References

50. Trr; http://www.ncbi.nlm.nih.gov/protein/NP_726773.2. Accessed 17 June 2016.

51. SET1; http://www.ncbi.nlm.nih.gov/protein/NP_001015221.1. Accessed 17 June 2016.

52. Trx; http://www.ncbi.nlm.nih.gov/protein/AAA29025.1. Accessed 17 June 2016.

Hypomethylated domain-enriched DNA motifs prepattern the accessible nucleosome organization in teleosts

Ryohei Nakamura[1], Ayako Uno[1], Masahiko Kumagai[1], Shinichi Morishita[2] and Hiroyuki Takeda[1*]

Abstract

Background: Gene promoters in vertebrate genomes show distinct chromatin features such as stably positioned nucleosome array and DNA hypomethylation. The nucleosomes are known to have certain sequence preferences, and the prediction of nucleosome positioning from DNA sequence has been successful in some organisms such as yeast. However, at gene promoters where nucleosomes are much more stably positioned than in other regions, the sequence-based model has failed to work well, and sequence-independent mechanisms have been proposed.

Results: Using DNase I-seq in medaka embryos, we demonstrated that hypomethylated domains (HMDs) specifically possess accessible nucleosome organization with longer linkers, and we reassessed the DNA sequence preference for nucleosome positioning in these specific regions. Remarkably, we found with a supervised machine learning algorithm, k-mer SVM, that nucleosome positioning in HMDs is accurately predictable from DNA sequence alone. Specific short sequences (6-mers) that contribute to the prediction are specifically enriched in HMDs and distribute periodically with approximately 200-bp intervals which prepattern the position of accessible linkers. Surprisingly, the sequence preference of the nucleosome and linker in HMDs is opposite from that reported previously. Furthermore, the periodicity of specific motifs at hypomethylated promoters was conserved in zebrafish.

Conclusion: This study reveals strong link between nucleosome positioning and DNA sequence at vertebrate promoters, and we propose hypomethylated DNA-specific regulation of nucleosome positioning.

Keywords: Nucleosome positioning, DNA methylation, DNA sequence, Vertebrate

Background

Eukaryotic genomes are organized into chromatin, a DNA–protein complex, together with epigenetic information such as nucleosome position, histone modification, and DNA methylation. A nucleosome is a basic packaging unit of chromatin consisting of 147 base pairs (bp) DNA wrapped around a histone octamer [1]. Positioning of nucleosomes affects accessibility of regulatory proteins to DNA and thereby influences gene transcription [2]. Histone modification and DNA methylation also play critical roles in transcriptional regulation, and regulatory DNA regions such as promoters and enhancers are characterized by specific histone modifications, DNA hypomethylation, and accessible nucleosome organization [3–7].

Using next generation sequencing techniques, many studies have attempted to identify the basic principle for nucleosome positioning and have found that nucleosomes have DNA sequence preference. For example, nucleosome formation tends to occur at 10-bp periodic repeat of AT/TA dinucleotides and also GC-rich sequences, whereas poly(dA:dT) sequences tend to evict nucleosomes and thus reside in the linker region [8, 9]. Indeed, a periodic DNA sequence pattern associated with nucleosome has been found in genomes [10]. Furthermore, genome-wide nucleosome mapping in yeast and *C. elegans* revealed that the position of nucleosomes on the genome is accurately predictable from

*Correspondence: htakeda@bs.s.u-tokyo.ac.jp
[1] Department of Biological Sciences, Graduate School of Science, The University of Tokyo, 7-3-1 Hongo, Bunkyo-ku, Tokyo 113-0033, Japan
Full list of author information is available at the end of the article

DNA sequences [11], suggesting a certain dependency of nucleosome positioning on local DNA sequences in these organisms. However, in more complex organisms such as vertebrates the prediction from DNA sequence has not been successful [12, 13]. These facts suggest that the sequence dependency of nucleosome positioning varies among species.

The promoter region is unique in the genome, because nucleosomes at gene promoters are known to be stably positioned and strongly phased, which is one of the widely conserved features of nucleosome organization in eukaryotes including vertebrates [13–18]. In spite of these characteristics, the prediction of nucleosome positions in promoter regions from DNA sequence has not been successful even in yeast [11–13], suggesting that nucleosome positioning in promoter regions relies on other rules. Indeed, a transacting factor-mediated mechanism has been proposed in the promoter region [8, 18, 19]. One exception reported so far is tetrahymena, in which nucleosome positioning downstream of TSSs coincides significantly with GC content [14]. However, the logic underlying nucleosome positioning at promoters remains elusive for other organisms.

In vertebrates, the majority of the genome is maintained methylated, and hypomethylated domains (HMDs) are predominantly found in the region around gene promoters [20]. HMDs are mostly enriched with specific histone modifications such as H3K4me and required for gene transcription [21–23]. Recent studies have utilized a supervised machine learning algorithm, the k-mer support vector machine (SVM), and showed that HMDs can be accurately predicted from DNA sequence alone in *Xenopus* embryos and that these HMD regions are highly enriched with specific k-mers [24]. Importantly, the link between epigenetic modifications and nucleosome positioning has been also reported [13, 18, 25, 26], and epigenetic modification could be one of the key factors which affect nucleosome positioning. Given that majority of gene promoters are overlapped with HMDs, vertebrate promoter regions are distinct from the rest of the genomic regions in terms of both epigenetic modification and DNA sequence composition. Thus, distinct mechanism for nucleosome positioning might exist in promoter regions.

Here, we investigated the nucleosome organization and the contribution of DNA sequences to nucleosome positioning in HMDs using the medaka (Japanese killifish). We found that the nucleosome linkers in HMDs are specifically accessible, and their positions can be precisely mapped using DNase I-seq in medaka embryos. The nucleosome linkers in HMDs are longer than typical ones in the methylated medaka genome, and the average nucleosome spacing changes sharply at the boundary

of HMDs (200 bp in HMDs and 180 bp in methylated regions). Unlike the previous notion, the nucleosome positioning within HMDs was found to be highly predictable from DNA sequence using k-mer SVM, suggesting that nucleosome positioning in HMDs depends significantly on its proximal linker sequence. Surprisingly, this sequence feature was opposite from the previously reported global sequence preference of nucleosome in yeast. Finally, the specific sequence occurrence in hypomethylated linkers was also observed in zebrafish, a distantly related teleost species. Taken together, we propose a novel epigenetic modification-dependent and sequence-based rule for nucleosome positioning at teleost promoters.

Results
HMD have specific nucleosome organization

We previously reported 15,145 HMDs containing at least 10 continuous low-methylated (methylation rate < 0.4) CpGs in the genome of medaka blastula embryos, and the majority (69%) of the HMDs are found in gene promoter regions [23]. To examine the nucleosome organization within the HMD, we made a map of accessible chromatin in the medaka blastula genome using DNase I-seq. DNase I preferentially digests accessible DNA, such as nucleosome linkers or nucleosome-depleted regions [27, 28]. By deep sequencing, 323 million reads generated by DNase I digestions were mapped to the medaka reference genome and 36,375 DNase I hypersensitive sites (DHSs) were identified using MACS2 software [29] by searching regions with significant enrichment (FDR $< 0.1\%$, fold enrichment > 5) of DNase I cleavage. As expected, DHSs were highly enriched in HMDs (Fig. 1a); 84.8% of HMDs contained at least one DHS, and 40.7% of DHSs are found in the HMD which constitutes only 3% of the blastula genome. Notably, the DNase I-seq pattern in HMDs showed the clear periodic pattern (Fig. 1b), suggesting that the DNase I cleavage pattern in the medaka blastula genome represents arrays of long and accessible nucleosome linkers that specifically exist in HMDs.

To examine if the periodic DNase I-seq pattern reflects the array of nucleosome linkers in HMDs and if the nucleosome linker length is specifically longer in HMDs than in methylated regions, we compared the periodic DNase I cleavage pattern with our previous MNase-seq data in medaka blastula embryos [30]. To clarify the difference in nucleosome organization between HMDs and methylated regions, the DNase I-seq peak summits that reside at the most end of the HMD were designated as the base position. As nucleosomes are known to show strong phasing especially downstream of TSSs [2, 30], we wanted to distinguish the change in nucleosome phasing at HMD boundaries from TSS-dependent phasing.

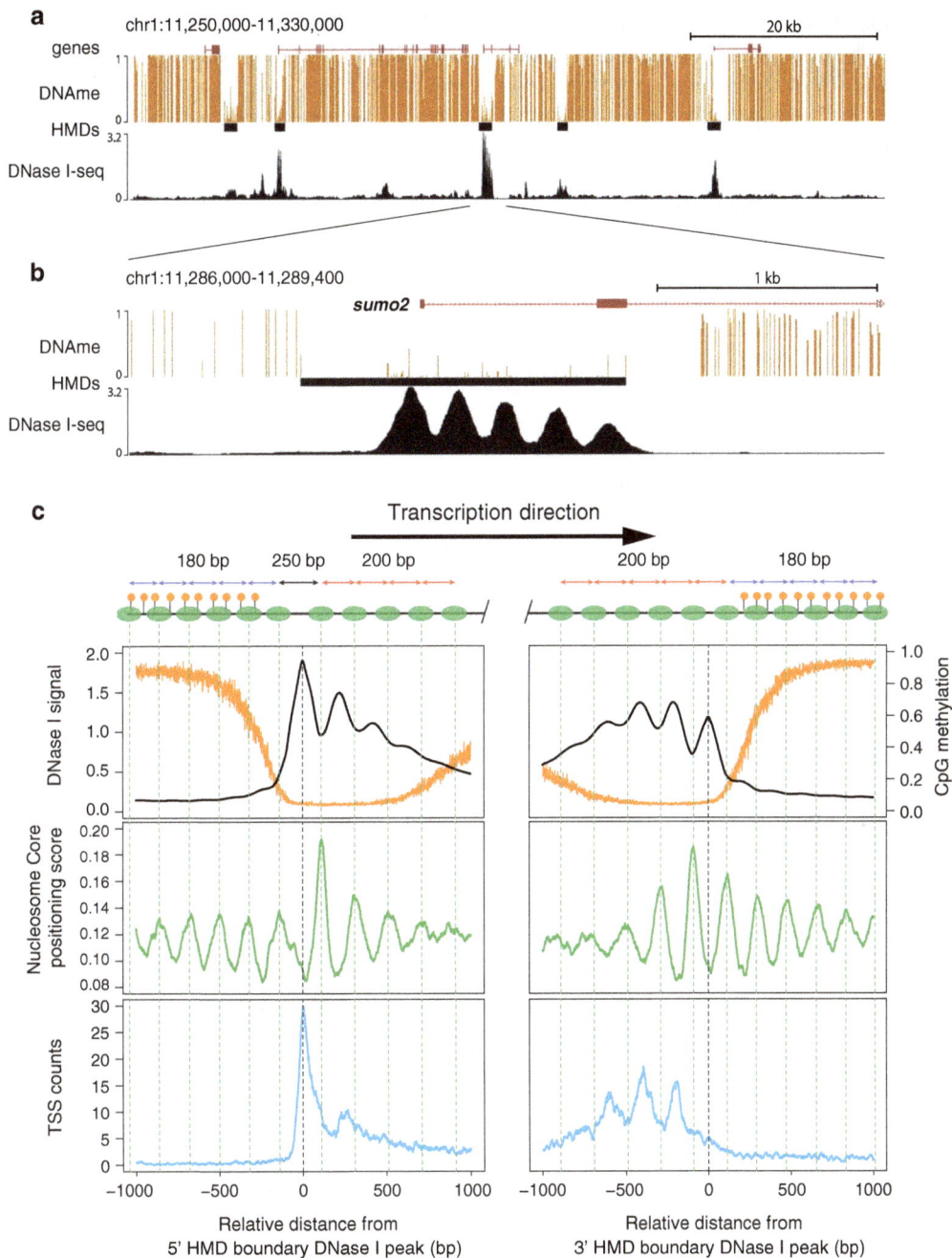

Fig. 1 DNase I-seq detects accessible nucleosome linkers within HMDs. **a** A representative genome browser view of DNA methylation, HMDs, and DNase I-seq pattern (signals per million reads) in medaka blastula embryos. Vertical line height of DNA methylation track indicates the ratio of methylated CpG. Black boxes represent HMDs. **b** A close-up view of single HMDs in (**a**). **c** Average profiles of DNase I-seq signal, DNA methylation, nucleosome core, and TSS counts around the accessible nucleosome linkers at the HMD boundaries. Vertical green dashed lines indicate the position of nucleosome core estimated from MNase-seq data. The top schema shows the position of nucleosomes (green ovals) and methylated CpGs (orange circle)

To this end, we oriented each HMD boundary by the direction of transcriptions from its nearest TSS (i.e., if the direction of the transcription was from the methylated side toward hypomethylated side, the boundary was classified as 5′ boundary, and 3′ boundary in the opposite case). First, we confirmed that the periodic pattern of DNase I-seq is inversely correlated with the nucleosome position estimated from the MNase-seq data (Fig. 1c; top and middle). In some cases, MNase-seq data could be affected by nonhistone DNA binding proteins [31]. However, we confirmed that the periodic patterns of DNase I-seq and MNase-seq are consistent with our previously published histone ChIP-seq pattern [23] (Additional file 1). These results indicate that the periodic DNase I-seq signals indeed reflected the nucleosome linkers in HMDs. Next, we observed that TSSs were most frequently found at the accessible linker located at the 5′ edge of HMDs, i.e., on the base position (Fig. 1c; bottom left). On the other hand, at 3′ boundary of HMDs, several peaks of TSS counts appear at linkers upstream of the base position (Fig. 1c; right bottom), probably reflecting TSSs at the 5′ boundary in short HMDs. Surprisingly, we found that the average spacing of nucleosomes changed clearly at the HMD boundary irrespective of the direction of transcription; in the methylated region, nucleosomes showed approximately 180-bp spacing, but in HMDs, the spacing was approximately 200 bp (Fig. 1c; Additional file 2). The spacing at 5′ HMD boundary was especially long (~250 bp), which is reminiscent of the fact that nucleosome-depleted region (NDR) exists at TSSs [2, 18]. Taken together, HMDs have distinct nucleosome organization, and our DNase I-seq data preferentially detect nucleosome linkers in HMDs that are longer (~200 bp) than typical linkers (~180 bp) in medaka embryos.

Prediction of nucleosome positioning by k-mer SVM

Since we precisely mapped the position of nucleosome linkers in each HMD, we then asked if specific DNA sequences can be correlated with the positioning of accessible linkers. k-mer-based DNA sequence analyses have been utilized to identify specific DNA elements [32]. We applied k-mer SVM, which finds a decision boundary that distinguishes the two sets of sequence data based on the frequency of all possible k-mers [33]. To discriminate linker sequences from nucleosome core sequences in HMDs, we extracted 100-bp sequences from DNase I peak summits in HMDs as positive (linker) data, and 100-bp sequences from the center regions between the two adjacent DNase I peak summits within HMDs for negative (core) data. Sequences on chromosome 8 were separated and used as test data, and the remaining sequences were used as training data. The performance of the k-mer SVM differed slightly between different k-mer length ($k = 2, 3$,

4, 5, 6, 7, 8), and we chose to use 6-mers for the further analyses, as this length produced high performance with minimized overfitting (Additional file 3). We refer to this trained SVM as SVM_{DNaseI}, as its purpose is to predict the DNase I-seq peaks (i.e., linkers) in HMDs. If nucleosome positioning in HMDs depends on specific DNA motifs, it should be predicted from DNA sequence. We calculated the SVM score for every 20 bp within HMDs on chromosome 8 and compared with DNase I-seq signal strength. Remarkably, SVM_{DNaseI} accurately predicted the DNase I pattern in HMDs (Fig. 2a left), and the correlation between the SVM_{DNaseI} score and actual DNase I-seq signal was significantly strong in each HMD (Fig. 2a right). This strong correlation was observed for the majority of HMDs on chromosome 8 (Fig. 2b). DNase I has been reported to have sequence preference [34–36], and thus the trained SVM might have been affected by this cleavage bias. In order to confirm that the SVM_{DNaseI} actually predicts nucleosome positioning, we performed ATAC-seq, an alternative method to map chromatin accessibility by Tn5 transposase [37], and compared with the SVM score. We found that the SVM_{DNaseI} score also showed significant correlation with ATAC-seq signal (Additional file 4). These results revealed that nucleosome positioning in HMDs is predictable from 6-mer distributions, suggesting that a sequence-based rule dominates in HMDs.

Specific 6-mers periodically distribute with 200-bp intervals in the linker regions of HMDs

The SVM outputs a weight for each k-mer which corresponds to the degree it contributes to the prediction [33] (Additional file 5). In this case, 6-mers with large positive weights were most frequently found in linker sequences, whereas those with large negative weights tended to be excluded from linkers but present in nucleosome core sequences. We noticed that the top positive 6-mers have larger absolute weights than top negative ones (Additional file 5), suggesting that a few number of specific 6-mers in linkers have strong contribution to nucleosome positioning. To test whether the high SVM-weight 6-mers appear periodically in a single HMD, we examined the distances between every pairs of top 10 high-weight 6-mers of SVM_{DNaseI} within HMDs. The histogram of all distances between the top 6-mer pairs showed clear enrichment at 200 bp (Fig. 2c), indicating

(See figure on next page.)

Fig. 2 Nucleosome positioning in HMDs is predictable by k-mer SVM. **a** Examples of prediction of nucleosome linkers (DNase I accessible regions) by k-mer SVM in HMDs on chromosome 8. Dark purple indicates the score higher than 0, light purple, lower than 0. Pearson's correlation and its P value between DNase I signal and SVM score for every 20 bp along the HMD are shown on the right. **b** A histogram of correlations for all HMDs on chromosome 8. Green and gray boxes represent the number of HMDs with and without significant correlation ($P < 0.05$), respectively. **c** A histogram of distances between top 10 SVM_{DNaseI}-weight 6-mers within HMDs. Distances shorter than 3 bp were excluded from the histogram

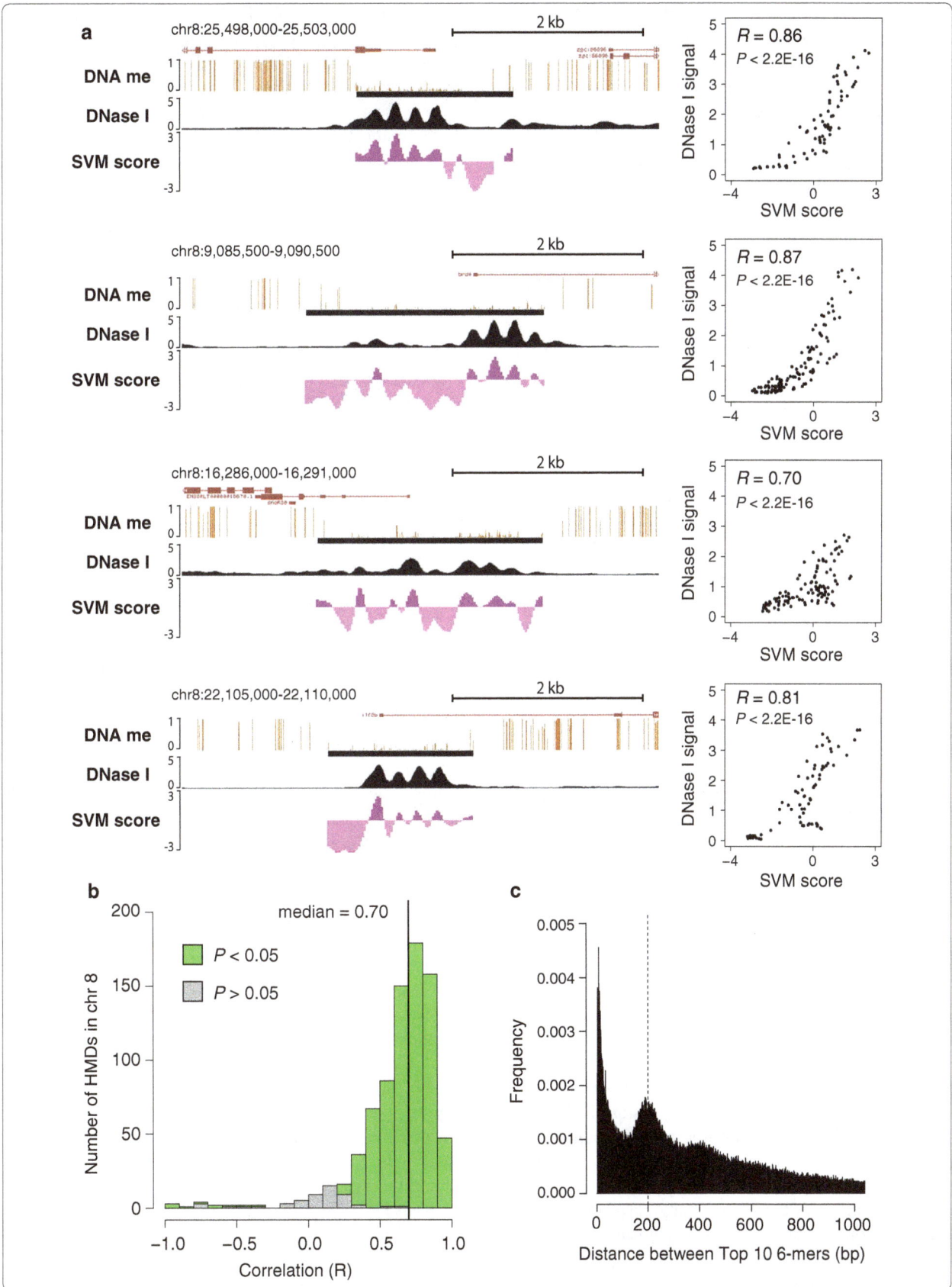

that those top 6-mers tend to distribute with approximately 200-bp intervals within a HMD.

We then examined the pattern of the SVM score around the HMD boundary and confirmed that the SVM score shows a periodical pattern with high levels at nucleosome linker regions specifically in HMDs (Fig. 3), suggesting that specific DNA motifs strongly contribute to the nucleosome positioning in HMDs. It is known that nucleosomes have specific sequence preference, poly(dA:dT) sequences for linkers and relatively GC-rich for nucleosome cores [8, 9]. Thus, the enrichment of specific 6-mers at the nucleosome linkers could be the result of distinct base compositions. To test this idea, we examined the distribution pattern of 6-mers with the highest SVM_{DNaseI}-weight (GCTAAC) and its reverse sequence (CAATCG) which is not reverse-complement but has the same base composition to highlight the importance of the base ordering in the motif. The highest SVM_{DNaseI}-weight 6-mer showed the clear periodic distribution pattern that is consistent with the position of linkers in HMDs, whereas the reverse sequence did not show such pattern (Fig. 3). These results suggest that specific DNA motifs, but not simple base composition, contribute to the formation of accessible nucleosome linkers. Furthermore, the SNP rate between the two closely related medaka species, Hd-rR and HNI [38–41], also showed a periodic pattern, indicating that nucleosome linker regions are highly conserved in HMDs (Fig. 3). This further suggests the importance of linker sequences. The eviction of nucleosomes from specific 6-mers could be caused by the binding of certain proteins to those specific sequences. However, the majority of high SVM-weight 6-mers do not show any similarity to known TF binding motifs (Additional file 6). Thus, intrinsic preference of the specific 6-mers for nucleosome linkers may exist in HMDs.

Previously, the global sequence preference of nucleosome has been proposed to predict in vivo genome-wide nucleosome positioning in yeast and *C. elegans* [11]. However, this model has limited performance when applied to human and zebrafish genomes [12, 13]. To test whether this model can be applied to the nucleosome positioning in HMDs, we calculated the Kaplan occupancy (expected nucleosome occupancy) around HMD boundaries. As shown in Fig. 3, the Kaplan occupancy showed the clear periodic pattern similar to that of the SVM score. This result was surprising, because the Kaplan occupancy is known to predict the nucleosome core position, but the SVM score correlates with the linker region in HMDs. Thus, nucleosomes in medaka HMDs have the sequence preference opposite to the global tendency in yeast.

Linker-specific 6-mers distribute preferentially in HMDs

We reasoned that the specific localization of high SVM-weight 6-mers is only observed in HMDs (Fig. 2) but not in the methylated region. To examine whether those 6-mers are actually enriched in HMDs, we trained *k*-mer SVM to discriminate HMD sequences from randomly selected methylated sequences and compared the contribution to the prediction of each 6-mers between HMDs and nucleosome linkers. We refer to this new trained SVM as SVM_{hypo}, as it is to predict the HMD. The performance of SVM_{hypo} was tested on HMDs and methylated sequences from chromosome 8, and the prediction quality was measured by calculating the area under the ROC curve (ROCauc). Consistent with the previous study [24], the SVM_{hypo} was able to distinguish HMD sequences from methylated sequences with high accuracy (Fig. 4a, b). Furthermore, we also measured 'precision and recall,' as it is a more reliable measure when positive and negative datasets are of unequal size. The precision–recall curve revealed that the SVM_{hypo} can distinguish HMD sequences from a 10× excess of methylated sequences (Fig. 4b). These results demonstrate that HMDs in blastula embryos are specifically enriched with a certain set of 6-mers. Intriguingly, the comparison between the SVM-weight of each 6-mer by SVM_{hypo} and SVM_{DNaseI} demonstrated that high SVM_{DNaseI}-weight 6-mers tended to have high SVM_{hypo} weight (i.e., the top 20 SVM_{DNaseI}-weight 6-mers had significantly high SVM_{hypo}-weights) (Fig. 4c). Thus, the 6-mers that contribute to the prediction of nucleosome linkers are preferentially distributed in HMDs and much less frequently present in methylated regions. This suggests that the sequence-based rule we propose is specific to the HMD, but should not be applicable to the methylated genomic region which constitutes the majority of the genome.

Taken together, accessible nucleosome organization in HMDs might uniquely depend on DNA sequence, which is directed by specific short sequences preferentially distributed with approximately 200-bp intervals in HMDs, longer than those in methylated regions (~180 bp) (Fig. 4d).

Similar sequence preference of nucleosome positioning in zebrafish HMDs

Finally, we tested whether the unique sequence preference of nucleosomes in HMDs also exists in other vertebrate species. We examined the sequence features of nucleosome core and linker regions in zebrafish by investigating the SVM_{DNaseI} score, together with the published data of methylome [42] and MNase-seq data in zebrafish embryos [13]. We applied the SVM_{DNaseI} trained with the medaka dataset to the zebrafish genome. As the DNase I-seq data were not available for blastula-stage zebrafish embryos, we were unable to determine the position of

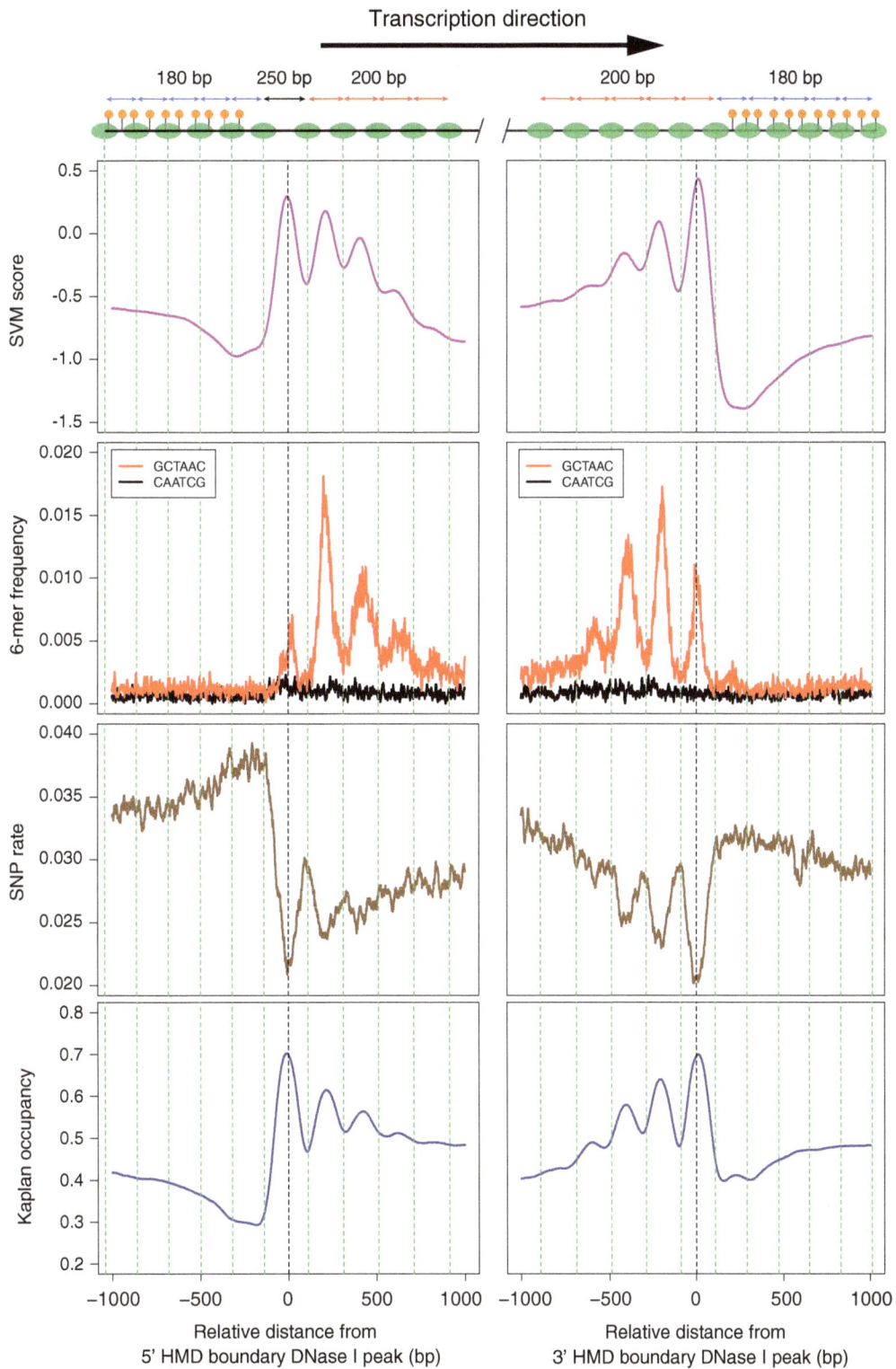

Fig. 3 Specific DNA 6-mers are enriched in accessible linkers. Average profiles of the SVM score, distribution frequency of 6-mers with highest SVM-weight (GCTAAC) and its reverse sequence (CAATCG), SNP rate, and Kaplan occupancy around the HMD boundaries. Vertical green dashed lines indicate the position of nucleosome core, and the top schema shows the position of nucleosomes (green ovals) and methylated CpGs (orange circle) (same as Fig. 1c)

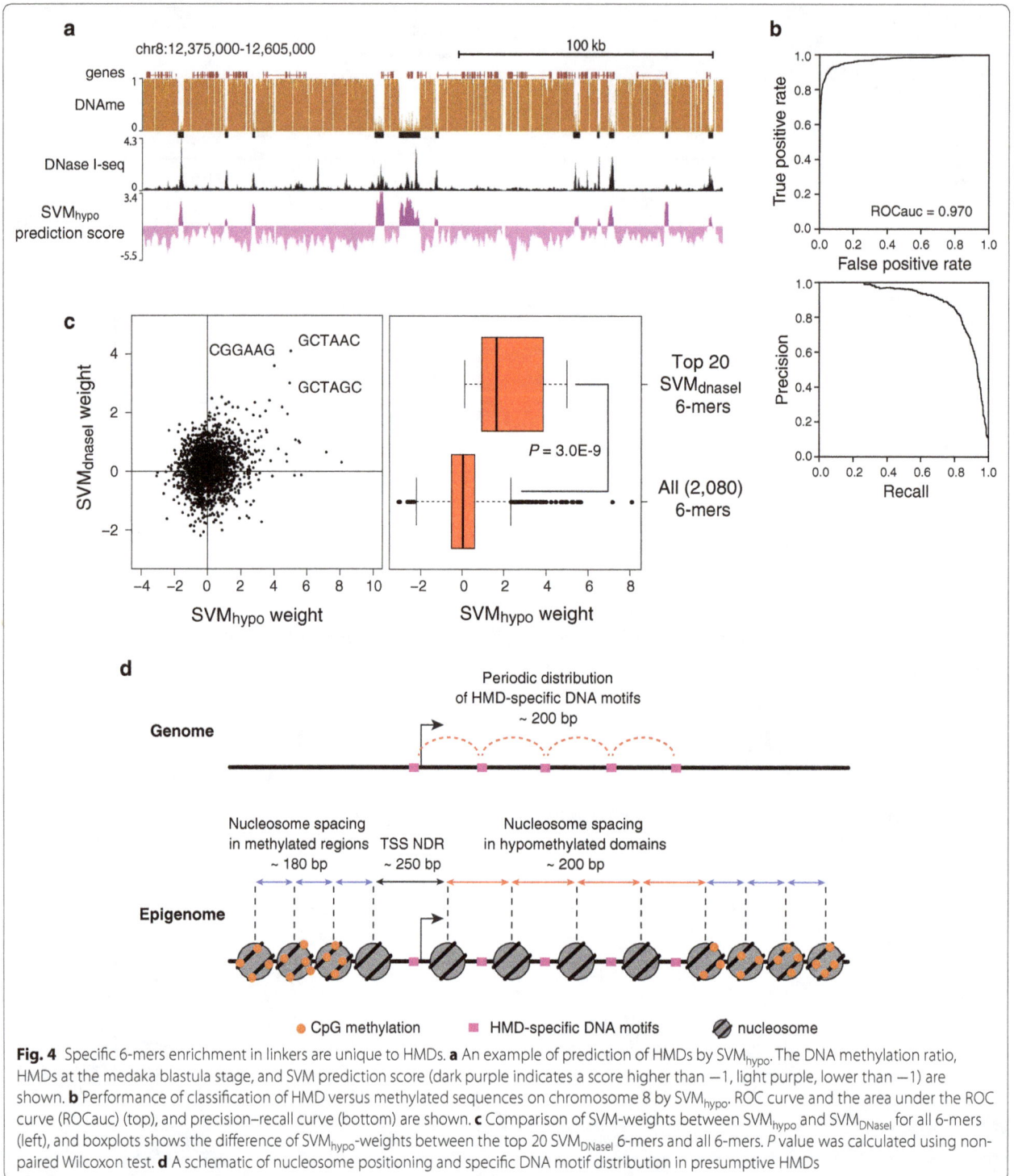

Fig. 4 Specific 6-mers enrichment in linkers are unique to HMDs. **a** An example of prediction of HMDs by SVM_{hypo}. The DNA methylation ratio, HMDs at the medaka blastula stage, and SVM prediction score (dark purple indicates a score higher than −1, light purple, lower than −1) are shown. **b** Performance of classification of HMD versus methylated sequences on chromosome 8 by SVM_{hypo}. ROC curve and the area under the ROC curve (ROCauc) (top), and precision–recall curve (bottom) are shown. **c** Comparison of SVM-weights between SVM_{hypo} and SVM_{DNaseI} for all 6-mers (left), and boxplots shows the difference of SVM_{hypo}-weights between the top 20 SVM_{DNaseI} 6-mers and all 6-mers. *P* value was calculated using non-paired Wilcoxon test. **d** A schematic of nucleosome positioning and specific DNA motif distribution in presumptive HMDs

accessible linker at zebrafish HMD boundaries like we did in medaka analyses. We therefore investigated the nucleosome pattern and SVM_{DNaseI} score only around the TSSs in HMDs and methylated regions. We found that in both medaka and zebrafish, nucleosome positions are phased and positioned around the TSSs that reside in HMDs, and that the SVM_{DNaseI} score was periodically high at linker regions (Fig. 5a, b left). By contrast, such periodicity was not observed for both nucleosome and SVM score around the methylated TSSs (Fig. 5a, b right).

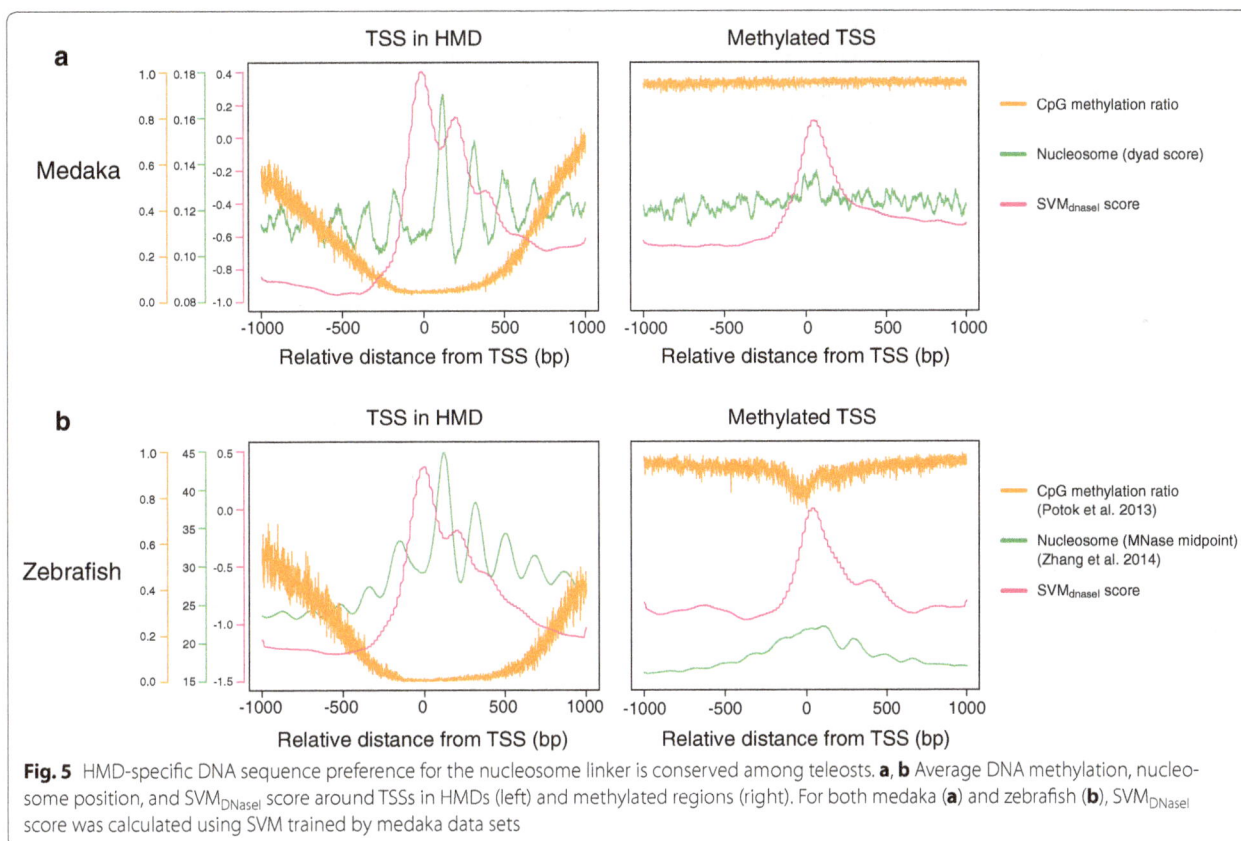

Fig. 5 HMD-specific DNA sequence preference for the nucleosome linker is conserved among teleosts. **a**, **b** Average DNA methylation, nucleosome position, and SVM$_{DNaseI}$ score around TSSs in HMDs (left) and methylated regions (right). For both medaka (**a**) and zebrafish (**b**), SVM$_{DNaseI}$ score was calculated using SVM trained by medaka data sets

These results suggest that the specific 6-mers occurrence at nucleosome linkers in the HMD is conserved between the two distantly related teleost species.

Discussion

Thus far, prediction of the nucleosome position on the basis of DNA sequence has not been successful in vertebrate genomes, in particular, gene promoter regions. In vertebrates, most gene promoters reside in HMDs, and in the present study, we reassessed the DNA sequence preference for nucleosome positioning in these specific regions. DNase I-seq was recently applied to genome-wide mapping of nucleosome positions in yeast and human [27], but in medaka embryos, DNase I was found to preferentially digest long linker DNA in HMDs. This feature allowed us to unveil the clear transition in nucleosome spacing length at the HMD boundary; from closed (180-bp interval in methylated) to open (200-bp in HMD) nucleosome organization. More importantly, with this precise map of linkers in HMDs, we identified the novel sequence-based rule that allows us to accurately predict the positions of nucleosomes in vertebrate HMDs harboring gene promoters. The 200-bp periodic occurrence of the predictable 6-mers accounts for longer spacing of nucleosomes in HMDs, and thereby promoters in

HMDs could maintain accessibility to regulatory proteins (Fig. 4d). In general, the majority of hypomethylated promoters persist throughout cell differentiation and sustain gene expression of housekeeping genes and early developmental genes [7, 43, 44]. Thus, DNA sequence directed long nucleosome linkers could contribute to their transcriptional regulation by constitutively maintaining accessible nucleosome organizations at those promoters. On the other hand, cell-type specifically hypomethylated promoters may not depend on the predictable 6-mers identified in blastula embryos, as they are activated by cell-type-specific transcription factors and epigenetic modifications. Notably, this HMD-specific rule was at least conserved among teleosts, as the similar tendency was observed in zebrafish which is evolutionarily long diverged from medaka. However, this rule holds true only in HMDs, and the co-occurrence of the specific 6-mers and nucleosome linker is not observed in methylated regions. Since the HMD constitutes only 3% of the entire genome, despite its crucial role in gene regulation, the HMD-specific rule could have been overlooked in previous genome-wide analyses.

The strong phasing of nucleosome positions downstream of TSSs is widely conserved among eukaryote genomes, but the degree of sequence contribution to the nucleosome positioning varies among species [11, 13, 14]. Surprisingly, the novel HMD-specific rule in medaka clearly contradicts the global sequence preference previously reported (poly(dA:dT) for linkers and GC-rich for nucleosome cores) [11]; the predictable short sequences enriched in medaka HMD linkers are relatively GC-rich, and the Kaplan occupancy, which was originally used to predict the global nucleosome occupancy in yeast, exhibits the opposite tendency in HMDs. At the moment, the reason for the reverse sequence preference of nucleosomes and the function of high SVM_{DNaseI}-weight 6-mers remain speculative. Those 6-mers may be intrinsically unfavorable for nucleosome formation in HMDs, although we cannot rule out the possibility that unknown proteins bind to those 6-mers and influence nucleosome positioning. To examine whether the 6-mers alone can direct nucleosome positioning, it would be informative to perform in vitro reconstitution of chromatin from histone octamers and naked medaka genomic DNA. Importantly, however, it has been reported in zebrafish that the strongly phased nucleosome array at gene promoters does not exist in early embryos, but appears during the zygotic genome activation (ZGA) stage, correlating with the emergence of H3K4me3, a histone modification specific to HMDs [13, 21, 45, 46]. This indicates that nucleosome positioning in promoter regions is not solely determined by DNA sequence, but may require specific chromatin environment (e.g., modifications such as H3K4me3 or binding of chromatin factors which function at the ZGA stage). Therefore, it is likely that the specific epigenetic environment override the normal sequence preference of nucleosome in HMDs.

Conclusion

In summary, although the molecular mechanisms by which identified short sequences are translated into nucleosome positioning remain elusive, the present study focusing on the HMD provides novel insights into a hypomethylated DNA-specific regulation of nucleosome positioning in the vertebrate genome.

Materials and methods
Fish strains

We used medaka d-rR strain as wild type. Medaka fishes were maintained and raised under standard condition.

DNase I-seq

DNase I-seq was performed as previously described [47] with modifications. 5000 d-rR strain medaka blastula embryos were dechorionated and dissociated by forcing the embryos through a 21G needle using a syringe, and cells were harvested by centrifugation at 500g for 5 min. After washing with PBS, cells were resuspended in 500 μl

of buffer A [15 mM Tris–HCl (pH 8.0), 15 mM NaCl, 60 mM KCl, 1 mM EDTA, 0.5 mM EGTA, 1 mM PMSF]. Cells were isolated using a cell-strainer (Falcon, 352235), centrifuged at 500g for 10 min, and resuspended in 1.5 ml of lysis buffer [buffer A with 0.1% IGEPAL CA-630]. After a 1-min incubation at 4 °C, nuclei were collected by centrifugation at 500g for 10 min. Nuclei were washed in buffer A, then resuspended in nuclear storage buffer [20 mM Tris–HCl pH 8.0, 75 mM NaCl, 0.5 mM EDTA, 50% (v/v) glycerol, 1 mM DTT, and 0.1 mM PMSF], and stored at −80 °C. For DNase I digestion, frozen nuclei were thawed on ice, washed in buffer A with 0.5 mM spermidine and 0.3 mM spermine, incubated for exactly 2 min at 37 °C in 3.5 ml of buffer D [1 volume of 10× DNase I digestion buffer with 9 volume of Buffer A] containing 480 U of DNase I. The reaction was stopped by adding stop buffer [50 mM Tris–HCl, pH 8.0, 100 mM NaCl, 0.1% SDS, 100 mM EDTA, 20 μg/ml RNase A, 0.5 mM spermidine, and 0.3 mM spermine] and proteinase K, and incubated at 55 °C overnight. Digested DNA was purified by phenol chloroform, sucrose fractionated, and fragments below 1 kb were collected, end-repaired, ligated with adapters compatible with the Illumina sequencing platform and sequenced as single-end tags on HiSeq 1500 platform (Illumina).

ATAC-seq

ATAC-seq was performed as previously described [37] with some modifications. Embryos were homogenized in PBS, and cells were harvested by centrifugation at 500g for 5 min. Approximately 5000 cells were used. After washing with PBS, cells were resuspended in 500 μl of cold lysis buffer [10 mM Tris–HCl pH7.4, 10 mM NaCl, 3 mM $MgCl_2$, 0.1% Igepal CA-630], centrifuged for 10 min at 500g, and supernatant was removed. Tagmentation reaction was performed as described previously [37] with Nextera Sample Preparation Kit (Illumina). After DNA was purified using MinElute kit (Qiagen), two sequential PCR were performed to enrich small DNA fragments. First, 9-cycle PCR were performed using indexed primers from Nextera Index Kit (Illumina) and KAPA HiFi HS ReadyMix, and amplified DNA was size selected (less than 500 bp) using AMPure XP beads. Then, a second 7-cycle PCR were performed using the same primer as the first PCR, and purified by AMPure XP beads. Libraries were sequenced using the Illumina HiSeq 1500 platform.

DNase I-seq and ATAC-seq data processing

The sequenced tags were aligned to the medaka reference genome by BWA [48], and tags with mapping quality larger than 20 were used for further analyses. Before the peak detection, each read position was shifted toward 5′

side with 50 bp. Then, MACS2 (version 2.0.10.20120913) [29] was used to identify regions that are significantly enriched (FDR 0.1%, fold enrichment > 5) with sequence tags (DHSs) using following options: –keep-dup all –nomodel –shiftsize 50 –q 0.01 –nolambda –call-summits –B –SPMR. For visualization and further analyses, signals per million reads data produced by MACS2 were used.

Nucleosome organization and sequence profiles around HMD boundaries

We used HMDs identified in the previous study [23]. To calculate the average chromatin profile around HMD boundaries, we needed to set the base position (position $x = 0$) in each HMD. For this, we first determined DNase I-seq peak summits that locate within HMDs and selected the summit nearest to the boundary (the first low-methylated CpG in the HMD) as the base position. Then, the boundaries were classified by the orientation relative to the direction of transcription from the nearest TSS. HMDs that have TSSs within 1 kb distance were used for this analysis. Kaplan nucleosome occupancy was calculated using the model previously reported [11]. SNP rate was calculated using the genome sequences of two medaka species, Hd-rR and HNI. SNPs identified in previous study [41] were used.

To estimate the average spacing of nucleosomes, we calculated the autocorrelation of nucleosome dyad score using acf function of R. The autocorrelation in HMDs and methylated regions were calculated using the average nucleosome dyad score (Fig. 1c, middle) at position $x = 0, ..., 1000$ and $x = -1000, ..., -100$, respectively, where $x = 0$ is the position of boundary DNase I-seq summit.

SVM for nucleosome linker and HMD prediction

We used the previously described method [33] for k-mer SVM. For training of $\mathrm{SVM_{DNaseI}}$ we first selected DNase I-seq peak summits within HMDs as the center of accessible nucleosome linkers. Then, we selected the center of two adjacent DNase I-seq peak summits as the nucleosome core position if the distance between the two summits was longer than 150 bp. From the linker (positive) and the core (negative) regions, 100-bp sequences were extracted. The sequences not on chromosome 8 were used for training, and those on chromosome 8 were used as test data to draw ROC and precision–recall curve. To test the performance of the prediction of DNase I pattern, we calculated the SVM score for each of the HMDs in chromosome 8 with sliding window of 100 bp with a step of 20 bp. Then, at each step, the average of overlapping windows was calculated. The correlation between

the average SVM score and DNase I signal level was calculated for each HMDs.

For the positive data set of $\mathrm{SVM_{DNA}}$, all HMD sequences below 3 kb were used. For the negative data set, ten copies of original HMD genome-coordinate set were randomly distributed on methylated regions using bedtools. HMD sequences and methylated sequences not on chromosome 8 were used for training, and those on chromosome 8 were used as test data (for Fig. 4b, c).

Motif analyses

TOMTOM [49] was used to search motifs similar to 6-mers. JASPAR Vertebrates and UniPROBE Mouse databases were used as target motifs.

Additional files

Additional file 1. The comparison between DNase I-seq pattern and histone ChIP-seq pattern. Average profiles of DNase I-seq signal (black), DNA methylation (orange), and H3K27ac ChIP-seq signal (blue) around the accessible nucleosome linkers at the HMD boundaries. Vertical green dashed lines indicate the position of nucleosome core estimated from MNase-seq data (see Fig. 1c). The top schema shows the position of nucleosomes (green ovals) and methylated CpGs (orange circle).

Additional file 2. The average nucleosome spacing in HMDs and methylated regions. The autocorrelation in HMDs and methylated regions were calculated using the average nucleosome dyad score (Fig. 1c, middle) for both 5′ (left) and 3′ (right) boundary regions.

Additional file 3. The performance of k-mer SVM for different k-mer length. ROC curve and the area under the ROC curve (auc) are shown for different k-mer length ($k = 2, 3, 4, 5, 6, 7, 8$).

Additional file 4. Validation of the performance of $\mathrm{SVM_{DNaseI}}$ by ATAC-seq. (A) An example of prediction of nucleosome linkers (DNase I accessible regions) by $\mathrm{SVM_{DNaseI}}$ in HMDs on chromosome 8. Dark purple indicates the score higher than 0, light purple, lower than 0. Pearson's correlation and its P value between ATAC-seq signal and $\mathrm{SVM_{DNaseI}}$ score for every 20 bp along the HMD are shown on the right. (B) A histogram of correlations between ATAC-seq signal and $\mathrm{SVM_{DNaseI}}$ score for all HMDs on chromosome 8. Blue and gray boxes represent the number of HMDs with and without significant correlation ($P < 0.05$), respectively.

Additional file 5. $\mathrm{SVM_{DNaseI}}$-weights for all 6-mers. All 6-mers are listed with SVM-weight.

Additional file 6. Known TF binding motifs similar to top 20 $\mathrm{SVM_{DNaseI}}$ 6-mers. Top 20 6-mers are listed with known TF motifs.

Authors' contributions

RN performed the experiments, analyzed data, and drafted the manuscript. HT and SM supervised the research and carried out revisions of the manuscript for important intellectual content. AU calculated SNP rate between Hd-rR and HNI strains. MK conducted sequencing of DNase I-seq. All authors read and approved the final manuscript.

Author details

[1] Department of Biological Sciences, Graduate School of Science, The University of Tokyo, 7-3-1 Hongo, Bunkyo-ku, Tokyo 113-0033, Japan. [2] Department of Computational Biology and Medical Sciences, Graduate School of Frontier Sciences, The University of Tokyo, 5-1-5 Kashiwanoha, Kashiwa 277-8562, Japan.

Acknowledgements
We thank Andrew Fire for critical reading of the manuscript.

Competing interests
The authors declare that they have no competing interests.

Funding
This research was supported by the Core Research for Evolutional Science and Technology (CREST) program of the Japan Science and Technology Agency (JST).

References

1. Richmond TJ, Davey CA. The structure of DNA in the nucleosome core. Nature. 2003;423(6936):145–50.
2. Jiang C, Pugh BF. Nucleosome positioning and gene regulation: advances through genomics. Nat Rev Genet. 2009;10(3):161–72.
3. Heintzman ND, Stuart RK, Hon G, Fu Y, Ching CW, Hawkins RD, Barrera LO, Van Calcar S, Qu C, Ching KA, et al. Distinct and predictive chromatin signatures of transcriptional promoters and enhancers in the human genome. Nat Genet. 2007;39(3):311–8.
4. Thurman RE, Rynes E, Humbert R, Vierstra J, Maurano MT, Haugen E, Sheffield NC, Stergachis AB, Wang H, Vernot B, et al. The accessible chromatin landscape of the human genome. Nature. 2012;489(7414):75–82.
5. Bird A. DNA methylation patterns and epigenetic memory. Genes Dev. 2002;16(1):6–21.
6. Rada-Iglesias A, Bajpai R, Swigut T, Brugmann SA, Flynn RA, Wysocka J. A unique chromatin signature uncovers early developmental enhancers in humans. Nature. 2011;470(7333):279–83.
7. Jones PA. Functions of DNA methylation: islands, start sites, gene bodies and beyond. Nat Rev Genet. 2012;13(7):484–92.
8. Struhl K, Segal E. Determinants of nucleosome positioning. Nat Struct Mol Biol. 2013;20(3):267–73.
9. Tillo D, Hughes TR. G + C content dominates intrinsic nucleosome occupancy. BMC Bioinform. 2009;10:442.
10. Knoch TA, Goker M, Lohner R, Abuseiris A, Grosveld FG. Fine-structured multi-scaling long-range correlations in completely sequenced genomes–features, origin, and classification. Eur Biophys J. 2009;38(6):757–79.
11. Kaplan N, Moore IK, Fondufe-Mittendorf Y, Gossett AJ, Tillo D, Field Y, LeProust EM, Hughes TR, Lieb JD, Widom J, et al. The DNA-encoded nucleosome organization of a eukaryotic genome. Nature. 2009;458(7236):362–6.
12. Gaffney DJ, McVicker G, Pai AA, Fondufe-Mittendorf YN, Lewellen N, Michelini K, Widom J, Gilad Y, Pritchard JK. Controls of nucleosome positioning in the human genome. PLoS Genet. 2012;8(11):e1003036.
13. Zhang Y, Vastenhouw NL, Feng J, Fu K, Wang C, Ge Y, Pauli A, van Hummelen P, Schier AF, Liu XS. Canonical nucleosome organization at promoters forms during genome activation. Genome Res. 2014;24(2):260–6.
14. Beh LY, Muller MM, Muir TW, Kaplan N, Landweber LF. DNA-guided establishment of nucleosome patterns within coding regions of a eukaryotic genome. Genome Res. 2015;25(11):1727–38.
15. Mavrich TN, Jiang C, Ioshikhes IP, Li X, Venters BJ, Zanton SJ, Tomsho LP, Qi J, Glaser RL, Schuster SC, et al. Nucleosome organization in the *Drosophila* genome. Nature. 2008;453(7193):358–62.
16. Saito TL, Hashimoto S, Gu SG, Morton JJ, Stadler M, Blumenthal T, Fire A, Morishita S. The transcription start site landscape of *C. elegans*. Genome Res. 2013;23(8):1348–61.
17. Nakatani Y, Mello CC, Hashimoto S, Shimada A, Nakamura R, Tsukahara T, Qu W, Yoshimura J, Suzuki Y, Sugano S, et al. Associations between nucleosome phasing, sequence asymmetry, and tissue-specific expression in a set of inbred Medaka species. BMC Genom. 2015;16:978.
18. Valouev A, Johnson SM, Boyd SD, Smith CL, Fire AZ, Sidow A. Determinants of nucleosome organization in primary human cells. Nature. 2011;474(7352):516–20.
19. Mavrich TN, Ioshikhes IP, Venters BJ, Jiang C, Tomsho LP, Qi J, Schuster SC, Albert I, Pugh BF. A barrier nucleosome model for statistical positioning of nucleosomes throughout the yeast genome. Genome Res. 2008;18(7):1073–83.
20. Suzuki MM, Bird A. DNA methylation landscapes: provocative insights from epigenomics. Nat Rev Genet. 2008;9(6):465–76.
21. Cedar H, Bergman Y. Linking DNA methylation and histone modification: patterns and paradigms. Nat Rev Genet. 2009;10(5):295–304.
22. Zhou VW, Goren A, Bernstein BE. Charting histone modifications and the functional organization of mammalian genomes. Nat Rev Genet. 2011;12(1):7–18.
23. Nakamura R, Tsukahara T, Qu W, Ichikawa K, Otsuka T, Ogoshi K, Saito TL, Matsushima K, Sugano S, Hashimoto S, et al. Large hypomethylated domains serve as strong repressive machinery for key developmental genes in vertebrates. Development. 2014;141(13):2568–80.
24. van Heeringen SJ, Akkers RC, van Kruijsbergen I, Arif MA, Hanssen LL, Sharifi N, Veenstra GJ. Principles of nucleation of H3K27 methylation during embryonic development. Genome Res. 2014;24(3):401–10.
25. Chodavarapu RK, Feng S, Bernatavichute YV, Chen PY, Stroud H, Yu Y, Hetzel JA, Kuo F, Kim J, Cokus SJ, et al. Relationship between nucleosome positioning and DNA methylation. Nature. 2010;466(7304):388–92.
26. Huff JT, Zilberman D. Dnmt1-independent CG methylation contributes to nucleosome positioning in diverse eukaryotes. Cell. 2014;156(6):1286–97.
27. Zhong J, Luo K, Winter PS, Crawford GE, Iversen ES, Hartemink AJ. Mapping nucleosome positions using DNase-seq. Genome Res. 2016;26(3):351–64.
28. Neph S, Vierstra J, Stergachis AB, Reynolds AP, Haugen E, Vernot B, Thurman RE, John S, Sandstrom R, Johnson AK, et al. An expansive human regulatory lexicon encoded in transcription factor footprints. Nature. 2012;489(7414):83–90.
29. Zhang Y, Liu T, Meyer CA, Eeckhoute J, Johnson DS, Bernstein BE, Nusbaum C, Myers RM, Brown M, Li W, et al. Model-based analysis of ChIP-Seq (MACS). Genome Biol. 2008;9(9):R137.
30. Sasaki S, Mello CC, Shimada A, Nakatani Y, Hashimoto S, Ogawa M, Matsushima K, Gu SG, Kasahara M, Ahsan B, et al. Chromatin-associated periodicity in genetic variation downstream of transcriptional start sites. Science. 2009;323(5912):401–4.
31. Chereji RV, Ocampo J, Clark DJ. MNase-sensitive complexes in yeast: nucleosomes and non-histone barriers. Mol Cell. 2017;65(3):565–77 **(e563)**.
32. Sievers A, Bosiek K, Bisch M, Dreessen C, Riedel J, Fross P, Hausmann M, Hildenbrand G. K-mer content, correlation, and position analysis of genome DNA sequences for the identification of function and evolutionary features. Genes (Basel). 2017;8(4):122.
33. Lee D, Karchin R, Beer MA. Discriminative prediction of mammalian enhancers from DNA sequence. Genome Res. 2011;21(12):2167–80.
34. He HH, Meyer CA, Hu SS, Chen MW, Zang C, Liu Y, Rao PK, Fei T, Xu H, Long H, et al. Refined DNase-seq protocol and data analysis reveals intrinsic bias in transcription factor footprint identification. Nat Methods. 2014;11(1):73–8.
35. Koohy H, Down TA, Hubbard TJ. Chromatin accessibility data sets show bias due to sequence specificity of the DNase I enzyme. PLoS ONE. 2013;8(7):e69853.
36. Lazarovici A, Zhou T, Shafer A, Dantas Machado AC, Riley TR, Sandstrom R, Sabo PJ, Lu Y, Rohs R, Stamatoyannopoulos JA, et al. Probing DNA shape and methylation state on a genomic scale with DNase I. Proc Natl Acad Sci U S A. 2013;110(16):6376–81.
37. Buenrostro JD, Giresi PG, Zaba LC, Chang HY, Greenleaf WJ. Transposition of native chromatin for fast and sensitive epigenomic profiling of open chromatin, DNA-binding proteins and nucleosome position. Nat Methods. 2013;10(12):1213–8.

38. Kasahara M, Naruse K, Sasaki S, Nakatani Y, Qu W, Ahsan B, Yamada T, Nagayasu Y, Doi K, Kasai Y, et al. The medaka draft genome and insights into vertebrate genome evolution. Nature. 2007;447(7145):714–9.
39. Setiamarga DH, Miya M, Yamanoue Y, Azuma Y, Inoue JG, Ishiguro NB, Mabuchi K, Nishida M. Divergence time of the two regional medaka populations in Japan as a new time scale for comparative genomics of vertebrates. Biol Lett. 2009;5(6):812–6.
40. Takeda H, Shimada A. The art of medaka genetics and genomics: what makes them so unique? Annu Rev Genet. 2010;44:217–41.
41. Uno A, Nakamura R, Tsukahara T, Qu W, Sugano S, Suzuki Y, Morishita S, Takeda H. Comparative analysis of genome and epigenome in closely related Medaka species identifies conserved sequence preferences for DNA hypomethylated domains. Zool Sci. 2016;33(4):358–65.
42. Potok ME, Nix DA, Parnell TJ, Cairns BR. Reprogramming the maternal zebrafish genome after fertilization to match the paternal methylation pattern. Cell. 2013;153(4):759–72.
43. Weber M, Hellmann I, Stadler MB, Ramos L, Paabo S, Rebhan M, Schubeler D. Distribution, silencing potential and evolutionary impact of promoter DNA methylation in the human genome. Nat Genet. 2007;39(4):457–66.
44. Xie W, Schultz MD, Lister R, Hou Z, Rajagopal N, Ray P, Whitaker JW, Tian S, Hawkins RD, Leung D, et al. Epigenomic analysis of multilineage differentiation of human embryonic stem cells. Cell. 2013;153(5):1134–48.
45. Ooi SK, Qiu C, Bernstein E, Li K, Jia D, Yang Z, Erdjument-Bromage H, Tempst P, Lin SP, Allis CD, et al. DNMT3L connects unmethylated lysine 4 of histone H3 to de novo methylation of DNA. Nature. 2007;448(7154):714–7.
46. Hu JL, Zhou BO, Zhang RR, Zhang KL, Zhou JQ, Xu GL. The N-terminus of histone H3 is required for de novo DNA methylation in chromatin. Proc Natl Acad Sci U S A. 2009;106(52):22187–92.
47. Sabo PJ, Kuehn MS, Thurman R, Johnson BE, Johnson EM, Cao H, Yu M, Rosenzweig E, Goldy J, Haydock A, et al. Genome-scale mapping of DNase I sensitivity in vivo using tiling DNA microarrays. Nat Methods. 2006;3(7):511–8.
48. Li H, Durbin R. Fast and accurate long-read alignment with Burrows–Wheeler transform. Bioinformatics. 2010;26(5):589–95.
49. Gupta S, Stamatoyannopoulos JA, Bailey TL, Noble WS. Quantifying similarity between motifs. Genome Biol. 2007;8(2):R24.

Histone isoform H2A1H promotes attainment of distinct physiological states by altering chromatin dynamics

Saikat Bhattacharya[1,4,6], Divya Reddy[1,4], Vinod Jani[5†], Nikhil Gadewal[3†], Sanket Shah[1,4], Raja Reddy[2,4], Kakoli Bose[2,4], Uddhavesh Sonavane[5], Rajendra Joshi[5] and Sanjay Gupta[1,4*]

Abstract

Background: The distinct functional effects of the replication-dependent histone H2A isoforms have been demonstrated; however, the mechanistic basis of the non-redundancy remains unclear. Here, we have investigated the specific functional contribution of the histone H2A isoform H2A1H, which differs from another isoform H2A2A3 in the identity of only three amino acids.

Results: H2A1H exhibits varied expression levels in different normal tissues and human cancer cell lines (H2A1C in humans). It also promotes cell proliferation in a context-dependent manner when exogenously overexpressed. To uncover the molecular basis of the non-redundancy, equilibrium unfolding of recombinant H2A1H-H2B dimer was performed. We found that the M51L alteration at the H2A–H2B dimer interface decreases the temperature of melting of H2A1H-H2B by ~ 3 °C as compared to the H2A2A3-H2B dimer. This difference in the dimer stability is also reflected in the chromatin dynamics as H2A1H-containing nucleosomes are more stable owing to M51L and K99R substitutions. Molecular dynamic simulations suggest that these substitutions increase the number of hydrogen bonds and hydrophobic interactions of H2A1H, enabling it to form more stable nucleosomes.

Conclusion: We show that the M51L and K99R substitutions, besides altering the stability of histone–histone and histone–DNA complexes, have the most prominent effect on cell proliferation, suggesting that the nucleosome stability is intimately linked with the physiological effects observed. Our work provides insights into the molecular basis of the non-redundancy of the histone H2A isoforms that are being increasingly reported to be functionally important in varied physiological contexts.

Keywords: Cancer, Chromatin, Differentiation, Histone, Nucleosome

Background

Histones are a class of highly conserved basic proteins that package the genome. The core histones are comprised of H2A, H2B, H3 and H4 which form the octameric protein core of the fundamental repeating unit of chromatin, the nucleosome. Around this core, ~147 bp of DNA is wrapped to form the nucleosome core particle (NCP) [1]. Further compaction of the chromatin is achieved with the aid of the linker histone H1 [2].

The canonical histone proteins are synthesized during the S-phase, and to meet up with their high demand during DNA replication, genes that encode them are present in clusters. There are three clusters of canonical histone genes present in humans at chromosome numbers 1 and 6. Notably, differences in the primary sequence are observed amongst the histone proteins encoded by these genes. For the sake of clarity, these are termed as the histone isoforms in this manuscript. In humans, there are 17 genes for H2A that code for 12 isoforms [3, 4]. Likewise,

*Correspondence: sgupta@actrec.gov.in
†Vinod Jani and Nikhil Gadewal contributed equally to this work
[1] Epigenetics and Chromatin Biology Group, Gupta Lab, Cancer Research Institute, Advanced Centre for Treatment, Research and Education in Cancer (ACTREC), Tata Memorial Centre, Kharghar, Navi Mumbai, MH 410210, India
Full list of author information is available at the end of the article

there are 13 genes for H2A in rats that code for 9 iso-forms (most are "predicted").

The histone isoform genes are named based on their identity and location in the genome. In the name of the gene, the first part refers to the cluster (HIST1—cluster 1, HIST2—cluster 2, HIST3—cluster 3), the second part of the gene name introduces the type of histone (H2A, H2B, H3, H4, H1), and the third part indicates the alphabetical order within each cluster (centromere distal to proximal). Therefore, HIST1H2AB refers to the second histone H2A gene in the histone cluster 1 and HIST2H2AB refers to the second histone H2A gene in the histone cluster 2. The proteins coded by these genes, however, were not referred to as systematically. Traditionally, the histone H2A isoforms were broadly classified into two categories, H2A.1 and H2A.2, based on the difference in their mobil-ity on AUT (acetic acid, urea, Triton X-100)–PAGE gels. The H2A isoforms that migrated slowly were collectively termed H2A.1 and the isoforms that migrated faster were collectively referred to as H2A.2 [5]. The difference in migration arises due to the L51M alteration in H2A. Leu-cine binds more Triton X, and hence, the H2A isoforms with L51 migrate slower than isoforms with M51 residue. However, as each of these two bands may be constituted of multiple proteins, this system of referring to isoforms can be misleading. Especially considering the growing evidence of the changes in the expression level of the iso-forms, a better way to name them would be to maintain consistency with their gene nomenclature. For example, the protein coded by HIST1H2AB will be referred to as H2A1B. If two genes code for the same protein as in the case of HIST1H2AB and HIST1H2AE, the protein will be referred to as H2A1B/E. Hence, in rats, the pro-teins H2A3, H2A4, H2A1F, H2A1K, H2A1H and H2A1C (H2AE-like, H2A1I, H2A1N) constitute the H2A.1 iso-forms, and the H2A2B, H2A2C and H2A2A3 proteins belong to H2A.2 isoforms (see Additional file 1: Figure S1 for the alignment).

The histone isoforms were considered function-ally redundant for a long time considering the similar-ity in their amino acid sequences. Interestingly, though, the H2A isoforms have been reported to be differen-tially expressed in a variety of physiological states. For instance, the proportion of the H2A.1 and H2A.2 iso-forms in rats has been shown to decrease during the course of development, differentiation and aging [6–8]. An earlier report from our laboratory revealed the over-expression of the H2A.1 isoforms during the sequential stages of rat hepatocellular carcinoma [9]. The expression level of the isoform H2A1C in humans has been reported to alter in pathological states. Expression of the H2A1C isoform was reported to be downregulated in chronic lymphocytic leukemia (CLL) and gall bladder cancer

cells [10, 11]. Interestingly, later on in a larger cohort of samples, H2A1C expression was conversely reported to be upregulated in CLL [12]. Also, H2A1C was found to be upregulated in non-small cell lung carcinoma [13]. The levels of H2A1C, in particular, have been reported to change in other diseases including human papillo-maviruses hyperplasia, AIDS and multiple sclerosis [14, 15]. Collectively, these reports demonstrate the altered expression of the H2A isoforms in different pathophysi-ological states. The question now is whether the observed changes are merely a consequence of the change in the state or these isoforms also contribute to the attainment of such states. One report that aims to address this ques-tion showed that specific knockdown of H2A1C leads to a marked increase in cell proliferation. This effect is not observed on depleting the other abundant isoforms like H2A1B/E [10]. However, how the histone isoforms impart their non-redundant effects remains unclear.

Here, we show that the expression level of the H2A1H/H2A1C isoform markedly varies in different tissues in addition to being generally upregulated in many cancer cell lines. We provide further evidence that H2A1H (encoded by HIST1H2AH, accession number: NM_001315492.1) provides a growth advantage to cells; however, this effect is context dependent. Importantly, with the help of in vitro and in silico studies, we demon-strate that H2A1H forms more stable nucleosomes than the H2A.2 isoform H2A2A3 (encoded by HIST2H2AA3, accession number: NM_001315493.1), and this is specu-lated to confer the non-redundant functionality. Our studies reveal that the highly similar histone isoforms can bring about changes in cell physiology by modulating chromatin dynamics.

Results

H2A1H/H2A1C expression level varies in cancer cell lines and amongst different normal tissues

Previously, we have reported the upregulation of H2A.1 isoforms during the progress of hepatocellular carcinoma (HCC) [9]. During the course of development of HCC, the animals were under the administration of NDEA. We wanted to see whether the increased expression of H2A.1 persists even without the influence of NDEA. To address this, a tumor was developed in the liver of Sprague–Dawley rat by feeding NDEA with drinking water. After the development of the tumor (105 days since the start of NDEA administration), a 3-mm^2 tumor tissue was excised and subcutaneously implanted in a NOD-SCID mice. The NOD-SCID mice were not fed with NDEA. Two weeks post-implantation, the animals were sacri-ficed and the developed tumor was excised. Analysis of the isolated histones from the tumor resolved onto AUT-PAGE showed a higher expression of H2A.1 isoforms compared to the normal liver (Fig. 1a). This suggests that H2A.1 upregulation is indeed a stable alteration that occurs during the process of tumorigenesis. The changes in H2A composition in HCC were further appreciated by performing reverse-phase HPLC of the extracted histones (Fig. 1b) (see Additional file 1: Figure S2 for the complete elution profile). The most prominent difference in the chromatogram of the control vs tumor histones is the distinct peak at around 84 ml elution volume (Fig. 1c, d). Mass spectrometry followed by peptide fingerprinting of the eluted fractions 84 and 85 ml revealed high scores for the H2A.1 isoforms (H2A1H, H2A3, H2A1C, H2A1K) with the maximum score obtained for H2A1H (Fig. 1e) (see Additional file 1: Figure S3 for the peptides detected in MS). We next performed real-time PCR to check the transcript levels of the histone isoforms in normal vs tumor liver tissues (see Additional file 1: Figure S4). We performed normalization to the widely used normalization control, glyceraldehyde phosphate dehydrogenase (GAPDH) gene (see Additional file 1: Figure S4a). The histone isoforms are synthesized during the S-phase of the cell cycle. As the cells in tumor tissues are more proliferative, to normalize for the overall changes in the histone content, we also performed normalization to histone H4 genes (see Additional file 1: Figure S4b). The primers for H4 genes were designed to pick up all the H4 transcripts. Irrespective of the normalization control used, we found that H2A1H was the most prominently upregulated H2A isoform.

In terms of the protein sequence, H2A1C in humans is the most similar to H2A1H of rat, differing in only the S16T substitution [see Additional file 1: Figure S7(c)]. The altered expression level of H2A1C has been reported in human cancers [10–13, 16–18]. Our observations in the rat hepatocellular carcinoma prompted us to investigate the expression level of H2A1C isoform in human transformed cell lines of the liver (HEPG2). We also included cell lines of skin (A431) and stomach (KATOIII, AGS) origin and their non-transformed immortalized counterparts, that is, HHL5 (liver), HACAT (skin) and HFE145 (stomach) in our study, as the expression level of H2A1C in these cell lines has not been previously reported. An increase in the relative expression of H2A1C was observed in HEPG2 and A431 (Fig. 2a, b). We did not find any significant changes in the levels of the isoform H2A2A3 (identical to rat H2A2A3). The two isoforms did not show any significant alteration in expression in both the transformed cell lines of the stomach with respect to their immortalized counterpart, that is, HFE145 (Fig. 2d). We also found upregulation of H2A1C in MCF7 consistent with a previously published report (Fig. 2c) [17].

We speculated that if H2A1H has some specific non-redundant function, then its expression may vary in different tissues. To test this hypothesis, the transcript level of H2A1H in different organs was compared. Marked variation in H2A1H level was observed. A very high level of H2A1H expression was observed in rectum (Fig. 2e). On the other hand, in the stomach and tongue tissues, the expression level was found to be particularly low (Fig. 2e). Isoform H2A2A3 exhibited much lesser variation in expression level (Fig. 2e). AUT-PAGE analysis of histones isolated from kidney, brain and liver demonstrates that the variations observed in transcript level of the H2A.1 isoform H2A1H is also reflected in protein expression (Fig. 2f). The brain showed an increased proportion of H2A.1, whereas kidney and liver have higher levels of H2A.2 isoforms (Fig. 2g).

H2A1H isoform is functionally non-redundant from the H2A2A3 isoform

The expression level of H2A.1 isoforms varies in different tissues, differentiation status, age and diseases. Based on our results, we wanted to test the effect of overexpressing H2A.1 isoform H2A1H on cell physiology. Two cell lines that are derived from the liver of NDEA-administered Sprague–Dawley rats were chosen for our studies: CL44 (pre-neoplastic), with an equimolar ratio of H2A.1 and H2A.2, and CL38 (neoplastic), in which H2A.1 is naturally elevated (see Additional file 1: Figure S5). By RT-PCR, we validated that the CL38 cells express higher levels of the H2A1H isoform. Localization of YFP-tagged H2A1H/H2A2A3 in CL38 cells suggested that both the isoforms are incorporated across the entire chromatin (see Additional file 1: Figure S5). By isolating histones from the CL38 cells exogenously overexpressing the isoforms [pcDNA3.1(+) vector] and resolving them on AUT-PAGE, we confirmed that the overexpression of H2A1H leads to its increased abundance in the chromatin (Fig. 3a, b).

A marked increase in proliferation was observed in the CL38 cells on exogenous overexpression of H2A1H (Fig. 3c). Similar effects were reflected in the colony formation assay, with H2A1H overexpressing colonies substantially larger (Fig. 3d, e). Associated upregulation in proliferation markers Ki67 and PCNA was also noted by qRT-PCR (Fig. 3f). To see the effect of the isoforms overexpression on the cycling of cells, we studied the cell cycle profile of G1-enriched H2A1H/H2A2A3 CL38 cells post-72-h serum release. Overexpression of H2A1H led to a discernible increase in the mitotic cell population (12%) compared to the vector control (4%) (Fig. 3g). We also observed an increase in the mitotic cell population with H2A2A3 overexpression (7%) compared to the vector control (4%). This was also reflected in the proliferation assays (Fig. 3c, d). No significant difference in the

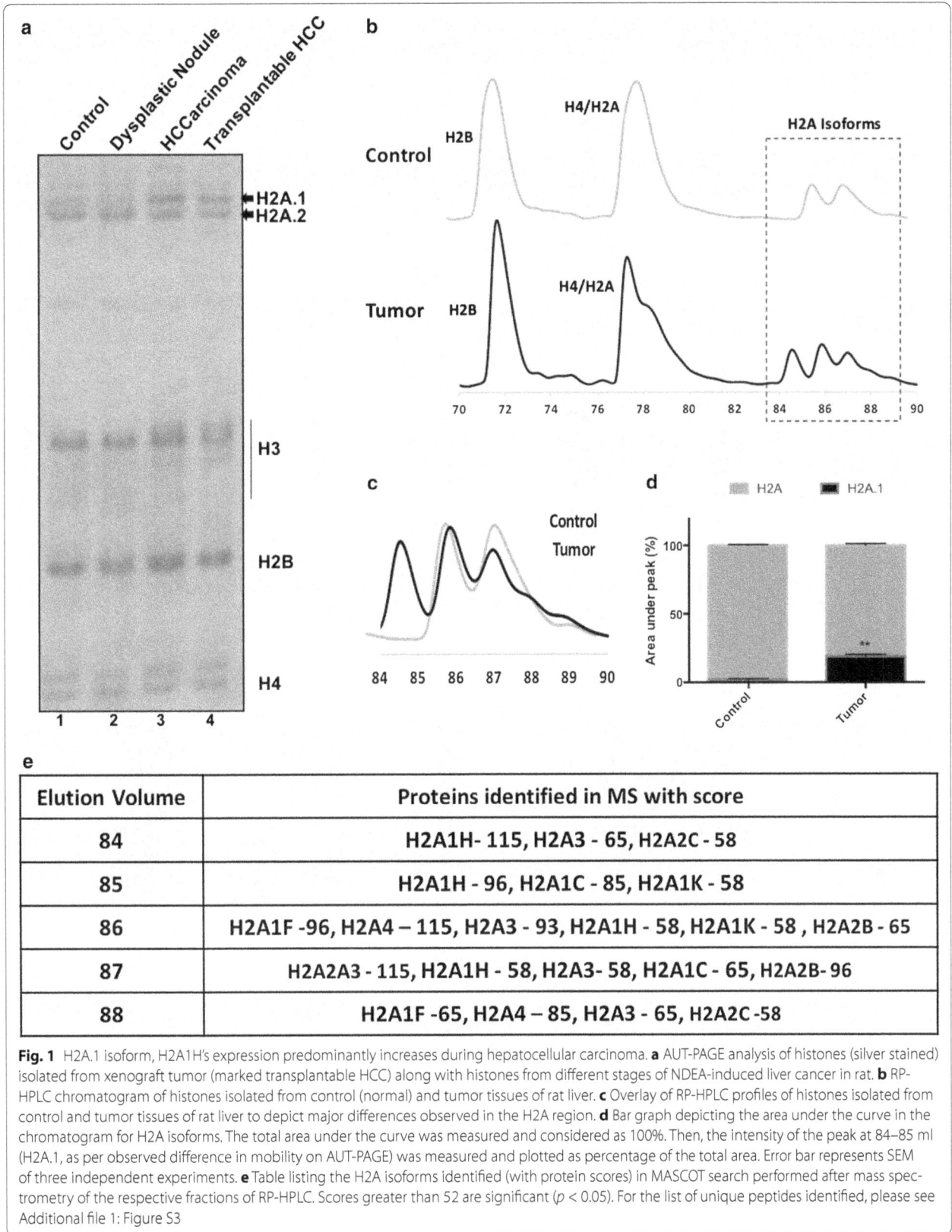

Fig. 1 H2A.1 isoform, H2A1H's expression predominantly increases during hepatocellular carcinoma. **a** AUT-PAGE analysis of histones (silver stained) isolated from xenograft tumor (marked transplantable HCC) along with histones from different stages of NDEA-induced liver cancer in rat. **b** RP-HPLC chromatogram of histones isolated from control (normal) and tumor tissues of rat liver. **c** Overlay of RP-HPLC profiles of histones isolated from control and tumor tissues of rat liver to depict major differences observed in the H2A region. **d** Bar graph depicting the area under the curve in the chromatogram for H2A isoforms. The total area under the curve was measured and considered as 100%. Then, the intensity of the peak at 84–85 ml (H2A.1, as per observed difference in mobility on AUT-PAGE) was measured and plotted as percentage of the total area. Error bar represents SEM of three independent experiments. **e** Table listing the H2A isoforms identified (with protein scores) in MASCOT search performed after mass spectrometry of the respective fractions of RP-HPLC. Scores greater than 52 are significant ($p < 0.05$). For the list of unique peptides identified, please see Additional file 1: Figure S3

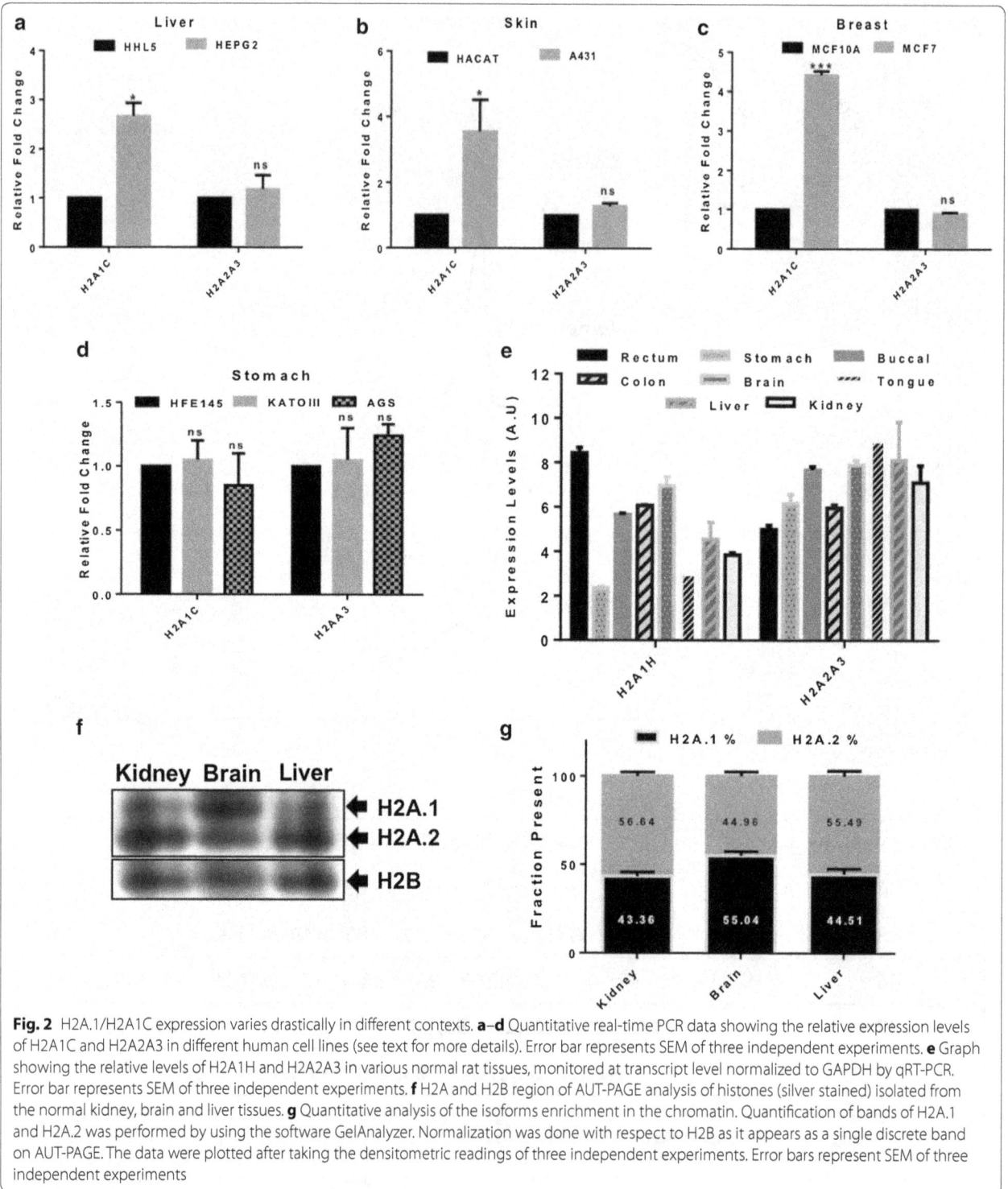

Fig. 2 H2A.1/H2A1C expression varies drastically in different contexts. **a–d** Quantitative real-time PCR data showing the relative expression levels of H2A1C and H2A2A3 in different human cell lines (see text for more details). Error bar represents SEM of three independent experiments. **e** Graph showing the relative levels of H2A1H and H2A2A3 in various normal rat tissues, monitored at transcript level normalized to GAPDH by qRT-PCR. Error bar represents SEM of three independent experiments. **f** H2A and H2B region of AUT-PAGE analysis of histones (silver stained) isolated from the normal kidney, brain and liver tissues. **g** Quantitative analysis of the isoforms enrichment in the chromatin. Quantification of bands of H2A.1 and H2A.2 was performed by using the software GelAnalyzer. Normalization was done with respect to H2B as it appears as a single discrete band on AUT-PAGE. The data were plotted after taking the densitometric readings of three independent experiments. Error bars represent SEM of three independent experiments

closure of the wound in scratch assays performed with CL38 cells on H2A1H overexpression was perceived (see Additional file 1: Figure S6a) in comparison with H2A2A3 overexpression. Notably, we did not observe any significant change in the proliferation of CL44 cells upon H2A1H overexpression (see Additional file 1: Figure S6b). Importantly, during liver regeneration post-partial hepatotectomy, H2A.1 expression was not found to alter [8]. Taken together, these data suggest that although H2A1H expression provides a growth advantage to cells,

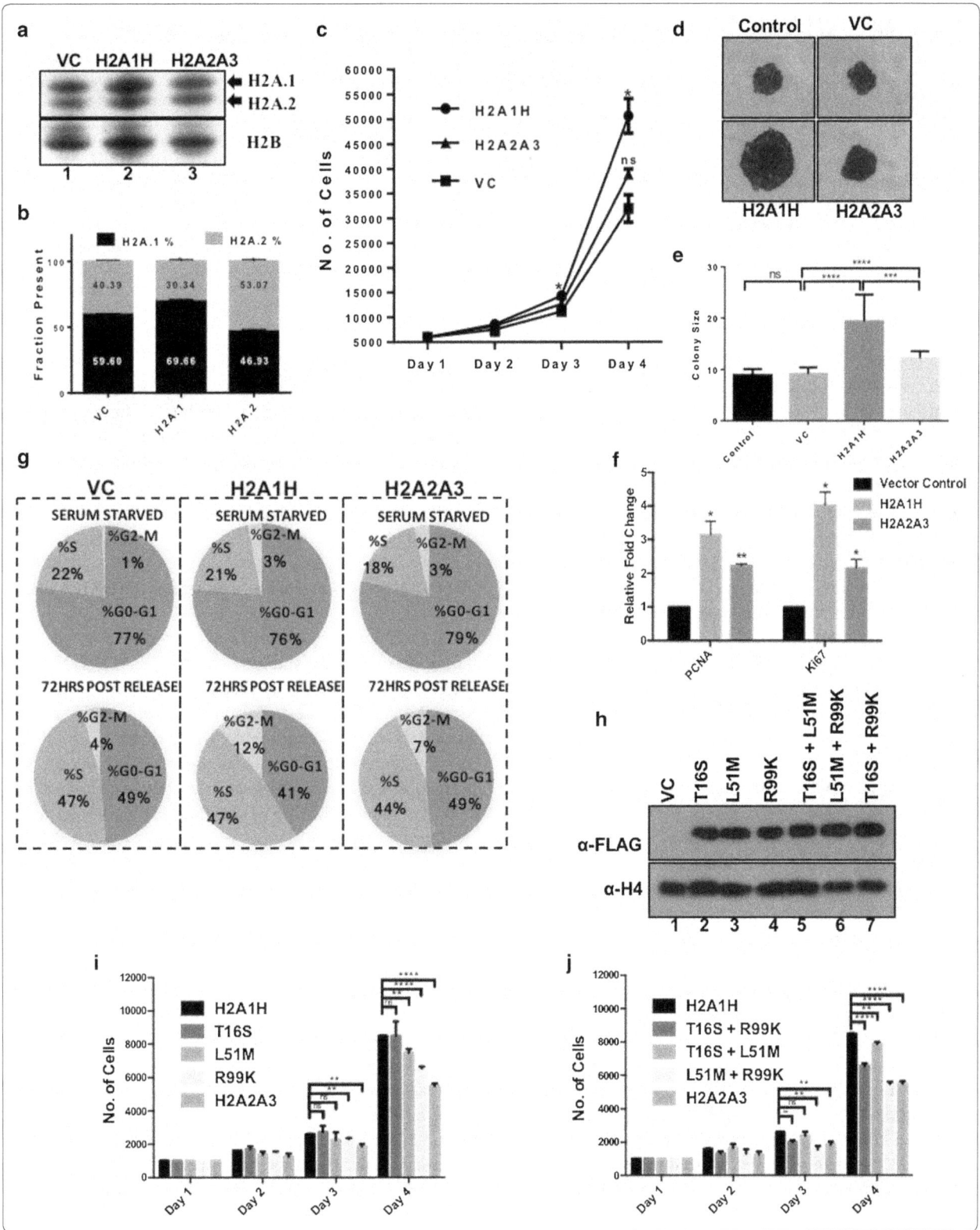

a

b

c

d

e

f

g

h

i

j

(see figure on previous page.)
Fig. 3 H2A1H overexpression leads to increase in cell proliferation. **a** AUT-PAGE analysis (silver stained) showing the enrichment of the H2A isoforms in chromatin upon their overexpression in CL38 cells. **b** Quantitative analysis of the isoforms enrichment in the chromatin. Quantification of bands of H2A.1 and H2A.2 was performed by using the software GelAnalyzer. Normalization was done with respect to H2B as it appears as a single discrete band on AUT-PAGE. The data were plotted after taking the densitometric readings of three independent experiments. Error bars represent SEM of three independent experiments. **c** Cell proliferation curves by MTT assay of H2A1H and H2A2A3 overexpressing CL38 cells in comparison with control CL38 cells. Error bars represent SEM of six independent experiments. **d** Colony formation assay of CL38 cells upon H2A1H and H2A2A3 over-expression. **e** Quantitative analysis of the colony sizes of 20 colonies each performed using ImageJ. Error bar represents SEM. **f** qRT-PCR for the cell proliferation markers Ki67 and PCNA on H2A1H and H2A2A3 overexpression normalized to 18S rRNA. Error bars represent SEM of three independent experiments. **g** Cell cycle analysis of the CL38 cells exogenously overexpressing H2A isoform post-serum starvation and release. **h** The expression level analysis of the CL38 cells expressing H2A1H single or double mutants with anti-FLAG antibody. **i** Bar graph depicting the proliferation of the CL38 cells expressing H2A1H single mutants by the MTT assay. Error bars represent SEM of 6 independent experiments. **j** Bar graph depicting the proliferation of the CL38 cells expressing H2A1H double mutants by MTT assay. Error bars represent SEM of six independent experiments. VC—vector control. H2A1H, H2A2A3 and their mutants in the figure are the genes cloned and expressed as FLAG tagged proteins in pcDNA3.1(+) vector

its expression is not always correlated with proliferation (discussed later).

Leu51 and Arg99 are important in conferring the non-redundant functionality to the H2A1H isoform

The H2A isoforms, H2A1H and H2A2A3, differ in three residues in their primary amino acid sequence (see Additional file 1: Figure S7a). To understand which residue(s) are important for the non-redundant functionality of H2A1H, we substituted the residues of H2A1H to the corresponding ones of H2A2A3. MTT assays performed with CL38 cells suggested that mutating R99K of H2A1H drastically reduced the pro-proliferative effect observed on its overexpression (Fig. 3i). Mutating L51M also negatively affected cell proliferation and had a synergistic effect when substituted alongside R99K (Fig. 3j). The assays were conducted with populations showing similar levels of overexpressed proteins to rule out possible variations resulting from any differences in the expression level (Fig. 3h). Notably, the 16th residue where the rat H2A1H and human H2A1C differ did not have any significant effect on the non-redundant effects of H2A1H in the assays performed by us (Fig. 3i, j).

Leu51 and Arg99 of H2A1H are present at important locations in the nucleosome and may potentially impact its stability

Our results show that the expression of H2A1H varies markedly in different states and it does have non-redundant functionality. Further, Leu51 and Arg99 contribute majorly in conferring the non-redundant functionality to the H2A1H isoform. We next wanted to address how the H2A1H isoform imparts its non-redundant functional effects.

We carried out in silico simulation of mononucleosome and looked for the interactions of the three differential residues between H2A1H and H2A2A3 in the nucleosome core particle (NCP). The 16th residue of

H2A is involved in interactions with the minor groove of DNA in the NCP, residue 51st lies in the dimer interface with H2B, and residue 99th of H2A interacts with the H4 tail in the octamer core (see Additional file 1: Figure S7b). Therefore, potentially the alterations at these residues can alter the stability of the nucleosome and its subcomplexes.

The H2A1H-H2B dimer is less stable than the H2A2A3-H2B dimer

To investigate the possibility discussed above, we compared the in vitro stability of the H2A1H-H2B with the H2A2A3-H2B dimer reconstituted using purified recombinant histones. Equilibrium unfolding of the reconstituted full-length H2A–H2B dimers, which was previously described [19], was used to perform the stability analysis. For details pertaining to the structural and stability characterization of the dimers, please refer to the "Methods" section. Once the equilibrium unfolding curves for both H2A1H-H2B and H2A2A3-H2B dimers were obtained, a comparative analysis of their stability was carried out (Fig. 4a). Co-plotting the Fapp (apparent fraction unfolded) of the H2A1H-H2B and H2A2A3-H2B dimers against the increasing temperature/denaturant concentration shows a hysteresis, suggestive of the difference in the propensity to unfold in response to the denaturant (Fig. 4a–c). The temperature of melting (Tm) for the H2A1H-H2B dimer was determined to be 50.04 °C, whereas that of the H2A2A3-H2B dimer was found to be higher by ~ 3 at 53.31 °C (Fig. 4d), suggesting that the former is less stable. The circular dichroism (CD) and the fluorescence data plotted in response to the increasing chemical denaturant concentration were in good agreement with each other. The $[urea]_{1/2}$ for the H2A1H-H2B dimer was obtained as 1.59 and 1.52 M, respectively, using the two methods. The $[urea]_{1/2}$ for the H2A2A3-H2B dimer was found to be 1.74 and 1.73 M with CD and fluorescence spectroscopy, respectively. Further,

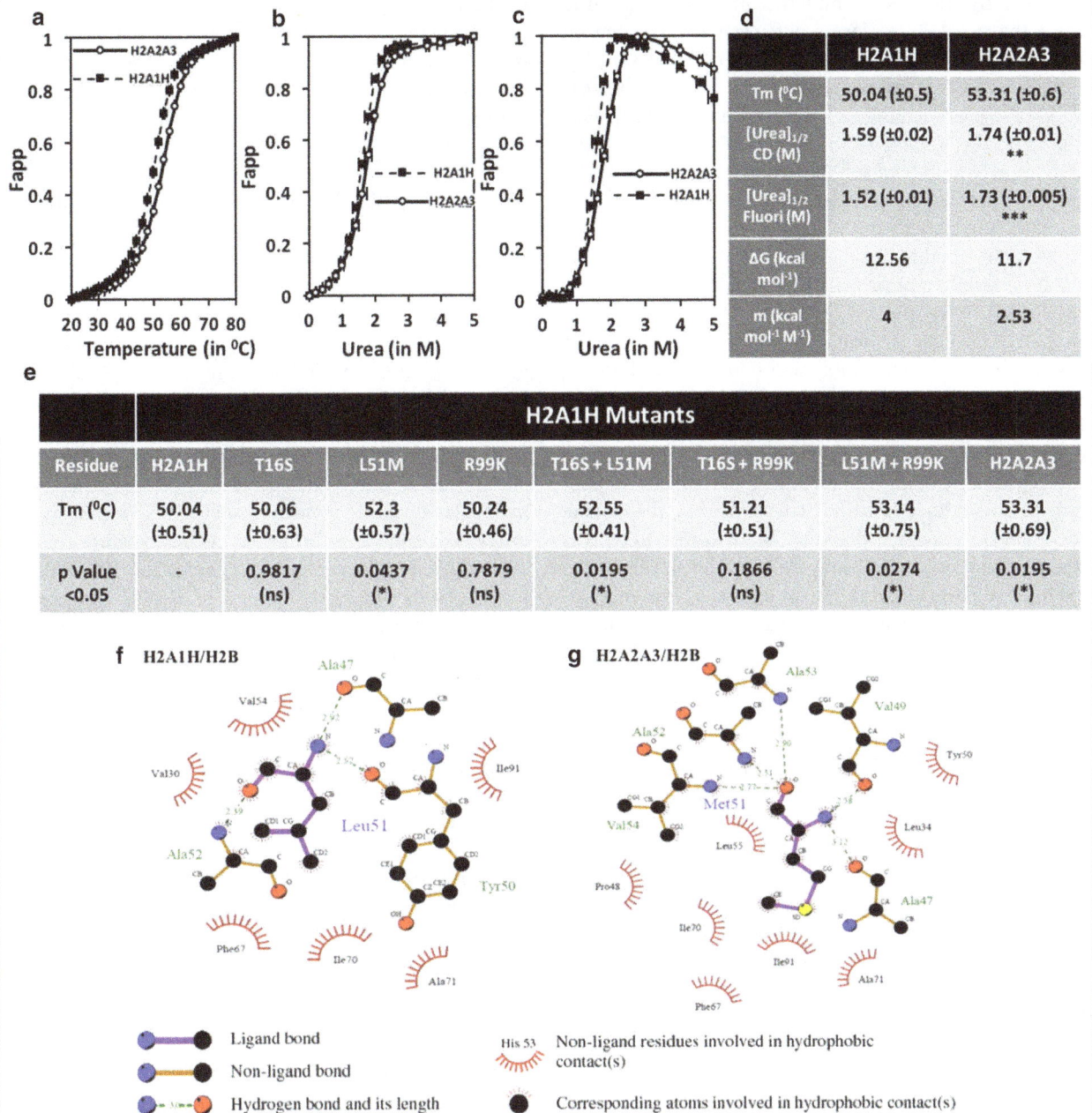

Fig. 4 H2A1H-H2B dimer is less stable than the H2A2A3-H2B dimer. **a, b** Apparent fraction unfolded (Fapp) obtained from the analysis of the CD spectra of H2A1H-H2B and H2A2A3-H2B monitored during thermal and urea denaturation. **c** Fapp obtained from the analysis of the fluorescence spectra of H2A1H-H2B and H2A2A3-H2B monitored during urea denaturation. Error bar represents SEM of six independent experiments. **d** Comparative determination of the various parameters obtained by the CD and fluorescence spectra of H2A1H-H2B and H2A2A3-H2B. **e** Comparison of temperature of melting (Tm) of various H2A1H single, double mutants and H2A2A3 with H2A1H. **f, g** Ligplots depicting the interaction of the 51st residue of both H2A1H and H2A2A3 in the H2A–H2B dimer interface

the m value obtained for the H2A1H-H2B dimer was 4 kcal mol^{-1} M^{-1} and that for the H2A2A3-H2B dimer was 2.53 kcal mol^{-1} M^{-1} (Fig. 4d) which are suggestive of the higher sensitivity of the H2A1H-H2B dimer to the denaturant concentration.

The L51M substitution in H2A at the dimer interface with H2B is primarily responsible for the differential stability

The stability of the H2A1H-H2B dimer was determined to be lower than of the H2A2A3-H2B dimer.

Subsequently, the effect of mutating the three residues in which the two H2A isoforms differ was investigated on the dimer stability by carrying out thermal denaturation with the reconstituted mutant dimers. Studies with the mutants suggest that the L51M alteration had the biggest impact on the stability of the dimers (Fig. 4e). Mutating L51M in H2A1H increased the Tm from 50.04 to 52.3 °C and that of H2A2A3 to M51L (H2A1H T16S + R99K) decreased the stability by 2.1 °C (Fig. 4e).

Leucine-to-methionine alteration at the 51st residue, which we found to be primarily responsible for the differential stability, has been suggested to be context dependent [20]. Although the van der Waals volume occupied by leucine is the same as for methionine, two opposing forces are at play when leucine-to-methionine substitution occurs. The substitution of methionine with leucine within the interior of a protein is expected to increase the stability because of both a more favorable solvent transfer term and the reduced entropic cost of holding the leucine side chain in a defined position. At the same time, this expected beneficial effect may be offset by steric factors due to the differences in the shape of leucine and methionine [20]. To understand the possible alteration in interactions on the incorporation of methionine, we carried out energy minimization of the structures. As depicted in the ligplots, the substitution L51M led to an increased number of hydrogen bonds and hydrophobic interactions that explains the higher stability observed in denaturation experiments (Fig. 4f, g). Altering the 16th and 99th residues in isolation did not have a major effect on the dimer stability; however, mutating R99K along with L51M had a synergistic effect on stabilizing the dimer by an additional increment in stability by ~ 0.8 °C (discussed in more detail in "Discussion" section).

The H2A1H isoform-containing nucleosomes are more stable owing to the formation of higher number of hydrogen bonds

To understand the importance of the alteration in dimer stability in the context of the chromatin, we investigated the effect of the incorporation of these isoforms on the nucleosome stability. Beyond 600 mM NaCl concentration, the nucleosome core particle starts losing its integrity as the histone H2A–H2B dimers start to irreversibly dissociate from the particle [21]. Hence, to compare the stability of the chromatin association of H2A1H-H2B and H2A2A3-H2B dimer, the chromatin was incubated in buffers of increasing ionic strength starting from 600 mM NaCl. Detectable levels of the H2A2A3 isoform (FLAG tagged) were obtained in the soluble fraction (supernatant post-centrifugation at 13,000g for 30 min, 4 °C) at a lower ionic strength (600 mM NaCl) compared to H2A1H (700 mM NaCl) (Fig. 5a). Analysis

of the chromatin fraction also indicated that the H2A1H isoform is more resistant to elution from the chromatin with increasing ionic strength compared to the H2A2A3 isoform (Fig. 5b).

To see whether the more stable association of H2A1H with chromatin is also reflected in its dynamics, we monitored the recovery of fluorescently tagged histone isoforms in a bleached region of the nucleus of CL38 cells (Fig. 5c). We documented that the distribution of both the isoforms is similar in the soluble and chromatin-bound fractions with undetectable levels in the soluble fraction (Fig. 5e). The percentage recovery of H2A1H after 1 h was markedly less (44.14%) compared to H2A2A3 (64.7%) (Fig. 5c, d) in the FRAP assay, suggesting that H2A1H is less dynamic than the H2A2A3 isoform.

To understand the basis of the increased stability of H2A1H-containing nucleosomes, we performed the molecular dynamic simulation (MDS). The convergence of the MD simulation in terms of the structure was calculated by the root mean square deviation (RMSD) with respect to the initial structure. The RMSD analysis was in agreement with the in vitro data with lower RMSD of H2A1H-containing system, suggesting that it forms more stable nucleosomes as compared to H2A2A3 (Fig. 5f). Corroboratively, the hydrogen bonding analysis shows that during the course of the simulation, H2A1H nucleosome has a higher number of hydrogen bonds (Fig. 5g). The RMSD of the octamer and DNA independently showed a similar trend (see Additional file 1: Figure S10).

Leu51 and Arg99 residues lead to the increased stability of H2A1H-containing nucleosomes as compared to H2A2A3-containing ones

We carried out site-directed mutagenesis of the isoforms followed by FRAP in CL38 cells to identify the important alteration(s) that is majorly responsible for the difference in chromatin dynamics of H2A1H and H2A2A3. The R99K substitution, which is involved in the interaction with the H4 tails in the NCP, independently brought about the most drastic increase (20%) in the dynamics of H2A1H followed by L51M (12%) (Fig. 6a, b). Mutating both the L51M and R99K together led to almost similar dynamics as observed for H2A2A3. Mutating only T16S did not have a significant impact on the H2A1H dynamics. However, a synergism was observed when residue T16S was mutated alongside L51M and R99K (discussed later).

We wanted to understand how the substitutions with very similar amino acids brought about the observed changes in the nucleosome stability. Analysis of the number of hydrogen bonds formed by the residues at the three positions with nearby residues throughout the simulation

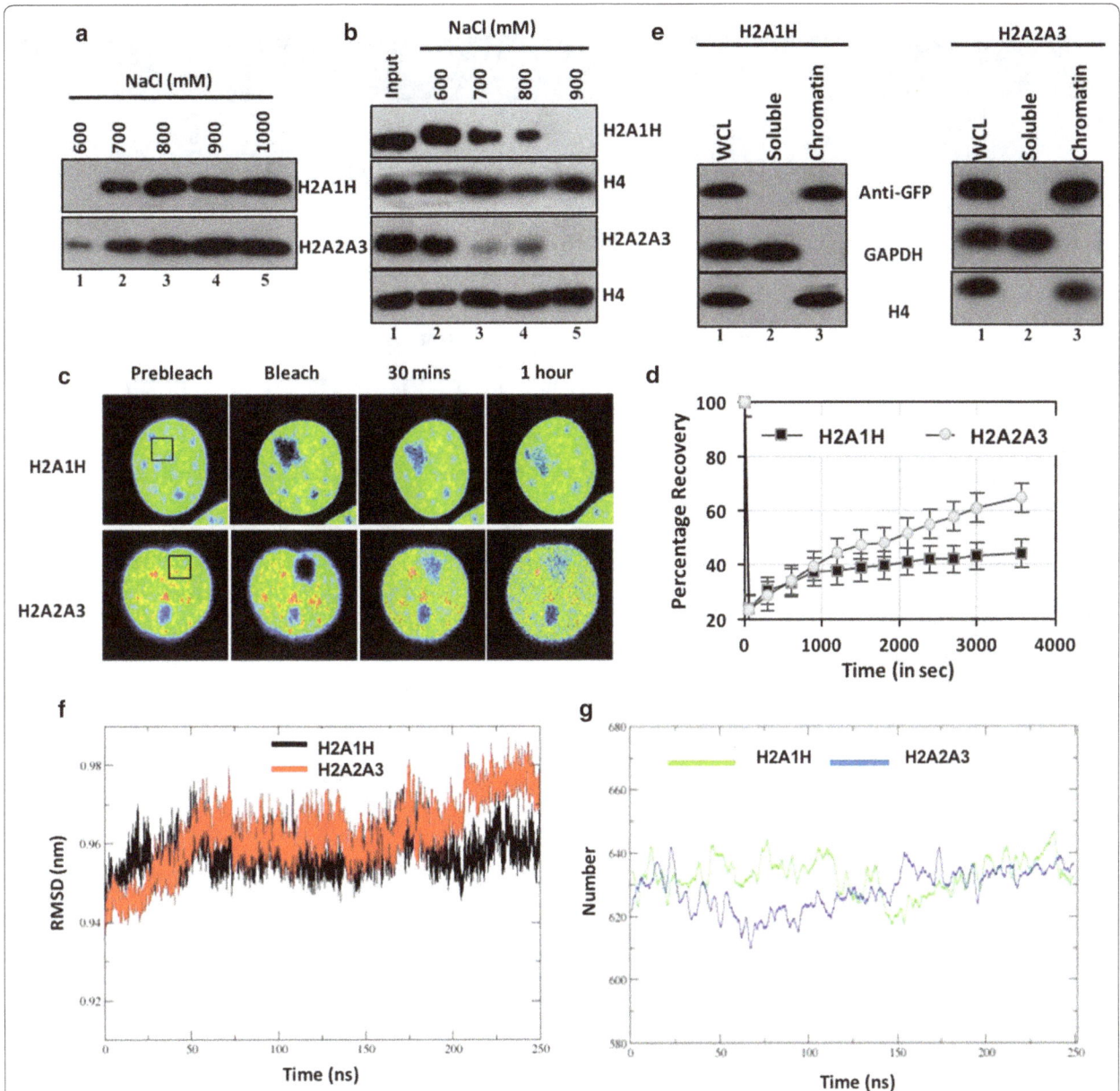

Fig. 5 H2A1H-containing nucleosome is more stable than the H2A2A3 nucleosome. **a, b** Levels of H2A1H and H2A2A3 in the soluble and chromatin fractions upon incubation of CL38 cells in buffers of increasing ionic strength. **c** FRAP assay performed with CL38 cells expressing YFP-tagged H2A1H and H2A2A3. Recovery was monitored for a period of 1 h. **d** Graph depicting the percentage recovery of YFP-H2A1H and YFP-H2A2A3 over a span of 4000 s. Error bar represents SEM of ten independent experiments. **e** Cellular fractionation of the CL38 cells followed by immunoblotting with the marked antibodies to determine the distribution of histones. **f** RMSD of H2A1H and H2A2A3 nucleosomes over a span of 250 ns of molecular dynamic simulation. **g** Hydrogen bond analysis of the H2A1H- and H2A2A3-containing nucleosome over the span of 250 ns of molecular dynamic simulation (MDS)

time of 250 ns was performed for both the H2A1H- and H2A2A3-containing nucleosomes. The data suggested that the 51st and the 99th residues majorly participate in the formation of hydrogen bonds with very less contribution from the 16th residue (Fig. 6c). Importantly, the

arginine at 99th position in H2A1H system forms more number of hydrogen bonds than lysine (Fig. 6c). The ligplots depicts the hydrogen and hydrophobic interactions between the 99th and nearby residues of H2A1H (Fig. 6d) and H2A2A3 systems (Fig. 6e).

a

	H2A1H Mutants						
Residue	H2A1H	T16S	L51M	R99K	T16S + L51M	T16S + R99K	L51M + R99K	H2A2A3
Recovery after 1hr	44.14 (±1.61)	46.63 (±2.0)	49.76 (±0.49)	53.19 (±1.18)	54.52 (±1.91)	55.70 (±2.24)	60.68 (±0.44)	64.70 (±1.33)

b

c

d

e

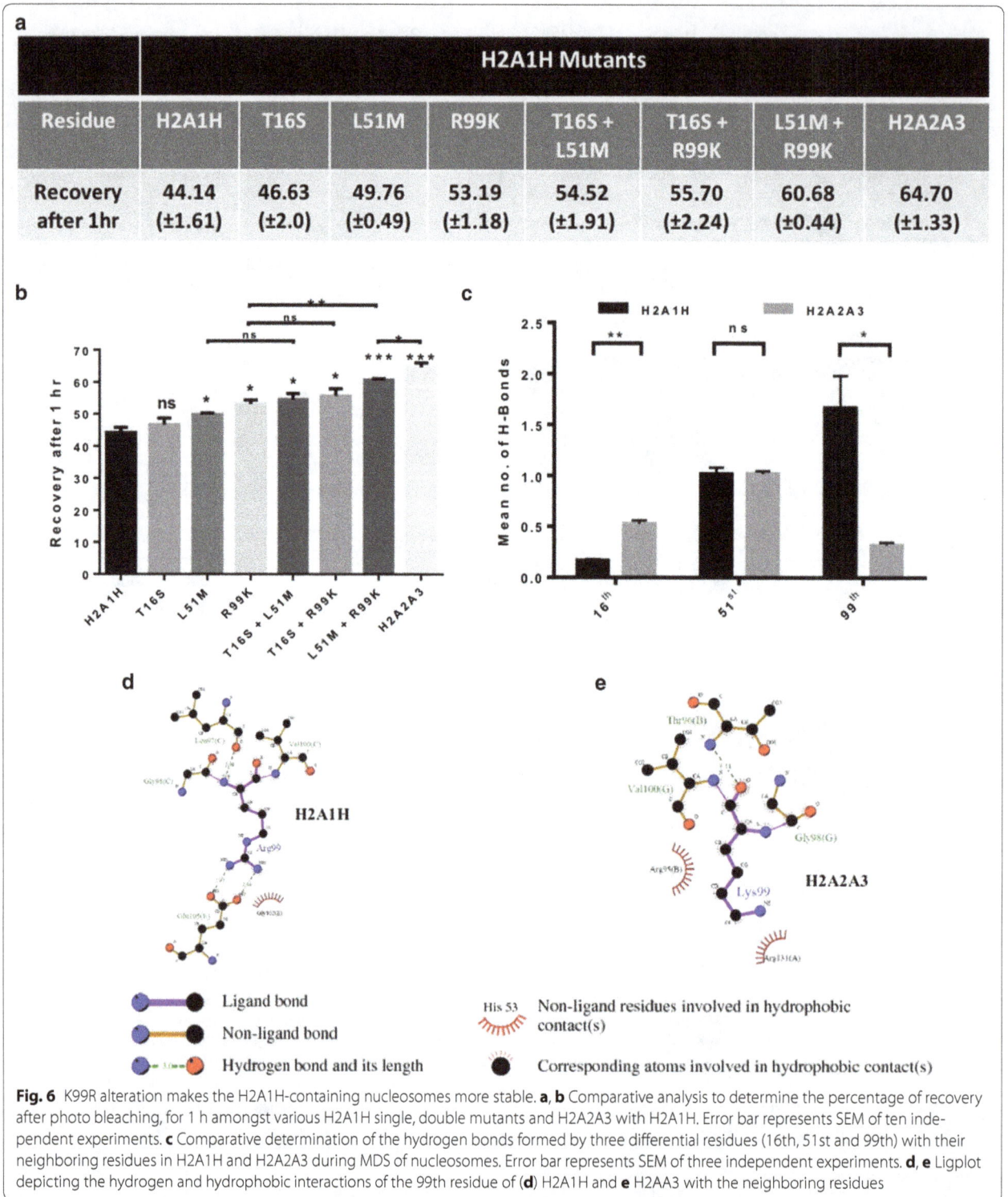

Fig. 6 K99R alteration makes the H2A1H-containing nucleosomes more stable. **a, b** Comparative analysis to determine the percentage of recovery after photo bleaching, for 1 h amongst various H2A1H single, double mutants and H2A2A3 with H2A1H. Error bar represents SEM of ten independent experiments. **c** Comparative determination of the hydrogen bonds formed by three differential residues (16th, 51st and 99th) with their neighboring residues in H2A1H and H2A2A3 during MDS of nucleosomes. Error bar represents SEM of three independent experiments. **d, e** Ligplot depicting the hydrogen and hydrophobic interactions of the 99th residue of (**d**) H2A1H and **e** H2AA3 with the neighboring residues

Principal component analysis suggests that H2A1H-containing nucleosome structures are better correlated

Next, the principal component analysis (PCA) was carried out to discriminate between relevant conformational changes in the protein structure from the background atomic fluctuations. The Fig. 7a(i) shows the cross-correlation plot for protein octamer for H2A1H and H2A2A3. In the H2A1H nucleosome, nearby interacting chains

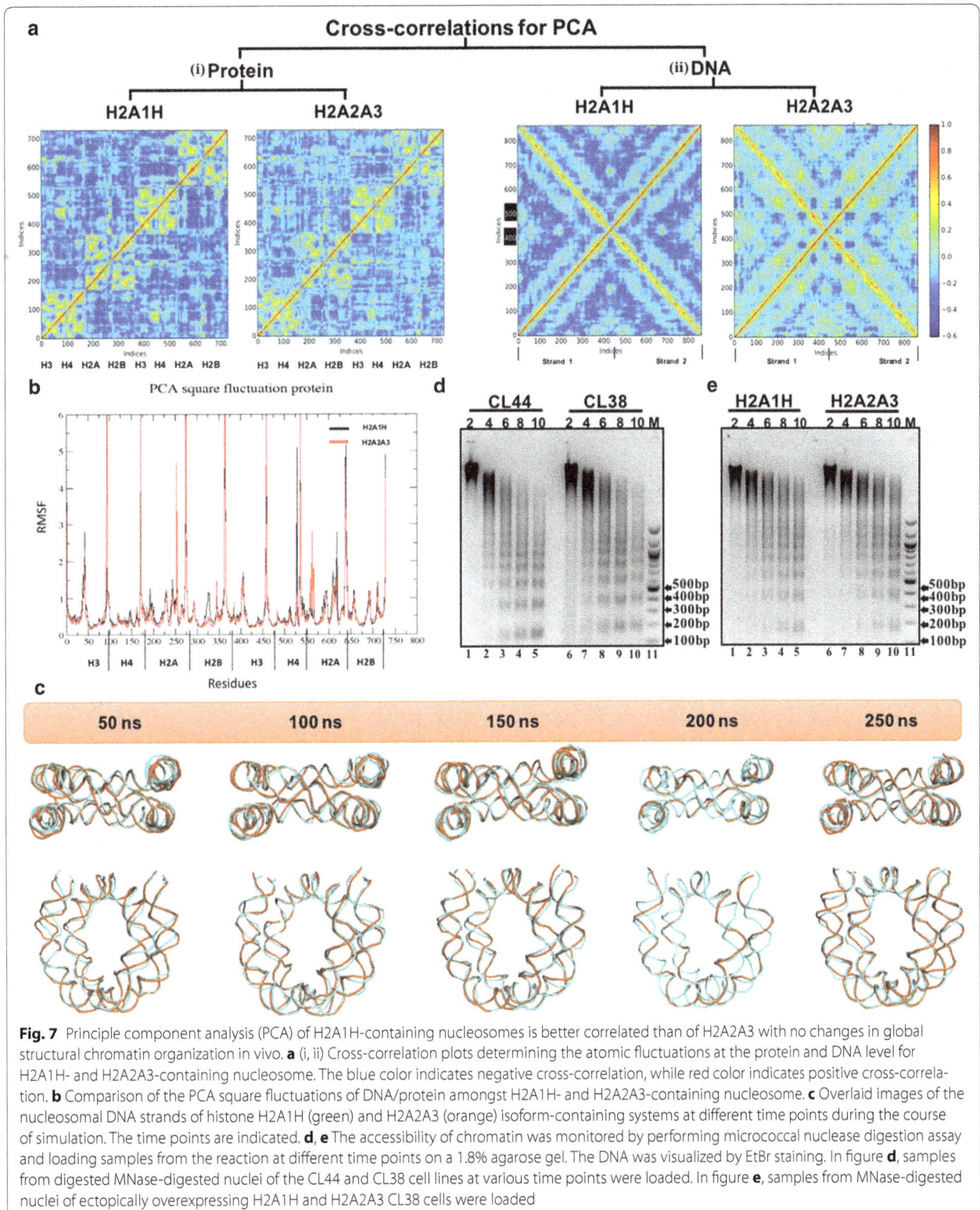

Fig. 7 Principle component analysis (PCA) of H2A1H-containing nucleosomes is better correlated than of H2A2A3 with no changes in global structural chromatin organization in vivo. **a** (i, ii) Cross-correlation plots determining the atomic fluctuations at the protein and DNA level for H2A1H- and H2A2A3-containing nucleosome. The blue color indicates negative cross-correlation, while red color indicates positive cross-correlation. **b** Comparison of the PCA square fluctuations of DNA/protein amongst H2A1H- and H2A2A3-containing nucleosome. **c** Overlaid images of the nucleosomal DNA strands of histone H2A1H (green) and H2A2A3 (orange) isoform-containing systems at different time points during the course of simulation. The time points are indicated. **d, e** The accessibility of chromatin was monitored by performing micrococcal nuclease digestion assay and loading samples from the reaction at different time points on a 1.8% agarose gel. The DNA was visualized by EtBr staining. In figure **d**, samples from digested MNase-digested nuclei of the CL44 and CL38 cell lines at various time points were loaded. In figure **e**, samples from MNase-digested nuclei of ectopically overexpressing H2A1H and H2A2A3 CL38 cells were loaded

show a positive correlation, while the distant regions are showing a negative correlation. Generally, a positive correlation is seen in nearby residues with a synchronous motion, whereas a negative correlation is observed between distantly interacting residues with asynchronous motion. Histones H3 and H4 together form a dimer; therefore, H3 shows a positive correlation for H4, while negative correlation for rest of the histone chains. Similarly, H2A shows a positive correlation for H2B. The pattern of correlation observed with H2A1H- or H2A2A3-containing nucleosome is the same for nearby chains; however, the correlation between H2A2A3 and H2B (system 2) is slightly less positive compared to H2A1H- and H2B-containing nucleosome (system 1). Also, in system 2 there is a less negative correlation between distant chains. Thus, comparing the cross-correlation data with the PCA square fluctuation (Fig. 7b) it can be seen that the negatively correlated motion between distant chains is providing a rigidity and stability to the H2A1H nucleosome. The cross-correlation of DNA [Fig. 7a(ii)] follows the same trend.

Incorporation of the H2A1H isoform does not impart structural alterations to the chromatin

The difference in the cross-correlation plot for the DNA of H2A1H- and H2A2A3-containing systems [Fig. 7a(ii)] prompted us to investigate whether there might be a structural alteration in the DNA on the incorporation of the H2A isoforms. Overlaying the structures of different time points of simulation suggested that there is no prominent structural alteration (Fig. 7c). To see whether there are any global changes in nucleosome spacing or chromatin accessibility, the chromatin of CL44 and CL38 cells was subjected to micrococcal nuclease (MNase) digestion. No structural alterations were discernible on resolving the digestion products on an agarose gel (Fig. 7d). Similarly, the digestion profile was virtually identical for the chromatins isolated from CL38 cells exogenously overexpressing the H2A1H/H2A2A3 isoforms (Fig. 7e), suggesting that the global chromatin structure and accessibility do not alter significantly on the incorporation of the H2A1H and H2A2A3 isoforms. However, more sensitive experiments are needed to be performed to rule out the possibility of very minute changes that might occur on the incorporation of the histone isoforms.

Discussion

The non-redundancy of the histone isoforms has made the understanding of the epigenetic regulations employed by cells more complex, nevertheless, interesting. Previous studies have attempted to elucidate the role of the H2A1C isoform in context of cancer [17]; however, insights into the basic non-redundant role of H2A isoforms, which may contribute to the attainment or persistence of a particular physiological or pathological state, remain poorly addressed. Earlier we had reported that the expression of H2A.1 isoforms increases in HCC

[9]. Considering the growing identification of a variety of H2A isoforms, we validated our earlier findings with the help of RP-HPLC. Further, we addressed the molecular basis of the functional non-redundancy of the histone H2A isoform H2A1H that is overexpressed in cancer.

We found the L51M alteration to have the most significant impact on the H2A–H2B dimer stability. The difference observed between the H2A1H-H2B and H2A2A3-H2B dimer stability is subtle compared to the change brought about by histone variants like H2A.Z [19]. This is consistent with previous reports where L to M replacements altered the protein stability only by 0.4–1.9 kcal/mol [22]. Possibly, the ubiquitous abundance of H2A isoforms in the genome, as opposed to variants, makes this difference significant to induce alterations in epigenetic regulation. Probably, the cell uses the histone variants to bring about major changes in gene regulation and has evolved the histone isoforms for subtle modulations of chromatin-mediated processes.

Interestingly, besides the involvement of the L51M alteration in determining H2A–H2B dimer stability, a synergistic effect was seen when residue R99K was mutated along with L51M. This was intriguing as the residue 99th is not present in the dimer interface. Arginine (in H2A1H) and lysine (in H2A2A3) are positively charged residues and play important roles in stabilizing proteins by forming ionic interactions and hydrogen bonds in the protein as well as with water [23]. Notably, the guanidinium group in arginine allows interactions in three possible directions through its three asymmetrical nitrogen atoms in contrast to only one direction of interaction allowed for lysine. Owing to this difference in geometry of the two amino acids, arginine might have a more stabilizing effect on proteins over lysine [20]. The presence of arginine in H2A1H probably stabilizes the H2A monomer more as compared to lysine in H2A2A3 which thermodynamically makes the H2A1H-H2B dimer less stable. Further, the ability of arginine to form a higher number of H-bonds compared to lysine is also reflected in our FRAP assay and MDS studies.

As discussed earlier, the altered stability of the H2A–H2B dimer will have its implications in the nucleosome stability. Previous MDS studies focussed on the histone octamer–DNA interactions revealed that the H2A–H2B dimer is the least stable part of the nucleosome and could make a significant contribution to the histone–DNA interaction dynamics [24]. We found that the H2A1H isoform gives rise to a more stable nucleosome although the H2A1H-H2B dimers were less stable. This is consistent thermodynamically as a less stable dimer would favor a more stable nucleosome. This is because the association of the H2A–H2B dimers with the nucleosome core particle (NCP) is a dynamic process. Therefore, there is an important equilibrium between the fully assembled NCP and partially unfolded NCPs in which the H2A–H2B dimers are less tightly bound or completely dissociated. The shift in this equilibrium will be affected by the overall entropy of the system, which, in turn, would depend on the free energy of the dissociated dimers. Therefore, the stability of the free H2A–H2B dimer will have consequences on the state of nucleosome assembly and its stability. A more stable H2A–H2B dimer should favor a more unfolded, dissociated state of the NCP. Similar to our observations, for the H2A.Z variant it was reported that the H2A.Z–H2B dimer was unstable as compared to the canonical H2A–H2B [19]; however, the nucleosome was found to be more stable [25].

A more stable nucleosome is expected to cause hindrance to chromatin-mediated processes like transcription, replication and repair. Previously, the HAR domain of H2A, which comprises of the residues 16–20 of the N-terminal tail, has been implicated in transcriptional repression owing to its ability to govern nucleosome dynamics by interacting with the minor groove of DNA [26]. Although the HAR domain was initially identified in yeast, it was later shown to be important in humans as well [17]. In addition, the S16A substitution at the HAR domain was found to disrupt its repressive ability [17]. Our data shows that the S16T substitution does not significantly alter nucleosome dynamics by itself. However, a synergism is observed when this substitution is carried out alongside alteration at 51st and 99th residues. This suggests that probably the presence of a serine at the 16th position instead of threonine favors the disassembly of the H2A–H2B dimer from the NCP; however, the interactions of the 51st and 99th residue are predominant in governing the nucleosome stability.

One very important aspect that collectively emerges from our study and the earlier reports is that the functional effects exhibited by the H2A isoforms might be context dependent, in terms of both the extent and the effect itself. For example, the pro-proliferative effect conferred by H2A1H was not observed in the pre-neoplastic CL44 cells. Notably, the human H2A1C isoform, which was initially reported to be downregulated in CLL, was shown to exhibit anti-proliferative effects [10]. However, in a later study with a higher number of samples, H2A1C levels were found to be higher in CLL patients compared to the samples from healthy individuals [12]. Further, the high expression of H2A1H seen in the brain cells, which are terminally differentiated and do not regenerate, suggests that the actual functional effect of H2A1H may also be context dependent and need not be always proliferation associated. Interestingly, the higher expression

of H2A1C has been seen in chemo-resistance in the pancreatic cancer cell lines [27]. It remains to be seen how overexpression of H2A1C might contribute to that.

Based on the discussion above, some of the questions that arise are: what determines the context in which the non-redundant functionality of the H2A isoforms is exhibited? And in those contexts, which are the genes that are regulated by a particular isoform? Difficulty in raising specific antibodies against the endogenous H2A1H and H2A2A3 proteins, which differ in only three residues that are well spaced apart, poses a technical challenge to address these questions. It is reasonable to hypothesize that other factors that contribute to the epigenetic landscape of cells and/or the differential PTMs that the histone isoform itself may undergo, determine the context in which the differential functional effects of the H2A isoforms are exhibited. Interestingly, Arg 99 of H2A has been shown to undergo methylation [28]. One study, which has tried to identify the genes in the particular context, shows that H2A1C isoform controls ER target genes in ER-positive breast cancer cell lines [17]. Interestingly, the deletion of the H2A N-terminal domain (Δ4-20) led to upregulation of only 248 genes [26]. Clearly, much remains to be understood of the correlation between H2A-mediated nucleosome stability and gene expression.

Conclusion

H2A1H-containing nucleosomes are more stable owing to the M51L and K99R substitutions that also have the most prominent effect on cell proliferation, suggesting that the nucleosome stability is intimately linked with the physiological effects observed. Possibly, the increased nucleosome stability resulting from H2A1H incorporation contributes to the contextual alteration in the global gene expression pattern that collectively promotes the attainment of different physiological states. This possibility of the non-redundant function, when extended to the plethora of the histone isoforms (H2A, 12 isoforms; H2B, 16 isoforms; H3, 6 isoforms; and H1, 6 isoforms), truly increases the complexity of the epigenome by many folds. Undoubtedly, such complexity is the necessity for multicellular organisms as the diversity in the epigenome plays a central role in cell-type-specific gene expression. This, in turn, leads to the specialized functions in thousands of cell types with the same genome.

Methods

Antibodies and reagents

Anti-FLAG antibody (Sigma-Aldrich, A8592), anti-GFP antibody (Roche, 11,814,460,001), anti-GAPDH antibody (Ambion, AM4300), anti-H4 (Millipore, 07-108) and oligos (Sigma-Aldrich) were used.

Animal handling and experiments

All the experiments were performed on male Sprague–Dawley rats (spp. *Rattus norvegicus*) or SCID mice after approval of the Institute Animal Ethics Committee, Advanced Centre for Treatment Research and Education in Cancer and the Committee for the Purpose of Control and Supervision on Animals, India, standards. Protocol to induce the sequential stages of liver carcinogenesis is as previously described [9].

AUT-PAGE

Core histones were applied horizontally to the top of a 15% AUT-PAGE and sealed using sealing buffer (1% w/v agarose, 0.75 mol/L potassium acetate, pH 4, 20% v/v glycerol and 0.001% pyronin Y). The gel was electrophoresed at a constant voltage of 200 V.

RP-HPLC

Reversed-phase separation was carried out on a C18 column (1.0 × 250 mm, 5 mm, 300 Å; Phenomenex). Mobile phases A and B consisted of water and acetonitrile, respectively, with 0.05% trifluoroacetic acid. The flow rate was 0.42 ml/min, and the gradient started at 20% B and increased linearly to 30% B in 2 min, to 35% B in 33 min, 55% B in 120 min and 95% B in 5 min. After washing with 95% B for 10 min, the column was equilibrated at 20% B for 30 min, and a blank was run between each sample injection.

Mass spectrometry

Histone spots of interest from AUT-PAGE and the fractions of RP-HPLC were subjected to matrix-assisted laser desorption/ionization mass spectrometry (MALDI-MS) using MALDI-TOF/TOF mass spectrometer (Bruker Daltonics Ultraflex II). In brief, gel pieces were washed, destained, reduced, alkylated and subjected to in-gel digestion, and HPLC fractions were subjected to in solution trypsin digestion. Mass spectra were acquired on reflector ion positive mode. Database searching for protein masses was carried out using MASCOT search engine (version 2.2.03) by comparing the peptide masses with those in the NCBInr protein database (database version: NCBInr_20080812.fasta) in Rattus species. The searches were carried out with trypsin digestion, one missed cleavage, fixed carbamidomethylation of cysteine residues and optional oxidation of methionine with 100 ppm mass tolerance for monoisotopic peptide masses.

Isolation of total RNA and PCR

Total RNA was extracted from cells as per the manufacturer's (Macherey-Nagel) instructions. It was further treated with DNaseI for 30 min at 72 °C to degrade any possible DNA contamination. RNA (2 μg) was subjected to reverse transcription using M-MLV reverse transcriptase and random hexamer primers according to the manufacturer's (Fermentas) instructions. cDNAs were then amplified with the corresponding gene-specific primer sets (see Additional file 1: Figure S11). For RT-PCR, PCR was conducted for 24 cycles using the condition of 30 s at 94 °C, 1 min at 58 °C and 1 min at 72 °C. The PCR products were analyzed on a 1% agarose gels containing 0.5 μg/ml ethidium bromide. For real-time PCR Syber-Green from Ambion was used. The reactions were performed and monitored using QuantStudio 12K Flex Real-Time PCR System.

Histone purification and dimerization

Histones were purified and the H2A–H2B dimers were reconstituted as previously described [29]. The dimers were purified by size exclusion chromatography using HiLoad 16/60 Superdex-200 gel filtration column (GE).

Equilibrium unfolding of dimers

The dimers were subjected to equilibrium unfolding which was monitored by observing both secondary and tertiary structure changes.

Secondary structure changes

Unfolding was observed in response to thermal and chemical denaturant by circular dichroism.

Thermal unfolding

Unfolding was carried out starting from 20 up to 80 °C with a 2 °C increment and an equilibration time of 3 min. The CD spectra of only three temperatures are plotted for clarity. Analysis of the thermal unfolding curves suggests that dip at 222 nm can serve as a good spectroscopic probe for monitoring secondary structure unfolding [see Additional file 1: Figure S8a(i)]. Further, the unfolding was completely reversible with no protein aggregation as suggested by the completely overlapping unfolding and refolding curves [see Additional file 1: Figure S8a(ii)]. The data obtained could be fit into two-state unfolding model for dimeric proteins with residual in the range of only ± 2 using IgorPro [see Additional file 1: Figure S8a(iii)].

Chemical unfolding

Urea-induced denaturation was also monitored with CD with an increment of 0.2 M urea concentration starting from 0 M, and like thermal denaturation, the dip at 222 nm in the CD spectra was used to plot the unfolding [see Additional file 1: Figure S8b(i)]. Initially, a titration up to 8 M urea was carried out; however, as the unfolding was complete in 3 M urea, subsequent titrations were performed with up to 5 M concentration of urea. The denaturation was completely reversible [see Additional file 1: Figure S8b(ii)]. Similar to the thermal unfolding data, the chemical denaturation data could be fit into the two-state unfolding model [see Additional file 1: Figure S8c(iii)].

Tertiary structure changes

To follow the tertiary structure unfolding, urea-induced denaturation monitored by fluorescence spectroscopy was performed.

Chemical unfolding

On carrying out urea-induced denaturation, there was a drop in the fluorescence intensity with the unfolding of proteins as expected because of the quenching of fluorescence of the tyrosines previously buried in the dimer interface [see Additional file 1: Figure S8c(i)]. The drop in the intensity of emission maxima at 305 nm could be used for monitoring and plotting denaturation as there was no apparent redshift [see Additional file 1: Figure S8c(i)]. The folding was reversible [see Additional file 1: Figure S8c(ii)]; however, the pre- and post-transition baselines in the urea denaturation curve had a positive slope as observed in previous reports [19]. However, to ensure that transitions were not missed during the unfolding process, denaturation was carried out with GdmCl as well. Similar pre- and post-transition baselines corroborated the urea denaturation data (see Additional file 1: Figure S9). The unfolding also showed a concentration dependence as is expected for a dimeric protein [see Additional file 1: Figure S8c(iii)] and could be fit into the two-state model of unfolding [see Additional file 1: Figure S8c(ii)] substantiating the data obtained for secondary structure unfolding.

Site-directed mutagenesis For making mutants for the study, site-directed mutagenesis was performed using the kit and guidelines given in the QuickChange™ Site-Directed Mutagenesis Kit from Stratagene. Oligos were procured from Sigma-Aldrich.

Data fitting The unfolding data were fit into the two-state model of unfolding as described previously [30].

FRAP assay H2A1H and H2A2A3 coding sequences were cloned into peYFPn1 (YFP at C-terminal) vector and transfected in CL38 cells. LSM510 Meta (Zeiss) microscope equipped with CO_2 and temperature maintenance accessories was used to carry out the studies. The nuclei was bleached (in a box of fixed area) using 488-nm laser set at 100% power, and the recovery in the region was monitored for 1 h. Images were taken at 30-s intervals for the first 15 min and then at a 5-min interval for the remaining

45 min to minimize photobleaching. Quantification of the recovery was done as described previously [31].

Molecular dynamics simulation All the simulations were performed using the Gromacs-4.6.5 software, with periodic boundary conditions. The particle mesh Ewald method was used to treat the long-range electrostatics, together with a cutoff of 1.2 nm for the short-range repulsive and attractive dispersion interactions, which were modeled via a Lennard–Jones potential. The Settle algorithm was used to constrain bond lengths and angles of water molecules and the P-Lincs for all other bond lengths. The time step of 2 fs was used for the entire system. The temperature was kept constant at 300 K by using the Nose–Hoover thermostat method. To control the pressure at 1 atmosphere, Parrinello–Rahman method was used. The following DNA sequence was used to model nucleosomes: ATCAATATCCACCTGCAGATTCTACCAA AAGTGTATTTGGAAACTGCTCCATCAAAAGGCAT GTTCAGCTGAATTCAGCTGAACATGCCTTTTGAT GGAGCAGTTTCCAAATACACTTTTGGTAGAATCT GCAGGTGGATATTGAT.

Cell line maintenance and synchronization The cells from the human origin were maintained in appropriate growth media depending on the line at 37 °C with 5% CO_2 supplemented with 10% FBS, 100 U/ml penicillin, 100 mg/ml streptomycin and 2 mM L-glutamine (Sigma). Cell lines CL38 and CL44 from rat liver origin were cultured in MEM (invitrogen) media with 10% FBS and were maintained at 37 °C with 5% CO_2.

For overexpression experiments, mammalian expression vectors with CMV promoters (pcDNA3.1, pcDNA3.1 FLAG HA or peYFPn1) were used. The coding sequence of H2A1H (NM_001315492.1) or H2A2A3 (NM_001315493.1) was cloned in frame. For generating stable lines, the CL38 and CL44 cells were transfected with vectors (empty or encoding gene of interest) using TurboFect (ThermoFisher). Stable populations were selected by adding G418 (Sigma-Aldrich) in the growth media.

For cell cycle experiments, cells were enriched in the early G1-phase by serum starvation (0.1% FBS) for 24 h. Cells were released from the arrest by supplementing the media with 10% FBS.

Cell cycle analysis Ethanol-fixed cells were washed twice with PBS and suspended in 500 μl of PBS with 0.1% Triton X-100 and 100 μg/ml of RNaseA followed by incubation at 37 °C for 30 min. After incubation, propidium iodide (25 μg/ml) was added followed by incubation at 37 °C for 30 min. DNA content analysis was carried out in a FACS-Calibur flow cytometer (BD Biosciences, USA). Cell cycle analysis was performed using the ModFit software from Verity house.

Histone isolation and immunoblot analysis   First, nuclei were isolated from cells. For this, the cell pellet was resuspended in 0.1 ml PBS in a microcentrifuge tube. To this suspension, 0.9 ml lysis solution (250 mM sucrose, 50 mM Tris–Cl pH 7.5, 25 mM KCl, 5 mM $MgCl_2$, 0.2 mM PMSF, 50 mM NaHSO3, 45 mM sodium butyrate, 10 mM β-ME and 0.2% v/v Triton X-100) was added. Tube was inverted several times and centrifuged for 15 min at 800 g, 4 °C. For nuclei isolation from tissues, the tissue was homogenized in hypotonic buffer (10 mM HEPES pH 7.5, 10 mM KCl, 0.2 mM EDTA, 0.1% NP40, 10% glycerol, 1 mM DTT) using Dounce homogenizer. The homogenate was overlayed on the same buffer containing 1.8 M sucrose and ultracentrifuged (20,000*g* for 2 h). The nuclear pellet obtained was subjected to histone extraction by acid extraction method by adding 0.3 ml of 0.2 M H_2SO_4. The tubes were vortexed thoroughly with intermittent incubation on ice. The tubes were then centrifuged at 13,000*g*, 4 °C for 30 min. The supernatant was transferred to a fresh tube without disturbing the pellet. The proteins in the supernatant were precipitated by adding 4 volumes of acetone and stored overnight at −20 °C. The tubes were then centrifuged at 13,000*g*, 4 °C for 10 min. The pellet was washed once in chilled acidified acetone (0.05 M HCl in 100% acetone) and once in chilled 100% acetone. Protein pellet was dried in vacuum centrifuge for 15 min. The pellet was resuspended in 0.1% β-ME at −20 °C. For immunoblotting, histones were resolved on 18% SDS–polyacrylamide gel, transferred to PVDF membrane and probed with antibodies. Signals were detected by using ECL plus detection kit (Millipore; Catalogue no. WBKLS0500).

MTT assay   Cell viability was quantified by its ability to reduce tetrazolium salt 3-(4,5-dimethylthiazole-2Υ)-2,5-diphenyl tetrasodium bromide (MTT) to colored formazan products. MTT reagent (5 mg/ml in PBS) was added to the cells at 1/10th volume of the medium to stain only the viable cells and incubated at 37 °C for 4 h. MTT solubilization buffer (0.01 M HCl, 10% SDS) of twofold volume was added to the cells, followed by incubation in the dark at 37 °C for 24 h. The absorbance was measured at 570 nm with Spectrostar Nano-Biotek, Lab Tech plate reader. Cell viability was expressed as the percentage of absorbance obtained in the control cultures.

Colony formation assay   The cells (*n* = 1000) were plated in triplicate in 60-mm tissue culture plates, and they were allowed to grow as a monolayer for 14 days. Cells were incubated in complete culture medium, with media changes after every 2–3 days. After 14 days, the cells were fixed with 4% paraformaldehyde for 1 h. The colonies were stained with 0.5% crystal violet (0.5 in 70% ethanol) for 1 h at room temperature, rinsed and air-dried. Surviving colonies with more than 50 cells were counted, and images were captured using a high-resolution Nikon D70 camera (Nikon, Tokyo, Japan). For quantification of the size of the colonies, ImageJ was used.

Wound healing assay   Cells were seeded at a high density, serum-starved for 16 h and wounded when the cells formed a confluent monolayer. Recovery of the wounds was recorded by using an inverted microscope equipped with CO_2 and temperature maintenance accessory for 20 h with images captured at 10-min interval.

MNase digestion assay   Nuclei containing 2 mM $CaCl_2$ were incubated for 2, 4, 6, 8 and 10 min with 5U MNase/mg of DNA at 37 °C in MNase digestion buffer (15 mM Tris–Cl pH 7.4, 15 mM NaCl, 2 mM $CaCl_2$, 60 mM KCl, 15 mM β-ME, 0.5 mM spermidine, 0.15 mM spermine, 0.2 mM PMSF, protease and phosphatase inhibitors). The digestion was stopped by adding equal volume of 2 × lysis buffer (0.6 M NaCl, 20 mM EDTA, 20 mM Tris–Cl pH 7.5, 1% SDS). MNase-digested samples were treated with RNaseA (100 μg/ml) for 30 min at 37 °C followed by proteinase K (80 μg/ml) treatment for 2 h at 50 °C. The samples were extracted sequentially with phenol, phenol/chloroform and chloroform followed by ethanol precipitation at −20 °C. The precipitated DNA was recovered by centrifugation at 10,000*g* for 20 min. The DNA pellet was washed, air-dried and dissolved in TE buffer, and its concentration was determined by A260/A280 absorbance. MNase-digested samples were resolved on 1.8% 1XTAE agarose gel electrophoresis with 0.5 μg/ml ethidium bromide.

Additional file

Additional file 1: Figure S1. Multiple alignment of all the H2A protein sequences in rat. **Figure S2**. RP-HPLC chromatogram of histones isolated from the rat liver tissue. **Figure S3.** Unique peptides identified for the H2A isoforms. **Figure S4**. Real-time PCR of H2A isoforms in the normal vs tumor liver tissues. **Figure S5**. H2A.1 and H2A.2 isoforms in the CL44 and CL38 cells. **Figure S6**. Migration and cell proliferation upon overexpression of the histone H2A isoforms. **Figure S7**. Sequence comparison of H2A isoforms. **Figure S8**. Approach for equilibrium unfolding analysis of histone dimers. **Figure S9**. Guanidine chloride induced denaturation of H2A-H2B dimer. **Figure S10**. RMSD of MDS. **Figure S11**. Primers.

Authors' contributions

Experiments were majorly performed by SB. He also contributed in manuscript writing. DR conducted cell-based assay. SS performed HPLC. VJ conducted the in silico experiments. NG carried out analysis of in silico data. RR carried out gel filtration experiments and constructed mutants. KB contributed majorly toward designing biophysics-oriented experiments and manuscript editing.

US and RJ designed the in silico experiments. SG conceived the idea, planned the experiments and wrote the manuscript. All authors read and approved the final manuscript.

Author details

[1] Epigenetics and Chromatin Biology Group, Gupta Lab, Cancer Research Institute, Advanced Centre for Treatment, Research and Education in Cancer (ACTREC), Tata Memorial Centre, Kharghar, Navi Mumbai, MH 410210, India. [2] Integrated Biophysics and Structural Biology Lab, Cancer Research Institute, Advanced Centre for Treatment, Research and Education in Cancer (ACTREC), Tata Memorial Centre, Kharghar, Navi Mumbai, MH 410210, India. [3] BTIS, Cancer Research Institute, Advanced Centre for Treatment, Research and Education in Cancer (ACTREC), Tata Memorial Centre, Kharghar, Navi Mumbai, MH 410210, India. [4] Homi Bhabha National Institute, Training School Complex, Anushakti Nagar, Mumbai, MH 400085, India. [5] Bioinformatics Group, Centre for Development of Advanced Computing (C-DAC), University of Pune Campus, Pune, MH 411007, India. [6] Stowers Institute for Medical Research, Kansas City, MO 64110, USA.

Acknowledgements

The authors are grateful to Dr. Vikram Gota and Murari Gurjar, Gota Lab, ACTREC, for assistance with reverse-phase HPLC.

Competing interests

The authors declare that they have no competing interests.

Funding

SB and DR were supported by CSIR fellowship. SS and RR were supported by ACTREC fellowship when the work was conducted. We are grateful to the Department of Biotechnology, India, and ACTREC for funding Gupta Lab.

References

1. Luger K, Mädera W, Richmond RK, Sargent DF, Richmond TJ. Crystal structure of the nucleosome core particle at 2.8 A resolution. Nature. 1997;389:251–60.
2. Finch JT, Klug A. Solenoidal model for superstructure in chromatin. Proc Natl Acad Sci USA. 1976;73:1897–901.
3. Marzluff WF, Gongidi P, Woods KR, Jin J, Maltais LJ. The human and mouse replication-dependent histone genes. Genomics. 2002;80:487–98.
4. Khare SP, Habib F, Sharma R, Gadewal N, Gupta S, Galande S. Histome—a relational knowledgebase of human histone proteins and histone modifying enzymes. Nucleic Acids Res. 2012;40:1–6.
5. Rogakou EP, Sekeri-Pataryas KE. Histone variants of H2A and H3 families are regulated during in vitro aging in the same manner as during differentiation. Exp Gerontol. 1999;34:741–54.
6. Piña B, Suau P. Changes in histones H2A and H3 variant composition in differentiating and mature rat brain cortical neurons. Dev Biol. 1987;123:51–8.

7. Vassilev AP, Rasmussen HH, Christensen EI, Nielsen S, Celis JE. The levels of ubiquitinated histone H2A are highly upregulated in transformed human cells: partial colocalization of uH2A clusters and PCNA/cyclin foci in a fraction of cells in S-phase. J Cell Sci. 1995;108(Pt 3):1205–15.
8. Tyagi M, Khade B, Khan SA, Ingle A, Gupta S. Expression of histone variant, H2A.1 is associated with the undifferentiated state of hepatocyte. Exp Biol Med (Maywood). 2014;1535370214531869. http://ebm.sagepub.com/content/early/2014/04/24/1535370214531869.full.
9. Khare SP, Sharma A, Deodhar KK, Gupta S. Overexpression of histone variant H2A.1 and cellular transformation are related in N-nitrosodi-ethylamine-induced sequential hepatocarcinogenesis. Exp Biol Med (Maywood). 2011;236:30–5.
10. Singh R, Mortazavi A, Telu KH, Nagarajan P, Lucas DM, Thomas-Ahner JM, et al. Increasing the complexity of chromatin: functionally distinct roles for replication-dependent histone H2A isoforms in cell proliferation and carcinogenesis. Nucleic Acids Res. 2013;41:9284–95.
11. Su X, Lucas DM, Zhang L, Xu H, Zabrouskov V, Davis ME, et al. Validation of an LC-MS based approach for profiling histones in chronic lymphocytic leukemia. Proteomics. 2009;9:1197–206.
12. Singh R, Harshman SW, Ruppert AS, Mortazavi A, Lucas DM, Thomas-Ahner JM, et al. Proteomic profiling identifies specific histone species associated with leukemic and cancer cells. Clin Proteom BioMed Central. 2015;12:22.
13. Zhang H-H, Zhang Z-Y, Che C-L, Mei Y-F, Shi Y-Z. Array analysis for potential biomarker of gemcitabine identification in non-small cell lung cancer cell lines. IntJ Clin Exp Pathol. 2013;6:1734–46.
14. Olivares I, Ballester A, Lombardia L, Dominguez O, López-Galíndez C. Human immunodeficiency virus type 1 chronic infection is associated with different gene expression in MT-4, H9 and U937 cell lines. Virus Res. 2009;139:22–31.
15. Singh MK, Scott TF, LaFramboise WA, Hu FZ, Post JC, Ehrlich GD. Gene expression changes in peripheral blood mononuclear cells from multiple sclerosis patients undergoing β-interferon therapy. J Neurol Sci. 2007;258:52–9.
16. Zhan F, Barlogie B, Arzoumanian V, Huang Y, Williams DR, Hollmig K, et al. Gene-expression signature of benign monoclonal gammopathy evident in multiple myeloma is linked to good prognosis. Blood. 2007;109:1692–700.
17. Su CH, Tzeng TY, Cheng C, Hsu MT. An H2A histone isotype regulates estrogen receptor target genes by mediating enhancer-promoter-3′-UTR interactions in breast cancer cells. Nucleic Acids Res. 2014;42:3073–88.
18. Kavak E, Unlü M, Nistér M, Koman A. Meta-analysis of cancer gene expression signatures reveals new cancer genes, SAGE tags and tumor associated regions of co-regulation. Nucleic Acids Res. 2010;38:7008–21.
19. Placek BJ, Harrison LN, Villers BM, Gloss LM. The H2A.Z/H2B dimer is unstable compared to the dimer containing the major H2A isoform. Protein Sci. 2005;14:514–22.
20. Lipscomb LA, Gassner NC, Snow SD, Eldridge AM, Baase WA, Drew DL, et al. Context-dependent protein stabilization by methionine-to-leucine substitution shown in T4 lysozyme. Protein Sci. 1998;7:765–73.
21. March R, April A. Nucleic acids research. Nucleic Acids Res. 1987;5:3987–96.
22. Gassner NC, Baase WA, Matthews BW. A test of the "jigsaw puzzle" model for protein folding by multiple methionine substitutions within the core of T4 lysozyme. Proc Natl Acad Sci USA. 1996;93:12155–8.
23. Sokalingam S, Raghunathan G, Soundrarajan N, Lee S-G. A Study on the effect of surface lysine to arginine mutagenesis on protein stability and structure using green fluorescent protein. PLoS ONE. 2012;7:e40410.
24. Ettig R, Kepper N, Stehr R, Wedemann G, Rippe K. Dissecting DNA-histone interactions in the nucleosome by molecular dynamics simulations of DNA unwrapping. Biophys J Biophys Soc. 2011;101:1999–2008.
25. Hoch DA, Stratton JJ, Gloss LM. Protein-protein Förster resonance energy transfer analysis of nucleosome core particles containing H2A and H2A.Z. J Mol Biol. 2007;371:971–88.
26. Parra MA, Wyrick JJ. Regulation of gene transcription by the histone H2A N-terminal domain. Mol Cell Biol. 2007;27:7641–8.
27. Zhang Y, Shi X. Department of General Surgery, Zhongda Hospital of Southeast University, Nanjing 210009, Jiangsu Province C. Expression of HIST1H2AC and HIST1H2BC genes in pancreatic cancer cell line Puta8988 treated with pemetrexed. World Chin J Dig 2009;14.

28. Zhang K, Tang H. Analysis of core histones by liquid chromatography-mass spectrometry and peptide mapping. J Chromatogr B Anal Technol Biomed Life Sci. 2003;783:173–9.

29. Tanaka Y, Tawaramoto-Sasanuma M, Kawaguchi S, Ohta T, Yoda K, Kurumizaka H, et al. Expression and purification of recombinant human histones. Methods. 2004;33:3–11.

30. Mann CJ, Matthews CR. Structure and stability of an early folding intermediate of *Escherichia coli* trp aporepressor measured by far-UV stopped-flow circular dichroism and 8-anilino-1-naphthalene sulfonate binding. Biochem Am Chem Soc. 1993;32:5282–90.

31. Wiedemann SM, Mildner SN, Boenisch C, Israel L, Maiser A, Matheisl S, et al. Identification and characterization of two novel primate-specific histone H3 variants, H3.X and H3.Y. J Cell Biol. 2010;190:777–91.

Chromatin organization changes during the establishment and maintenance of the postmitotic state

Yiqin Ma and Laura Buttitta*

Abstract

Background: Genome organization changes during development as cells differentiate. Chromatin motion becomes increasingly constrained and heterochromatin clusters as cells become restricted in their developmental potential. These changes coincide with slowing of the cell cycle, which can also influence chromatin organization and dynamics. Terminal differentiation is often coupled with permanent exit from the cell cycle, and existing data suggest a close relationship between a repressive chromatin structure and silencing of the cell cycle in postmitotic cells. Heterochromatin clustering could also contribute to stable gene repression to maintain terminal differentiation or cell cycle exit, but whether clustering is initiated by differentiation, cell cycle changes, or both is unclear. Here we examine the relationship between chromatin organization, terminal differentiation and cell cycle exit.

Results: We focused our studies on the *Drosophila* wing, where epithelial cells transition from active proliferation to a postmitotic state in a temporally controlled manner. We find there are two stages of G_0 in this tissue, a flexible G_0 period where cells can be induced to reenter the cell cycle under specific genetic manipulations and a state we call "robust," where cells become strongly refractory to cell cycle reentry. Compromising the flexible G_0 by driving ectopic expression of cell cycle activators causes a global disruption of the clustering of heterochromatin-associated histone modifications such as H3K27 trimethylation and H3K9 trimethylation, as well as their associated repressors, Polycomb and heterochromatin protein 1 (HP1). However, this disruption is reversible. When cells enter a robust G_0 state, even in the presence of ectopic cell cycle activity, clustering of heterochromatin-associated modifications is restored. If cell cycle exit is bypassed, cells in the wing continue to terminally differentiate, but heterochromatin clustering is severely disrupted. Heterochromatin-dependent gene silencing does not appear to be required for cell cycle exit, as compromising the H3K27 methyltransferase *Enhancer of zeste*, and/or HP1 cannot prevent the robust cell cycle exit, even in the face of normally oncogenic cell cycle activities.

Conclusions: Heterochromatin clustering during terminal differentiation is a consequence of cell cycle exit, rather than differentiation. Compromising heterochromatin-dependent gene silencing does not disrupt cell cycle exit.

Keywords: *Drosophila*, Cell cycle exit, E2F, Histone modifications, Heterochromatin binding proteins

Background

Cellular differentiation is the acquisition of cell-type specific characteristics, driven by changes in gene expression. Changes in gene expression are largely controlled by transcription factors, which can be facilitated or impeded by chromatin modifications, binding site accessibility and chromatin organization. A reciprocal relationship exists between chromatin organization, modification and gene expression, and several studies have shown that chromatin organization and modifications can change during differentiation. For example, during neural differentiation silenced genes move to repressive compartments in the nucleus [1–3]. In certain contexts of differentiation, global nuclear compartments can become dramatically

*Correspondence: buttitta@umich.edu
Department of Molecular, Cellular and Developmental Biology,
University of Michigan, Ann Arbor, MI 48109, USA

reorganized to facilitate specialized functions [4]. At a more local level, chromatin modifiers can be recruited to specific genes involved in differentiation to facilitate their expression and limit the expression of genes involved in other cell-type programs that must be kept off [5]. Thus, dynamic changes in chromatin organization and modification can have critical consequences on proper differentiation during development.

There is also an intimate relationship between the cell cycle and chromatin organization and modifications. Chromatin in actively cycling cells is highly dynamic. During S-phase, new histones are incorporated onto nascent DNA requiring re-establishment of histone modifications [6]. During mitosis, nuclear organization including intra- and interchromosomal contacts is lost and many chromatin modifiers are ejected from chromatin to facilitate proper chromosome condensation and segregation [7, 8]. In addition, the activity of histone modifiers can be regulated in a cell cycle-dependent manner [9–14]. During differentiation cells often transition from rapid proliferation to slower cycling, which can be followed by cell cycle exit or entry into G_0 coordinated with terminal differentiation. Thus, the modification and organization of chromatin in the nucleus can be impacted by the differentiation process itself, but also by the changes in cell cycle dynamics during differentiation. For example, chromatin compacts and heterochromatin clusters as cells in the embryo cycle more slowly and become lineage restricted [15]. In *Drosophila* loci within constitutive heterochromatin show increased association in terminally differentiated postmitotic cells [16] and facultative heterochromatin-forming Polycomb bodies cluster as cells differentiate and the cell cycle slows during embryogenesis [17]. Methods such as inducing developmental arrest have been used in attempt to disentangle the influence of cell cycle changes from differentiation process [16], but these approaches cannot fully uncouple terminal differentiation from the accompanying cell cycle exit and it has remained unclear whether changes in heterochromatin clustering and dynamics are due to differentiation, the accompanying cell cycle changes, or both. The influence of cell cycle changes during differentiation adds a layer of complexity to our understanding of the relationship between chromatin organization and modifications and differentiation.

Here we directly address the relationship between heterochromatin organization, chromatin modification and cell cycle exit using the temporally controlled cell cycle exit in the *Drosophila* wing [18–20]. In our experiments, we take advantage of tools that can effectively uncouple cell cycle exit and differentiation to ask whether heterochromatin clustering is a consequence of cell cycle exit or differentiation. In addition, we examine changes in

chromatin modifications caused by the delay of cell cycle exit and examine the impact of disrupting heterochromatin-dependent gene silencing on cell cycle exit.

Methods

Fly stocks and genetics

Disruption of G0 in the posterior wing

w/y, w, hs-FLP; en-GAL4, UAS-GFP/UAS-E2F1, UAS-DP; tub-gal80TS/+

w/y, w, hs-FLP; en-GAL4, UAS-GFP/UAS-CycD, UAS-Cdk4; tub-gal80TS/UAS-E2F1, UAS-DP

w/y, w, hs-FLP; en-GAL4, UAS-GFP/+; tub-gal80TS/UAS-CycE, UAS-Cdk2

w/y, w, hs-FLP; en-GAL4, UAS-GFP/UAS-CycD, UAS-Cdk4; tub-gal80TS/+

Disruption of G0 in clones

w/y, w, hs-FLP; tub > CD2 > GAL4, UAS-GFP/UAS-CycD, UAS-Cdk4; tub-gal80TS/UAS-E2F1, UAS-DP

Disruption of H3K27me3

w/y, v; en-GAL4, UAS-GFP/+; tub-gal80TS/UAS-E(z)^{RNAi} (Bloomington 33659)

Disruption of HP1

w/y, v; en-GAL4, UAS-GFP/+; tub-gal80TS/UAS-Su(var)205^{RNAi} (Bloomington 33400)

Disruption of HP1 with Y10C

w/w, Y10C; en-Gal4, UAS-RFP/+; +/UAS-Su(var)205^{RNAi} (Bloomington 33400)

All the crosses containing gal80^{TS} were maintained in 18 °C to suppress Gal4 in early development. To disrupt G_0 with cell cycle regulators, white prepupae were collected and shifted to 28 °C to indicated time points. For $E(z)$ knockdown experiments, L3 larva were shifted from 18 to 28 °C to induce $E(z)$ RNAi. For *HP1* knockdown, crosses were kept in 28 °C after egg laying (AEL). For clonal expression of cell cycle regulators, animals were heat-shocked in 37 °C for 8 min during 48–72 h AEL and then kept in 18 °C. White prepupae were collected and shifted to 28 °C to indicated time points. All timings are adjusted according to the equivalent development at 25 °C as described previously [18].

Immunostaining

Imaginal disks or pupal wings were dissected in $1 \times$ PBS and fixed in 4% paraformaldehyde/$1 \times$ PBS for 30 minutes. Samples were washed twice in $1 \times$ PBS, 0.1% Triton X, 10 min each and incubated in PAT ($1 \times$ PBS, 0.1% Triton X-100 and 1% BSA) for 10 min for larval tissues and 3×20 min for pupal tissues. Samples were then incubated with primary antibodies for 4 h or 4 °C overnight followed by three washes and secondary antibodies at room temperature for 4 h or 4 °C overnight. Primary antibodies used in this study include: Anti-phospho-Ser10 histone H3, 1:2000 rabbit (Millipore #06-570) or mouse (Cell Signaling #9706); Anti-GFP, 1:1000 chicken (Life Technologies A10262) or 1:1000 rabbit (Life Technologies A11122); Anti-pH2Av, 1:100 mouse (DSHB, UNC93-5.2.1); Anti-H3K27me3, 1:500 rabbit (Millipore #07-449); Anti-HP1, 1:250 mouse (DSHB, C1A9); Anti-H2Av, 1:500 rabbit (Active Motif #39715); Anti-H3, 1:500 mouse (Cell Signaling #3638); Anti-H3ac, 1:500 rabbit (Millipore #06-599); Anti-H3K4me3, 1:500 rabbit (Millipore #07-473); Anti-H3K9me3, 1:500 rabbit (Millipore #07-523) or (Active Motif #39161); Anti-H3K27ac, 1:500 rabbit (Abcam ab4729); Anti-H4ac, 1:500 rabbit (Millipore #06-866); Anti-H4K16ac, 1:500 rabbit (Millipore #07-329); Anti-H4K20me3, 1:500 mouse (Abcam ab78517); Anti-E2F, 1:500 guinea pig (kindly provided by Dr. Terry L. Orr-Weaver); Anti-Ubx, 1:250 mouse (DSHB, FP3.38); Anti-D1, 1:200 guinea pig (kindly provided by Dr. Yukiko Yamashita). DNA was labeled by 1 µg/ml DAPI in 1 × PBS, 0.1% Triton X for 10 min. F-actin was stained using 1:100 rhodamine–phalloidin (Invitrogen; R415) in 1 × PBS for 4 h.

Microscopy and image quantification

Images were taken with a 100 × oil objective on a Leica SP5 confocal with a system optimized z-section of 0.13 µm. Three-dimensional reconstructions were performed using the "3D viewer" function in Leica LAS AF software. Images of whole pupal wings in Figs. 2 and 7 were obtained using a Leica DMI6000B epifluorescence system. All adjustments of brightness or contrast were applied to the entire image in Adobe Photoshop and performed equally with equal threshold values across control and experiment samples.

For integrated intensity quantifications, we used maximum projections of 12 continuous z-sections of confocal images. We developed a toolkit in MATLAB (Release 2015b) that automatically segments nuclei and foci within nuclei and integrates the pixel intensities with the help of the Advocacy and Research Support, U. Michigan LSA-IT. To identify nuclei, images were smoothed using a circular averaging filter through the fspecial and imfilter function of MATLAB. Next, a watershed algorithm was applied to segment nuclei from the background, and nuclei were masked using local maxima with an h-maxima transform. Thresholds were manually set and checked for each image to accurately delineate nuclei. GFP positive versus negative was established using an intensity threshold for the GFP channel. Integrated intensities for all nuclei were exported to Excel. Segmentation and measurement of foci followed

a similar process for foci within the defined nuclear regions. In brief, foci were segmented using a watershed algorithm and then further measured for pixel intensity and number, which was used for foci area and intensity measurements.

Fluorescent in situ hybridization (FISH)

Alexa-488 probes against the rDNA internal transcribed spacer (ITS) region and Cy3 probes against AACAC repetitive satellite sequences were kindly provided by Dr. Yukiko Yamashita. For FISH, fixed tissues were treated with 2 mg/ml Rnase A in 1 × PBS, 0.1% Triton X at 37 °C for 10 min, and rinsed in 2 × SSC/1 mM EDTA/0.2% Tween 20. Then tissues were incubated in 2 × SSC/1 mM EDTA/0.2% Tween 20 solution with increasing formamide concentration from 20, 40 to 50% for 15 min to 30 min. Finally, tissues were incubated in 100 µl hybridization solution with 50 µl formamide, 20 µl 50% dextran sulfate, 20 µl 2 × SSC/1 mM EDTA/0.2% Tween 20 and 10 µl of 10 µm probe for 15 min at 91 °C and left at 37 °C overnight. Quantification of size for rDNA loci area was carried out using our customized MATLAB toolkit. For the quantification of AACAC satellite to the chromocenter, we used a single 0.13-µm z-section with the strongest FISH signal and measured the relative distance of the center of the FISH signal to the brightest Dapi-stained region and corrected for the total nuclear radius using the Leica LAS AF software.

Flow cytometry

FACS was performed on dissociated wings to measure DNA content on an Attune Cytometer (Life Technologies) as described [21].

RNA interference

Kc167 cells were kindly provided by Dr. K. Cadigan and cultured as described [22]. For RNA interference, cells were placed with concentration of 1 million/ml and starved in serum-free medium with 10 µg/ml double-strand RNA (dsRNA) for 4–6 h, then 10% serum medium was added to the culture, and cells were collected for staining 3 days after serum medium addition. dsRNA was synthesized with T7 Megascript Kit (Ambion). T7 primers used in this study:

T7-Wee-fwd, TAATACGACTCACTATAGGGATGAC TTTGACAAGGACAC;
T7-Wee-rev, TAATACGACTCACTATAGGATCTAG TCGATTGACGCATT;
T7-Myt1-fwd, TAATACGACTCACTATAGGAATTG CACGACGACAAACAC;
T7-Myt1-rev, TAATACGACTCACTATAGGTGTCC AGATGGATGAGATTC;

T7-Myt1-fwd2, TAATACGACTCACTATAGGACAA
CAATCTGAACCGAAGC;

T7-Myt1-rev2, TAATACGACTCACTATAGGTGGA
GCCATATACCTCGAAT;

T7-GFP-fwd, TAATACGACTCACTATAGGCATGT
GGTCTCTCTTTTCGT

T7-GFP-rev, TAATACGACTCACTATAGGGGGCAC
AAATTTTCTGTCAG

T7-CycB-fwd TAATACGACTCACTATAGGGGCGT
TTTTGCGTTCGAATT

T7-CycB-rev, TAATACGACTCACTATAGGCAATT
GCAAGTACGTGCGTT.

Western blots

Western blots were performed on staged fly wings using Bio-Rad TGX precast 4–20% gels and high-sensitivity ECL reagents (Thermo) to detect HRP-conjugated secondary antibodies [23]. Mouse anti-α-tubulin (1:1000, DSHB, AA4.3) was used as a loading control. Blot signals were detected and quantified with FluorChem M digital system from ProteinSimple.

Results

Heterochromatin clusters as proliferation slows and cells differentiate

The impact of the cell cycle on heterochromatin clustering during cellular differentiation has not been resolved. Specifically, how the transition from a proliferative to a postmitotic state impacts global chromatin organization in *Drosophila* is unclear. To examine this, we immunostained for various chromatin marks and chromatin binding proteins in wild-type *Drosophila* wings at three stages with distinct proliferation parameters. We examined quickly proliferating second instar larval (L2) wings, slowly proliferative wandering third instar larva (L3) and postmitotic 28 h pupal wings (Fig. 1A). Cells of the L2 wing region examined have a cell doubling time (CDT) of about 10 h, while cells of the same region in L3 wings have a longer CDT of 15 h. By 28 h after puparium formation (APF) during metamorphosis, cells of the wing

blade have entered G0 and are permanently postmitotic [18, 24, 25]. We examined the histone modification H3K27me3 associated with facultative heterochromatin, H3K9me3, HP1 and the AT-rich repetitive sequence binding protein D1 associated with constitutive heterochromatin and the euchromatin-associated modification H3K4me3 (Fig. 1A). The immunofluorescence (IF) signals for H3K27me3, H3K9me3 and D1 were weakest at the L2 stage, but increased at the L3 and pupal stages, and clustered into larger and more intense, distinct foci in the slower cycling tissues (Fig. 1A). In *Drosophila* cells, the chromocenter, containing constitutive heterochromatin such as clustered centromeres, can be easily visualized as a DAPI-bright region within the nucleus [26]. We confirmed the co-localization of the chromocenter with D1 staining, which binds centromeric satellite repeats, and also co-localized with the centromeric histone Cenp-A (not shown) [27]. H3K9Me3 and HP1 label heterochromatin foci partially overlapping and adjacent to the DAPI-bright region [28]. H3K27Me3 labels distinct foci throughout the nucleus associated with facultative heterochromatin and represents Polycomb repressive complex 2 (PRC2) binding and formation of Polycomb group (PcG) clusters or foci [29–31]. By contrast, H3K4Me3 broadly localizes throughout the chromatin, does not form distinct foci and is excluded from the centromeric and pericentromeric regions (Fig. 1A).

To automatically detect and measure heterochromatin foci parameters such as intensity and number for a large number of nuclei, we developed a custom MATLAB App (described in supplemental methods) that uses DAPI staining to mask individual nuclei followed by foci segmentation and measurement. We measured clustering of heterochromatin foci as a function of the integrated intensity for each focus (the sum of intensities for all pixels in a focus) [17, 31]. This automated approach allowed us to examine the distribution of heterochromatin foci at a single cell level, across hundreds to thousands of nuclei, sampled from multiple wings for each experiment in an unbiased manner. We found that heterochromatin

---

(See figure on next page.)

Fig. 1 Heterochromatin clustering increases as the cell cycle slows and cells differentiate. **A** Wings of the indicated developmental stages were immunostained for the indicated chromatin modifications and chromatin binding proteins. **B** As the cell cycle slows down and cells differentiate, the distribution of heterochromatin-associated foci shifts toward larger, brighter foci indicating increased clustering. Coalescence is quantified as the total intensity of individual focus within 129–448 nuclei at each developmental stage. **C**, **D** Kc cells treated with dsRNA against GFP, *wee/myt* and *cycB* were immunostained for the indicated chromatin modifications and fluorescence intensity was quantified. **E**, **F** Fluorescent in situ hybridization (FISH) against the rDNA ITS region was performed on wings of the indicated stages. rDNA foci coalesce and condense in postmitotic cells. **G**, **H** FISH against the AACAC pericentromeric satellite repeats was performed on wings of the indicated stages and the distance to the center of the DAPI-bright chromocenter was measured. The distance decreases in postmitotic cells indicating increased condensation of heterochromatin. **I** A box plot of the RNA $\log_2$-fold changes compared to proliferative L3 for each time point is shown. **J** A line plot of average FAIREseq signal across all accessible chromatin for the indicated stages is shown. The accessibility of regulatory elements is similar in cycling and postmitotic wings. Scale bars = 2 µm. *P* values were determined by an unpaired *t* test *< 0.05; **< 0.01; ***< 0.001; ****< 0.0001

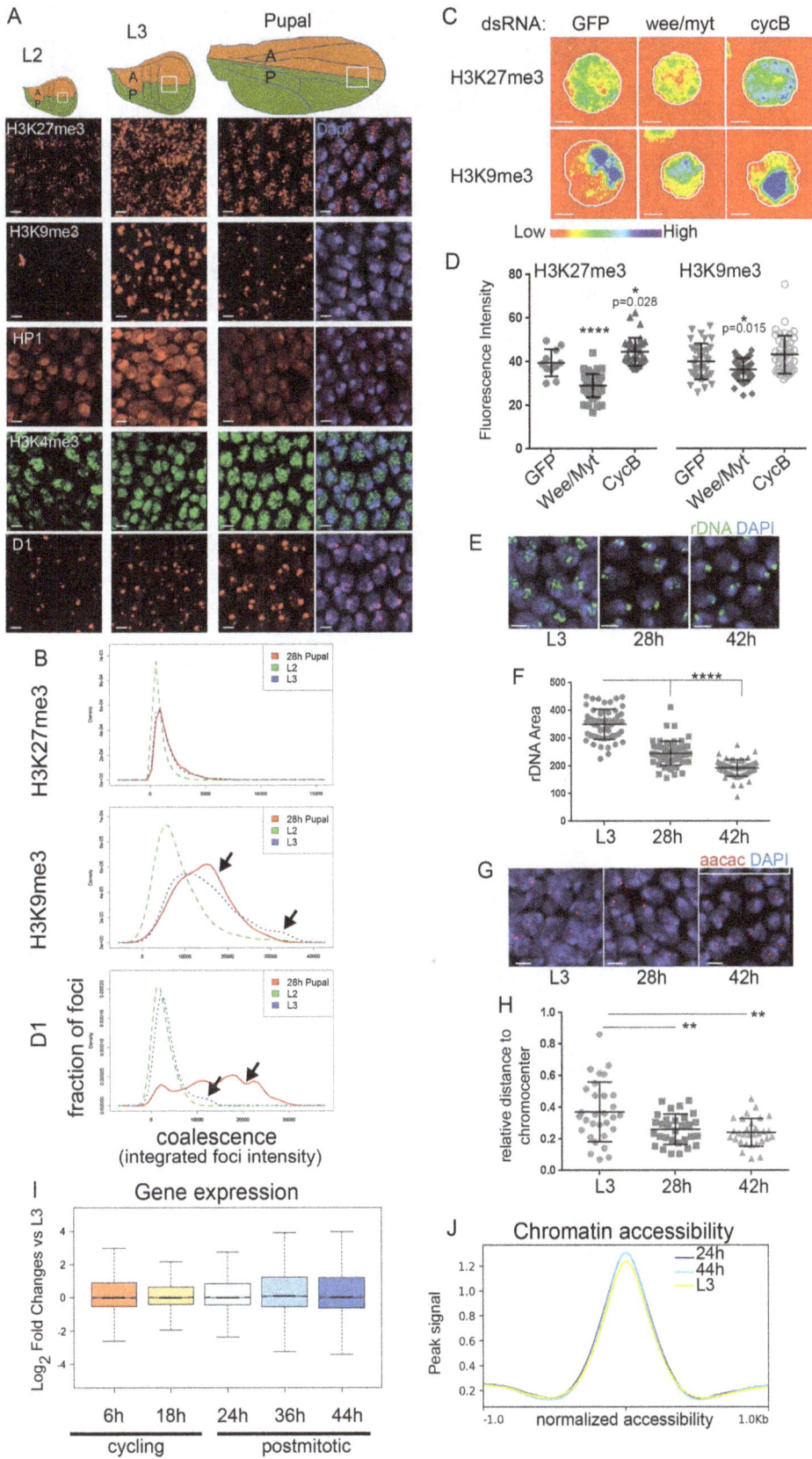

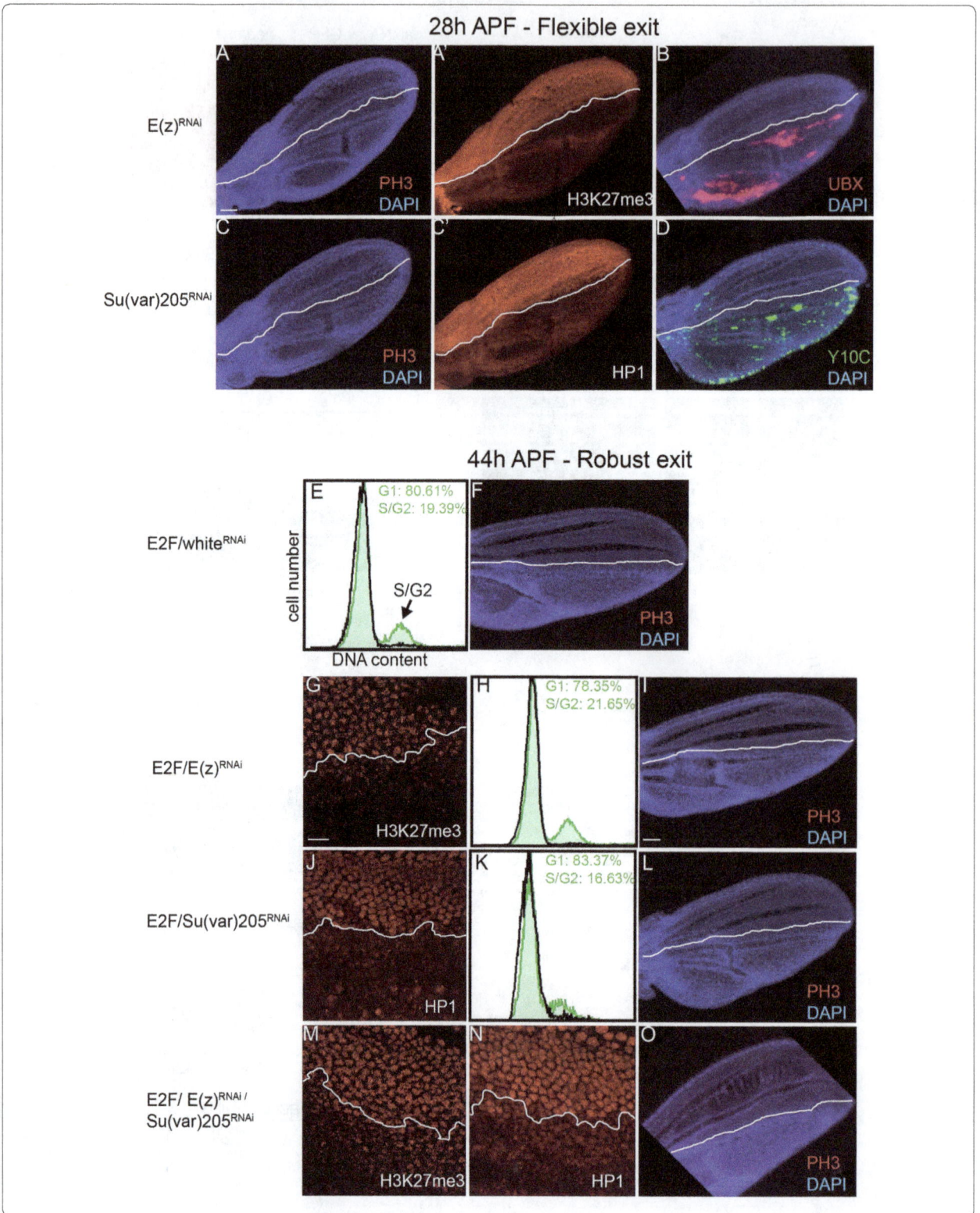

Fig. 2 Heterochromatin-dependent gene silencing is not required for cell cycle exit. **A, B** RNAi to *E(z)* was expressed in the posterior wing from the L3 stage until the indicated time points in metamorphosis. Postmitotic wings at 26–28 h were examined for mitoses as indicated by phospho-Ser10-histone H3, PH3 (**A**), H3K27me3 (**A′**) and de-repression of Ubx (**B**). **C** RNAi to *Su(var)205* (the gene encoding HP1) was overexpressed in the posterior wing, and postmitotic tissues were immunostained for PH3 and HP1 (**C′**). These conditions led to loss of HP1 and disrupted heterochromatin-mediated silencing of the Y10C reporter (**D**). Control RNAi (to the *white* gene), *E(z)* and/or *Su(var)205* was expressed in the posterior wing in combination with E2F from the start of metamorphosis. Postmitotic wings at 42–44 h were dissected and examined for H3K27me3 (**G, M**), HP1 (**J, N**) and PH3 (**F, I, L, O**). **E, H, K** Flow cytometry was also performed to measure cells that enter S and G2 phases. Green trace indicates cells from the posterior wing expressing the indicated transgenes. Black trace: control non-expressing anterior wing cells. Reduced heterochromatin gene silencing does not compromise G0 even in the presence of high E2F activity. Scale bars = 50 μm **A–D, F, I, L** and **O**; 10 μm **G, J, M** and **N**

clustering increased as the cell cycle slowed and stopped during L3 and pupal stages (Fig. 1B). We noted a dramatic increase in H3K9Me3 and HP1 staining at the L3 stage, which may reflect a developmentally controlled stage-specific increase in this modification/reader pair.

To distinguish whether an increase in heterochromatin clustering is due to the changes in the cell cycle, we turned to *Drosophila* cell culture. In *Drosophila* Kc cells, the overall cell doubling time is controlled by the negative and positive regulators of the G2/M transition Wee/Myt and Cyclin B, respectively [22]. We sped up the cell cycle by reducing Wee/Myt1 activity via RNAi or slowed the cell cycle using RNAi to *cyclin B*. Slowing the cell cycle increased the clustering and intensity of H3K27Me3 and H3K9Me3 compared to controls exposed to RNAi to GFP (Fig. 1C, D).

The increased clustering of heterochromatin could be due to chromatin condensation and compaction. To examine chromatin condensation, we performed fluorescent in situ hybridization (FISH) using probes against the internal transcribed spacer region between the 18S RNA and 28S rDNA loci, which are tandemly repeated on the X, and measured the total rDNA area before and after cell cycle exit in the wing [32, 33]. In proliferating L3 wings, the rDNA is extended. The rDNA becomes more compact as cells enter G0 at 24–28 h APF and condenses further as G0 is maintained at 42 h APF (Fig. 1E, F). The changes in the rDNA locus suggest chromatin condensation increases in prolonged G0. To verify that compaction is not specific to the rDNA locus on the X, we also performed FISH to the pericentromeric satellite repeat AACAC on chromosome II and measured the distance of the signal to the chromocenter (Fig. 1G, H). The distance of the pericentromeric heterochromatin to the chromocenter also decreased suggesting that heterochromatin condensation, coalescence and compaction occur throughout the nucleus after cell cycle exit.

An increase in chromatin clustering could be correlated with a global reduction in gene expression when cells become postmitotic [34]. To test whether global gene expression is reduced in postmitotic wings, we examined an RNAseq time course of gene expression

from proliferating to postmitotic stages [22]. We found the global gene expression levels to be similar in proliferating and postmitotic tissues (Fig. 1I); however, since RNAseq reveals steady-state mRNA levels, changes in RNA Pol II could still occur. Transcriptional shutoff upon quiescence in yeast is associated with a repressive chromatin structure and reduced chromatin accessibility [34]. Therefore, we also compared the global changes in chromatin accessibility between proliferating and postmitotic wings through Formaldehyde-Assisted Identification of Regulatory Elements (FAIRE)-seq [35] (Fig. 1J). Consistent with the global gene expression profile, we found no obvious changes in the average level of chromatin accessibility in cycling versus postmitotic tissue. This suggests that clustering of heterochromatin as cells exit the cell cycle does not cause global changes in genome accessibility or steady-state mRNA levels during differentiation and cell cycle exit.

Compromising heterochromatin-dependent gene silencing does not disrupt cell cycle exit

We have shown that heterochromatin clustering increases with entry into G0. Heterochromatin clustering is associated with increased target gene silencing [31] and has been suggested to repress cell cycle gene expression to facilitate cell cycle exit in mammalian muscle and neurons [36–38]. To test whether heterochromatin-dependent gene silencing promotes cell cycle exit in Drosophila wings, we compromised the H3K27me3 methyltransferase *E(z)* and/or the H3K9Me3 binding protein HP1. As *E(z)* and HP1 perform many functions during development, we turned to an inducible system with RNAi to alter gene function after embryogenesis. We used the *engrailed*-Gal4 driver with a temperature sensitive Gal80 (*en*-Gal4/Gal80^TS) to turn on UAS-driven expression of dsRNAs to *E(z)* and *HP1* in the posterior wing from the early L1 and L3 stages, respectively. We then dissected wings at 24–28 h APF and stained for the mitotic marker phosphorylated phospho-Ser-10 histone H3 (PH3) to determine whether cells in the posterior wing delayed or bypassed cell cycle exit. We saw no effect of *E(z)* or *HP1* reduction on cell cycle exit despite a clear loss of H3K27Me3 and HP1 in the posterior wing (Fig. 2A–C). We further confirmed that our knockdowns effectively compromised heterochromatin-dependent gene silencing in the wing, by examining de-repression of the Polycomb target Ultrabithorax (UBX) and the HP1-silenced Y10C GFP reporter (Fig. 2B, D). Recent work has suggested Polycomb (Pc) can repress certain targets independent of $E(z)$ [29]. We therefore also directly inhibited *Pc* by RNAi, but observed no effect on cell cycle exit despite de-repression of UBX in the wing (Additional file 1: Fig. S2).

Compromising heterochromatin-dependent gene silencing does not disrupt or delay cell cycle exit on its own, but we wondered whether it may sensitize cells to other perturbations that compromise cell cycle exit. We have previously shown that activation of various cell cycle regulators, including the cell cycle transcription factor complex E2F/DP (hereafter referred to as E2F), can cause 1–2 extra cell cycles in the pupa wing between 24 and 36 h APF followed by a delayed entry into G0 at 36 APF (Additional file 2: Fig. S3). We refer to the 24–36 h APF period as flexible cell cycle exit or "flexible G0" which is followed by a more difficult to disrupt "robust G0" after 36 h. We co-expressed E2F with RNAi to $E(z)$ and/or HP1 to examine whether loss of heterochromatin-dependent gene silencing can further delay cell cycle exit in the presence of high E2F activity. However, inhibition of $E(z)$, *HP1* or $E(z)$ + *HP1* together did not further compromise cell cycle exit in the presence of high E2F activity (Fig. 2G–O). Altogether our results demonstrate that compromising heterochromatin-dependent gene silencing does not disrupt cell cycle exit in the *Drosophila* wing.

Delaying cell cycle exit disrupts heterochromatin clustering and chromosome compaction
Constitutive and facultative heterochromatin clusters in postmitotic wings. To examine whether compromising cell cycle exit affects clustering, we used the system described above to express E2F in the posterior pupal wing to drive 1–2 extra cell cycles and delay exit from 24 to 36 h APF. We immunostained for the heterochromatin-associated histone modifications H3K27me3, H3K9me3 and H4K20me3 at 26–28 h APF, a time point when E2F induces abundant mitoses in the posterior

wing (Additional file 2: Fig. S3). We compared the clustering of the chromatin marks in the unperturbed anterior to the posterior wing. When cell cycle exit is delayed, all three modifications appear more diffuse throughout the nucleus and heterochromatin clustering is disrupted (Fig. 3A–M). To determine whether E2F altered the total abundance of the modified histones, we performed semi-quantitative western blots on 28-h pupal wings. With E2F expression, total levels of H3 were increased, consistent with additional S-phases leading to replication-coupled canonical histone production [39, 40]. However, the ratio of modified H3 to total H3 was relatively unchanged or even slightly increased when cell cycle exit was delayed (Additional file 3: Fig. S1). This may be because E2F activity also increases the expression of several PRC2 components ($E(z)$, *esc*, $Su(z)12$) and $Su(var)3-9$ as well as several other histone modifying enzymes (Additional file 4: Table S1), a feature conserved with mammalian E2Fs [41]. Thus, delaying cell cycle exit increases new histone production, but the histone modification rate is maintained by a coordinated increase in the expression of the modifying enzymes.

Delaying cell cycle exit disrupts the localization of heterochromatin-associated proteins
To determine whether delaying cell cycle exit also affected the localization of proteins associated with heterochromatin, we examined HP1, D1 and Polycomb using a Pc-GFP fusion protein [17]. We observed a more diffuse localization and a reduction in the clustering of these heterochromatin-associated proteins when cell cycle exit was compromised (Fig. 4A–M). This was also accompanied by a reduction in heterochromatin condensation, as assessed by the distance of the AACAC satellite to the chromocenter (Fig. 4N).

The accumulation of PRC1 components such as Pc into large foci or Pc bodies is important for target gene repression [17, 31]. Since E2F expression disrupts Pc clustering we examined whether increased E2F activity can disrupt the repression of Pc target genes [42–44]. We selected 12 high-confidence Pc target genes predicted not to be direct E2F targets based upon published genome-wide E2F complex binding in *Drosophila* [45]. We examined their expression upon E2F activation in pupal wings at 24 and 36 h APF using our previously published array data

---

(See figure on next page.)
Fig. 3 Heterochromatin clustering is disrupted when G0 is compromised. E2F was co-expressed with GFP in the posterior wing (boundary indicated by a white line) from the start of metamorphosis (0 h APF) to delay cell cycle exit. At 26–28 h wings were dissected and immunostained for the indicated chromatin modifications and DAPI to label nuclei (**A–I**). **J–M** Fluorescence signal were measured for 485–848 nuclei for each chromatin modification. The distribution of overall fluorescence intensity (**J**), foci number per nucleus section (**K**) and individual focus intensity (**L, M**) all indicate that delaying cell cycle exit disrupts heterochromatin clustering in wing cell nuclei. Scale bars = 10 μm in **A–I** except for anterior (A) and posterior (P) zoomed images where the bar = 2 μm (e.g., B$_A$, B$_P$). *P* values were determined by an unpaired *t* test. *< 0.05; **< 0.01; ***< 0.001; ****< 0.0001

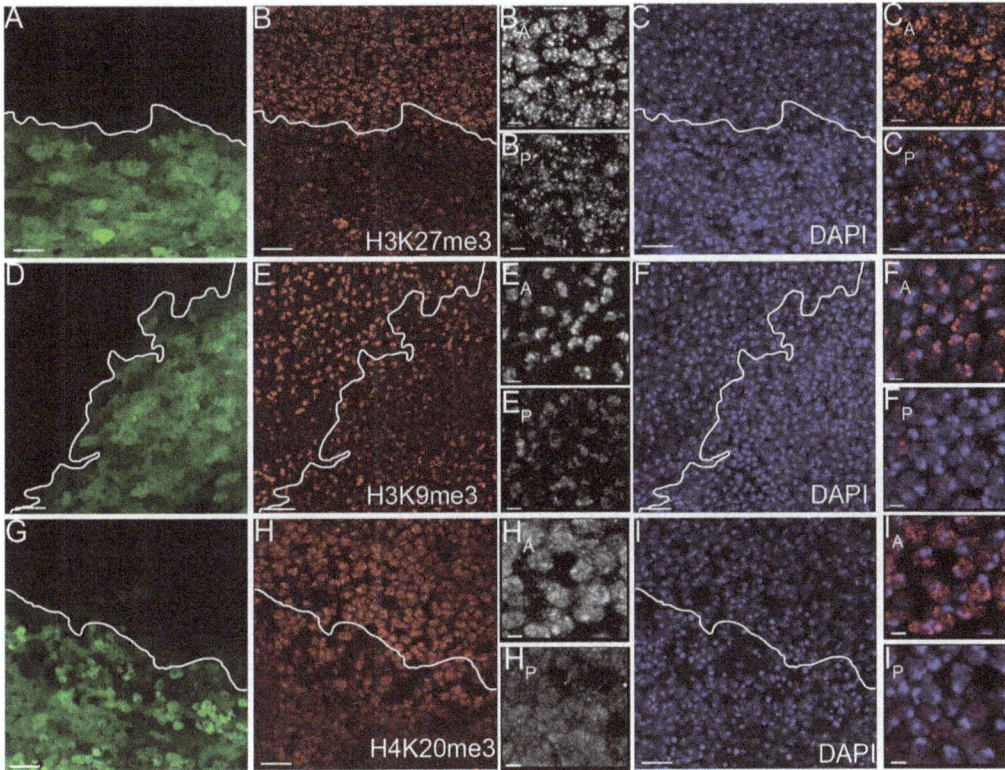

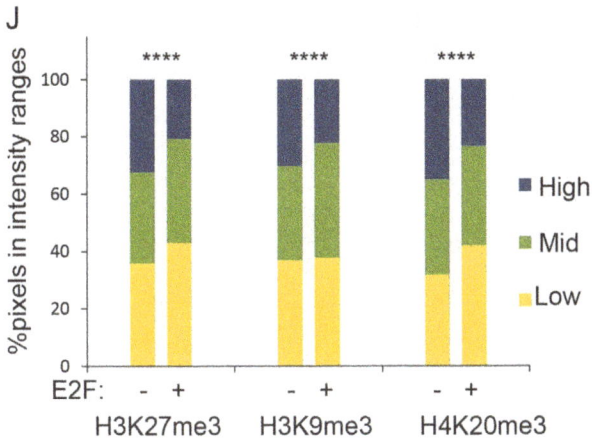

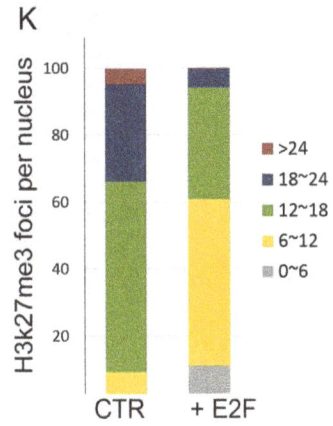

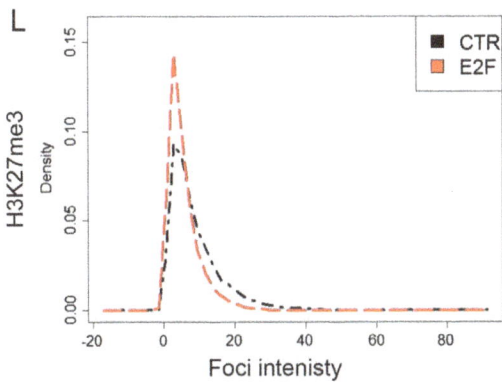

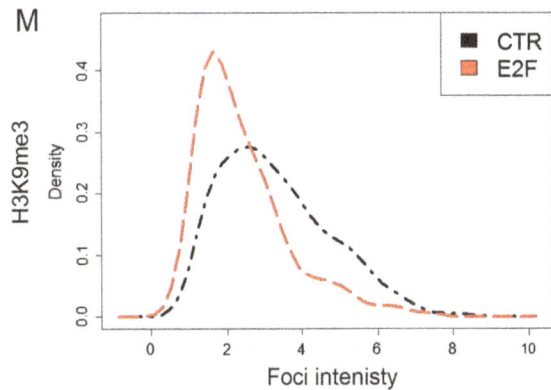

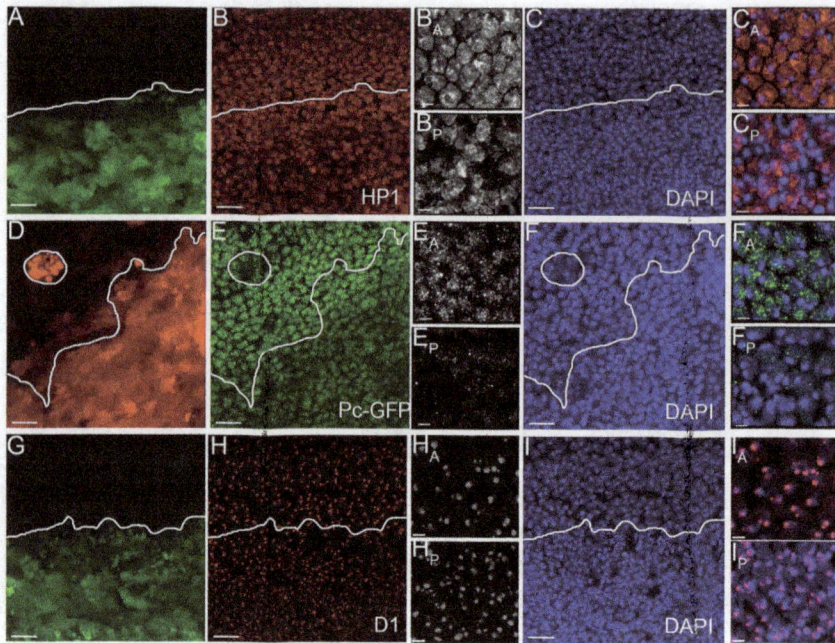

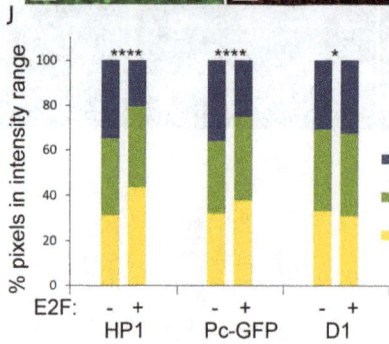

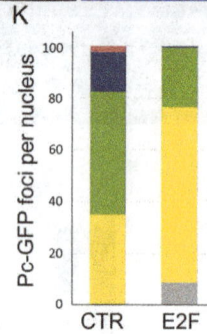

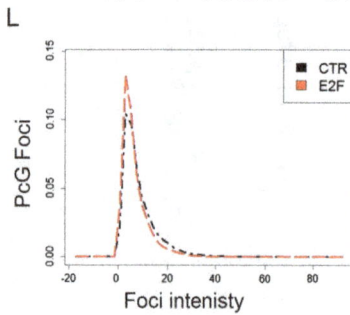

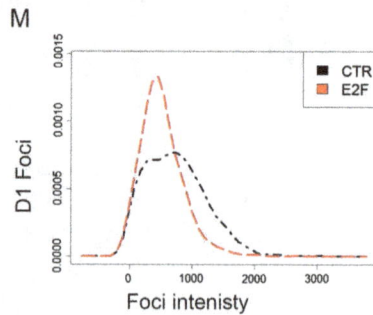

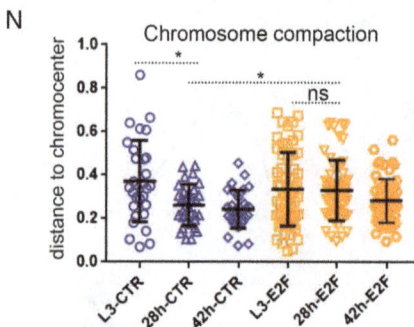

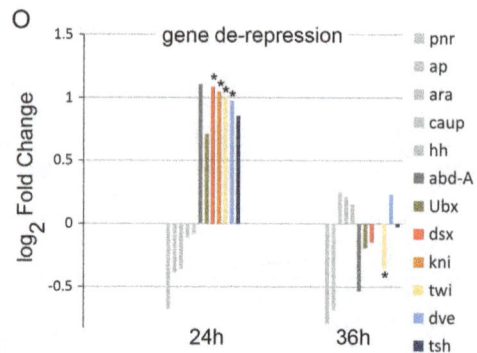

Fig. 4 Compromising G0 disrupts D1, HP1 and Polycomb body clustering and leads to partial de-repression of select PcG targets. E2F was co-expressed with GFP or RFP in the posterior wing to delay cell cycle exit. At 26–28 h wings were dissected and immunostained for the indicated heterochromatin binding proteins and DAPI to label nuclei (**A–I**). **J** Overall fluorescence intensities were measured for 319–1270 nuclei for each chromatin modification. The distribution of individual focus intensity (**L, M**), foci number per nucleus section (**K**) all indicate that delaying cell cycle exit disrupts heterochromatin clustering and formation of large Polycomb bodies in wing cell nuclei. **N** Chromosome compaction was measured using the distance of the AACAC repeats to the chromocenter. When cell cycle exit is delayed, chromosome compaction is also compromised. For **J, N**, P values were determined by an unpaired t test. *< 0.05; **< 0.01; ***< 0.001; ****< 0.0001. **O** Microarray analysis revealed specific PcG target genes that become temporarily de-repressed in wings expressing E2F at 24 h when cell cycle exit is delayed. P values were determined by ANOVA *< 0.05. Scale bars = 10 µm in **A–I** except for anterior (A) and posterior (P) zoomed images where the bar = 2 µm (e.g., B_A, B_P)

[46]. We found four Pc targets, *dsx, kni, twi* and *dve* to be reproducibly de-repressed 1.97–2.12-fold specifically at 24 h APF, during the window of time that cell cycle exit is delayed. This suggests that delaying cell cycle exit can partially compromise Pc-dependent gene silencing. E2F activity similarly impacts heterochromatin-dependent gene silencing at the pericentromeric heterochromatin, with the loss of *e2f1* increasing gene silencing by position effect variegation and an increase in E2F activity de-repressing variegated gene expression [47].

In our experiments to delay cell cycle exit, E2F is over-expressed for 28 h, which includes the final 1–2 normal cell cycles in the pupa wing as well as 1–2 extra cell cycles based upon lineage tracing [18, 46]. We therefore asked whether expression of E2F within only the final cell cycle during terminal differentiation is adequate to disrupt heterochromatin clustering. We used temperature shifts to limit the expression of E2F to a 12 h window within the final cell cycle in the pupa and observed a similar disruption of heterochromatin clustering (Additional file 5: Fig. S4). We also observed similar effects on heterochromatin clustering when cell cycle exit was delayed by expression of other cell cycle regulators such as CycE/Cdk2 or CycD/Cdk4 (Additional file 6: Fig. S5). This demonstrates that heterochromatin clustering in differentiating cells can be disrupted by a single extra cell cycle and that this effect is not specific to E2F overexpression.

Histone modifications associated with de-condensation are upregulated upon G0 disruption

Compromising G0 leads to the disruption of heterochromatin clustering and chromatin condensation (Fig. 4). H3K27ac and H4K16ac are associated with open chromatin such as active enhancers and origins [48–51], and H4K16ac can suppress the formation of higher-order chromatin structure [52]. We therefore examined whether these histone modifications were affected by delaying cell cycle exit with E2F overexpression. Indeed, during the delay of cell cycle exit, we observed dramatic increase in the levels of these two histone marks throughout the nucleus (Fig. 5A–D). However, other histone modifications associated with active chromatin were not

affected, such as H3K4me3, pan H3 and H4 acetylation (Fig. 5E–J). Thus, an increase of H3K27ac and H4K16ac could contribute to the compromised chromatin condensation and disruption of heterochromatin clustering observed when cell cycle exit is delayed.

Heterochromatin clustering is restored when cells enter a robust G0 state

Delaying cell cycle exit disrupts heterochromatin clustering; however, this is reversible. When we examined wings at 42–46 h, a time point when cells enter a robust G0 state refractory to E2F activation, heterochromatin clustering is either partially or completely restored (Fig. 6). Interestingly, levels of H3K27me3 and HP1 became higher in robust G0 after cell cycle exit is delayed (Fig. 6A, D). This could be due to the E2F-dependent upregulation of E(z) and Su(var)3-9, which may indicate an expansion of heterochromatin in differentiating cells that enter a robust G0. Consistent with this idea we also observe an increase in the H2A variant H2Av in E2F-expressing cells in robust G0 (Fig. 6F) and an upregulation of several components of the NuA4 complex responsible for incorporation of H2Av (Additional file 4: Table S1). Heterochromatin expansion is associated with senescence, suggesting delaying cell cycle exit with E2F overexpression could induce oncogenic stress or senescence-like features [53, 54]. Consistent with this, ectopic E2F in the wing induced multiple genes associated with senescence in mammals during robust G0 (Additional file 7: Table S2) and led to a widespread increase in phosphorylated H2Av, a hallmark of E2F-induced replication stress and DNA damage in *Drosophila* (Fig. 6E) [55].

Heterochromatin clustering during terminal differentiation is a consequence of cell cycle exit, rather than differentiation

Heterochromatin clustering becomes restored at the robust G0 phase in the wing as terminal differentiation proceeds. But whether differentiation or cell cycle arrest restores the heterochromatin clustering remains unclear. We previously demonstrated that the robust G0 state in the wing can be bypassed by co-expression

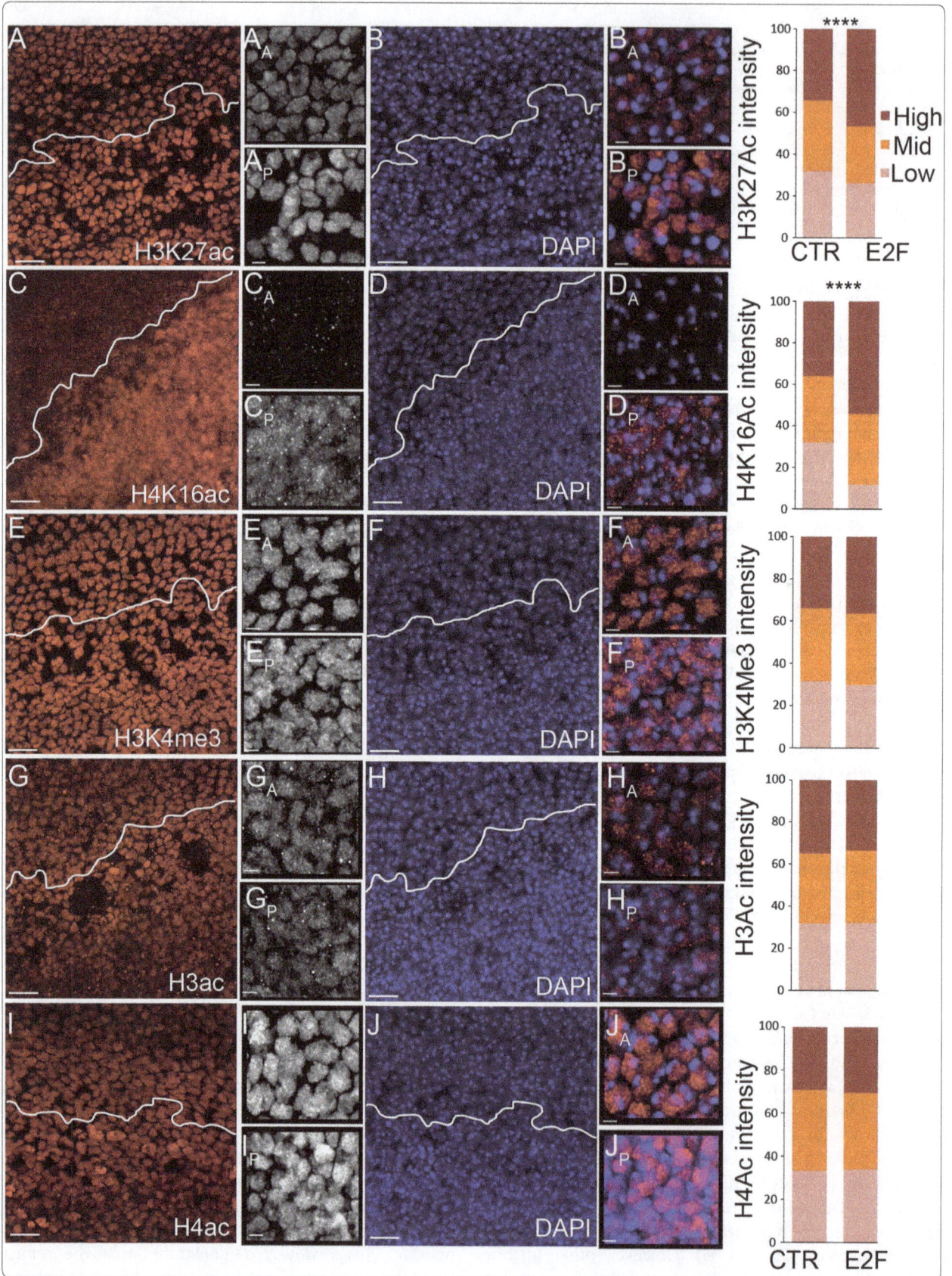

Fig. 5 Specific histone modifications associated with gene activation are increased when flexible G0 is compromised. E2F was expressed in the posterior wing to delay cell cycle exit. At 26–28 h wings were dissected and immunostained for the indicated histone modifications and DAPI to label nuclei (**A–J**). The anterior–posterior boundary is indicated by a white line. The distribution of staining intensity in 217–1312 nuclei, binned into three ranges, is shown at right. Compromising flexible G0 specifically increases H3K27ac and H4K16ac. P values were determined by an unpaired t test; **** < 0.0001. Scale bars $= 10$ μm in **A–J** except for anterior (A) and posterior (P) zoomed images where the bar $= 2$ μm (e.g., B_A, B_P)

of E2F + CycD/Cdk4 [18]. Under these conditions cells in the wing continue cycling past 48 h APF, yet physical hallmarks of wing terminal differentiation such as cuticle secretion and wing hair formation proceed after 36 h and adult wings form. This condition effectively uncouples cell cycle exit from terminal differentiation in the wing, with actively dividing cells forming actin-rich wing hairs and developing adult cuticle (Fig. 7A–E). We took advantage of this dividing-yet-differentiated context to ask whether heterochromatin clustering requires cell cycle exit. We immunostained 42 h wings expressing E2F + CycD/Cdk4 for H3K27Me3, H3K9Me3 and HP1 and found that clustering of facultative and constitutive heterochromatin was dramatically disrupted. We quantified facultative heterochromatin foci and found H3K27Me3 forming fewer, smaller and less intense foci (Fig. 7F–K). By contrast, HP1 levels became extremely high, with a diffuse localization throughout the nucleus (Fig. 7H, J), similar to the effects of E2F on HP1 at robust G0 (Fig. 6). These results demonstrate that heterochromatin clustering is a consequence of cell cycle exit rather than terminal differentiation. In addition, terminal differentiation can proceed despite a visibly significant disruption of heterochromatin organization.

Discussion

The relationship between heterochromatin clustering and differentiation

A number of studies have documented increased clustering and condensation of heterochromatin as cells differentiate [reviewed in 56]. In this study we reveal a substantial effect of the cell cycling status on heterochromatin clustering independent of differentiation. Heterochromatin clustering increases as the cell cycle slows and cells exit the cell cycle. By delaying or bypassing cell cycle exit in terminally differentiating cells, we show that the highly clustered state of heterochromatin in postmitotic cells is a consequence of cell cycle exit rather than the process of terminal differentiation. Importantly, we show that differentiation still proceeds even when cell cycle exit is prevented and heterochromatin clustering is severely disrupted (Fig. 7). We suggest this is because disrupting heterochromatin clustering has only limited effects on the expression of specific heterochromatin-repressed genes in the context of the *Drosophila* wing

(Fig. 4) and minimal effects on the terminal differentiation gene expression program. Indeed, we show that cell cycle exit can proceed normally in the *Drosophila* wing even when heterochromatin-dependent gene silencing is directly compromised (Fig. 2). Altogether this demonstrates that the increased heterochromatin clustering observed during differentiation is a consequence rather than a cause of cell cycle exit and raises questions regarding the function of increasing very long-range heterochromatin interactions and heterochromatin clustering in differentiation.

What is the function of heterochromatin clustering?

When we delay or bypass cell cycle exit, we visibly disrupt heterochromatin clustering. We find that this leads to very mild effects on the expression of only a small number of Polycomb target genes (Fig. 4), and we did not find significant de-repression of genes that are located in or near constitutive heterochromatin [57] (not shown). Our result is consistent with recent work showing that compromising some types of PcG clustering seems to have limited and selective effects on Polycomb target gene silencing [31]. However, the minimal effect of disrupting cell cycle exit on heterochromatin-dependent gene silencing is somewhat unexpected as the E2F1 gene was one of the early identified modifiers of position effect variegation (PEV), which is thought to be due to heterochromatin-dependent gene silencing through association with constitutive heterochromatin [47, 58, 59]. This suggests either the PEV assay is highly sensitive to even mild or selective changes in heterochromatin-dependent gene silencing or that this assay reads out changes in the chromatin state that are different from the silencing of the endogenous genes we examined. Indeed, there are additional possible functions for heterochromatin clustering beyond heterochromatin-dependent gene silencing. For example, heterochromatin clustering could facilitate DNA damage repair in postmitotic cells, which downregulate many DNA repair genes when they exit the cell cycle and become more reliant on error prone NHEJ [reviewed in 60]. Sequestration of heterochromatin may prevent inappropriate interactions and fusions. It has also been proposed that sequestration of heterochromatin could lead to an increased efficiency of gene activation for very highly expressed genes by reducing the availability of possible binding sites for specific

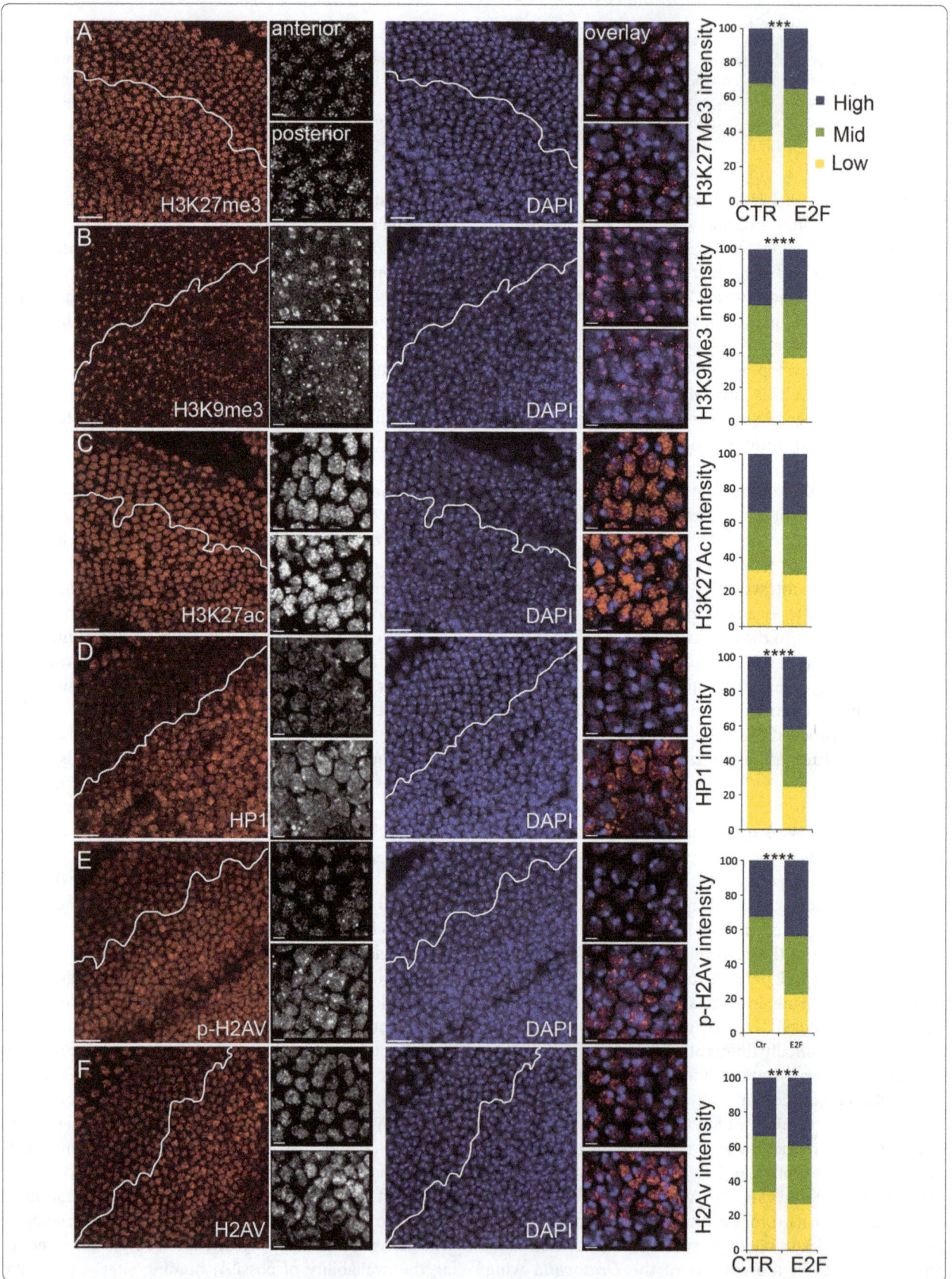

(See figure on previous page.)

Fig. 6 Robust G0 restores heterochromatin clustering and shares features with senescence. E2F was expressed in the posterior wing to delay cell cycle exit. At 42–44 h, when cells are in robust G0, wings were dissected and immunostained for the indicated histone modifications or chromatin binding proteins and DAPI to label nuclei (**A–F**). The distribution of staining intensity in 509–1185 nuclei, binned into three ranges, is shown at right. Robust G0 in the presence of high E2F increases H3K27Me3, HP1 and pH2Av, chromatin marks associated with senescence. P values were determined by an unpaired t test; ****< 0.0001; ***< 0.001. Scale bars = 10 μm in **A–F** except for anterior (A) and posterior (P) zoomed images where the bar = 2 μm (e.g., B_A, B_P).

transcription factors [56] or could facilitate the formation of transcription factories [61].

A number of other studies also describe changes in the abundance of specific chromatin modifications associated with entry into or exit from G0. For example, H3K9Me3 and H3K27Me3 accumulate in postmitotic, differentiated cardiac muscle [37], while H4K16Ac and H3K27Ac increase in activated B cells exiting G0 [62]. While our data from the *Drosophila* wing suggest clustering of H3K9Me3 and H3K27Me3 domains during cell cycle exit rather than obvious changes in total levels, we do observe a strong upregulation of H4K16 acetylation when G0 is delayed by E2F activation, a situation similar to cell cycle reentry from G0. We also observe a decrease in H4K20Me3 when G0 is compromised, similar to what has been reported for quiescent human fibroblasts [63]. While H4K16Ac was not specifically measured in the fibroblast study, H4K20 methylation and H4K16 acetylation are antagonistic marks [64] and H4K16 acetylation can de-compact nucleosomes in vitro, although whether this also occurs in vivo has been questioned [52, 65]. We suggest some aspects of chromatin remodeling, such as compaction and coalescence of heterochromatin (which may be tied to H4K16/20 dynamics) are shared among different contexts of cell cycle exit/reentry, while other chromatin changes associated with G0 entry/exit may be more cell-type specific.

Why does delaying or bypassing cell cycle exit disrupt heterochromatin clustering in interphase?

Our experiments effectively separate cell cycle exit from terminal differentiation to reveal that heterochromatin clustering is a consequence of cell cycle exit. Furthermore, heterochromatin clustering can be disrupted within a single cell cycle (Additional file 5: Fig. S4), suggesting progression through one round of S or M-phase is sufficient to disrupt heterochromatin clustering and long-range interactions. The effects we observe are not due to the dilution of chromatin marks by incorporation of new histones in S-phase, since we do not see changes in all histone marks (e.g., H3K4 methylation) or reduced global levels of histone marks in proliferating versus postmitotic cells (Additional file 3: Fig. S1). Indeed, when global levels of chromatin marks in actively proliferating

fibroblasts were quantified and compared to fibroblasts held in G0 under contact inhibition for 14d, the majority of histone modifications did not exhibit significantly different levels [63]. This is likely because the levels of many histone modifiers are upregulated by positive cell cycle regulators through E2F transcriptional activity (Additional file 4: Table S1) which effectively coordinates increased histone modification with increased production and incorporation in S-phase.

Overexpression of E2F could have effects on chromatin modifications and condensation through sequestration or indirect inhibition of RB-family proteins via increased Cyclin/Cdk expression. RB associates with chromatin modifying complexes that promote facultative and constitutive heterochromatin formation [66–68]. RB also impacts chromosome condensation and cohesin levels at pericentromeric heterochromatin [69–71]. However, in our experiments during robust exit, E2F levels and transcriptional targets remain high while heterochromatin clustering and chromatin marks are restored (Fig. 6). This suggests that even if RB is inhibited by overexpression of E2F, the eventual entry into robust G0 somehow restores heterochromatin organization and chromatin modifications independent of RB.

During mitosis most transcription factors and chromatin modifiers are ejected from chromatin and higher-order architecture is lost [7]. This together with mitotic spindle assembly leads to the loss of long-range interactions and interchromosomal associations. These interactions are then restored, even in the presence of high E2F activity once cells engage additional mechanisms to exit the cell cycle during the robust G0 phase [46]. Our findings are in agreement with previous studies showing that the motion of heterochromatin domains and Polycomb bodies become more constrained as the cell cycle slows and cells exit the cell cycle [16, 17]. We suggest that constrained motion combined with increased self-association or polymerization likely leads to the coalescence of heterochromatin after cell cycle exit.

Coalescence of heterochromatin can be driven by heterochromatin-bound proteins that self-associate such as HP1. HP1 has recently been shown to undergo phase separation to form liquid droplets that fuse when interphase becomes longer during *Drosophila* embryogenesis

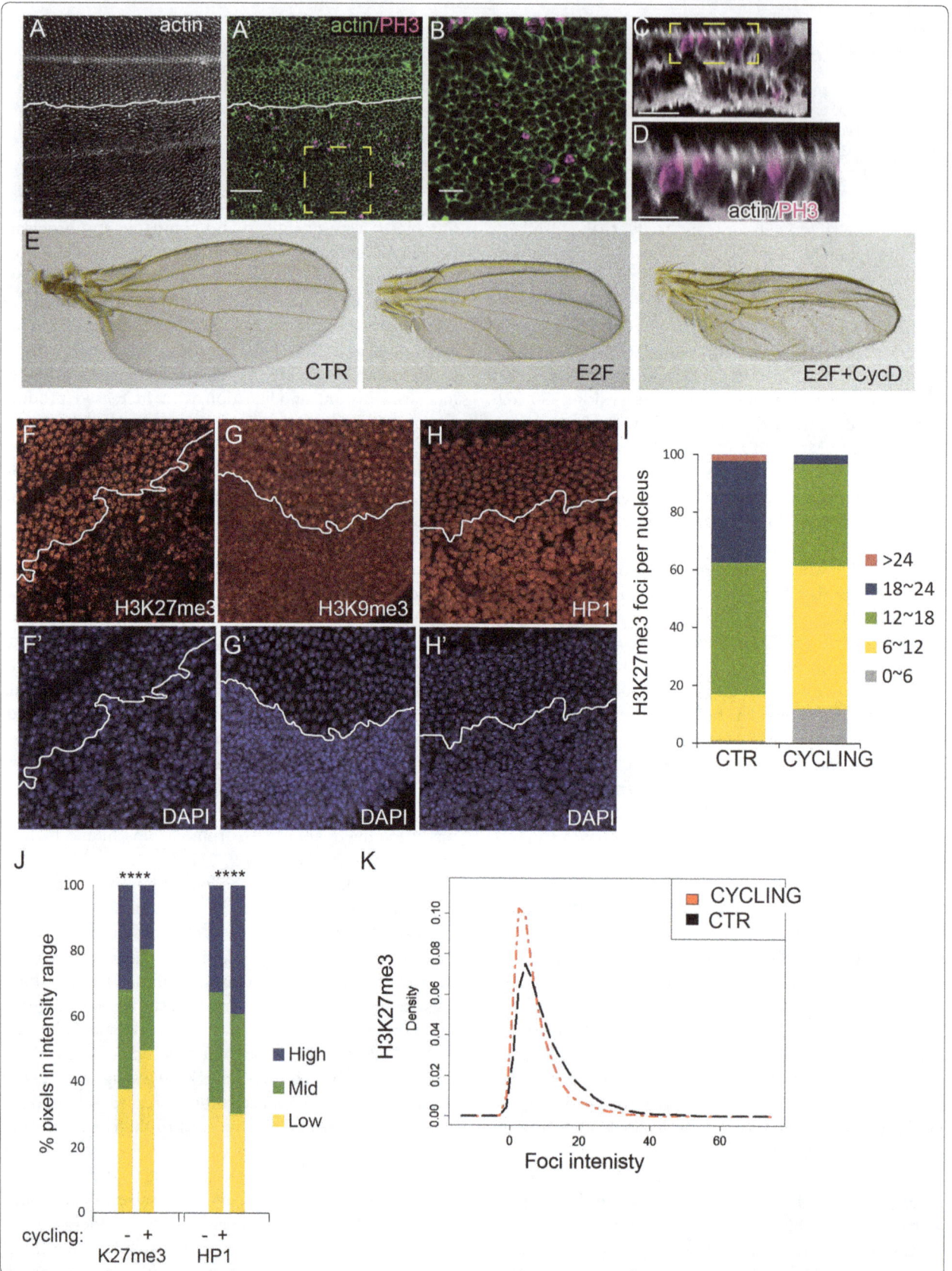

(See figure on previous page.)
Fig. 7 Heterochromatin clustering during terminal differentiation is a consequence of cell cycle exit. CycD, Cdk4 and E2F were co-expressed in the posterior wing to bypass robust cell cycle exit without preventing terminal differentiation. The anterior–posterior boundary is indicated by a white line. (**A–D**) Pupal wings at 42–44 h were dissected stained for actin and PH3 to label mitoses. Mitoses are evident in cells generating wings hairs, a hallmark of wing terminal differentiation (**E**) and the wings generate intact adult wing cuticle. **C**, **D** show optical cross sections (x/z) of wings reveal PH3 and actin-rich hairs in the same section. **F–K** CycD, Cdk4 and E2F expression in the posterior wing prevents G0 entry and disrupts proper localization of heterochromatin-associated histone modifications and HP1. **J** The distribution of staining intensity in 474–1191 nuclei, binned into three ranges is shown. The reduced foci number (**I**) and intensity (**K**) indicate compromised clustering of H3K27me3 containing chromatin when entry into G0 is prevented. P values were determined by an unpaired t test; ****< 0.0001. Scale bars = 10 μm in **A**, 5 μm in **B** and 2.5 μm in **D**

[72]. These droplets have been suggested to form diffusion barriers to limit heterochromatin access to factors involved in transcription such as TFIIB [73]. As the droplets fuse and mature during longer interphases, an immobile HP1 fraction forms. In our experiments to bypass cell cycle exit, we may limit the coalescence and maturation of HP1 droplets without preventing HP1 binding to H3K9me3. This could explain why dramatic effects on HP1 clustering may have minimal effects on gene de-repression. Alternatively, the role for HP1 clustering after cell cycle exit may largely affect silencing of transposons and piRNA clusters, an intriguing possibility to be addressed in future studies [74].

Conclusions
Heterochromatin clusters as cell exit the cell cycle and terminally differentiate. Delaying or preventing cell cycle exit disrupts heterochromatin clustering and globally alters chromatin modifications. Heterochromatin clustering during terminal differentiation is a consequence of cell cycle exit, rather than differentiation. Compromising heterochromatin-dependent gene silencing does not disrupt cell cycle exit.

Additional files

Additional file 1: Figure S2. Compromising PRC1 does not delay cell cycle exit. RNAi to Pc or white (as a control) was expressed in the posterior wing from the L3 stage and postmitotic wings at 26–28 h were examined for mitoses as indicated by PH3 and effective knockdown of PRC1 function by de-repression of the PRC1 target gene Ubx. Flow cytometry was also performed to measure cells that enter S and G2 phases. Green trace indicates cells from the posterior wing expressing the indicated transgenes. Black trace: control non-expressing anterior wing cells. Compromising PRC1 activity does not delay cell cycle exit. Scale bars = 100 μm.

Additional file 2: Figure S3. Two stages of G0 in differentiating wings. E2F was expressed in the posterior wing to delay cell cycle exit. 28 h and 42 h APF pupal tissues were dissected and immunostained for PH3 (to label mitoses) and E2F1. The anterior/posterior boundary is specified by the white line. Overexpression of E2F delays entry into G0 until 36 h. At 42 h cells expressing high E2F1 are postmitotic (in robust G0). CycD/Cdk4 + E2F expression in the posterior wing is able to bypass the robust G0 to promote continued cycling, as shown by abundant mitoses (PH3) at 42 h. Bar = 50 μm.

Additional file 3: Figure S1. Global levels of histone modifications do not dramatically change at cell cycle exit. (A-D) Quantitative western blots were performed on wings of the indicated stages to assess the levels of modified or total histone H3 or HP1. Control (Ctrl) and E2F samples are from 28 h postmitotic wings overexpressing GFP or E2F respectively. Total H3K9me3, H3K27Me3, and HP1 levels do not dramatically change with cell cycle exit, however they increase with E2F expression. Modifications associated with active chromatin, H3K4Me3 and H3K27Ac also do not dramatically change with cell cycle exit, but increase upon E2F expression.

Additional file 4: Table S1. Chromatin modifiers/organizers/remodelers that are upregulated upon E2F1/DP expression in pupal wings.

Additional file 5: Figure S4. Clustering of heterochromatin can be disrupted within one cell cycle. E2F was overexpressed in the posterior wing from 10 h APF. 12 h later (within approximately one cell cycle) tissues were immunostained for indicated histone modifications. The posterior region is labeled by the expression of GFP and the anterior/posterior boundary is specified by the white line. The distribution of staining intensity in 1112–1339 nuclei, binned into three ranges, is shown at bottom. E2F disrupts heterochromatin clustering within one cell cycle. P values were determined by an unpaired t test; ****< 0.0001.

Additional file 6: Figure S5. Delaying cell cycle exit disrupts heterochromatin. (A) CycE/Cdk2 or CycD/Cdk4 complexes were overexpressed in the posterior wing from 0 h APF. The anterior/posterior boundary is indicated by the white line. At 28 h (flexible G0) or 42 h APF (robust G0) pupal tissues were dissected and immunostained for the indicated histone modifications. (B) The distribution of staining intensity from 492 to 976 nuclei, binned into three ranges, is shown. Wings expressing E2F or CycD/Cdk4 to delay cell cycle exit were stained for mitoses (PH3) and the mitotic index at 27 h was quantified for the posterior compartment (C-D). The degree of heterochromatin disruption correlates with the number of cells cycling. P-values were determined by an unpaired t test; ****P value < 0.0001.

Additional file 7: Table S2. Genes associated with senescence that are upregulated during robust G0 in the presence of ectopic E2F1/DP.

Abbreviations
G0: gap zero, exit from the cell cycle; G1: first gap phase of the cell cycle; S-phase: DNA synthesis phase; rDNA: ribosomal DNA; ITS: internal transcribed spacer; FISH: fluorescent in situ hybridization; L2: second larval instar; L3: third larval instar; h APF: hours after puparium formation; CDT: cell doubling time; DAPI: 4′,6-diamidino-2-phenylindole; HP1: heterochromatin protein 1; H3K9: histone H3 lysine 9; H3K4: histone H3 lysine 4; H3K27: histone H3 lysine

27; H4K20: histone H4 lysine 20; H4K16: histone H4 lysine 16; H2Av: histone H2A variant; Me3: trimethylation; Ac: acetylation; PcG: Polycomb group; Pc: Polycomb; PRC1, PRC2: Polycomb repressive complexes 1 and 2; RNAi: RNA interference; RNAseq: high-throughput sequencing of cDNA libraries made from mRNA; FAIREseq: formaldehyde-assisted identification of regulatory elements with high-throughput sequencing; Gal80TS: temperature-sensitive Gal80; E(z): enhancer of zeste; Ubx: ultrabithorax; E2F/DP: E2 factor, transcription factor complex with dimerization partner; CycE, CycD: cyclins E or D; cdk: cyclin-dependent kinase.

Authors' contributions
YM performed all experiments. YM and LB conceived of the project, analyzed the data and wrote the manuscript. Both authors read and approved the final manuscript.

Acknowledgements
A special thanks to Abbey Roelofs (University of Michigan, Advocacy and Research Support, LSA-IT) for developing the automatic quantification toolkit in MATLAB. We thank Dr. Yukiko Yamashita (University of Michigan, Ann Arbor) for sharing FISH reagents and protocols as well as D1 antibody. We thank Dr. Keith Maggert (Texas A&M University) for sharing the Y10C reporter and Dr. Terry Orr-Weaver (Massachusetts Institute of Technology) for sharing the E2F1 antibody. Additional antibodies were obtained from Developmental Studies Hybridoma Bank (DSHB), created by the NICHD of the NIH and maintained at The University of Iowa. Stocks obtained from the Bloomington Drosophila Stock Center (NIH P40OD018537) were used in this study.

Competing interests
The authors declare that they have no competing interests.

Funding
This work in the Buttitta Lab was supported by the NIH (R21AG047931) and the American Cancer Society (RSG-15-161-01-DDC).

References
1. Peric-Hupkes D, Meuleman W, Pagie L, Bruggeman SW, Solovei I, Brugman W, Graf S, Flicek P, Kerkhoven RM, van Lohuizen M, et al. Molecular maps of the reorganization of genome-nuclear lamina interactions during differentiation. Mol Cell. 2010;38(4):603–13.
2. Kohwi M, Lupton JR, Lai SL, Miller MR, Doe CQ. Developmentally regulated subnuclear genome reorganization restricts neural progenitor competence in Drosophila. Cell. 2013;152(1–2):97–108.
3. Clowney EJ, LeGros MA, Mosley CP, Clowney FG, Markenskoff-Papadimitriou EC, Myllys M, Barnea G, Larabell CA, Lomvardas S. Nuclear aggregation of olfactory receptor genes governs their monogenic expression. Cell. 2012;151(4):724–37.
4. Solovei I, Kreysing M, Lanctot C, Kosem S, Peichl L, Cremer T, Guck J, Joffe B. Nuclear architecture of rod photoreceptor cells adapts to vision in mammalian evolution. Cell. 2009;137(2):356–68.
5. Asp P, Blum R, Vethantham V, Parisi F, Micsinai M, Cheng J, Bowman C, Kluger Y, Dynlacht BD. Genome-wide remodeling of the epigenetic landscape during myogenic differentiation. Proc Natl Acad Sci USA. 2011;108(22):E149–58.
6. Alabert C, Bukowski-Wills JC, Lee SB, Kustatscher G, Nakamura K, de Lima Alves F, Menard P, Mejlvang J, Rappsilber J, Groth A. Nascent chromatin capture proteomics determines chromatin dynamics during DNA replication and identifies unknown fork components. Nat Cell Biol. 2014;16(3):281–93.
7. Naumova N, Imakaev M, Fudenberg G, Zhan Y, Lajoie BR, Mirny LA, Dekker J. Organization of the mitotic chromosome. Science. 2013;342(6161):948–53.
8. Kadauke S, Blobel GA. Mitotic bookmarking by transcription factors. Epigenet Chromatin. 2013;6(1):6.
9. Rice JC, Nishioka K, Sarma K, Steward R, Reinberg D, Allis CD. Mitotic-specific methylation of histone H4 Lys 20 follows increased PR-Set7 expression and its localization to mitotic chromosomes. Genes Dev. 2002;16(17):2225–30.
10. Liu W, Tanasa B, Tyurina OV, Zhou TY, Gassmann R, Liu WT, Ohgi KA, Benner C, Garcia-Bassets I, Aggarwal AK, et al. PHF8 mediates histone H4 lysine 20 demethylation events involved in cell cycle progression. Nature. 2010;466(7305):508–12.
11. Bracken AP, Pasini D, Capra M, Prosperini E, Colli E, Helin K. EZH2 is downstream of the pRB-E2F pathway, essential for proliferation and amplified in cancer. EMBO J. 2003;22(20):5323–35.
12. Chen S, Bohrer LR, Rai AN, Pan Y, Gan L, Zhou X, Bagchi A, Simon JA, Huang H. Cyclin-dependent kinases regulate epigenetic gene silencing through phosphorylation of EZH2. Nat Cell Biol. 2010;12(11):1108–14.
13. Kaneko S, Li G, Son J, Xu CF, Margueron R, Neubert TA, Reinberg D. Phosphorylation of the PRC2 component Ezh2 is cell cycle-regulated and up-regulates its binding to ncRNA. Genes Dev. 2010;24(23):2615–20.
14. Wei Y, Chen YH, Li LY, Lang J, Yeh SP, Shi B, Yang CC, Yang JY, Lin CY, Lai CC, et al. CDK1-dependent phosphorylation of EZH2 suppresses methylation of H3K27 and promotes osteogenic differentiation of human mesenchymal stem cells. Nat Cell Biol. 2011;13(1):87–94.
15. Ahmed K, Dehghani H, Rugg-Gunn P, Fussner E, Rossant J, Bazett-Jones DP. Global chromatin architecture reflects pluripotency and lineage commitment in the early mouse embryo. PLoS ONE. 2010;5(5):e10531.
16. Thakar R, Csink AK. Changing chromatin dynamics and nuclear organization during differentiation in Drosophila larval tissue. J Cell Sci. 2005;118(Pt 5):951–60.
17. Cheutin T, Cavalli G. Progressive polycomb assembly on H3K27me3 compartments generates polycomb bodies with developmentally regulated motion. PLoS Genet. 2012;8(1):e1002465.
18. Buttitta LA, Katzaroff AJ, Perez CL, de la Cruz A, Edgar BA. A double-assurance mechanism controls cell cycle exit upon terminal differentiation in Drosophila. Dev Cell. 2007;12(4):631–43.
19. Schubiger M, Palka J. Changing spatial patterns of DNA replication in the developing wing of Drosophila. Dev Biol. 1987;123(1):145–53.
20. Milan M, Campuzano S, Garcia-Bellido A. Cell cycling and patterned cell proliferation in the Drosophila wing during metamorphosis. Proc Natl Acad Sci USA. 1996;93(21):11687–92.
21. Flegel K, Sun D, Grushko O, Ma Y, Buttitta L. Live cell cycle analysis of Drosophila tissues using the attune acoustic focusing cytometer and Vybrant DyeCycle violet DNA stain. J Vis Exp. 2013;75:e50239.
22. Guo Y, Flegel K, Kumar J, McKay DJ, Buttitta LA. Ecdysone signaling induces two phases of cell cycle exit in Drosophila cells. Biol Open. 2016;5(11):1648–61.
23. Sun D, Buttitta L. Protein phosphatase 2A promotes the transition to G0 during terminal differentiation in Drosophila. Development. 2015;142(17):3033–45.
24. Johnston LA, Sanders AL. Wingless promotes cell survival but constrains growth during Drosophila wing development. Nat Cell Biol. 2003;5(9):827–33.
25. Reis T, Edgar BA. Negative regulation of dE2F1 by cyclin-dependent kinases controls cell cycle timing. Cell. 2004;117(2):253–64.
26. Riddle NC, Minoda A, Kharchenko PV, Alekseyenko AA, Schwartz YB, Tolstorukov MY, Gorchakov AA, Jaffe JD, Kennedy C, Linder-Basso D, et al. Plasticity in patterns of histone modifications and chromosomal proteins in Drosophila heterochromatin. Genome Res. 2011;21(2):147–63.
27. Aulner N, Monod C, Mandicourt G, Jullien D, Cuvier O, Sall A, Janssen S, Laemmli UK, Kas E. The AT-hook protein D1 is essential for Drosophila melanogaster development and is implicated in position-effect variegation. Mol Cell Biol. 2002;22(4):1218–32.
28. Figueiredo ML, Philip P, Stenberg P, Larsson J. HP1a recruitment to promoters is independent of H3K9 methylation in Drosophila melanogaster. PLoS Genet. 2012;8(11):e1003061.

29. Loubiere V, Delest A, Thomas A, Bonev B, Schuettengruber B, Sati S, Martinez AM, Cavalli G. Coordinate redeployment of PRC1 proteins suppresses tumor formation during Drosophila development. Nat Genet. 2016;48(11):1436–42.

30. Gonzalez I, Mateos-Langerak J, Thomas A, Cheutin T, Cavalli G. Identification of regulators of the three-dimensional polycomb organization by a microscopy-based genome-wide RNAi screen. Mol Cell. 2014;54(3):485–99.

31. Wani AH, Boettiger AN, Schorderet P, Ergun A, Munger C, Sadreyev RI, Zhuang X, Kingston RE, Francis NJ. Chromatin topology is coupled to Polycomb group protein subnuclear organization. Nat Commun. 2016;7:10291.

32. Lavoie BD, Hogan E, Koshland D. In vivo requirements for rDNA chromosome condensation reveal two cell-cycle-regulated pathways for mitotic chromosome folding. Genes Dev. 2004;18(1):76–87.

33. Freeman L, Aragon-Alcaide L, Strunnikov A. The condensin complex governs chromosome condensation and mitotic transmission of rDNA. J Cell Biol. 2000;149(4):811–24.

34. McKnight JN, Boerma JW, Breeden LL, Tsukiyama T. Global promoter targeting of a conserved lysine deacetylase for transcriptional shutoff during quiescence entry. Mol Cell. 2015;59(5):732–43.

35. Uyehara CM, Nystrom SL, Niederhuber MJ, Leatham-Jensen M, Ma Y, Buttitta LA, McKay DJ: Hormone-dependent control of developmental timing through regulation of chromatin accessibility. Genes Dev. 2017;31(9):862–75.

36. Blais A, van Oevelen CJ, Margueron R, Acosta-Alvear D, Dynlacht BD. Retinoblastoma tumor suppressor protein-dependent methylation of histone H3 lysine 27 is associated with irreversible cell cycle exit. J Cell Biol. 2007;179(7):1399–412.

37. Sdek P, Zhao P, Wang Y, Huang CJ, Ko CY, Butler PC, Weiss JN, Maclellan WR. Rb and p130 control cell cycle gene silencing to maintain the postmitotic phenotype in cardiac myocytes. J Cell Biol. 2011;194(3):407–23.

38. Panteleeva I, Boutillier S, See V, Spiller DG, Rouaux C, Almouzni G, Bailly D, Maison C, Lai HC, Loeffler JP, et al. HP1alpha guides neuronal fate by timing E2F-targeted genes silencing during terminal differentiation. EMBO J. 2007;26(15):3616–28.

39. Ma T, Van Tine BA, Wei Y, Garrett MD, Nelson D, Adams PD, Wang J, Qin J, Chow LT, Harper JW. Cell cycle-regulated phosphorylation of p220(NPAT) by cyclin E/Cdk2 in Cajal bodies promotes histone gene transcription. Genes Dev. 2000;14(18):2298–313.

40. Zhao J, Kennedy BK, Lawrence BD, Barbie DA, Matera AG, Fletcher JA, Harlow E. NPAT links cyclin E-Cdk2 to the regulation of replication-dependent histone gene transcription. Genes Dev. 2000;14(18):2283–97.

41. Muller H, Bracken AP, Vernell R, Moroni MC, Christians F, Grassilli E, Prosperini E, Vigo E, Oliner JD, Helin K. E2Fs regulate the expression of genes involved in differentiation, development, proliferation, and apoptosis. Genes Dev. 2001;15(3):267–85.

42. Schwartz YB, Kahn TG, Nix DA, Li XY, Bourgon R, Biggin M, Pirrotta V. Genome-wide analysis of Polycomb targets in Drosophila melanogaster. Nat Genet. 2006;38(6):700–5.

43. Tolhuis B, de Wit E, Muijrers I, Teunissen H, Talhout W, van Steensel B, van Lohuizen M. Genome-wide profiling of PRC1 and PRC2 Polycomb chromatin binding in Drosophila melanogaster. Nat Genet. 2006;38(6):694–9.

44. Schuettengruber B, Cavalli G. Recruitment of polycomb group complexes and their role in the dynamic regulation of cell fate choice. Development. 2009;136(21):3531–42.

45. Korenjak M, Anderssen E, Ramaswamy S, Whetstine JR, Dyson NJ. RBF binding to both canonical E2F targets and noncanonical targets depends on functional dE2F/dDP complexes. Mol Cell Biol. 2012;32(21):4375–87.

46. Buttitta LA, Katzaroff AJ, Edgar BA. A robust cell cycle control mechanism limits E2F-induced proliferation of terminally differentiated cells in vivo. J Cell Biol. 2010;189(6):981–96.

47. Seum C, Spierer A, Pauli D, Szidonya J, Reuter G, Spierer P. Position-effect variegation in Drosophila depends on dose of the gene encoding the E2F transcriptional activator and cell cycle regulator. Development. 1996;122(6):1949–56.

48. Ho JW, Jung YL, Liu T, Alver BH, Lee S, Ikegami K, Sohn KA, Minoda A, Tolstorukov MY, Appert A, et al. Comparative analysis of metazoan chromatin organization. Nature. 2014;512(7515):449–52.

49. Schwaiger M, Stadler MB, Bell O, Kohler H, Oakeley EJ, Schubeler D. Chromatin state marks cell-type- and gender-specific replication of the Drosophila genome. Genes Dev. 2009;23(5):589–601.

50. Liu J, McConnell K, Dixon M, Calvi BR. Analysis of model replication origins in Drosophila reveals new aspects of the chromatin landscape and its relationship to origin activity and the prereplicative complex. Mol Biol Cell. 2012;23(1):200–12.

51. Shlyueva D, Stampfel G, Stark A. Transcriptional enhancers: from properties to genome-wide predictions. Nat Rev Genet. 2014;15(4):272–86.

52. Shogren-Knaak M, Ishii H, Sun JM, Pazin MJ, Davie JR, Peterson CL. Histone H4-K16 acetylation controls chromatin structure and protein interactions. Science. 2006;311(5762):844–7.

53. Di Micco R, Sulli G, Dobreva M, Liontos M, Botrugno OA, Gargiulo G, dal Zuffo R, Matti V, d'Ario G, Montani E, et al. Interplay between oncogene-induced DNA damage response and heterochromatin in senescence and cancer. Nat Cell Biol. 2011;13(3):292–302.

54. Narita M, Nunez S, Heard E, Narita M, Lin AW, Hearn SA, Spector DL, Hannon GJ, Lowe SW. Rb-mediated heterochromatin formation and silencing of E2F target genes during cellular senescence. Cell. 2003;113(6):703–16.

55. Davidson JM, Duronio RJ. S phase-coupled E2f1 destruction ensures homeostasis in proliferating tissues. PLoS Genet. 2012;8(8):e1002831.

56. Meister P, Mango SE, Gasser SM. Locking the genome: nuclear organization and cell fate. Curr Opin Genet Dev. 2011;21(2):167–74.

57. Smith CD, Shu S, Mungall CJ, Karpen GH. The Release 5.1 annotation of Drosophila melanogaster heterochromatin. Science. 2007;316(5831):1586–91.

58. Harmon B, Sedat J. Cell-by-cell dissection of gene expression and chromosomal interactions reveals consequences of nuclear reorganization. PLoS Biol. 2005;3(3):e67.

59. Csink AK, Henikoff S. Genetic modification of heterochromatic association and nuclear organization in Drosophila. Nature. 1996;381(6582):529–31.

60. Iyama T, Wilson DM 3rd. DNA repair mechanisms in dividing and non-dividing cells. DNA Repair (Amst). 2013;12(8):620–36.

61. Van Bortle K, Corces VG. Nuclear organization and genome function. Annu Rev Cell Dev Biol. 2012;28:163–87.

62. Kieffer-Kwon KR, Nimura K, Rao SSP, Xu J, Jung S, Pekowska A, Dose M, Stevens E, Mathe E, Dong P et al. Myc regulates chromatin decompaction and nuclear architecture during B cell activation. Mol Cell. 2017;67(4):566–78.

63. Evertts AG, Manning AL, Wang X, Dyson NJ, Garcia BA, Coller HA. H4K20 methylation regulates quiescence and chromatin compaction. Mol Biol Cell. 2013;24(19):3025–37.

64. Nishioka K, Rice JC, Sarma K, Erdjument-Bromage H, Werner J, Wang Y, Chuikov S, Valenzuela P, Tempst P, Steward R, et al. PR-Set7 is a nucleosome-specific methyltransferase that modifies lysine 20 of histone H4 and is associated with silent chromatin. Mol Cell. 2002;9(6):1201–13.

65. Taylor GC, Eskeland R, Hekimoglu-Balkan B, Pradeepa MM, Bickmore WA. H4K16 acetylation marks active genes and enhancers of embryonic stem cells, but does not alter chromatin compaction. Genome Res. 2013;23(12):2053–65.

66. Ishak CA, Marshall AE, Passos DT, White CR, Kim SJ, Cecchini MJ, Ferwati S, MacDonald WA, Howlett CJ, Welch ID, et al. An RB-EZH2 complex mediates silencing of repetitive DNA sequences. Mol Cell. 2016;64(6):1074–87.

67. Gonzalo S, Blasco MA. Role of Rb family in the epigenetic definition of chromatin. Cell Cycle. 2005;4(6):752–5.

68. Isaac CE, Francis SM, Martens AL, Julian LM, Seifried LA, Erdmann N, Binne UK, Harrington L, Sicinski P, Berube NG, et al. The retinoblastoma protein regulates pericentric heterochromatin. Mol Cell Biol. 2006;26(9):3659–71.

69. Longworth MS, Herr A, Ji JY, Dyson NJ. RBF1 promotes chromatin condensation through a conserved interaction with the Condensin II protein dCAP-D3. Genes Dev. 2008;22(8):1011–24.

70. Manning AL, Longworth MS, Dyson NJ. Loss of pRB causes centromere dysfunction and chromosomal instability. Genes Dev. 2010;24(13):1364–76.

71. Longworth MS, Dyson NJ. pRb, a local chromatin organizer with global possibilities. Chromosoma. 2010;119(1):1–11.

72. Strom AR, Emelyanov AV, Mir M, Fyodorov DV, Darzacq X, Karpen GH. Phase separation drives heterochromatin domain formation. Nature. 2017;547(7662):241–5.

73. Larson AG, Elnatan D, Keenen MM, Trnka MJ, Johnston JB, Burlingame AL, Agard DA, Redding S, Narlikar GJ. Liquid droplet formation by HP1al-pha suggests a role for phase separation in heterochromatin. Nature. 2017;547(7662):236–40.

74. Penke TJ, McKay DJ, Strahl BD, Matera AG, Duronio RJ. Direct interrogation of the role of H3K9 in metazoan heterochromatin function. Genes Dev. 2016;30(16):1866–80.

Site-specific regulation of histone H1 phosphorylation in pluripotent cell differentiation

Ruiqi Liao[1] and Craig A. Mizzen[1,2]* (ID)

Abstract

Background: Structural variation among histone H1 variants confers distinct modes of chromatin binding that are important for differential regulation of chromatin condensation, gene expression and other processes. Changes in the expression and genomic distributions of H1 variants during cell differentiation appear to contribute to phenotypic differences between cell types, but few details are known about the roles of individual H1 variants and the significance of their disparate capacities for phosphorylation. In this study, we investigated the dynamics of interphase phosphorylation at specific sites in individual H1 variants during the differentiation of pluripotent NT2 and mouse embryonic stem cells and characterized the kinases involved in regulating specific H1 variant phosphorylations in NT2 and HeLa cells.

Results: Here, we show that the global levels of phosphorylation at H1.5-Ser18 (pS18-H1.5), H1.2/H1.5-Ser173 (pS173-H1.2/5) and H1.4-Ser187 (pS187-H1.4) are regulated differentially during pluripotent cell differentiation. Enrichment of pS187-H1.4 near the transcription start site of pluripotency factor genes in pluripotent cells is markedly reduced upon differentiation, whereas pS187-H1.4 levels at housekeeping genes are largely unaltered. Selective inhibition of CDK7 or CDK9 rapidly diminishes pS187-H1.4 levels globally and its enrichment at housekeeping genes, and similar responses were observed following depletion of CDK9. These data suggest that H1.4-S187 is a *bona fide* substrate for CDK9, a notion that is further supported by the significant colocalization of CDK9 and pS187-H1.4 to gene promoters in reciprocal re-ChIP analyses. Moreover, treating cells with actinomycin D to inhibit transcription and trigger the release of active CDK9/P-TEFb from 7SK snRNA complexes induces the accumulation of pS187-H1.4 at promoters and gene bodies. Notably, the levels of pS187-H1.4 enrichment after actinomycin D treatment or cell differentiation reflect the extent of CDK9 recruitment at the same loci. Remarkably, the global levels of H1.5-S18 and H1.2/H1.5-S173 phosphorylation are not affected by these transcription inhibitor treatments, and selective inhibition of CDK2 does not affect the global levels of phosphorylation at H1.4-S187 or H1.5-S18.

Conclusions: Our data provide strong evidence that H1 variant interphase phosphorylation is dynamically regulated in a site-specific and gene-specific fashion during pluripotent cell differentiation, and that enrichment of pS187-H1.4 at genes is positively related to their transcription. H1.4-S187 is likely to be a direct target of CDK9 during interphase, suggesting the possibility that this particular phosphorylation may contribute to the release of paused RNA pol II. In contrast, the other H1 variant phosphorylations we investigated appear to be mediated by distinct kinases and further analyses are needed to determine their functional significance.

Keywords: Histone H1, Phosphorylation, Embryonic stem cell, Pluripotency factors, Cell differentiation, Cyclin-dependent kinase (CDK), CDK2, CDK7, CDK9, P-TEFb

*Correspondence: cmizzen@life.uiuc.edu
[1] Department of Cell and Developmental Biology, University of Illinois at Urbana Champaign, B107 Chemistry and Life Sciences Building, MC-123 601 S. Goodwin Ave., Urbana, IL 61801, USA
Full list of author information is available at the end of the article

Background

The H1 family of linker histones is important chromatin architectural proteins that bind linker DNA and facilitate higher order chromatin folding [1–3]. Human cells differentially express 11 genes encoding non-allelic amino acid sequence variants of H1 [4, 5]. These share a common tripartite structure in which a conserved globular domain (GD) is flanked by a short N-terminal domain (NTD) and a longer C-terminal domain (CTD) [6]. Fluorescence recovery after photobleaching (FRAP) microscopy of H1 variant-GFP fusions suggests that the association of H1 with chromatin is highly dynamic in vivo [7, 8] and that variation in CTD structure between individual H1 variants is major determinants of differences in their chromatin binding affinities and dynamics [9, 10].

Several types of post-translational modifications, including phosphorylation, methylation and acetylation, have been identified in H1 variants, with phosphorylation being particularly abundant [11–13]. Analyses of synchronized cells from multiple organisms suggest that H1 phosphorylation increases progressively during interphase before peaking transiently during mitosis [14–19]. However, these early studies did not identify which sites in individual H1 variants are phosphorylated during interphase and mitosis. More recently, mass spectrometry has enabled the precise identification of H1 phosphorylation sites. Phosphorylation at both cyclin-dependent kinase (CDK) consensus motifs and non-CDK sites has been detected [20–23]. Interphase H1 phosphorylation occurs predominantly, if not exclusively, at serine-containing CDK motifs (SPXZ, X = any amino acid, Z = K or R). Mitotic H1 phosphorylation occurs at these same SPXZ sites together with phosphorylation that occurs exclusively during mitosis at TPXZ CDK motifs and some non-CDK sites [21, 23–28]. S18, S173 and S189 were identified as sites of interphase phosphorylation of H1.5 [21, 24], but the significance of these phosphorylations remains unclear [13].

Our laboratory identified H1.2-S173, H1.4-S172 and H1.4-S187 as the predominant sites of interphase phosphorylation of H1.2 and H1.4 in HeLa S3 cells. Analyses with antisera that are highly specific for H1.4 phosphorylation at S187 (pS187-H1.4) suggest that pS187-H1.4 is enriched at sites of transcription by RNA polymerases I and II [23], but how the levels and chromatin distribution of pS187-H1.4 are regulated is unknown. Similarities in the nuclear and nucleolar staining patterns obtained with our antisera to pS187-H1.4 and pS173-H1.2/5 suggest that enrichment of the latter may also correlate with transcription [23], and this is supported by evidence that staining with pS173-H1.2/5 antisera raised by another laboratory colocalizes extensively with Br-UTP labeling of nascent RNA and to a lesser extent with EdU labeling of replicating chromatin, in HeLa and HEK293 cells [24]. However, other interphase phosphorylations may be functionally distinct since staining for pS18-H1.5 showed little colocalization with either Br-UTP or EdU labeling [24]. The data available currently suggest that the impact of interphase S18 phosphorylation in the H1.5 NTD differs from that of the transcription-associated roles proposed for interphase phosphorylations in the CTDs of H1.2, H1.4 and H15. This suggests, in turn, that different kinases may mediate interphase phosphorylation at different sites in H1 variants. Although CDK2 and CDK9 have been implicated as possible interphase H1 kinases [29–33], whether they display site-specificity among H1 variants in vivo is not known.

Recent evidence suggests that individual H1 variants play significant roles in embryonic stem (ES) cell differentiation. Analyses of the abundance of the mRNAs encoding seven somatic H1 variants suggest that they are differentially expressed during the differentiation of human ES cells (hESCs) in vitro [34]. H1 triple-knockout mouse ES cells (mESCs) depleted of H1c, H1d and H1e (H1.2-4) display global alterations in chromatin structure and an impaired capacity for differentiation compared to wild-type mESCs [35, 36]. Comparison of the genomic distribution of H1.5 in hESCs versus differentiated cell lines suggests that H1.5 becomes enriched at specific gene family clusters during cellular differentiation [37]. Although the mechanisms involved have yet to be defined, these findings support an emerging view that individual H1 variants differ in their impact on cellular differentiation. Metazoan H1 variants differ conspicuously from each other in possessing different numbers of sites for interphase phosphorylation that vary in their relative location and amino acid sequence context. Thus, differences in the expression, genomic distribution and interphase phosphorylation dynamics between individual H1 variants may enable them to play distinct roles during ES cell differentiation. However, data on H1 variant phosphorylation dynamics during cell differentiation has not been reported. Here, we show that the global levels of interphase phosphorylation at H1.5-S18 and H1.2/H1.5-S173 decrease to different extents when pluripotent human NTERA-2/D1 (NT2) cells differentiate. Similar decreases in the global levels of pS18-H1.5 and pS173-H1.2/5 occur when mESCs differentiate, but pS173-H1.2/5 levels are reduced more extensively than in NT2 cells. The global level of pS187-H1.4 also decreased when NT2 cells differentiated, but pS187-H1.4 levels at the transcription start sites (TSSs) of pluripotency factor genes fell to a disproportionately greater extent that correlated with their diminished expression. In contrast, the association of pS18-H1.5 and pS173-H1.2/5 with pluripotency factor gene TSSs in differentiated cells tended to increase or decrease, respectively, but these changes did not reach statistical significance. pS18-H1.5, pS173-H1.2/5 and pS187-H1.4 were associated with different extents with housekeeping gene TSSs in undifferentiated cells, and this did not change significantly upon cell differentiation. Contrary to evidence suggesting that CDK2 is involved in interphase H1 phosphorylation, selective inhibition of CDK7 or CDK9, but not CDK2, significantly decreased the global levels of pS187-H1.4, and inhibition of CDK7 or CDK9 rapidly diminished the association of pS187-H1.4 with specific genes in HeLa cells. The likelihood that CDK9 mediates interphase H1.4-S187 phosphorylation was further supported by their colocalization at gene promoters in both HeLa and NT2 cells in reciprocal re-ChIP analyses and by manipulation of the levels of chromatin-associated CDK9. Depletion of CDK9 with siRNA significantly decreased the global level of pS187-H1.4 in HeLa cells, and the level of pS187-H1.4 at housekeeping genes reflected the extent to which CDK9 association was reduced at these sites. In contrast, actinomycin D rapidly induced the enrichment of both CDK9 and pS187-H1.4 at housekeeping genes. Remarkably, none of the inhibitor treatments affected the global levels of pS18-H1.5 or pS173-H1.2/5. Taken together, our data suggest that CDK9 is the predominant interphase kinase for H1.4-S187, but not for H1.5-S18 or H1.2/H1.5-S173 in both HeLa and NT2 cells, and that interphase phosphorylation of H1 variants is regulated in a site-specific and gene-specific fashion that may confer specialized roles to individual H1 variants in chromatin processes.

Methods
Cell culture and differentiation
NTERA-2/D1 (NT2) human embryonal testicular teratocarcinoma cells and W4/129S6 mESCs were obtained from Dr. Fei Wang (UIUC). NT2 cells were grown in DMEM + 10% FBS and subcultured by scraping. Differentiation was induced by dissociating cells with trypsin, followed by seeding at a density of 1×10^6 cells per T-75 flask or 1.33×10^4 cells/cm^2 in DMEM, supplemented with 10% FBS and 10 μM all-trans retinoic acid (RA). mESCs were maintained in DMEM with high glucose, supplemented with 15% FBS (ES-Cult FBS, Stemcell Technologies, 06952), 0.1 mM non-essential amino acids (Gibco), 1 mM sodium pyruvate (Gibco), 0.1 mM β-mercaptoethanol (Sigma), 2 mM L-glutamine (GlutaMAX, Gibco), 1000 U/mL LIF (Nacalai USA, NU0012) and 1× penicillin/streptomycin (Gibco) on gelatin-coated plates. The cells were fed with fresh medium daily and subcultured in new gelatin-coated plates by trypsinization every other day. Differentiation was induced by seeding cells in ES cell medium without LIF. HeLa cells were grown in DMEM + 10% FBS and subcultured by

trypsinization. Cells were treated with flavopiridol (NIH AIDS Reagent Program), NU-6140 (Tocris), actinomycin D (Fisher), α-amanitin (Cayman), triptolide (Tocris) or THZ1 (ApexBio) dissolved in DMSO to selectively inhibit RNA Pol II or CDK activities as described in the figure legends.

Histone preparation and chromatography
Crude histones were extracted from isolated nuclei with 0.4 N H$_2$SO$_4$ as described previously [38]. Crude H1 was prepared by 5% perchloric acid fractionation of crude histones and recovered by precipitation with 20% (w/v, final concentration) trichloroacetic acid (TCA). Hydrophobic interaction chromatography was performed using a 4.6 mm ID × 100 mm PolyPROPYL A column (PolyLC Inc.) and a multistep linear gradient from buffer A [2.5 M (NH$_4$)$_2$SO$_4$ in 50 mM sodium phosphate, pH 7.0] to buffer B [1.0 M (NH$_4$)$_2$SO$_4$ in 50 mM sodium phosphate, pH 7.0]. Fractions were collected by time, and proteins were recovered by precipitation with 20% TCA.

siRNA transfection
siRNA transfection was performed using Lipofectamine RNAiMAX (Invitrogen) according to manufacturer's protocol. Cells were seeded the day before transfection in order to reach 60–80% confluency at the time of transfection. Transfection reagent and siRNA were diluted in Opti-MEM (Gibco), mixed, and incubated for 5 min at room temperature. The complexes were then added directly to cell cultures, and the cells were harvested 72 h later.

Immunoblotting
Whole-cell lysates or histone extracts were electrophoresed in 15% polyacrylamide gels (6% gel for Pol II blots) containing SDS, transferred to a PVDF membrane and blocked with 5% milk powder in TBS for 1 h at room temperature. The blocked membrane was then incubated with primary antibody at 4 °C overnight, washed with TBST, and incubated with secondary antibody conjugated with HRP (Amersham, NA-931 or NA-934) for 1 h at room temperature, washed with TBST again, developed with chemiluminescence reagents (Thermo, SuperSignal West Pico Chemiluminescent Substrate) and images recorded with a series of lengthening exposures on X-ray films. Bitmap images were generated from selected films using a flatbed scanner and densitometry performed using ImageJ https://imagej.nih.gov/ij/.

The pS18-H1.5 antisera were generated by immunizing rabbits with a synthetic phosphopeptide (CPVEK-phosphoserine-PAKK) conjugated to maleimide-activated keyhole limpet hemocyanin (Thermo Fisher Scientific) using standard procedures. Pan antisera to H1.0 and H1.5

were generated by immunizing rabbits with full-length recombinant human H1.0 or H1.5 as described previously [23]. The antisera to pS187-H1.4, pS173-H1.2/5 and pan-H1.4 (UI-100) have been described previously [23]. The antisera against other histones and pluripotency markers were obtained from Abcam: H1.2 (ab4086), H1.5 (ab24175), histone H3 (ab1791), Oct4 (ab19857), Sox2 (ab97959), Nanog (ab21624). The commercial antibody against pS18-H1.5 (61107) was purchased from Active Motif. The commercial antisera against H1.0 (sc-56695), Pol II (sc-899) and CDK9 (sc-13130, sc-484) were purchased from Santa Cruz Biotechnology, Inc. The antibodies against phosphorylated CTD of RNA Pol II were purchased from Abcam: Pol II pS2 (ab5095), Pol II pS5 (ab5131).

For peptide competition assays, 10 µg of non-phosphorylated (CPVEKSPAKK) or phosphorylated (CPVEKpSPAKK) H1.5-S18 peptides (Genscript) was incubated with 2 µL of primary antisera in a small volume (500 µL) of PBS for 2 h at room temperature. These mixtures were then further diluted in TBST for use in immunoblotting.

ChIP and re-ChIP

Chromatin immunoprecipitation (ChIP) experiments were performed as described previously with minor modifications [23]. Cells were cross-linked by adding formaldehyde directly to cultures (1% final) and incubating for 8 min at room temperature. 125 mM final glycine was added, and cultures were incubated for 10 min at room temperature. Cells were then washed twice with cold PBS, scraped, and resuspended in ChIP lysis buffer with protease and phosphatase inhibitors. Chromatin was sheared to ~1 kb mean length by repeated cycles of sonication in a 4 °C water bath using a Bioruptor (Diagenode). After centrifuging at $18,000 \times g$ for 10 min, the supernatants were diluted tenfold with ChIP dilution buffer. Aliquots representing $1-2 \times 10^6$ cells in 1.0 ml final volume were used for each pull down. Samples were incubated with specific antibodies [15 µL pS187-H1.4, 30 µL pS173-H1.2/5, 10 µL pS18-H1.5 (Active Motif) or 20 µL CDK9 (Santa Cruz)] at 4 °C overnight. Immunocomplexes were incubated with 50 µL BSA-blocked protein G Dynabeads (Invitrogen) for 4 h at 4 °C, collected using a magnetic rack, and washed sequentially with ChIP wash buffer I, II, III and twice with TE. Beads were eluted twice with 200 µL 1% SDS in 0.1 M NaHCO₃ at 65 °C for 10 min. The combined eluates were made 200 mM NaCl (final), incubated at 65 °C overnight to reverse cross-links, digested with 50 µg/ml RNase A at 37 °C for 30 min, and then digested with 50 µg/ml proteinase K at 50 °C for 1 h. The DNA fragments were then purified by phenol/chloroform extraction, recovered by ethanol precipitation using 20 µg glycogen as a carrier, and dissolved in 50 µL

of deionized water. For re-ChIP assays, immunoprecipitations from the first ChIP were washed sequentially as described above. The immunocomplexes were eluted with 10 mM DTT in TE at 37 °C for 30 min, diluted 20 times with ChIP dilution buffer and then immunoprecipitated with the second antibody using standard ChIP protocol. ChIP products were quantitated by real-time PCR using SYBR Green master mix (Applied Biosystems) and the primers listed in Additional file 1: Table S1.

Results
Site-specific changes in global H1 phosphorylation during cell differentiation

We have generated a collection of highly specific antisera, raised against synthetic phosphopeptides, which recognize phosphorylation at single sites that are unique to individual human H1 variants or are shared between just two variants. We have also raised "pan" antisera against individual full-length recombinant human H1 variants that specifically recognize the intended variant regardless of whether it is phosphorylated or not. The former provide a relative measure of the levels of phosphorylation at defined sites between samples, whereas the latter can be used to confirm that equivalent amounts of a particular H1 variant, regardless of phosphorylation status, are present in the samples being compared. The specificity of our antisera to pS173-H1.2/5, pS187-H1.4 and pan-H1.4 has been described previously [23]. The specificity of our antisera to pan-H1.0, pan-H1.5 and pS18-H1.5 is shown in Additional file 2: Figure S1. We used these antisera and commercially available reagents in immunoblotting to monitor the relative expression and phosphorylation of H1 variants in NT2 cells during seven days of retinoic acid (RA)-induced differentiation. RA induces pluripotent NT2 cells to differentiate along a neural lineage [39, 40]. For comparison, we also analyzed the spontaneous differentiation of pluripotent mouse embryonic stem cells (mESCs) after the removal of leukemia inhibitory factor (LIF) [41, 42]. The sequences of the phosphopeptide antigens used to generate the pS173-H1.2/5, pS187-H1.4 and pS18-H1.5 antisera are completely conserved in the respective mouse H1 variants, and these antisera display the same apparent affinity and specificity for the respective phosphorylated H1 variants in analyses of murine and human samples. Our antisera raised against recombinant human H1 variants are specific for the corresponding mouse H1 variant, but some bind the mouse protein with lower apparent affinity compared to the human (see below).

Control analyses revealed that expression of the Oct-4, Sox-2 and Nanog transcription factors all decreased significantly in NT2 cells after RA addition, confirming differentiation and the loss of pluripotency in cultures during the treatment interval (Fig. 1a). Similarly, Oct-4 and Sox-2 expression in mESCs dropped markedly within 1 week after LIF was removed (Fig. 1b). Comparing the expression and phosphorylation of individual H1 variants relative to histone H3 expression in NT2 cells on days 0

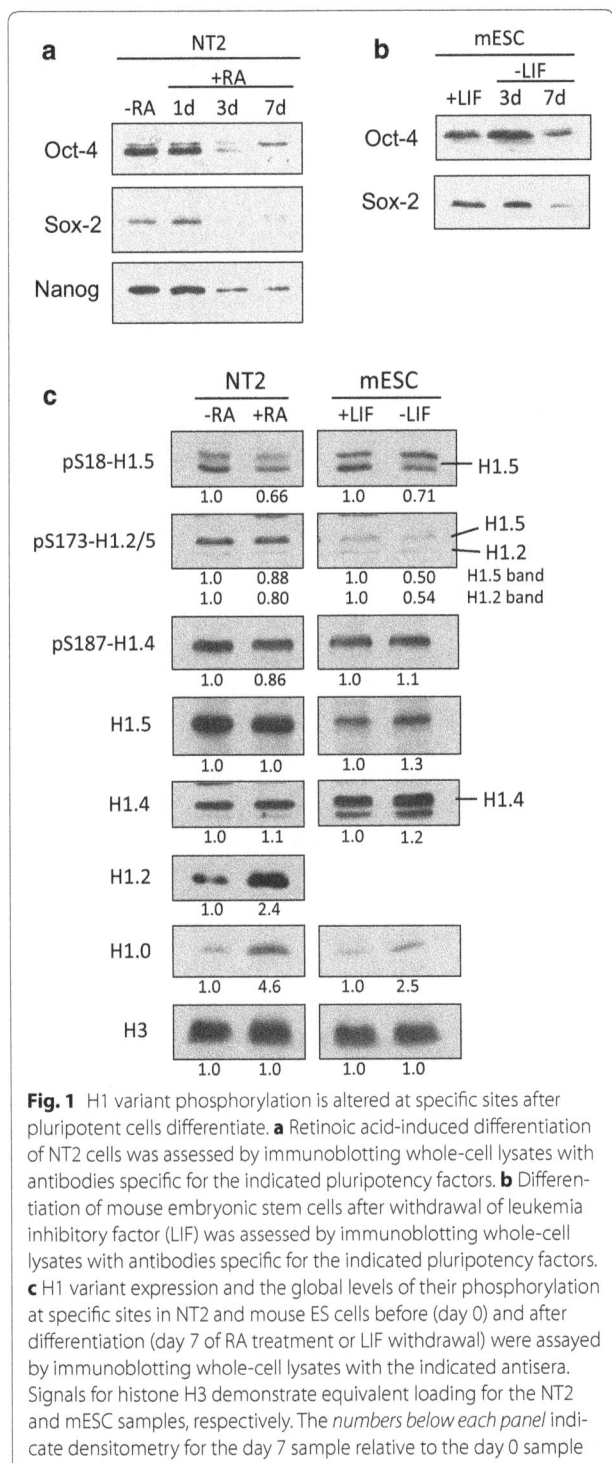

Fig. 1 H1 variant phosphorylation is altered at specific sites after pluripotent cells differentiate. **a** Retinoic acid-induced differentiation of NT2 cells was assessed by immunoblotting whole-cell lysates with antibodies specific for the indicated pluripotency factors. **b** Differentiation of mouse embryonic stem cells after withdrawal of leukemia inhibitory factor (LIF) was assessed by immunoblotting whole-cell lysates with antibodies specific for the indicated pluripotency factors. **c** H1 variant expression and the global levels of their phosphorylation at specific sites in NT2 and mouse ES cells before (day 0) and after differentiation (day 7 of RA treatment or LIF withdrawal) were assayed by immunoblotting whole-cell lysates with the indicated antisera. Signals for histone H3 demonstrate equivalent loading for the NT2 and mESC samples, respectively. The *numbers below each panel* indicate densitometry for the day 7 sample relative to the day 0 sample

and 7 of RA treatment revealed changes in the global levels of phosphorylation at some sites that did not correlate with changes in H1 variant expression and were presumably related to changes in H1 phosphorylation regulation during the transition to a longer cell cycle in differentiated cells [43]. The level of pS18-H1.5 was markedly lower, and pS173-H1.5 was reduced to a lesser extent, in differentiated NT2 cells even though H1.5 expression was not altered (Fig. 1c). In contrast, the levels of pS187-H1.4 and pS173-H1.2 were reduced in differentiated NT2 cells even though H1.2 and H1.4 expression was increased. Notably, the data suggest that the global levels of phosphorylation at each of these four sites changed in a site-specific fashion. Expression of H1.0 was also upregulated in differentiated NT2 cells, consistent with previous evidence that H1.0 expression is induced during differentiation of NT2 and human ES cells [34]. Although these authors did not detect changes in the relative levels of other H1 variant proteins despite significant changes in mRNA levels, our data suggest that expression of H1.2 and H1.4 also increases when NT2 cells differentiate.

We used hydrophobic interaction chromatography (HIC) as an antibody-independent approach to analyze crude H1 from pluripotent and differentiated NT2 cells. HIC is a convenient alternative to two-dimensional approaches for monitoring H1 variant modification dynamics that provides partial or complete resolution of H1 variants and their phosphorylated forms [23]. HIC analyses revealed that pluripotent NT2 cells express H1.2, H1.3, H1.4 and H1.5 predominantly, and that higher levels of phosphorylation affect H1.5 compared to the other variants. H1.4 and H1.2 were phosphorylated to lesser extents, while phosphorylation of H1.3 was not readily detectable (Additional file 3: Figure S2). Following the addition of RA, the level of H1.5 monophosphorylation (H1.5-1p) appeared to remain stable for the first three days but then decreased markedly between days 3 and 7 of RA treatment. Previous analyses of H1 variant phosphorylation in several cell lines suggest that H1.5-1p is comprised of a mixture of molecules monophosphorylated at S18 (pS18-H1.5) or S173 (pS173-H1.5) [21, 24]. Thus, the decrease in the H1.5-1p peak in HIC is consistent with the decreased signals observed for pS18-H1.5 and pS173-H1.5 in immunoblotting (Fig. 1c). RA treatment also led to diminished H1.4 monophosphorylation (H1.4-1p), but the kinetics differed from those of H1.5-1p. Loss of H1.4-1p was apparent after one day of RA treatment and appeared to decrease somewhat further over the following 6 days. Since monophosphorylation of H1.4 occurs exclusively or predominantly at S187 (pS187-H1.4) [23], the decrease observed for the H1.4-1p peak in HIC is consistent with the reduction in signal observed for pS187-H1.4 in immunoblotting. In contrast, the level of H1.2 monophosphorylation appeared to remain constant throughout the treatment interval (Additional file 3: Figure S2), while the peak for non-phosphorylated H1.2 seemed to increase. Therefore, the ratio of H1.2-1p relative to H1.2-0p was reduced after 7d of RA-induced differentiation, which is consistent with the decrease of pS173-H1.2 detected on blots. Together, the HIC and immunoblotting results suggest that the relative levels of global interphase phosphorylation at pS18-H1.5, pS173-H1.2, pS173-H1.5 and pS187-H1.4 decrease in a site-specific fashion during the differentiation of pluripotent NT2 cells.

Some, but not all, of these changes were observed when mES cells differentiated. In particular, the level of pS18-H1.5 in differentiated mESC was also markedly decreased (Fig. 1c). However, greater reductions in the phosphorylation of both H1.2-S173 and H1.5-S173 were apparent when mESC differentiated compared to NT2 cells. In contrast to NT2 cells, differentiation led to an increase in the level of pS187-H1.4 in mESC, providing further evidence that interphase H1 phosphorylation is regulated in a cell type-specific and site-specific fashion. Moreover, whereas the expression of H1.0 and H1.2 increased with differentiation in NT2 cells, the expression of H1.0, H1.4 and H1.5 increased upon mESC differentiation (Fig. 1c). We were unable to monitor H1.2 expression in mESC because the commercial antisera we used does not detect murine H1.2. Although stronger signals were obtained for total H1.5 in NT2 cells compared to mESCs, this appears to reflect weaker binding of mouse H1.5 by our pan-H1.5 antiserum due to species-specific amino acid sequence differences rather than a marked difference in abundance. Stained gels suggest that H1.5 is present at similar levels in both cell types (data not shown). Our findings that the expression of H1.0, H1.4 and H1.5 increases upon LIF withdrawal is partially consistent with a previous report that the expression of H1.0, H1.2, H1.3 and H1.4 increased when a different mESC line was differentiated to form embryoid bodies [36]. Taken together, the data presented here and findings from previous reports suggest that enhanced H1.0 expression and diminished interphase phosphorylation of H1.5-S18 and H1.5-S173 are common features of pluripotent cell differentiation. Additional changes in H1 variant expression and phosphorylation may also occur depending on the cell type and the differentiation pathway.

Enrichment of pS187-H1.4 at promoters is associated with pluripotency factor gene transcription

Since the global levels of pS18-H1.5, pS173-H1.2/5 and pS187-H1.4 appeared to decrease by different extents contemporaneously with the downregulation of pluripotency factor expression when NT2 cells differentiated,

we used ChIP-qPCR to investigate whether changes in the association of these phosphorylated forms at pluripotency genes might contribute to their transcriptional regulation. As shown in Additional file 4: Figure S3, pS187-H1.4 is enriched at the promoters of the active ACTG1 and GAPDH genes in HeLa cells, but not at the promoter of the repressed MYOD1 gene or an intergenic region within the chromosome 1p35.3 band, consistent with our prior evidence that enrichment of pS187-H1.4 may serve as a general marker for transcriptionally active chromatin [23]. pS187-H1.4 was enriched at the transcription start site (TSS) of pluripotency genes (Fig. 2a) and housekeeping genes (Fig. 2b) in pluripotent NT2 cells. Upon differentiation, the level of pS187-H1.4 at the TSSs of pluripotency factor genes fell markedly, whereas the levels at housekeeping gene TSSs either increased or decreased, but these changes were not statistically significant. Although the levels of pS187-H1.4 decrease globally during NT2 cell differentiation (Fig. 1; Additional file 3: Figure S2), the latter finding implies that pS187-H1.4 is selectively diminished at pluripotency factor gene promoters during NT2 cell differentiation. This notion is further supported by the fact that the magnitude of the loss of pS187-H1.4 at all four pluripotency factor promoters was greater than the loss detected at the global level. Similarly, we found that pS173-H1.2/5 association with the TSSs of pluripotency genes, but not housekeeping genes, decreased after differentiation, although the changes were not statistically significant (Fig. 2a). Since the pS173-H1.2/5 antibody detects phosphorylation on both H1.2 and H1.5, it is unclear whether the decrease at pluripotency gene TSSs in differentiated NT2 cells is due to changes in the association of both pS173-H1.2 and pS173-H1.5 or the selective loss of one form over the other. However, since the global level of pS173-H1.2 was much lower than that of pS173-H1.5 in both pluripotent and differentiated NT2 cells (Fig. 1; Additional file 3: Figure S2), it is possible that the decrease in ChIP signal reflects selective depletion of pS173-H1.5 from pluripotency gene promoters in differentiated cells.

Although the decrease in the global level of pS18-H1.5 in differentiated NT2 cells was notable (Fig. 1; Additional file 3: Figure S2), the level of pS18-H1.5 at the promoters we assessed did not change significantly (Fig. 2a, b). This finding is consistent with prior evidence that pS18-H1.5 is not enriched in active chromatin [24]. Taken together, the results in Fig. 2 suggest that enrichment of pS187-H1.4, and possibly pS173-H1.2 and pS173-H1.5, at gene promoters correlates with their transcription, whereas dephosphorylation or depletion of these forms correlates with their repression. Our results also suggest that the association of interphase phosphorylation at the H1.5-S18 NTD site with transcription may differ, in general,

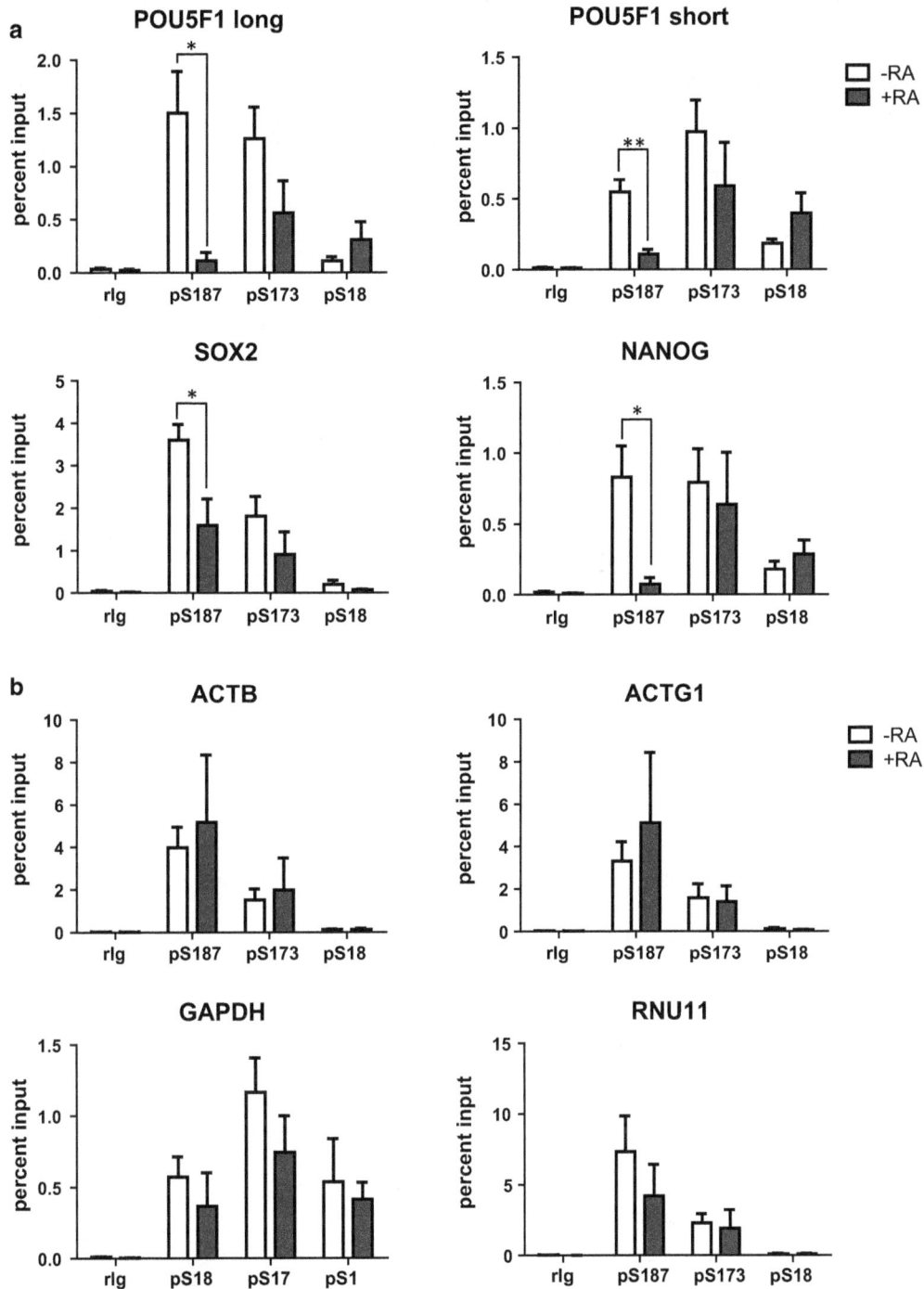

Fig. 2 Changes in the levels of phosphorylated H1 variants at pluripotency factor gene promoters correlate with their reduced expression in differentiated NT2 cells. **a** The levels of pS187-H1.4, pS173-H1.2/5 and pS18-H1.5 at the transcription start sites of pluripotency factor genes in NT2 cells before and after 7 days of RA treatment were assessed by ChIP-qPCR. **b** The levels of pS187-H1.4, pS173-H1.2/5 and pS18-H1.5 at the transcription start sites of housekeeping genes in NT2 cells before and after 7 days of RA treatment were assessed by ChIP-qPCR. Negative control ChIP assays employed non-immune rabbit immunoglobulin (rIg) in place of primary antisera. Custom antisera were used for pS187-H1.4 and pS173-H1.2/5 ChIP. Commercial antisera (Active Motif) was used for pS18-H1.5 ChIP. The data are expressed as the percent relative to input DNA (mean ± s.e.m., *$p < 0.05$; **$p < 0.01$)

from that of the H1.2/H1.5-S173 and H1.4-S187 CTD sites.

Interphase phosphorylation of H1.4-S187 and H1.5-S18 are regulated disparately

The evidence that phosphorylation at the H1.5-S18 NTD site may impact transcription differently than phosphorylation at the H1.4-S187 CTD site prompted us to investigate how these phosphorylations are regulated. Both CDK2 and CDK9 have been implicated previously in interphase H1 phosphorylation, but the evidence linking either kinase to the phosphorylation of specific sites in individual H1 variants in vivo is ambiguous [29–33]. Consequently, we assessed the global levels of pS18-H1.5 and pS187-H1.4 in pluripotent NT2 cells after selective inhibition of CDK2 with NU6140 or CDK9 with

flavopiridol (FLVP) [44–46]. Much of the CDK9 present in cells is associated with cyclin T in the P-TEFb complex that phosphorylates the CTD of RNAP II, DRB sensitivity inducing factor (DSIF), and negative elongation factor (NELF) to release promoter-proximal paused RNAP II and activate transcriptional elongation [47, 48]. Remarkably, the levels of pS18-H1.5 were unaffected during 24 h of NU6140 treatment (Fig. 3a). pS18-H1.5 levels were also stable during the initial 3 h of FLVP treatment, but a significant decrease was apparent after 8 h of treatment. Since the delayed onset of diminished pS18-H1.5 may be attributable to inhibition of cell cycle progression by FLVP [49], our data suggest it is unlikely that either CDK2 or CDK9 is involved in regulating H1.5-S18 interphase phosphorylation. Control experiments confirmed that our preparations of NU6140 and FLVP were active and

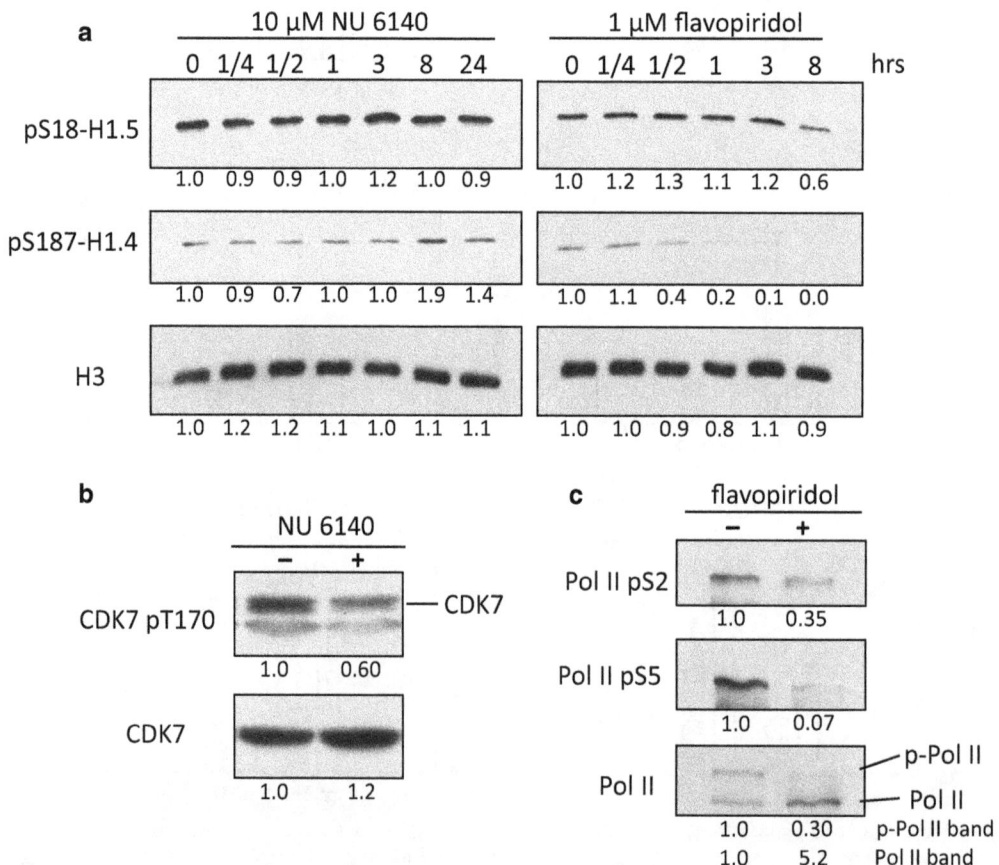

Fig. 3 The activity of CDK9, but not CDK2, is required for H1.4-S187 phosphorylation. **a** Pluripotent NT2 cells were treated for increasing intervals with 10 μM NU6140 or 1 μM flavopiridol to preferentially inhibit CDK2 and CDK9, respectively. The global abundance of pS18-H1.5 and pS187-H1.4 was assessed by immunoblotting whole-cell lysates. The blot for histone H3 serves as a loading control. The 0 h time points provide a solvent control (DMSO). Commercial antisera (Active Motif) was used to detect pS18-H1.5. **b** HeLa cells were treated with DMSO or 10 μM NU6140 for 1 h. The loss of CDK2-mediated phosphorylation of CDK7-T170 [50] was assessed by immunoblotting whole-cell lysates with the indicated antisera. **c** HeLa cells were treated with DMSO or 1 μM flavopiridol for 1 h. Altered phosphorylation in the CTD heptad repeats of RNAP II [51] was assessed by immunoblotting whole-cell lysates with the indicated antisera. The *numbers below each panel* indicate densitometry for the treated samples relative to the respective control sample

attenuated the phosphorylation of CDK7 and the RNAP II-CTD, known targets of CDK2 and CDK9, respectively (Fig. 3b, c) [50, 51]. Similarly, pS187-H1.4 levels remained unchanged during the first 3 h of NU6140 treatment, but were increased after 8 and 24 h of treatment, suggesting that CDK2 also does not directly regulate H1.4-S187 interphase phosphorylation. In contrast, FLVP treatments as short as 30 min markedly reduced the global levels of pS187-H1.4, suggesting that H1.4-S187 is a *bona fide* CDK9 substrate or that CDK9 is otherwise involved in regulating the global levels of this modification.

Inhibiting CDK7 or CDK9 rapidly diminishes global and gene-specific levels of pS187-H1.4

Although FLVP inhibits CDK9 preferentially [45], inhibition of additional kinases by FLVP [52–54] could reduce pS187-H1.4 levels and bias the identification of CDK9 as a major kinase for H1.4-S187. We investigated this possibility using THZ1, a recently developed inhibitor that exhibits extraordinary selectivity for CDK7 due to a novel mechanism of inhibition that involves covalent binding to a cysteine residue outside of the kinase domain that is unique to CDK7 [55]. As a subunit of both the CDK-activating kinase (CAK) and TFIIH multi-protein complexes, CDK7 is involved in regulating cell cycle progression and transcription [56]. TFIIH affects transcription via several mechanisms, including phosphorylating the heptapeptide repeats of the RNAP II-CTD at Ser5 and Ser7 during initiation and promoter clearance, and phosphorylating Thr186 in the T-loop of CDK9 to activate phosphorylation of RNAP II-CTD heptads at Ser2 and the release of paused RNAP II by the P-TEFb complex [57–59]. We reasoned that if H1.4-S187 was a *bona fide* CDK9 substrate, selective inhibition of CDK7 by THZ1 should impair CDK9 (P-TEFb) activation and lead to decreased levels of pS187-H1.4. Initial experiments revealed that THZ1 reduced global levels of pS187-H1.4 in HeLa cells in a dose- and time-dependent manner (Fig. 4a). We then compared the effects of treating HeLa cells for just one hour with 1 µM THZ1 or 1 µM FLVP. Both drugs significantly decreased the phosphorylation of the RNAP II-CTD and markedly suppressed global pS187-H1.4 levels, with FLVP eliciting a greater reduction than THZ1 (Fig. 4b). Remarkably, the level of pS18-H1.5 was not affected by either treatment. Both THZ1 and FLVP led to slight reductions in the level of pS173-H1.2, but these may have been attributable to reduced expression of H1.2. Both treatments also led to increases in the level of pS173-H1.5 that did not appear to be attributable to increased H1.5 expression. These data provide additional evidence that different kinases are involved in regulating interphase phosphorylation of H1.4-S187 versus H1.5-S18, H1.2-S173 and H1.5-S173, but differences in the rates of phosphoryl turnover at these sites could also be involved.

One hour treatments with THZ1 or FLVP markedly decreased the levels of pS187-H1.4 at the promoters and bodies of three housekeeping genes (Fig. 4c). However, as we found for global pS187-H1.4 levels, the decreases observed for FLVP were larger than those for THZ1. Moreover, the effect of THZ1 was more pronounced at these promoters compared to the gene bodies. Given the evidence that CDK9 activity is regulated by CDK7 [55, 59], our data support models in which CDK7 indirectly regulates pS187-H1.4 levels at promoters by controlling CDK9 (P-TEFb) activation, whereas CDK9 (P-TEFb) or other FLVP-sensitive kinases associated with elongating RNAP II mediate H1.4-S187 phosphorylation in gene bodies [60]. However, our data do not exclude the possibility that H1.4-S187 may also be a *bona fide* substrate for CDK7.

Depletion of CDK9 decreases the global and gene-specific levels of pS187-H1.4

To confirm that the effects observed for THZ1 or FLVP treatment were due to inhibition of CDK9, we assessed how depletion of CDK9 affected pS187-H1.4 levels in HeLa cells. Despite overloading of the CDK9-depleted sample for the repeat shown in Fig. 5 (compare pan-H3 levels), immunoblots revealed that very little CDK9 perdured after 72 h of siRNA treatment, and this was associated with a marked reduction in the global level of pS187-H1.4 (Fig. 5a). In contrast, the global levels of pS18-H1.5, pS173-H1.2 and pS173-H1.5 were all increased in the CDK9-depleted sample, but this may reflect the enhanced loading of this sample as noted above. These data, together with the results from the kinase inhibitor treatments (Figs. 3, 4), strongly suggest that interphase H1.4-S187 phosphorylation depends on the activity of CDK9, whereas other kinases mediate interphase phosphorylation of H1.5-S18, H1.2-S173 and H1.5-S173.

CDK9 depletion led to striking reductions in the association of both CDK9 and pS187-H1.4 with the promoters and bodies of the ACTB and ACTG1 genes that were not observed in the MYOD1 promoter and intergenic negative control regions (Fig. 5b). However, CDK9 depletion did not have significant effects on the binding of CDK9 and pS187-H1.4 at the GAPDH promoter or gene body. These data provide further evidence that CDK9 phosphorylates H1.4-S187, and that the level of pS187-H1.4 at genes is directly related to the extent of co-enrichment of CDK9. They also suggest that the extent of CDK9 dynamics at different genes may vary considerably.

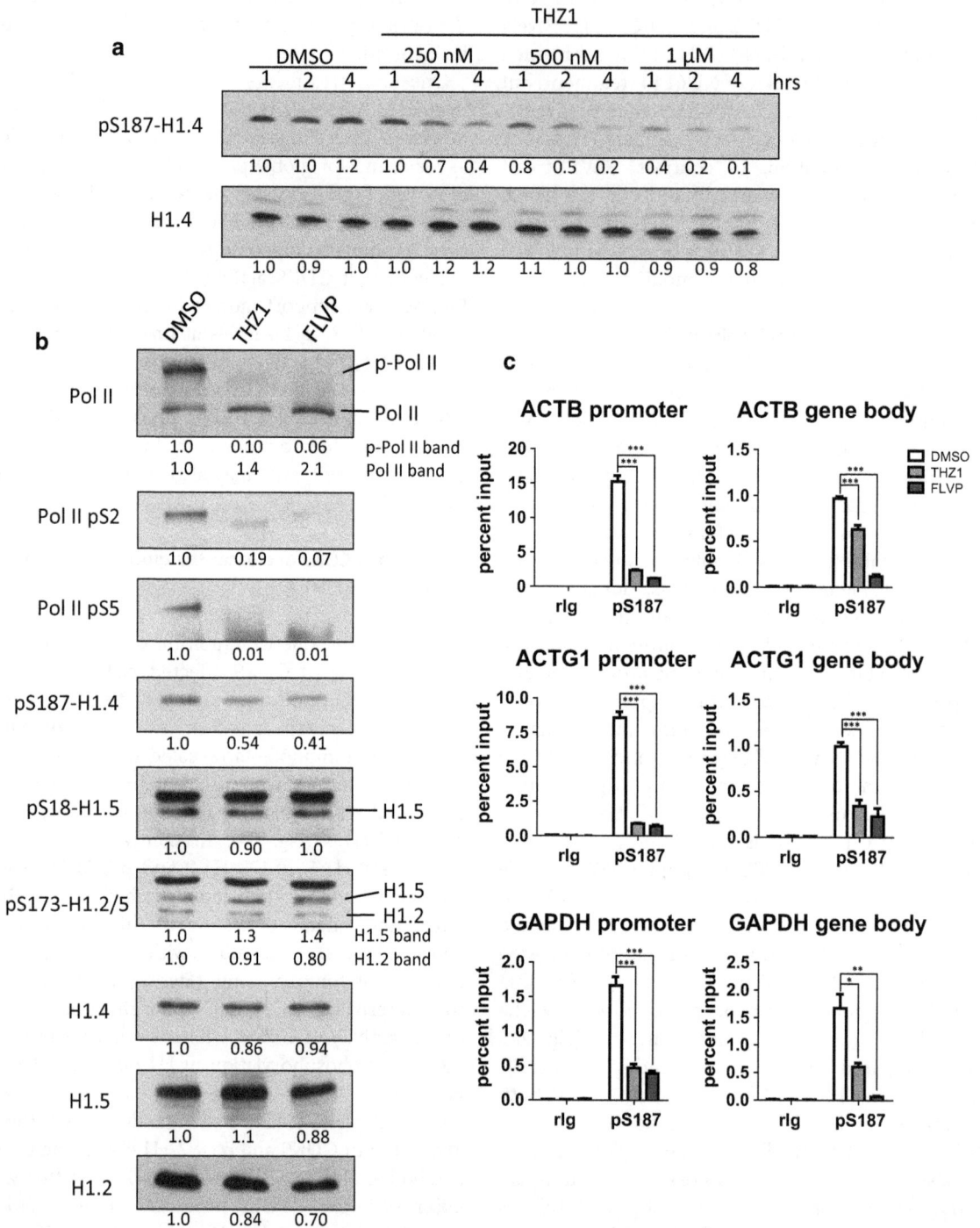

Fig. 4 Selective inhibition of CDK7 or CDK9 diminishes pS187-H1.4 levels globally and at specific genes. **a** HeLa cells were treated with DMSO (solvent control) or increasing amounts of THZ1 to selectively inhibit CDK7 for 1, 2 or 4 h. The global abundance of pS187-H1.4 and H1.4 was assessed by immunoblotting whole-cell lysates. **b** HeLa cells were treated with DMSO (solvent control), 1 μM THZ1 or 1 μM FLVP for 1 h to selectively inhibit CDK7 and CDK9, respectively. Immunoblotting of whole-cell lysates with the indicated antisera was used to monitor the abundance of phosphorylated forms of RNAP II and selected H1 variants. The *numbers below each panel* indicate densitometry for the treated samples relative to the control sample. **c** HeLa cells were treated as in (**b**) and the levels of pS187-H1.4 at the promoters or gene bodies of housekeeping genes were assessed by ChIP-qPCR. Negative control ChIP assays employed non-immune rabbit immunoglobulin (rIg) in place of primary antisera. The data are expressed as percent relative to input DNA (mean ± s.e.m., *$p < 0.05$; **$p < 0.01$; ***$p < 0.001$)

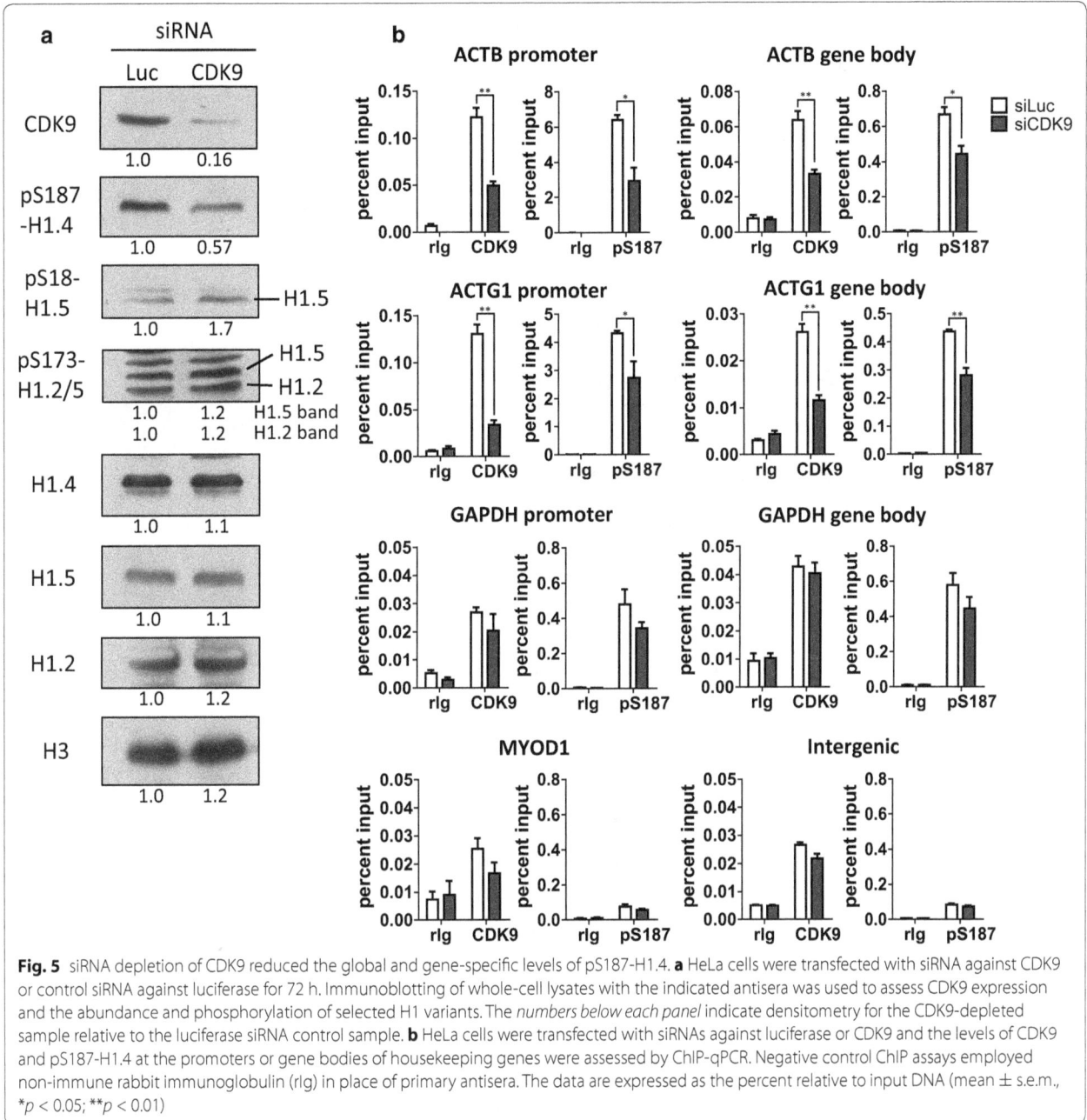

Fig. 5 siRNA depletion of CDK9 reduced the global and gene-specific levels of pS187-H1.4. **a** HeLa cells were transfected with siRNA against CDK9 or control siRNA against luciferase for 72 h. Immunoblotting of whole-cell lysates with the indicated antisera was used to assess CDK9 expression and the abundance and phosphorylation of selected H1 variants. The *numbers below each panel* indicate densitometry for the CDK9-depleted sample relative to the luciferase siRNA control sample. **b** HeLa cells were transfected with siRNAs against luciferase or CDK9 and the levels of CDK9 and pS187-H1.4 at the promoters or gene bodies of housekeeping genes were assessed by ChIP-qPCR. Negative control ChIP assays employed non-immune rabbit immunoglobulin (rIg) in place of primary antisera. The data are expressed as the percent relative to input DNA (mean ± s.e.m., *p < 0.05; **p < 0.01)

Phosphorylation at S187-H1.4 is not dependent on RNAP II progression

Our data on gene-specific pS187-H1.4 dynamics during NT2 cell differentiation (Fig. 2), following short-term inhibition of CDK7 or CDK9 (Fig. 4), and following CDK9 depletion (Fig. 5), suggest that H1.4-S187 phosphorylation by CDK9 could be involved in mechanisms that facilitate transcription by promoting initiation, elongation, or both processes. However, the possibility remains that enrichment of pS187-H1.4 at active genes is mediated by RNAP II-associated kinases other than CDK7 or CDK9. Consequently, we assessed how inhibiting transcription with drugs that are not kinase inhibitors affected the global and gene-specific levels of pS187-H1.4 (Fig. 6). Brief (1 h) exposure of HeLa cells to α-amanitin (50 μM), actinomycin D (500 nM) or triptolide (200 nM) elicited small or no changes in the expression of H1.2, H1.4 and H1.5, and the global levels pS18-H1.5, pS173-H1.2, pS173-H1.5 and pS187-H1.4 (Fig. 6a). Actinomycin D induced global RNAP II CTD hyperphosphorylation,

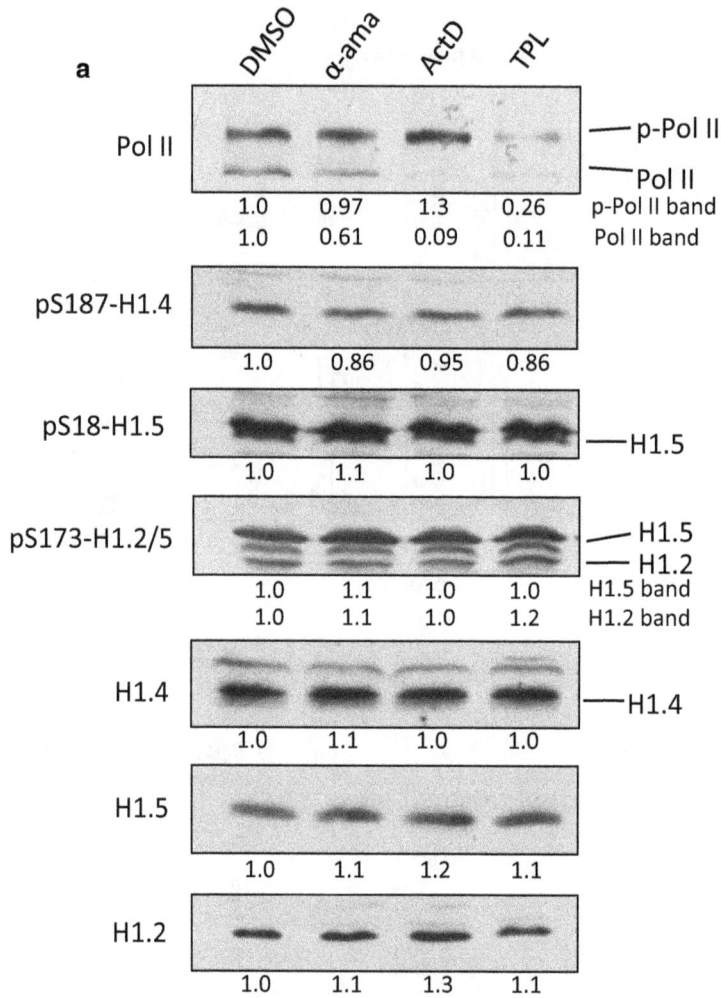

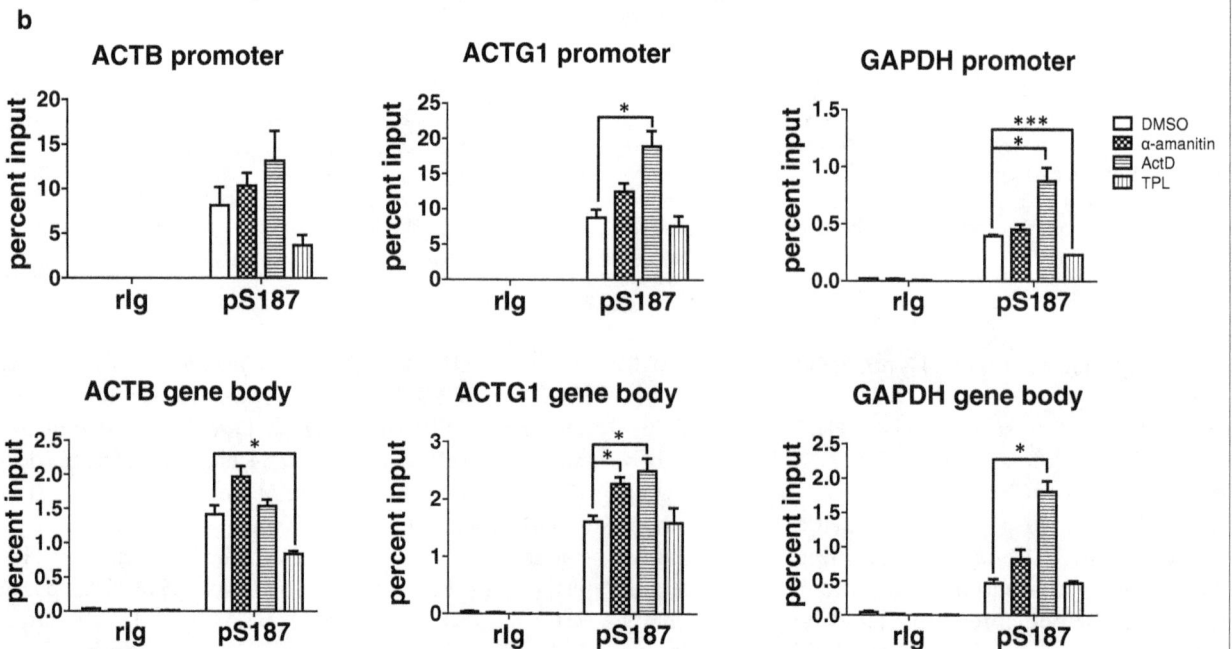

Fig. 6 pS187-H1.4 levels are independent of RNAP II progression. **a** HeLa cells were treated with DMSO (solvent control), 50 μM α-amanitin, 500 nM actinomycin D or 200 nM triptolide for 1 h. The abundance and phosphorylation of RNA polymerase II, H1.2, H1.4 and H1.5 were analyzed by immunoblotting whole-cell lysates with the indicated antisera. The *numbers below each panel* indicate densitometry for the treated samples relative to the DMSO control sample. **b** The levels of pS187-H1.4 at the promoters and gene bodies of housekeeping genes in HeLa cells treated with DMSO (solvent control), 50 μM α-amanitin, 500 nM actinomycin D or 200 nM triptolide for 1 h were assessed by ChIP-qPCR. Negative control ChIP assays employed non-immune rabbit immunoglobulin (rIg) in place of primary antisera. The data are expressed as percent relative to input DNA (mean ± s.e.m., *$p < 0.05$; **$p < 0.01$; ***$p < 0.001$)

consistent with prior evidence that actinomycin D and other treatments that promote P-TEFb release from 7SK snRNP complexes enhance the accumulation of CTD-hyperphosphorylated RNAP II [61, 62]. Triptolide dramatically reduced the levels of both phosphorylated and non-phosphorylated RNAP II, as expected from prior evidence that TPL inhibits the helicase activity of TFIIH by covalently binding the XPB subunit and rapidly induces proteasome-dependent degradation of the large subunit (RPB1) of RNAPII [63]. α-Amanitin reduced the level of non-phosphorylated RNAP II, but not the hyperphosphorylated forms, consistent with prior evidence that α-amanitin induces hyperphosphorylation of RNAP II CTD and RPB1 degradation [61, 64].

Remarkably, actinomycin D treatment led to a significant accumulation of pS187-H1.4 at the promoters and the bodies of ACTG1 and GAPDH genes compared to the control sample (Fig. 6b). α-Amanitin increased the level of pS187-H1.4 slightly at all of the gene regions assessed compared to control cells, but these differences were statistically significant only for the body of the ACTG1 gene. In contrast, TPL caused significant decreases in the level of pS187-H1.4 at the GAPDH promoter and the body of the ACTB gene. Taken together, the data in Fig. 6 suggest that the global and genic levels of pS187-H1.4 are not dependent on RNAP II progression, arguing against the possibility that interphase H1.4-S187 phosphorylation is mediated predominantly by kinases other than CDK9 (or possibly CDK7).

pS187-H1.4 is co-enriched with CDK9 at specific gene loci

The dependency of global and genic pS187-H1.4 levels on CDK9 expression (Fig. 5) and the enhanced enrichment of pS187-H1.4 at housekeeping genes induced by actinomycin D (Fig. 6) both suggest that the levels of pS187-H1.4 at genomic loci depend on the extent of CDK9 recruitment to those loci. We used sequential or "re-ChIP" analyses to test this hypothesis and discovered that pS187-H1.4 and CDK9 were co-enriched to a much greater extent on active housekeeping gene chromatin fragments in HeLa cells compared to fragments derived from the intergenic negative control region or the inactive MYOD1 promoter (Fig. 7a).

Similarly, pS187-H1.4 and CDK9 were extensively co-enriched on active pluripotency factor and housekeeping gene chromatin fragments compared to the intergenic region or the inactive MYOD1 promoter in NT2 cells (Fig. 7b).

We also assessed how treatments that alter the chromatin distribution of CDK9 affected the enrichment of pS187-H1.4 at genes. Actinomycin D rapidly induced the accumulation of CDK9 at both the promoters and bodies of the ACTB, ACTG1 and GAPDH genes in HeLa cells (Fig. 8a), and this was accompanied by corresponding increases in the level of pS187-H1.4 in every case. Similarly, differentiation of NT2 cells with RA led to marked reductions in the association of CDK9 with the Oct-4 and Nanog promoters which correlated with significant decreases in the level of pS187-H1.4 at these same sites (Fig. 8b). In contrast, CDK9 binding in differentiated cells was increased at the ACTG1 and GAPDH promoters, and to a lesser extent at the ACTB promoter, and this was accompanied by non-significant reductions in the association of pS187-H1.4. Taken together, the data from our re-ChIP analyses (Fig. 7), the effect of ActD on CDK9 and pS187-H1.4 levels at housekeeping genes in HeLa cells (Fig. 8a), and the correspondence of CDK9 and pS187-H1.4 levels at pluripotency factor promoters in pluripotent and differentiated NT2 cells (Fig. 8b), suggest that CDK9 mediates H1.4-S187 phosphorylation in vivo and that the levels of pS187-H1.4 at loci reflect the extent of CDK9 recruitment to those sites. However, this simple relationship was not evident when housekeeping genes in pluripotent and differentiated NT2 cells were compared (Fig. 8b). Potential explanations for this difference include the possibility that appreciable amounts of inactive CDK9 (e.g., chromatin–associated 7SK complexes) contribute to the ChIP signal for housekeeping gene promoters in differentiated NT2 cells, whereas active P-TEFb predominates in ActD-treated cells due to its release from 7SK complexes or that pS187-H1.4 dephosphorylation is regulated differently in these distinct cellular contexts. Nonetheless, the data in Fig. 8b provide striking evidence for differences in the mechanisms that regulate the levels of CDK9 and pS187-H1.4 at specific genes.

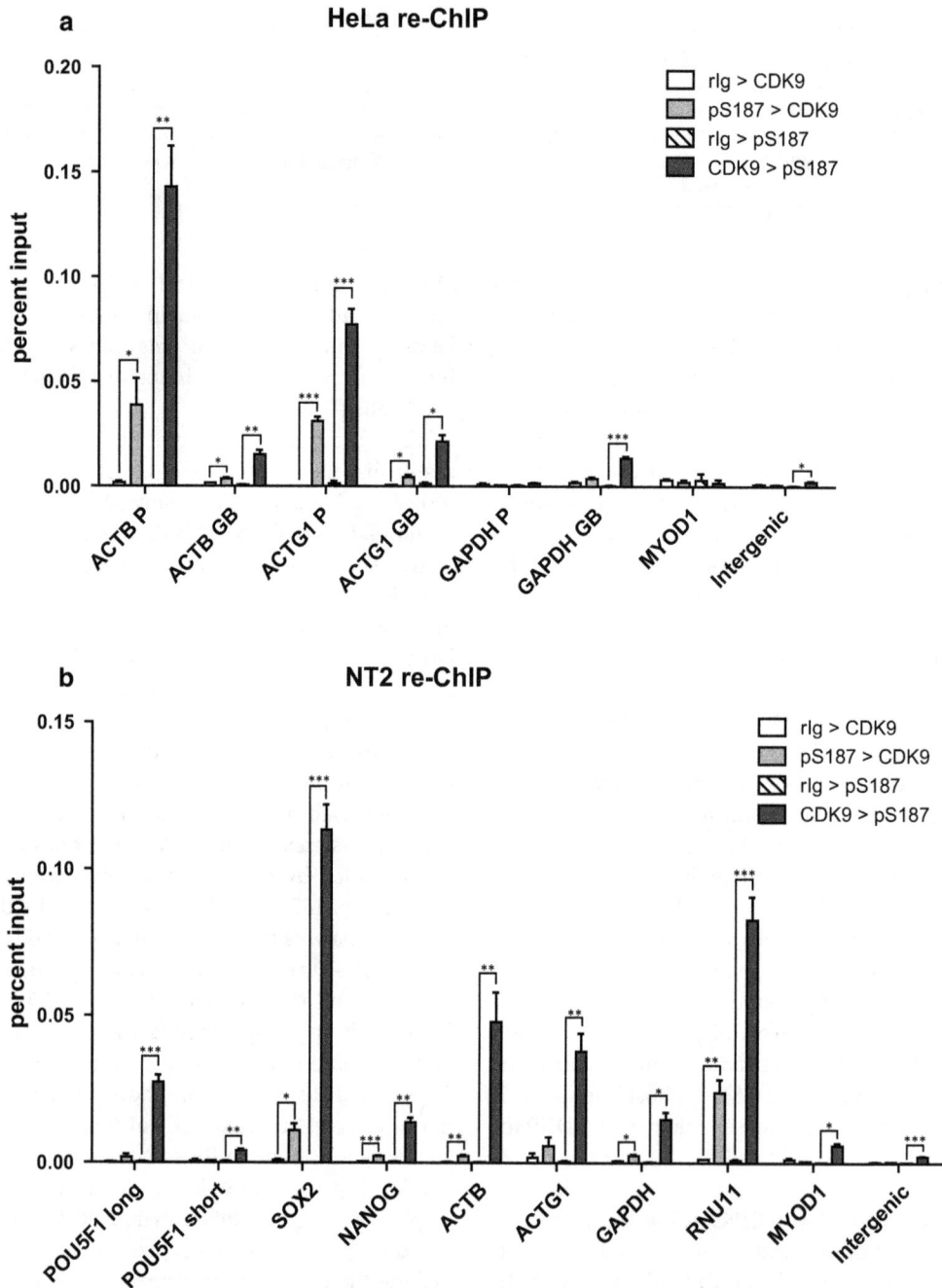

Fig. 7 Co-enrichment of CDK9 and pS187-H1.4 at selected genes. **a** re-ChIP assays were performed on asynchronous HeLa cells using CDK9 antibody in the first ChIP followed by pS187-H1.4 antiserum in the second ChIP or vice versa. The co-enrichment of CDK9 and pS187-H1.4 at selected promoters and bodies of housekeeping genes was assessed by qPCR. **b** re-ChIP assays were performed on pluripotent NT2 cells as in (**a**). The co-enrichment of CDK9 and pS187-H1.4 at the promoters of pluripotency genes and housekeeping genes was assessed by qPCR. Negative control re-ChIP assays employed non-immune rabbit IgG (rIg) in place of primary antisera for the first ChIP. The data are expressed as percent relative to input DNA (mean ± s.e.m., *$p < 0.05$; **$p < 0.01$; ***$p < 0.001$)

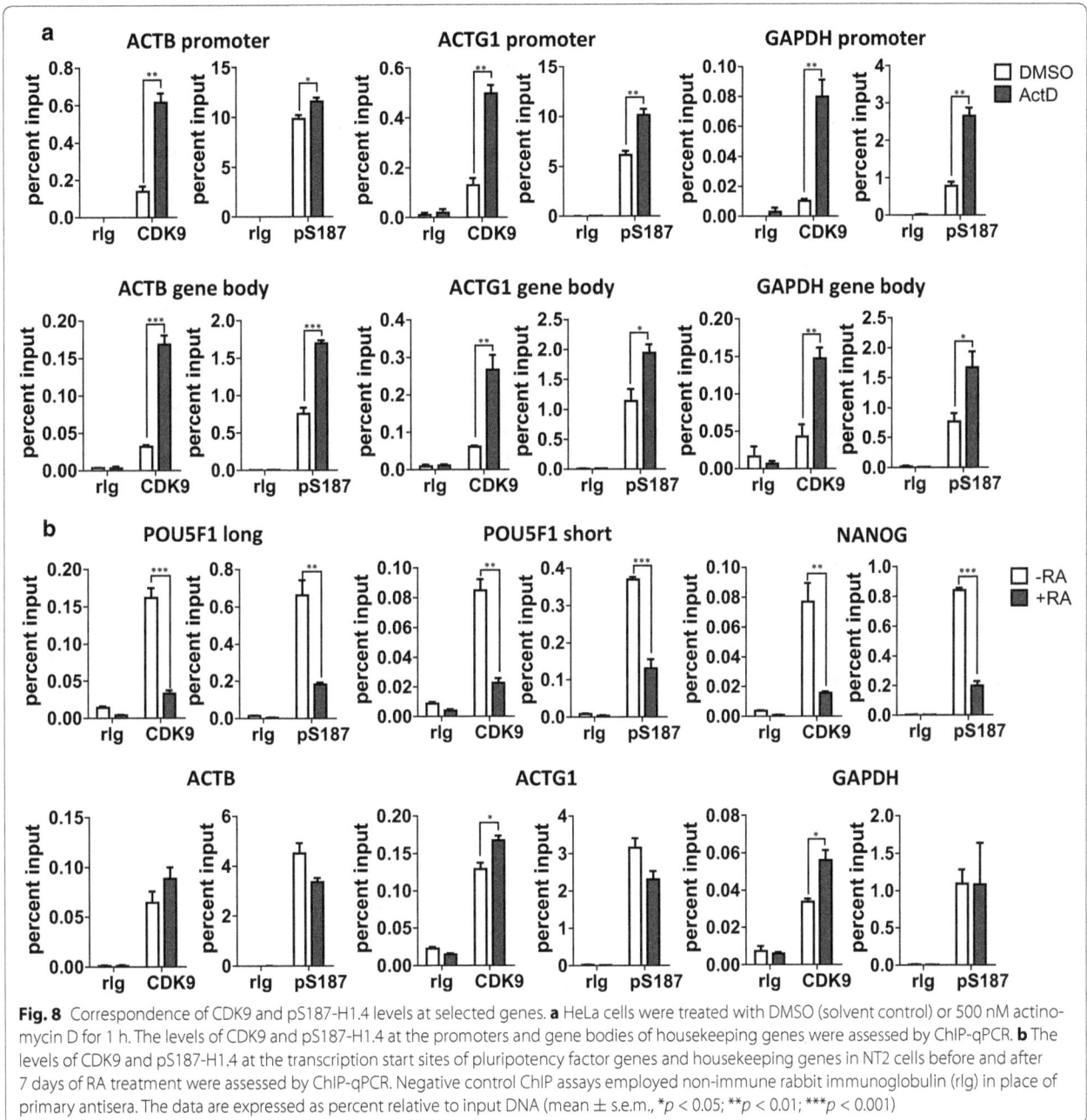

Fig. 8 Correspondence of CDK9 and pS187-H1.4 levels at selected genes. **a** HeLa cells were treated with DMSO (solvent control) or 500 nM actinomycin D for 1 h. The levels of CDK9 and pS187-H1.4 at the promoters and gene bodies of housekeeping genes were assessed by ChIP-qPCR. **b** The levels of CDK9 and pS187-H1.4 at the transcription start sites of pluripotency factor genes and housekeeping genes in NT2 cells before and after 7 days of RA treatment were assessed by ChIP-qPCR. Negative control ChIP assays employed non-immune rabbit immunoglobulin (rIg) in place of primary antisera. The data are expressed as percent relative to input DNA (mean ± s.e.m., *$p < 0.05$; **$p < 0.01$; ***$p < 0.001$)

Discussion

Cell differentiation has site-specific effects on H1 variant interphase phosphorylation

Multiple differences in epigenomic landscape and a more open chromatin structure distinguish pluripotent cells from their differentiated counterparts [65–67]. Differentiated cells have increased heterochromatin content over ES cells [68–70], consistent with evidence that repressive histone marks like H3K9me3 and H3K27me3 are distributed more widely in differentiated cells compared to stem cells [71]. Conversely, H3 and H4 acetylation are more prevalent in pluripotent ES cells than differentiated cells [67], and H3K9ac is enriched at a greater fraction of promoters in stem cells compared to differentiated cells [72]. The global abundance of multi-acetylated H4 diminishes during ES cell differentiation, with ES cell-specific and embryoid body-specific peaks of multi-acetylated H4 present at genes that are differentially expressed in these cell types [73]. The work presented here provides novel evidence that changes in the abundance and distribution of interphase phosphorylated forms of H1 variants are also involved in the changes in nuclear structure and gene activity that accompany cell differentiation. Phosphorylation decreases the affinity of H1-GFP fusions for chromatin binding and enhances their mobility in vivo [9]. Thus, the higher global levels of phosphorylation that we show are present at several interphase sites in pluripotent NT2 and mES cells may contribute to the weaker chromatin binding of endogenous H1 variants and the hyperdynamic mobility of H1-GFP fusions in ES cells compared to differentiated cells [74].

Significant changes in cell cycle regulation occur during pluripotent cell differentiation. Whereas S phase cells are predominant in asynchronous cultures of pluripotent NT2 cells, this fraction decreases and the proportion of G1 phase cells increases following differentiation (data not shown) [75, 76]. The levels of interphase H1 phosphorylation are expected to decrease globally with differentiation since analyses of crude H1 (i.e., the sum of the H1 variants expressed) have revealed that phosphorylation increases progressively over interphase before reaching maximum levels at mitosis [14–19]. However, it is unclear whether the progressive increase in H1 phosphorylation during interphase detected in earlier work was due to increasing levels of phosphorylation at a fixed number of sites, increased numbers of sites becoming phosphorylated, or a combination of these alternatives. Recent evidence supports the former possibility, suggesting that the majority of sites of interphase phosphorylation of H1 variants are established during G1 phase. A clear increase in the levels of phosphorylation at all interphase sites of H1.2, H1.3, H1.4 and H1.5 was observed during S phase, and this pattern was largely preserved in G2/M cells with additional hyperphosphorylated forms being identified [77]. These findings suggest that the increased abundance of G1 cells after differentiation should lead to uniform reductions in phosphorylation across interphase sites unless they are regulated in a variant-specific or site-specific fashion. Our findings that H1.2, H1.4 and H1.5 are more highly phosphorylated in asynchronous cultures of undifferentiated NT2 cells than they are after differentiation is consistent with the presence of a higher proportion of S phase cells. However, the preferential loss of pS18-H1.5 observed during NT2 cell differentiation, and the preferential loss of pS173-H1.2/5 combined with elevated pS187-H1.4 observed during mESC differentiation argue that interphase phosphorylation levels are not uniformly reduced across the distinct types of sites present in different H1 variants during cell differentiation as a consequence of the shift in cell cycle distribution. Although little is known about the numbers and nature of the pathways that regulate interphase H1 phosphorylation, our findings demonstrate that they act

in conjunction with changes in the differential expression of H1 variants to regulate the phosphorylation levels on individual H1 variants in a site-specific fashion.

Cell differentiation affects the genomic distributions of interphase phosphorylated H1 variants

Embryonic stem cells are pluripotent cells derived from the inner cell mass of the early embryo that are able to self-renew and differentiate into all three germ layers [78, 79]. A regulatory network involving transcription factors such as OCT4, SOX2 and NANOG is essential for the maintenance of pluripotency in ES cells [79, 80]. The binding of RA to RAR/RXR heterodimers at enhancers containing retinoic acid response elements (RARE) activates the transcription of RA-regulated primary response genes, including transcription factors that activate or repress their target genes [81, 82]. RA-induced repression of pluripotency factor genes is not mediated directly by RAR/RXR. Increased expression of the orphan nuclear receptor GCNF induced by RA represses OCT4 expression by promoting GCNF binding to the OCT4 promoter [83, 84]. The binding of GCNF recruits the DNA methyltransferase DNMT3 to the OCT4 promoter, facilitating its methylation and transcriptional repression. H3K9/K14 acetylation and H3K4 methylation at the OCT4 promoter decreases during differentiation, while H3K9 and H3K27 methylation levels remain constant [85]. Our data indicate that a loss of pS187-H1.4 enrichment, and possibly pS173-H1.2/5, at pluripotency factor promoters is part of the mechanism of their repression by RA in NT2 cells. In contrast, pS18-H1.5 appears to become modestly enriched at these loci when they are repressed. These differentiation-induced changes in phosphorylated H1 variant distribution appear to be gene-specific since significant changes were not detected for pS187-H1.4, pS173-H1.2/5 or pS18-H1.5 at several housekeeping genes. The elevated expression of H1.0 after cell differentiation may also facilitate the repression of genes important for self-renewal since ChIP analyses suggest that the association of HA-tagged H1.0 with pluripotency genes is enhanced in differentiated cells compared to pluripotent cells [34]. It may be significant in this regard that H1.0 lacks predicted sites for interphase phosphorylation [13].

Regulation of interphase H1 phosphorylation

Our data imply that enrichment of pS187-H1.4 at the promoters of pluripotency factor genes in pluripotent cells, and at the promoters of housekeeping genes (and presumably other types of genes) in both pluripotent and differentiated cells, is associated with their transcriptional activity. Interphase H1 phosphorylation has been implicated in facilitating transcription in a gene-specific fashion in several model systems. ChIP analyses with an antisera that recognizes multiple H1 phosphorylations suggest that H1 phosphorylation levels correlate directly with the hormone-inducibility and transcriptional competency of the murine mammary tumor virus (MMTV) promoter [29, 86]. This is consistent with subsequent work showing that pS187-H1.4 is enriched at hormone-response elements shortly after hormone stimulation, suggesting that this pS187-H1.4 enrichment facilitates the transcription of their target genes [23]. However, little is known about the mechanisms responsible for the targeted enrichment of pS187-H1.4 or other interphase phosphorylated forms of H1.

CDK2 has been suggested to be an interphase H1 kinase, but the evidence is equivocal. FRAP analyses suggested that the mobility of GFP-H1.4, but not GFP-M1-5 (all five S/TPXZ sites in H1.4 mutated to APXZ), was positively correlated with the level of CDK2 activity, and this difference was minimized upon CDK2 inhibition [30]. However, later work showed that three out of the five CDK motifs investigated are TPXZ motifs that are phosphorylated exclusively during mitosis [23]. Inhibitors thought to be selective for CDK2 such as CVT-313 and roscovitine decreased global levels of phosphorylated H1 and its association with the MMTV promoter [29], and roscovitine and olomoucine diminished H1 phosphorylation at replication foci [87], but several issues affect these studies. Both employed antisera raised against phosphorylated *Tetrahymena* macronuclear H1, which lacks SPXZ motifs [88], to detect phosphorylated H1 even though the sites in metazoan H1 variants that are recognized by this antisera have not been established. Moreover, CVT-313, roscovitine and olomoucine are now known to inhibit additional CDKs with similar potency [89–92], and prolonged treatments with these drugs can arrest cell cycle progression [91, 93] and consequently affect H1 phosphorylation levels indirectly. Here, we used a time course experiment to show that treatments as long as 24 h with NU6140, an inhibitor with greater selectivity for CDK2 [44], did not reduce the global level of pS187-H1.4 or pS18-H1.5 in NT2 cells, providing strong evidence that CDK2 is not involved in regulating phosphorylation of H1 at these two sites.

Multiple lines of evidence presented here strongly implicate CDK9 as the predominant interphase kinase for H1.4-S187 in NT2 and HeLa cells. Brief (one hour or less) treatments with the preferential CDK9 inhibitor FLVP significantly diminish the global level of pS187-H1.4 in both NT2 cells (Fig. 3) and HeLa cells (Fig. 4), without diminishing the levels of pS18-H1.5 or pS173-H1.2/5 (Fig. 4). Short FLVP treatment essentially abolished pS187-H1.4 enrichment at housekeeping genes in HeLa cells (Fig. 4). Depletion of CDK9 in HeLa cells significantly decreased global H1.4-S187 phosphorylation

and the association of pS187-H1.4 with housekeeping genes, while the global levels of pS18-H1.5 and pS173-H1.2/5 were slightly increased (Fig. 5). Moreover, re-ChIP analyses show that CDK9 and pS187-H1.4 are specifically co-enriched on housekeeping gene chromatin fragments recovered from HeLa and NT2 cells and also on pluripotency factor chromatin fragments recovered from NT2 cells (Fig. 7). Taken together, these findings provide strong evidence that CDK9 mediates H1.4-S187 phosphorylation which is consistent with prior evidence that H1 interacts with P-TEFb in vivo [31]. Remarkably, our findings also imply that the H1.5-S18 and H1.2/5-S173 interphase phosphorylations are mediated by distinct kinases. However, our data do not exclude the possibility that H1.4-S187 may be targeted by additional interphase kinases. Brief, highly selective inhibition of CDK7 by THZ1 also decreased the levels of pS187-H1.4 globally and at specific gene loci (Fig. 4), but further work is required to determine the extent to which this reflects the possible direct phosphorylation of H1.4-S187 or is secondary to diminished activation of CDK9 [57]. Our data are consistent with a role for CDK7 via either mechanism.

Given the roles of CDK7 and CDK9 in regulating RNAP II transcription [47, 57], the formal possibility remains that the effects of CDK7 and CDK9 inhibition on H1 phosphorylation are secondary to transcription deficits elicited by these drugs. We suggest that this is unlikely since we limited inhibitor treatments to just one hour. Furthermore, our findings that inhibition of RNAP II transcription with α-amanitin, actinomycin D or triptolide had relatively minor effects on global pS187-H1.4 levels (Fig. 6) compared to kinase inhibition by THZ1 and FLVP (Fig. 4) also argue against this possibility and suggest that H1.4-S187 phosphorylation does not depend directly on RNAP II progression. In fact, the enhancements in global RNAP II CTD phosphorylation and the association of pS187-H1.4 at gene promoters and bodies elicited by brief treatment with ActD (Fig. 6) provide further evidence that CDK9 is a *bona fide* interphase kinase for H1.4-S187 since ActD induces the release of active P-TEFb from 7SK snRNA complexes [61, 94, 95]. Moreover, we show that CDK9 and pS187-H1.4 are co-enriched at housekeeping gene promoters and gene bodies in ActD-treated HeLa cells but are co-depleted at pluripotency factor gene promoters after NT2 cells differentiate (Fig. 8).

Our collective data suggest that recruitment of existing active P-TEFb complexes, or P-TEFb released from 7SK complexes, to sites of transcription results in the accumulation and enrichment of pS187-H1.4 that facilitates RNAP II transcription by promoting the transient dissociation of H1.4 or possibly by other mechanisms. A variety of mechanisms have been implicated in releasing P-TEFb from 7SK complexes, including post-translational modification of 7SK snRNP components and direct interactions with RNA binding proteins or transcriptional regulators [96, 97]. The RNA splicing factor SRSF2 (also known as SC-35) was found to be part of the 7SK complex assembled at gene promoters. The binding of SRSF2 to promoter-associated nascent RNA triggers the coordinated release of SRSF2 and P-TEFb from the 7SK complex, reminiscent of the mechanism used by HIV Tat/TAR to activate the transcription of HIV genes [98]. The promoter-bound DEAD-box RNA helicase DDX21 was found to be recruited to promoters of RNAP II-transcribed genes encoding ribosomal proteins and snoRNAs and promote their activation by releasing P-TEFb from the 7SK snRNP in a helicase-dependent manner [99]. These different mechanisms for releasing P-TEFb from 7SK complexes share the common feature that the activation of P-TEFb occurs on chromatin where transcription and pre-mRNA processing occurs. Our findings suggest that H1.4-S187 phosphorylation at gene promoters/ TSSs occurring conjointly with phosphorylation of the RNAP II-CTD, DSIF and NELF by P-TEFb may facilitate transcriptional elongation by promoting the release of promoter-proximal paused RNAP II, but further work is required to distinguish between this and other possible consequences of H1.4-S187 phosphorylation.

Recent data suggest a possible mechanism for our observations that the levels of CDK9 and pS187-H1.4 at pluripotency factor gene promoters are reduced by differentiation (Figs. 2, 8b). BRD4 and acetylated histone H4 are enriched at the OCT4 enhancer and promoter regions in ESCs, and their enrichment decreases upon differentiation and embryoid body formation [73]. BRD4 binding to acetylated H4 mediates CDK9 recruitment at pluripotency factor genes such as OCT4 that are associated with super-enhancers in ESCs. Inhibiting chromatin binding by BRD4 with a specific bromodomain inhibitor impaired CDK9 recruitment at the OCT4 enhancer and promoter regions and reduced the expression of pluripotency genes [100]. These findings suggest that reduced BRD4 enrichment at OCT4 and other pluripotency genes leads to less CDK9 recruitment and consequently diminished H1.4-S187 phosphorylation following cell differentiation.

Functional diversity of H1 variants

An obvious difference that distinguishes the amino acid sequences of metazoan H1 variants from each other is the number and positions of SPXZ sites they possess for interphase phosphorylation. Human H1.1 and H1.3 each have one at the same relative position (S183 or S189). H1.2 also has one, but at a different position (S173). H1.4 has two (S172 and S187) and H1.5 has three (S18, S173 and S189). Other human H1 variants either lack SPXZ sites or they occur at non-conserved positions [13]. These differences in SPXZ motif conservation may enable phosphorylation to impart different functions to individual H1 variants by altering their conformation, DNA binding and other properties in specific ways [101]. Our evidence that pS187-H1.4, but not pS18-H1.5, is preferentially associated with active genes (vide infra) [23, 24] is consistent with this possibility. The differences we observed in the regulation of phosphorylation at H1.4-S187 versus H1.2-S173, H1.5-S173 and H1.5-S18 also support this hypothesis. The involvement of CDK7 and CDK9 revealed here in regulating H1.4-S187 phosphorylation may contribute to the unique requirement of H1.4 among H1.0 and H1.1-H1.5 for the viability of human breast cancer cells [102]. Moreover, these differences in phosphorylation regulation, together with the changes we observed for interphase phosphorylation of H1.2, H1.4 and H1.5 during pluripotent cell differentiation, suggest that these pathways are likely to play distinct roles in developmental regulation of gene expression.

Conclusions

In this study, we demonstrated that interphase phosphorylation of H1 variants is reduced preferentially at specific sites during pluripotent cell differentiation and that the enrichment of pS187-H1.4, but not pS18-H1.5, is positively correlated with transcription. Brief inhibition of CDK9 or CDK7 with specific inhibitors, or depletion of CDK9, decreased global and gene-specific levels of pS187-H1.4. Remarkably, inhibiting RNAP II transcription with actinomycin D significantly enhanced co-enrichment of pS187-H1.4 and CDK9 at housekeeping genes. In contrast, the global levels of pS18-H1.5 or pS173-H1.2/5 are not affected by any of the above treatments, suggesting that these interphase phosphorylations are regulated by different mechanisms. Our findings suggest that H1.4-S187 is a *bona fide* substrate for CDK9 that is phosphorylated during transcriptional activation, and that reductions in the levels of CDK9 and pS187-H1.4 at pluripotency factor gene promoters contribute to their repression of during cell differentiation. Interphase phosphorylation at other sites in H1 variants appears to be regulated by different kinase pathways, and this is expected to contribute to functional diversity among different H1 variants.

Additional files

Additional file 1: Table S1. List of ChIP-qPCR primers.

Additional file 2: Figure S1. Validations of custom H1 antibodies that have not been published previously. *A*, Recombinant H1 variants were analyzed by immunoblotting with our custom antisera for H1.5 and H1.0. *B*, PCA-extracted crude H1 from WI-38 VA-13 cells and a mixture of recombinant H1.5 and H1.0 were analyzed by immunoblotting with our custom antisera against pS18-H1.5. *C*, Antisera against pS18-H1.5 was mock-treated (none) or preadsorbed with pS18-H1.5 antigen peptide (pS18) or the corresponding non-phosphorylated peptide (S18) prior to immunoblotting with HeLa whole-cell lysate. *D*, HeLa and WI-38 VA-13 whole-cell lysates (WCL) and PCA-extracted crude H1 (PCAS) from WI-38 VA-13 cells were analyzed by immunoblotting with our custom antisera against H1.5 and H1.0.

Additional file 3: Figure S2. H1 variant expression and phosphorylation in differentiating NT2 cells. Crude H1 in acid extracts of nuclei isolated from NT2 cells induced to differentiate with 10 µM retinoic acid for 0, 1, 3 or 7 days was fractionated by hydrophobic interaction chromatography (HIC). Eluate absorbance at 214 nm (*Y* axis) is plotted relative to time (*X* axis) for equivalent portions of each separation. The relative elution positions of H1.2, H1.3, H1.4 and H1.5 and the phosphorylation stoichiometry of their major interphase forms, as characterized previously [23], are indicated above the 0 day trace. H1.0 coelutes as a broad peak that overlaps with both phosphorylated and non-phosphorylated H1.5 (data not shown).

Additional file 4: Figure S3. pS187-H1.4 is preferentially enriched in active chromatin. *A*, HeLa cells were treated with DMSO or 500 nM ActD for 1 h and the levels of CDK9 and pS187-H1.4 at the promoters of ACTG1, GAPDH or MYOD1 and one intergenic region were assessed by ChIP-qPCR. *B*, HeLa cells were treated with DMSO or 1 µM FLVP for 1 h and the levels of CDK9 and pS187-H1.4 at the promoters of ACTG1, GAPDH or MYOD1 and one intergenic region were assessed by ChIP-qPCR. Negative control ChIP assays employed non-immune rabbit IgG (rIg) in place of primary antisera for the first ChIP. The data are expressed as percent relative to input DNA (mean ± s.e.m., *: $p < 0.05$, **: $p < 0.01$, ***: $p < 0.001$).

Abbreviations
ActD: actinomycin D; CAK: CDK-activating kinase; CDK: cyclin-dependent kinase; ChIP: chromatin immunoprecipitation; CTD: C-terminal domain; ESC: embryonic stem cell; FLVP: flavopiridol; LIF: leukemia inhibitory factor; P-TEFb: positive transcription elongation factor b; RA: retinoic acid; TPL: triptolide; TSS: transcription start site.

Authors' contributions
RL conducted all of the experiments, analyzed the results, prepared the figures and wrote a draft of the manuscript. CAM conceived the project, interpreted the results and revised the paper. Both authors read and approved the final manuscript.

Author details
[1] Department of Cell and Developmental Biology, University of Illinois at Urbana Champaign, B107 Chemistry and Life Sciences Building, MC-123 601 S. Goodwin Ave., Urbana, IL 61801, USA. [2] Institute for Genomic Biology, University of Illinois at Urbana Champaign, Urbana, IL 61801, USA.

Competing interests
The authors declare that they have no competing interests.

Funding
This work was supported by Grants from the National Science Foundation (MCB-0821893) and the UIUC Campus Research Board (RB13053 and RB15254).

References
1. Allan J, Hartman PG, Crane-Robinson C, Aviles FX. The structure of histone H1 and its location in chromatin. Nature. 1980;288(5792):675–9.
2. Carruthers LM, Bednar J, Woodcock CL, Hansen JC. Linker histones stabilize the intrinsic salt-dependent folding of nucleosomal arrays: mechanistic ramifications for higher-order chromatin folding. Biochemistry. 1998;37(42):14776–87.
3. Bednar J, Horowitz RA, Grigoryev SA, Carruthers LM, et al. Nucleosomes, linker DNA, and linker histone form a unique structural motif that directs the higher-order folding and compaction of chromatin. Proc Natl Acad Sci USA. 1998;95(24):14173–8.
4. Happel N, Doenecke D. Histone H1 and its isoforms: contribution to chromatin structure and function. Gene. 2009;431(1–2):1–12.
5. Harshman SW, Young NL, Parthun MR, Freitas MA. H1 histones: current perspectives and challenges. Nucleic Acids Res. 2013;41(21):9593–609.
6. Chapman GE, Hartman PG, Bradbury EM. Studies on the role and mode of operation of the very-lysine-rich histone H1 in eukaryote chromatin. Eur J Biochem. 1976;61(1):69–75.
7. Misteli T, Gunjan A, Hock R, Bustin M, et al. Dynamic binding of histone H1 to chromatin in living cells. Nature. 2000;408:877–81.
8. Lever MA, Th'ng JPH, Sun X, Hendzel MJ. Rapid exchange of histone H1.1 on chromatin in living human cells. Nature. 2000;408(6814):873–6.
9. Hendzel MJ, Lever MA, Crawford E, Th'ng JPH. The C-terminal domain is the primary determinant of histone H1 binding to chromatin in vivo. J Biol Chem. 2004;279(19):20028–34.
10. Th'ng JPH, Sung R, Ye M, Hendzel MJ. H1 family histones in the nucleus: control of binding and localization by the C-terminal domain. J Biol Chem. 2005;280(30):27809–14.
11. Wood C, Snijders A, Williamson J, Reynolds C, et al. Post-translational modifications of the linker histone variants and their association with cell mechanisms. FEBS J. 2009;276(14):3685–97.
12. Izzo A, Schneider R. The role of linker histone H1 modifications in the regulation of gene expression and chromatin dynamics. Biochim Biophys Acta. 2016;1859(3):486–95.
13. Liao R, Mizzen CA. Interphase H1 phosphorylation: regulation and functions in chromatin. Biochim Biophys Acta. 2016;1859(3):476–85.
14. Bradbury EM, Inglis RJ, Matthews HR, Sarner N. Phosphorylation of very-lysine-rich histone in *Physarum polycephalum*. Eur J Biochem. 1973;33(1):131–9.
15. Bradbury EM, Inglis RJ, Matthews HR. Control of cell division by very lysine rich histone (F1) phosphorylation. Nature. 1974;247(5439):257–61.
16. Hohmann P, Tobey RA, Gurley LR. Cell-cycle-dependent phosphorylation of serine and threonine in Chinese hamster cell F1 histones. Biochem Biophys Res Commun. 1975;63(1):126–33.
17. Hohmann P, Tobey RA, Gurley LR. Phosphorylation of distinct regions of f1 histone. Relationship to the cell cycle. J Biol Chem. 1976;251(12):3685–92.
18. Ajiro K, Borun TW, Shulman SD, McFadden GM, et al. Comparison of the structures of human histones 1A and 1B and their intramolecular phosphorylation sites during the HeLa S-3 cell cycle. Biochemistry. 1981;20(6):1454–64.
19. Ajiro K, Borun TW, Cohen LH. Phosphorylation states of different histone 1 subtypes and their relationship to chromatin functions during the HeLa S-3 cell cycle. Biochemistry. 1981;20(6):1445–54.
20. Garcia BA, Busby SA, Barber CM, Shabanowitz J, et al. Characterization of phosphorylation sites on histone H1 isoforms by tandem mass spectrometry. J Proteome Res. 2004;3(6):1219–27.
21. Sarg B, Helliger W, Talasz H, Forg B, et al. Histone H1 phosphorylation occurs site-specifically during interphase and mitosis: identification

of a novel phosphorylation site on histone H1. J Biol Chem. 2006;281(10):6573–80.

22. Wiśniewski JR, Zougman A, Krüger S, Mann M. Mass spectrometric mapping of linker histone H1 variants reveals multiple acetylations, methylations, and phosphorylation as well as differences between cell culture and tissue. Mol Cell Proteomics. 2007;6(1):72–87.

23. Zheng Y, John S, Pesavento JJ, Schultz-Norton JR, et al. Histone H1 phosphorylation is associated with transcription by RNA polymerases I and II. J Cell Biol. 2010;189(3):407–15.

24. Talasz H, Sarg B, Lindner HH. Site-specifically phosphorylated forms of H1.5 and H1.2 localized at distinct regions of the nucleus are related to different processes during the cell cycle. Chromosoma. 2009;118(6):693–709.

25. Happel N, Stoldt S, Schmidt B, Doenecke D. M phase-specific phosphorylation of histone H1.5 at threonine 10 by GSK-3. J Mol Biol. 2009;386(2):339–50.

26. Daujat S, Zeissler U, Waldmann T, Happel N, et al. HP1 binds specifically to Lys26-methylated histone H1.4, whereas simultaneous Ser27 phosphorylation blocks HP1 binding. J Biol Chem. 2005;280(45):38090–5.

27. Hergeth SP, Dundr M, Tropberger P, Zee BM, et al. Isoform-specific phosphorylation of human linker histone H1.4 in mitosis by the kinase Aurora B. J Cell Sci. 2011;124(10):1623–8.

28. Chu CS, Hsu PH, Lo PW, Scheer E, et al. Protein kinase A-mediated Serine 35 phosphorylation dissociates histone H1.4 from mitotic chromosome. J Biol Chem. 2011;286(41):35843–51.

29. Bhattacharjee RN, Banks GC, Trotter KW, Lee HL, et al. Histone H1 phosphorylation by Cdk2 selectively modulates mouse mammary tumor virus transcription through chromatin remodeling. Mol Cell Biol. 2001;21(16):5417–25.

30. Contreras A, Hale TK, Stenoien DL, Rosen JM, et al. The dynamic mobility of histone H1 is regulated by cyclin/CDK phosphorylation. Mol Cell Biol. 2003;23(23):8626–36.

31. O'Brien SK, Cao H, Nathans R, Ali A, et al. P-TEFb kinase complex phosphorylates histone H1 to regulate expression of cellular and HIV-1 genes. J Biol Chem. 2010;285(39):29713–20.

32. O'Brien SK, Knight KL, Rana TM. Phosphorylation of histone H1 by P-TEFb is a necessary step in skeletal muscle differentiation. J Cell Physiol. 2012;227(1):383–9.

33. Deterding LJ, Bunger MK, Banks GC, Tomer KB, et al. Global changes in and characterization of specific sites of phosphorylation in mouse and human histone H1 isoforms upon CDK inhibitor treatment using mass spectrometry. J Proteome Res. 2008;7(6):2368–79.

34. Terme JM, Sese B, Millan-Arino L, Mayor R, et al. Histone H1 variants are differentially expressed and incorporated into chromatin during differentiation and reprogramming to pluripotency. J Biol Chem. 2011;286(41):35347–57.

35. Fan YH, Nikitina T, Zhao J, Fleury TJ, et al. Histone H1 depletion in mammals alters global chromatin structure but causes specific changes in gene regulation. Cell. 2005;123(7):1199–212.

36. Zhang Y, Cooke M, Panjwani S, Cao K, et al. Histone H1 depletion impairs embryonic stem cell differentiation. PLoS Genet. 2012;8(5):e1002691.

37. Li J-Y, Patterson M, Mikkola HKA, Lowry WE, et al. Dynamic distribution of linker histone H1.5 in cellular differentiation. PLoS Genet. 2012;8(8):e1002879.

38. Pesavento JJ, Yang H, Kelleher NL, Mizzen CA. Certain and progressive methylation of histone H4 at lysine 20 during the cell cycle. Mol Cell Biol. 2008;28(1):468–86.

39. Andrews PW. Retinoic acid induces neuronal differentiation of a cloned human embryonal carcinoma cell line in vitro. Dev Biol. 1984;103(2):285–93.

40. Schwartz CM, Spivak CE, Baker SC, McDaniel TK, et al. NTera2: a model system to study dopaminergic differentiation of human embryonic stem cells. Stem Cells Dev. 2005;14(5):517–34.

41. Niwa H, Ogawa K, Shimosato D, Adachi K. A parallel circuit of LIF signalling pathways maintains pluripotency of mouse ES cells. Nature. 2009;460(7251):118–22.

42. Hirai H, Karian P, Kikyo N. Regulation of embryonic stem cell self-renewal and pluripotency by leukaemia inhibitory factor. Biochem J. 2011;438:11–23.

43. White J, Dalton S. Cell cycle control of embryonic stem cells. Stem Cell Reviews and Reports. 2005;1(2):131–8.

44. Pennati M. Potentiation of paclitaxel-induced apoptosis by the novel cyclin-dependent kinase inhibitor NU6140: a possible role for survivin down-regulation. Mol Cancer Ther. 2005;4(9):1328–37.

45. Chao SH, Fujinaga K, Marion JE, Taube R, et al. Flavopiridol inhibits P-TEFb and blocks HIV-1 replication. J Biol Chem. 2000;275(37):28345–8.

46. Chao SH, Price DH. Flavopiridol inactivates P-TEFb and blocks most RNA polymerase II transcription in vivo. J Biol Chem. 2001;276(34):31793–9.

47. Price DH. P-TEFb, a cyclin-dependent kinase controlling elongation by RNA polymerase II. Mol Cell Biol. 2000;20(8):2629–34.

48. Wang S, Fischer P. Cyclin-dependent kinase 9: a key transcriptional regulator and potential drug target in oncology, virology and cardiology. Trends Pharmacol Sci. 2008;29(6):302–13.

49. Gallorini M, Cataldi A, Giacomo V. Cyclin-dependent kinase modulators and cancer therapy. Biodrugs. 2012;26(6):377–91.

50. Garrett S, Barton WA, Knights R, Jin P, et al. Reciprocal activation by cyclin-dependent kinases 2 and 7 is directed by substrate specificity determinants outside the T loop. Mol Cell Biol. 2001;21(1):88–99.

51. Ni Z, Schwartz BE, Werner J, Suarez J-R, et al. Coordination of transcription, RNA processing, and surveillance by P-TEFb kinase on heat shock genes. Mol Cell. 2004;13(1):55–65.

52. Sedlacek HH. Mechanisms of action of flavopiridol. Crit Rev Oncol Hematol. 2001;38(2):139–70.

53. Albert TK, Rigault C, Eickhoff J, Baumgart K, et al. Characterization of molecular and cellular functions of the cyclin-dependent kinase CDK9 using a novel specific inhibitor. Br J Pharmacol. 2014;171(1):55–68.

54. Yin T, Lallena MJ, Kreklau EL, Fales KR, et al. A novel CDK9 inhibitor shows potent antitumor efficacy in preclinical hematologic tumor models. Mol Cancer Ther. 2014;13(6):1442–56.

55. Kwiatkowski N, Zhang T, Rahl PB, Abraham BJ, et al. Targeting transcription regulation in cancer with a covalent CDK7 inhibitor. Nature. 2014;511(7511):616–20.

56. Malumbres M. Cyclin-dependent kinases. Genome Biol. 2014;15(6):122.

57. Fisher RP. Secrets of a double agent: CDK7 in cell-cycle control and transcription. J Cell Sci. 2005;118(22):5171–80.

58. Glover-Cutter K, Larochelle S, Erickson B, Zhang C, et al. TFIIH-associated Cdk7 kinase functions in phosphorylation of C-terminal domain Ser7 residues, promoter-proximal pausing, and termination by RNA polymerase II. Mol Cell Biol. 2009;29(20):5455–64.

59. Larochelle S, Amat R, Glover-Cutter K, Sanso M, et al. Cyclin-dependent kinase control of the initiation-to-elongation switch of RNA polymerase II. Nat Struct Mol Biol. 2012;19(11):1108–15.

60. Wang W, Yao X, Huang Y, Hu X, et al. Mediator MED23 regulates basal transcription in vivo via an interaction with P-TEFb. Transcription. 2013;4(1):39–51.

61. Cassé C, Giannoni F, Nguyen VT, Dubois M-F, et al. The transcriptional inhibitors, actinomycin D and α-amanitin, activate the HIV-1 promoter and favor phosphorylation of the RNA polymerase II C-terminal domain. J Biol Chem. 1999;274(23):16097–106.

62. Nguyen VT, Kiss T, Michels AA, Bensaude O. 7SK small nuclear RNA binds to and inhibits the activity of CDK9/cyclin T complexes. Nature. 2001;414(6861):322–5.

63. Titov DV, Gilman B, He Q-L, Bhat S, et al. XPB, a subunit of TFIIH, is a target of the natural product triptolide. Nat Chem Biol. 2011;7(3):182–8.

64. Nguyen VT, Giannoni F, Dubois M-F, Seo S-J, et al. In vivo degradation of RNA polymerase II largest subunit triggered by α-amanitin. Nucleic Acids Res. 1996;24(15):2924–9.

65. Gaspar-Maia A, Alajem A, Meshorer E, Ramalho-Santos M. Open chromatin in pluripotency and reprogramming. Nat Rev Mol Cell Biol. 2011;12(1):36–47.

66. Niwa H. Open conformation chromatin and pluripotency. Genes Dev. 2007;21(21):2671–6.

67. Meshorer E, Misteli T. Chromatin in pluripotent embryonic stem cells and differentiation. Nat Rev Mol Cell Biol. 2006;7(7):540–6.

68. Park S-H, Park SH, Kook M-C, Kim E-Y, et al. Ultrastructure of human embryonic stem cells and spontaneous and retinoic acid-induced differentiating cells. Ultrastruct Pathol. 2004;28(4):229–38.

69. Efroni S, Duttagupta R, Cheng J, Dehghani H, et al. Global transcription in pluripotent embryonic stem cells. Cell Stem Cell. 2008;2(5):437–47.

70. Schaniel C, Ang YS, Ratnakumar K, Cormier C, et al. Smarcc1/Baf155 couples self-renewal gene repression with changes in chromatin structure in mouse embryonic stem cells. Stem Cells. 2009;27(12):2979–91.

71. Hawkins RD, Hon GC, Lee LK, Ngo Q, et al. Distinct epigenomic landscapes of pluripotent and lineage-committed human cells. Cell Stem Cell. 2010;6(5):479–91.

72. Krejci J, Uhlirova R, Galiova G, Kozubek S, et al. Genome-wide reduction in H3K9 acetylation during human embryonic stem cell differentiation. J Cell Physiol. 2009;219(3):677–87.

73. Gonzales-Cope M, Sidoli S, Bhanu NV, Won KJ, et al. Histone H4 acetylation and the epigenetic reader Brd4 are critical regulators of pluripotency in embryonic stem cells. BMC Genom. 2016;17(1):95.

74. Meshorer E, Yellajoshula D, George E, Scambler PJ, et al. Hyperdynamic plasticity of chromatin proteins in pluripotent embryonic stem cells. Dev Cell. 2006;10(1):105–16.

75. Spinella MJ, Freemantle SJ, Sekula D, Chang JH, et al. Retinoic acid promotes ubiquitination and proteolysis of cyclin D1 during induced tumor cell differentiation. J Biol Chem. 1999;274(31):22013–8.

76. Fu W-Y, Wang JH, Ip NY. Expression of Cdk5 and its activators in NT2 cells during neuronal differentiation. J Neurochem. 2002;81(3):646–54.

77. Green A, Sarg B, Green H, Lonn A, et al. Histone H1 interphase phosphorylation becomes largely established in G(1) or early S phase and differs in G(1) between T-lymphoblastoid cells and normal T cells. Epigenetics Chromatin. 2011;4:15.

78. O'Shea KS. Self-renewal vs. differentiation of mouse embryonic stem cells. Biol Reprod. 2004;71(6):1755–65.

79. Niwa H. How is pluripotency determined and maintained? Development. 2007;134(4):635–46.

80. Boyer LA, Lee TI, Cole MF, Johnstone SE, et al. Core transcriptional regulatory circuitry in human embryonic stem cells. Cell. 2005;122(6):947–56.

81. Soprano DR, Teets BW, Soprano KJ. Role of retinoic acid in the differentiation of embryonal carcinoma and embryonic stem cells. Vitam Horm. 2007;75:69–95.

82. Gudas LJ, Wagner JA. Retinoids regulate stem cell differentiation. J Cell Physiol. 2011;226(2):322–30.

83. Gu P, LeMenuet D, Chung AC, Mancini M, et al. Orphan nuclear receptor GCNF is required for the repression of pluripotency genes during retinoic acid-induced embryonic stem cell differentiation. Mol Cell Biol. 2005;25(19):8507–19.

84. Wang H, Wang X, Xu X, Kyba M, et al. Germ cell nuclear factor (GCNF) represses Oct4 expression and globally modulates gene expression in human embryonic stem (hES) cells. J Biol Chem. 2016;291(16):8644–52.

85. Sato N, Kondo M, Arai K. The orphan nuclear receptor GCNF recruits DNA methyltransferase for Oct-3/4 silencing. Biochem Biophys Res Commun. 2006;344(3):845–51.

86. Lee HL, Archer TK. Prolonged glucocorticoid exposure dephosphorylates histone H1 and inactivates the MMTV promoter. EMBO J. 1998;17(5):1454–66.

87. Alexandrow MG, Hamlin JL. Chromatin decondensation in S-phase involves recruitment of Cdk2 by Cdc45 and histone H1 phosphorylation. J Cell Biol. 2005;168(6):875–86.

88. Lu MJ, Dadd CA, Mizzen CA, Perry CA, et al. Generation and characterization of novel antibodies highly selective for phosphorylated linked histone H1 in Tetrahymena and HeLa cells. Chromosoma. 1994;103(2):111–21.

89. Meijer L, Borgne A, Mulner O, Chong JP, et al. Biochemical and cellular effects of roscovitine, a potent and selective inhibitor of the cyclin-dependent kinases cdc2, cdk2 and cdk5. Eur J Biochem. 1997;243(1–2):527–36.

90. Schang LM, Bantly A, Knockaert M, Shaheen F, et al. Pharmacological cyclin-dependent kinase inhibitors inhibit replication of wild-type and drug-resistant strains of herpes simplex virus and human immunodeficiency virus type 1 by targeting cellular, not viral, proteins. J Virol. 2002;76(15):7874–82.

91. Brooks EE, Gray NS, Joly A, Kerwar SS, et al. CVT-313, a specific and potent inhibitor of CDK2 that prevents neointimal proliferation. J Biol Chem. 1997;272(46):29207–11.

92. Veselý J, Havlíček L, Strnad M, Blow JJ, et al. Inhibition of cyclin-dependent kinases by purine analogues. Eur J Biochem. 1994;224(2):771–86.

93. Alessi F, Quarta S, Savio M, Riva F, et al. The cyclin-dependent kinase inhibitors olomoucine and roscovitine arrest human fibroblasts in G1 phase by specific inhibition of CDK2 kinase activity. Exp Cell Res. 1998;245(1):8–18.

94. Yang Z, Zhu Q, Luo K, Zhou Q. The 7SK small nuclear RNA inhibits the CDK9/cyclin T1 kinase to control transcription. Nature. 2001;414(6861):317–22.

95. Bensaude O. Inhibiting eukaryotic transcription: which compound to choose? How to evaluate its activity? Transcription. 2011;2(3):103–8.

96. Quaresma AJ, Bugai A, Barboric M. Cracking the control of RNA polymerase II elongation by 7SK snRNP and P-TEFb. Nucleic Acids Res. 2016;44(16):7527–39.

97. McNamara RP, Bacon CW, D'Orso I. Transcription elongation control by the 7SK snRNP complex: Releasing the pause. Cell Cycle. 2016;15(16):2115–23.

98. Ji X, Zhou Y, Pandit S, Huang J, et al. SR proteins collaborate with 7SK and promoter-associated nascent RNA to release paused polymerase. Cell. 2013;153(4):855–68.

99. Calo E, Flynn RA, Martin L, Spitale RC, et al. RNA helicase DDX21 coordinates transcription and ribosomal RNA processing. Nature. 2015;518(7538):249–53.

100. Di Micco R, Fontanals-Cirera B, Low V, Ntziachristos P, et al. Control of embryonic stem cell identity by BRD4-dependent transcriptional elongation of super-enhancer-associated pluripotency genes. Cell Rep. 2014;9(1):234–47.

101. Roque A, Ponte I, Suau P. Interplay between histone H1 structure and function. Biochim Biophys Acta. 2016;1859(3):444–54.

102. Sancho M, Diani E, Beato M, Jordan A. Depletion of human histone H1 variants uncovers specific roles in gene expression and cell growth. PLoS Genet. 2008;4(10):e1000227.

Differential DNA methylation and lymphocyte proportions in a Costa Rican high longevity region

Lisa M. McEwen[1], Alexander M. Morin[1], Rachel D. Edgar[1], Julia L. MacIsaac[1], Meaghan J. Jones[1], William H. Dow[2], Luis Rosero-Bixby[3], Michael S. Kobor[1] and David H. Rehkopf[4*]

Abstract

Background: The Nicoya Peninsula in Costa Rica has one of the highest old-age life expectancies in the world, but the underlying biological mechanisms of this longevity are not well understood. As DNA methylation is hypothesized to be a component of biological aging, we focused on this malleable epigenetic mark to determine its association with current residence in Nicoya versus elsewhere in Costa Rica. Examining a population's unique DNA methylation pattern allows us to differentiate hallmarks of longevity from individual stochastic variation. These differences may be characteristic of a combination of social, biological, and environmental contexts.

Methods: In a cross-sectional subsample of the Costa Rican Longevity and Healthy Aging Study, we compared whole blood DNA methylation profiles of residents from Nicoya ($n = 48$) and non-Nicoya (other Costa Rican regions, $n = 47$) using the Infinium HumanMethylation450 microarray.

Results: We observed a number of differences that may be markers of delayed aging, such as bioinformatically derived differential CD8+ T cell proportions. Additionally, both site- and region-specific analyses revealed DNA methylation patterns unique to Nicoyans. We also observed lower overall variability in DNA methylation in the Nicoyan population, another hallmark of younger biological age.

Conclusions: Nicoyans represent an interesting group of individuals who may possess unique immune cell proportions as well as distinct differences in their epigenome, at the level of DNA methylation.

Keywords: DNA methylation, Epigenetics, Immune aging, Longevity, Biodemography, Epigenetic age

Background

Aging is a complex biological process that progressively leads to physiological decline and an increased risk of mortality. The genetic component of life span is approximated to be less than 30%, leaving the remainder to be determined by environmentally and socially influenced factors such as diet, exposure to infection, and lifestyle choices [1, 2]. While the mechanistic regulation of these non-genetic influences is poorly understood, previous work has suggested that epigenetic processes may be tightly interwoven with biological aging [3].

Epigenetics generally refers to the study of altered chromatin states, such as modifications to DNA and the proteins involved in its packaging and regulation. To date, DNA methylation (DNAm) is the most commonly studied epigenetic mark in human populations, as recent advances in technology have allowed for the inexpensive high-throughput measurement of >400,000 CpG sites across the genome. There are many other studied epigenetic processes, such as post-translational histone modifications, histone variants, and noncoding RNAs; however, these modifications have been more of a focus in model organism research and cancer biology.

*Correspondence: drehkopf@stanford.edu
[4] Division of General Medical Disciplines, Department of Medicine, School of Medicine, Stanford University, 1070 Arastradero Road, Suite 300, Palo Alto, CA 94304, USA
Full list of author information is available at the end of the article

DNAm is one type of epigenetic modification that impacts how genes are expressed and thus can have important phenotypic and functional consequences for an organism. Unlike the DNA sequence itself, DNAm is changeable through environmental influences over an individual's life course. DNAm involves the covalent addition of a methyl group to the 5' carbon of cytosine nucleotides, most often in the context of CpG dinucleotides (cytosine–phosphate–guanine). Genetic variants, such as single nucleotide polymorphisms (SNPs), can affect DNAm at nearby CpG sites, called methylation quantitative trait loci (mQTL) [4]. DNAm also varies in association with environmental and behavioral factors, such as diesel exposure and smoking. Additionally, variability in DNAm accumulates across one's entire life span, discussed in detail below. DNAm patterns are associated with altered gene activity. Shifts in DNAm levels may follow a change in gene expression, or may act in the recruitment of methylation-dependent transcription factors that regulate transcriptional machinery. Understanding the effect of environmental influences on DNAm is important for unraveling the intricate regulation of genes and possible functional consequences of these alterations.

Age-related DNAm encompasses at least two distinct phenomena. First, specific CpGs associated with chronological age have been identified and, in some cases, replicated in several human populations. These age-related DNAm signatures can be either tissue specific or occur across several tissue types [5]. The epigenetic clock is a tool based on CpGs that change with age. Epigenetic clocks are DNAm-based markers of biological age, either confined to a single tissue or consistently accurate across tissues [6–8]. Deviations from these epigenetic age estimates, referred to as a measure of age acceleration, have been associated with an increased risk of all-cause mortality, time until death, and frailty [9–11]. Second, variability increases with age due to stochastic non-site-specific changes in DNAm, a process referred to as epigenetic drift [12].

It is critical to address cell-type heterogeneity when investigating DNAm patterns in tissues containing mixed cell types. Not only does cell type change over one's life span but it is also the primary source of variation in DNAm across healthy individuals. DNAm profiles obtained from identical cell types, but separate individuals, show higher similarity than two different cell types from the same individual [13]. Given that isolating DNA from a single cell type is not always feasible or that cell count information is sometimes not available, bioinformatics-based methods have been developed to estimate cell-type proportions using DNAm profiles in blood and brain [14, 15]. The blood-based predictions are closely correlated with complete blood count measures, thus suggesting the validity of these methods to derive accurate blood count information bioinformatically [16]. It is also worth noting that measures of epigenetic age acceleration specific to whole blood have been defined to account for age-induced changes to cell-type proportions [6]. These measures are integral when analyzing DNAm from whole blood, as the proportion of certain blood cell types, such as CD8+ memory and naïve T cells, change with age [17].

Aging research in humans commonly investigates the unique biological and lifestyle characteristics of individuals surviving to old age [18]. An alternative approach, used in the current study design, is to examine the underlying biology of longevity by examining a population characterized as having a particularly high old-age life expectancy [19], the Nicoya region of Costa Rica, and comparing it to the rest of Costa Rica which has moderately lower life expectancy. By averaging out the stochastic variation in aging among individuals within each geographic region of the country, this approach offers a way to identify contextual (rather than individual) differences associated with healthy aging and longevity. While the method of examining area-based determinants of health and longevity has received substantial attention in biomedical research [20], a lack of appropriate data sources have limited its application in understanding the biological mechanisms of longevity.

The Nicoya peninsula of Costa Rica has been characterized by exceptionally high longevity, providing an intriguing framework to explore the relationship between DNAm and aging [21]. Mortality rates among elderly Costa Ricans in Nicoya are substantially lower than in the rest of Costa Rica, with individuals in Nicoya being some of the most long lived in the world. The relative mortality rate of Nicoya as compared to similar age cohorts in the rest of Costa Rica is 0.80. This advantage remains significant after statistical control for level of education and type of health insurance [21]. The Nicoyan advantage is particularly evident in cardiovascular disease, despite the fact that risk factors like smoking, physical activity and systolic blood pressure are similar throughout Costa Rica. One key indicator of the Nicoyan advantage is longer knee height—an anthropometric biomarker that is associated with early childhood environment [22]. Nicoyans also have lower BMI, waist circumference, and, among men, lower levels of HbA1c, glucose, triglycerides and total/HDL cholesterol ratio [21]. We have previously shown that leukocyte telomere length, an aging biomarker, also has more favorable (longer) levels among Nicoyans compared to individuals in the rest of Costa Rica [23].

In this study, we examined DNAm in a nationally representative population of Costa Ricans, investigating potential biological differences that may help explain the higher longevity observed in Nicoyans as compared to other Costa Ricans. We focused on DNAm, as this epigenetic mark can be modified by environmental influences, has the potential to regulate gene expression, and most importantly, has an established relationship with aging across the mammalian life span. Using genome-wide DNAm patterns to predict blood cell-type composition, we determined differential estimated proportions of age-related immune cells. While we did not observe differences in epigenetic age acceleration, we did find significantly decreased DNAm variability in Nicoyans. Finally, we identified DNAm patterns unique to Nicoyans, at both genomic regions and specific CpG sites. Understanding DNAm patterns between Nicoyans and other Costa Ricans (non-Nicoyans) will offer new insights both into the role of DNAm in aging and perhaps help to illuminate why Nicoyans have among the longest old-age life expectancies.

Results
Cohort characteristics and DNA methylation data
We examined a subset of samples from the Costa Rican Study on Longevity and Healthy Aging (CRELES), a longitudinal, nationally representative, and probabilistic sample of close to 3000 adults aged 60 years and over that were collected mostly in 2005, with over-sampling of older ages [24]. We assayed DNAm profiles of 48 Nicoyans (longevity group) and 47 non-Nicoyans (control group). In order to maximize statistical power for our age-based hypothesis, we randomly sampled half of the individuals between the ages of 60 and 75 and the other half aged 95 and above, selecting an equal number in each age category among Nicoyans and non-Nicoyans to have an age-matched sample. Table 1 shows the mean characteristics of these populations. Nicoyans tend to have lower levels of education and lower wealth than non-Nicoyans, but are similar on observed health-related characteristics. We obtained DNAm profiles from whole blood using the Infinium HumanMethylation450 (450k) array, a genome-wide microarray that quantifies DNAm at over 485,000 sites. We applied data quality controls to remove poor performing probes, probes that hybridized the XY chromosomes, and probes predicted to cross-hybridize [25]. Our final dataset for subsequent analyses consisted of 441,109 sites.

Nicoyans had fewer estimated CD8+ memory and more naïve T cells than non-Nicoyans
Whole blood is a heterogeneous tissue containing certain cell types that change with age, with age-related decreases occurring in CD8+ T, CD4+ T and B lymphocytes, and the greatest increases in natural killer

Table 1 Cohort characteristics (means and percents), Nicoyans and non-Nicoyans

Characteristics	Nicoya (n = 48)	Non-Nicoya (n = 47)
Age (mean in years)	83 (14)	85 (16)
Female (%)	57	55
Low education (%)	80	68
Low wealth (%)	35	21
Currently smoke (%)	4	6
Systolic blood pressure (mean mmHg)	139 (23)	140 (25)
Diastolic blood pressure (mean mmHg)	78 (12)	78 (13)
Body mass index (mean)	24 (7.1)	25 (5.8)

Standard deviations are shown in parenthesis. Low education is not completing primary school. The wealth index was based on a simple count of ten goods and conveniences in the household, ranging from running water and a toilet to having a cloth washer and a car. Low wealth is having six or fewer of these items. Systolic and diastolic blood pressure and body mass index were measured at the time of interview. Currently smoking was assessed through questionnaire. Further details on survey measures are available elsewhere [45]

cells and monocytes [26]. To assess these differences in our cohort, we performed a previously described blood cell-type deconvolution [14] by using the DNAm profiles of each sample to estimate the proportions of granulocytes, natural killer cells, CD8+ T lymphocytes, CD4+ T lymphocytes, monocytes, and B lymphocytes. We found that Nicoyans had a significantly lower proportion of estimated CD8+ T cells when compared to non-Nicoyans (Kruskal–Wallis p value = 0.0038) (Fig. 1a). We also observed that Nicoyans had a higher mean level of estimated granulocyte proportions, although only reaching borderline significance (Kruskal–Wallis p value = 0.0486). It is important to note that we did not focus on the blood composition as a whole, as we were primarily interested in specific age-related cell-type trends.

To further investigate the differential proportion of estimated CD8+ T cells, we applied a more detailed cell deconvolution tool that provides an expanded estimation of CD8+ T naïve cells (CD8+ CD45RA+ CCR7+) and CD8+ T memory cells (CD8+ CD28− CD45RA−) [6]. CD8+ T memory cells typically increase with age, and CD8+ T naïve cells generally decrease with age through thymic involution [27]. Given that these measures are proportional and highly correlated, it is statistically appropriate to assess the ratio of CD8+ T naïve cells to CD8+ T memory cells (Additional file 1: Figure S1). Using this approach, we found a significant difference between Nicoyans and non-Nicoyans (Kruskal–Wallis p value = 0.0135). We observed a greater abundance of predicted CD8+ T naïve cells in Nicoyans and a lower abundance of estimated CD8+ T memory cells in Nicoyans, as

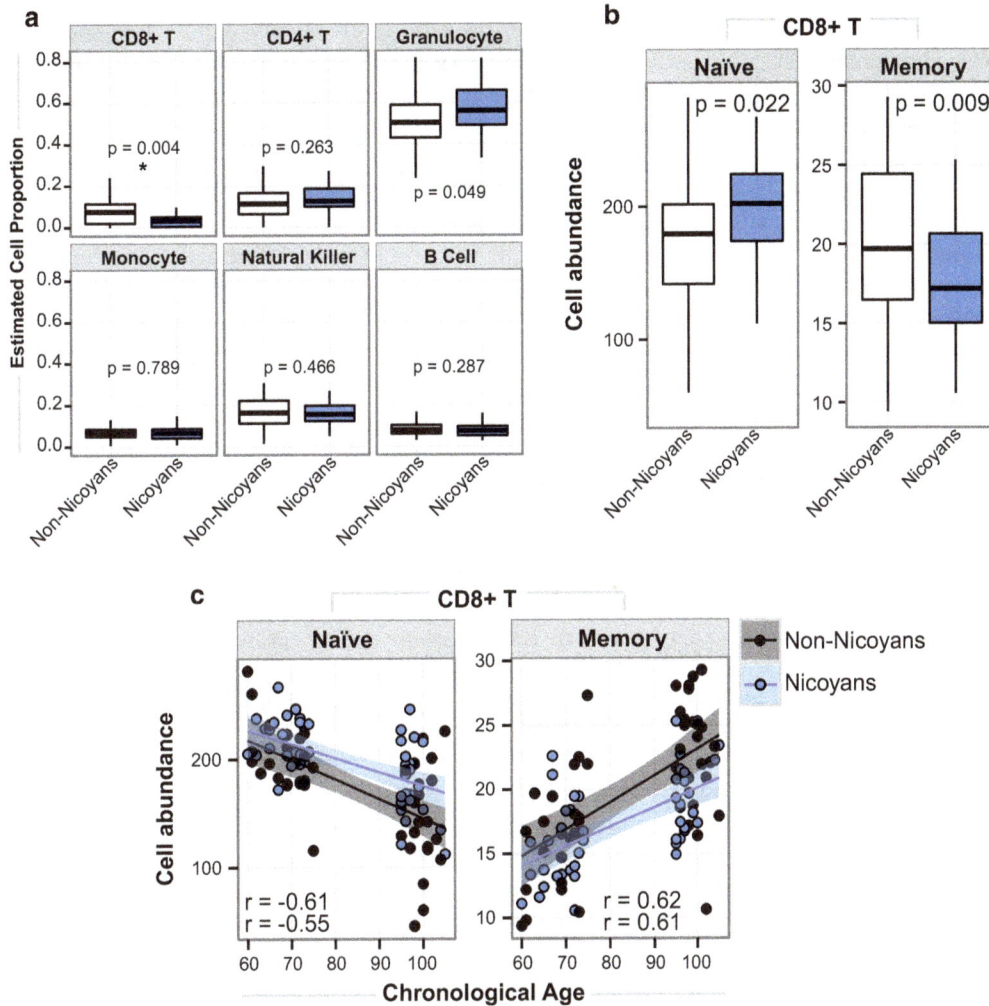

Fig. 1 Nicoyans had differential CD8+ naïve and memory T cell abundance levels. **a** Box plots demonstrating bioinformatically derived white blood cells in Nicoyans and non-Nicoyans. Cell proportions estimated using the Houseman method. *Blue*: Nicoyans, *white*: non-Nicoyans. The *p* value is derived from a nonparametric group comparison test using Kruskal–Wallis. **b** Box plots illustrating the relationship between the bioinformatically derived CD8+ naïve T cell and CD8+ memory T cell across Nicoyans and non-Nicoyans. **c** Scatter plots of chronological age plotted against each CD8+ naïve T cells and CD8+ memory T cells abundance level for each sample. CD8+ naïve T cell show a decrease with age, whereas CD8+ memory T cells increase with chronological age. Pearson's *r* coefficients derived from log-transformed age correlated with each respective cell-type level. *Blue*: Nicoyans, *black*: non-Nicoyans. Line of best fit shown with 95% confident intervals shaded in respective group color. The scale of cell abundance is a measure from a bioinformatically derived prediction of the respective cell types using flow-sorted counts from other datasets to infer cellular proportions of that specific isolated cell type based on the DNA methylation data [9, 46, 47]

compared to non-Nicoyans (Fig. 1b). These trends were suggestive of a younger immune cell profile in Nicoyans.

Pearson's correlation coefficients were computed to assess the relationship between chronological age (log transformed) and the estimated proportion of each CD8+ T cell type (Fig. 1c). We observed a negative correlation between chronological age and CD8+ T naïve proportion (Nicoyans and non-Nicoyans; Pearson's $r = -0.55$ and -0.61, respectively). As expected, there was a positive correlation between age and CD8+ T memory cells (Nicoyans and non-Nicoyans; Pearson's $r = 0.61$, and 0.61, respectively; Fig. 1c), demonstrating a known immunological aging trend, but here based on epigenetic data. We found no significant difference in the regression slopes of chronological age on either CD8+ T cell type, when comparing Nicoyans to non-Nicoyans ($p > 0.50$); while we did observe differences in mean levels of both estimated CD8+ T cell types, we did not see any group difference in the nature of how these cell types changed across age.

Epigenetic age did not differ between Nicoyans and non-Nicoyans

Having established differences in estimated CD8+ T cell populations between the two groups, we next examined DNAm age as an established metric of biological aging. We examined DNAm age of Nicoyans compared to non-Nicoyans using three epigenetic clocks, which provided measures of biological age using DNAm levels at different groups of CpG sites (Fig. 2a) [6–8]. Across all samples, we found correlations between DNAm age and chronological age (Horvath: Pearson's $r = 0.86$, Hannum: Pearson's $r = 0.85$, Weidner: Pearson's $r = 0.86$). However, we found no significant difference between Nicoyans and non-Nicoyans in terms of DNAm age (as calculated by each clock), while adjusting for chronological age (ANOVA $p > 0.30$, 95% CI for the Horvath clock was -6.3 to 3.8 years). We did, however, observe a mean difference of -6.9 years between epigenetic age and chronological age for all samples, suggesting that Costa Ricans, inclusive of both Nicoyans and non-Nicoyans, may on average be epigenetically younger than their chronological age (Fig. 2b). Furthermore, we reduced our data to only centenarians (age ≥ 100 years old), and the mean absolute difference between DNAm age and chronological age was -12.7 years.

We further examined biological age by assessing two other recently defined measures of age acceleration: intrinsic and extrinsic epigenetic acceleration, which are independent and dependent of blood cell-type proportions, respectively. We did not find any significant difference between Nicoyans and non-Nicoyans in any of these acceleration measures (ANOVA $p \geq 0.5$, Additional file 2: Figure S2).

DNA methylation variability was lower in Nicoyans

After assessing whether any differences in epigenetic age existed, we next investigated epigenetic drift between Nicoyans and non-Nicoyans. We hypothesized that lower variability in Nicoyans would be representative of a biologically younger profile based on the epigenetic drift phenomenon where stochastic variation in DNAm occurs with age. We calculated the interquantile range (IQR) at each CpG site (90th–10th percentile) represented on the 450k array to account for outliers and found a significant variability difference between Nicoyans and non-Nicoyans (Wilcoxon signed-rank test; $p < 2.2 \times 10^{-16}$), with a lower level of total mean DNAm variation in Nicoyans (Fig. 3a). Furthermore, we assessed the level of DNAm variability across individuals between 60 and 80 years old and individuals >80 years old, in Nicoyans and non-Nicoyans. We found a lower DNAm variability in the younger group, in both populations. However, there was a greater difference between the non-Nicoyan old and young age groups compared to the Nicoyans respective age groups (non-Nicoya: IQR mean difference = 0.0065, $p < 2.2 \times 10^{-16}$, Nicoya: IQR mean difference = 0.0043, $p = 1.79 \times 10^{-10}$). Lastly, when a β value IQR threshold of $\geq 5\%$ was applied to only variable sites, we found 129,971 and 146,047 variable sites in the >80-year-old range of Nicoyans and non-Nicoyans, respectively. For the younger age range, we found 116,038 and 120,711 sites that had greater than 5% IQR in Nicoyans and non-Nicoyans, respectively (Fig. 3b). In summary, we found a lower degree of DNAm variability was associated with Nicoyans for both age groups, 60–80 years old and >80 years old, when compared to non-Nicoyans.

Unique region-based differential methylation in Nicoyans

We next investigated differentially methylated regions (DMRs) between Nicoyans and non-Nicoyans to identify unique epigenetic signatures that may be associated with the longevity observed in Nicoya. Using the R package 'DMRcate,' we found that in the comparison of Nicoyans and non-Nicoyans, 20 DMRs containing three or more CpG sites passed a false discovery rate of ≤ 0.05, as well as an arbitrary biological cutoff of a β value $\geq 5\%$ between groups for at least one CpG site in each DMR (Table 2; Fig. 4; Additional file 3: Figure S3). Age, sex, and blood cell-type proportions were used as covariates. DMRs were associated with genes based on the closest proximity to a transcription start site (TSS). The mean length

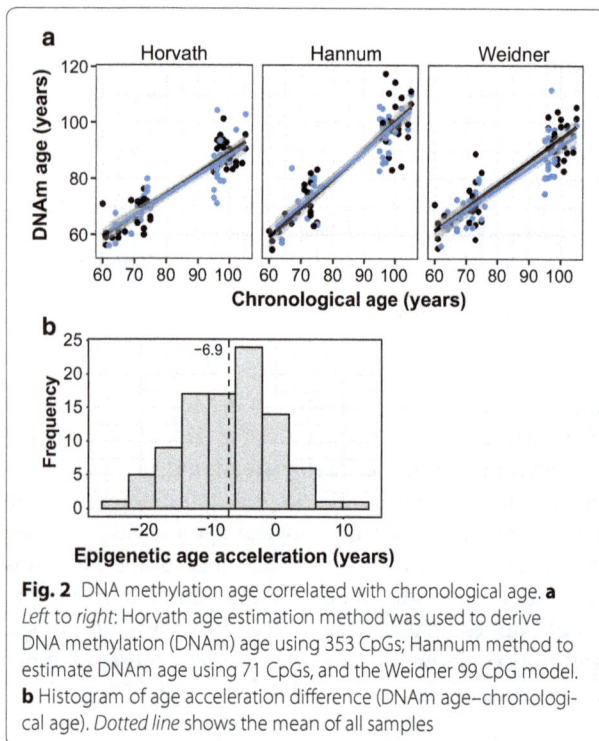

Fig. 2 DNA methylation age correlated with chronological age. **a** *Left to right*: Horvath age estimation method was used to derive DNA methylation (DNAm) age using 353 CpGs; Hannum method to estimate DNAm age using 71 CpGs, and the Weidner 99 CpG model. **b** Histogram of age acceleration difference (DNAm age–chronological age). *Dotted line* shows the mean of all samples

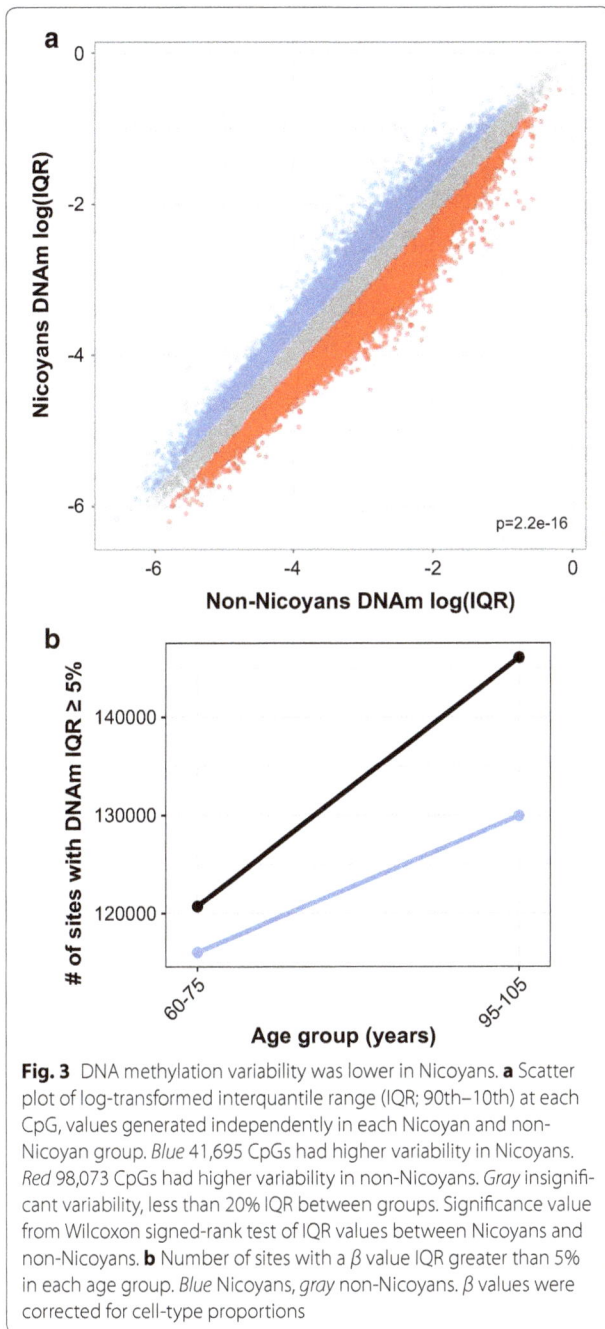

Fig. 3 DNA methylation variability was lower in Nicoyans. **a** Scatter plot of log-transformed interquantile range (IQR; 90th–10th) at each CpG, values generated independently in each Nicoyan and non-Nicoyan group. *Blue* 41,695 CpGs had higher variability in Nicoyans. *Red* 98,073 CpGs had higher variability in non-Nicoyans. *Gray* insignificant variability, less than 20% IQR between groups. Significance value from Wilcoxon signed-rank test of IQR values between Nicoyans and non-Nicoyans. **b** Number of sites with a β value IQR greater than 5% in each age group. *Blue* Nicoyans, *gray* non-Nicoyans. β values were corrected for cell-type proportions

LOC100128885 (uncharacterized) (3 CpGs). The majority of DMRs were found in intergenic regions (8/20) with an average number of four CpGs per DMR and average length of 267 bp. Four DMRs were found in the body (intragenic) of the following genes: Glutamate-rich protein 1 (*ERICH1*) (8 CpGs), Hook Microtubule Tethering Protein 2 (*HOOK2*) (4 CpGs), GATA2 Antisense RNA 1 (*GATA2-AS1*) (3 CpGs), and C15orf26 (uncharacterized) (4 CpGs). Two DMRs were found in the 3′end regions of Mitochondrial Ribosomal Protein L21 (*MRPL21*) (3 CpGs) and BolA family member 3 (*BOLA3*) (5 CpGs). The average absolute difference in DNAm between Nicoyans and non-Nicoyans was 7.9% when assessing the single CpG site for each DMR that showed the greatest difference between groups. When DNAm was averaged across the DMR per group, the mean β value difference was 5.9%.

Site-specific differential methylation in Nicoyans and technical verification

To complement the region-specific analysis, we also assessed DNAm differences at the site-by-site level to identify any single sites that were differentially methylated in Nicoyans as compared to non-Nicoyans. By investigating site-specific changes, we did not limit ourselves to highly represented genomic regions on the 450k array. We performed an epigenome-wide association study using a linear model of population group regressed on M values at each CpG site, with sex, age, and blood cell-type proportions included as covariates (Additional file 4: Figure S4). We observed nine CpGs below a nominal p value significance threshold of $p \leq 5 \times 10^{-7}$ (q value < 0.022, Table 3) that were differentially methylated between Nicoyans and non-Nicoyans. After applying a biological threshold similar to the DMRcate analysis, four CpG sites passed our significance criteria: cg02853387 (*DSCAML1*), cg02438481 (*C6orf123*), cg13979274 (*OR10H1*), cg26107275 (*BC042649*). Not surprisingly, the four significant CpG sites did not overlap with our DMR findings, as these single CpGs identified through the site-by-site analysis were either: (1) found in a genomic region with no proximal array probes, or (2) nearby CpGs sites were not significantly correlated. The nearest neighboring array-CpG probe was greater than 1 kb away for three out of the four significant CpG sites (cg02853387, cg13979274, and cg26107275). While for cg02438481, the closest array-CpG (cg00788354) was 324 bp away, it had a Pearson's correlation of $r < 0.10$ and was not significantly differentially methylated between groups.

We performed pyrosequencing, a targeted DNAm sequencing technology, to verify that our single differential DNAm results were reproducible using an independent platform. We designed assays to measure

of the DMRs was 411 bp, with the shortest being 76 bp and the longest being 1416 bp. On average, out of the 20 DMRs observed, there were seven CpG sites per region, with a range from 3 to 16. Six DMRs were found within 1500 bp of a TSS associated with the following genes' promoter region: Nudix Hydrolase 12 (*NUDT12*) (6 CpGs), Vault-RNA-2 (*VTRNA2-1*) (16 CpGs), Peptidase M20 Domain Containing 1 (*PM20D1*) (8 CpGs), Active BCR-Related (*ABR*) (3 CpGs), tRNA-Leu (5 CpGs), and

Table 2 Differentially methylated genomic regions between Nicoyans and non-Nicoyans

Genomic sequence[a]	# CpGs	Min FDR	Max Δβ	Mean Δβ	Associated gene	DMR length (bp)	Genomic region	Gene ontology
chr5:102898223-102898733	6	1.0E−11	−0.07	−0.05	NUDT12	511	Promoter	NAD + diphosphatase activity; NADH pyrophosphatase activity
chr4:132896266-132897018	6	4.9E−08	−0.09	−0.06	–	753	Intergenic	–
chr5:180643432-180643713	3	1.9E−05	−0.05	−0.05	–	282	Intergenic	–
chr7:155832831-155832992	3	2.9E−04	0.06	0.05	–	162	Intergenic	–
chr11:68658383-68658836	3	7.3E−04	−0.09	−0.08	MRPL21	454	3′ end	Poly(A) RNA binding, ribosomal protein
chr6:6894084-6894182	3	5.3E−4	0.06	0.05	–	99	Intergenic	–
chr7:64034943-64035529	3	2.4E−03	0.07	0.06	LOC100128885	593	Promoter	–
chr12:130707332-130707407	3	2.6E−03	0.07	0.05	–	76	Intergenic	Wnt-protein binding
chr3:128215433-128216122	3	1.7E−03	−0.06	−0.05	GATA2-AS1	690	Intragenic	Enhancer sequence-specific binding, chromatin/transcription factor/C2H2 zinc finger binding
chr5:4292415-42924694	4	2.6E−03	0.07	0.05	–	480	Intergenic	–
chr1:152161237-152162507	8	2.1E−06	−0.07	−0.05	–	113	Intergenic	–
chr16:3988694-3988869	3	1.9E−02	−0.07	−0.05	–	176	Intergenic	–
chr6:28446794-28447115	5	1.3E−02	0.06	0.06	tRNA-Leu	322	Promoter	–
chr19:12876846-12877188	4	3.6E−02	−0.14	−0.11	HOOK2	343	Intragenic	Kinase modulator; membrane traffic protein
chr17:1133546-1133706	3	3.5E−02	−0.07	−0.05	ABR	161	Promoter	Transcription factor; protein binding; guanyl-nucleotide exchange factor
chr15:81410745-81411066	4	2.3E−02	0.09	0.06	C15orf26	322	Intragenic	–
chr2:74357527-74358223	5	2.0E−02	0.08	0.06	BOLA3	697	3′ end	Production of iron-sulfur clusters
chr1:205818956-205819609	8	2.2E−02	−0.08	−0.07	PM20D1	654	Promoter	Deacetylase, metalloprotease
chr8:599525-600940	8	2.7E−04	−0.12	−0.05	ERICH1	1416	Intragenic	–
chr5:135415693-135416613	16	1.1E−03	−0.09	−0.06	VTRNA2-1	921	Promoter	Potential tumor suppressor

DMR differentially methylated region

Δβ: (Delta beta; absolute mean or max difference of β values between groups (Nicoya–non-Nicoya)

[a] Genome coordinates from Human Genome GRCh37/hg19 Assembly

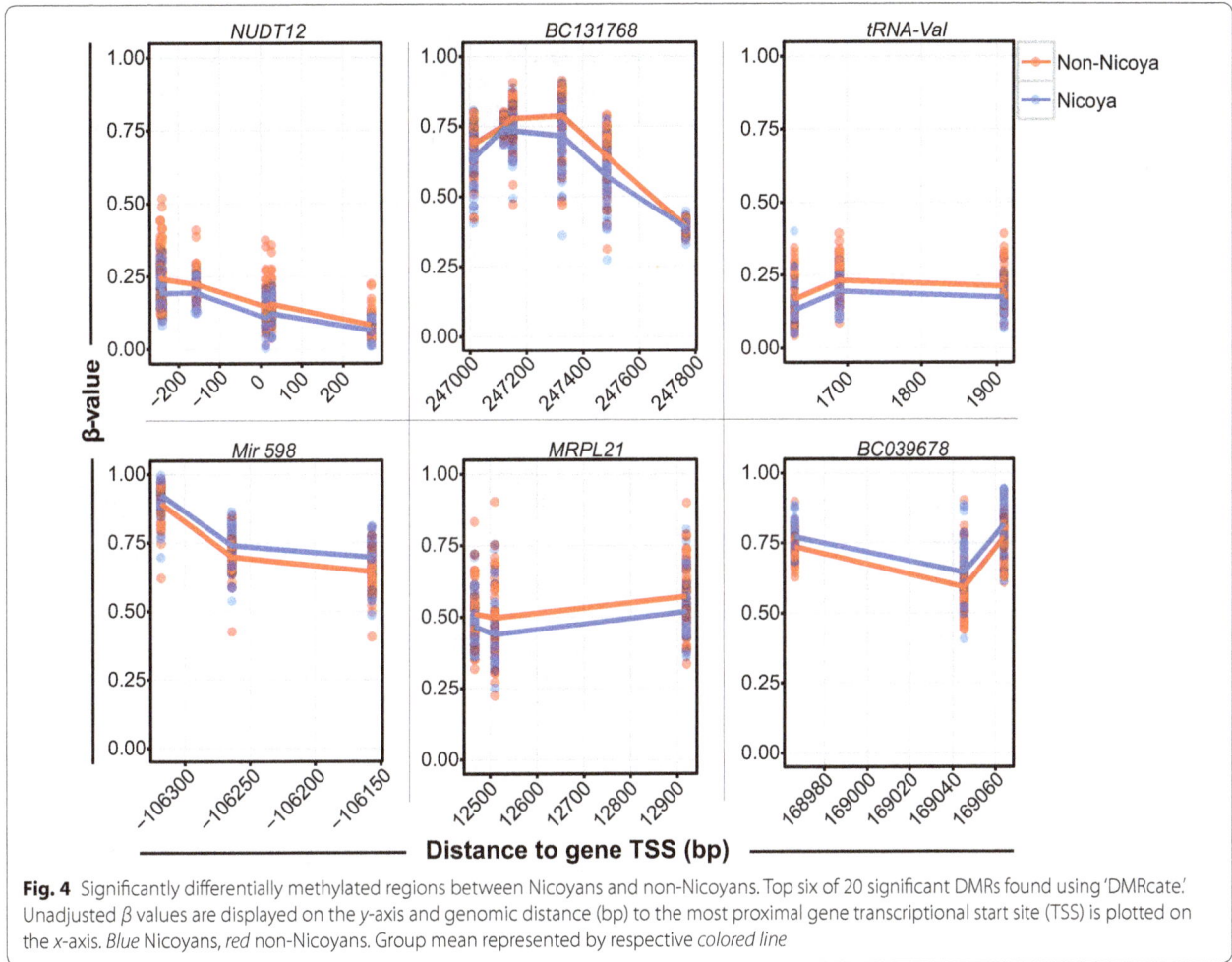

Fig. 4 Significantly differentially methylated regions between Nicoyans and non-Nicoyans. Top six of 20 significant DMRs found using 'DMRcate'. Unadjusted β values are displayed on the y-axis and genomic distance (bp) to the most proximal gene transcriptional start site (TSS) is plotted on the x-axis. *Blue* Nicoyans, *red* non-Nicoyans. Group mean represented by respective *colored line*

Table 3 Characteristics of four biologically and statistically significant DNA methylation sites between Nicoyans and non-Nicoyans

Probe ID	Pyrosequencing		450k array			Chr	Distance to closest TSS (bp)	Genomic region: gene	Gene ontology
	Δβ	p value	Δβ	p value	q value				
cg02853387	0.06	7.9E−05	0.07	1.3E−07	0.012	11	183072	Intragenic: intron 3 of *DSCAML1*	Protein homodimerization activity
cg02438481	0.08	1.9E−05	0.08	1.6E−07	0.012	6	−1861	Intergenic: C6orf123	–
cg13979274	0.06	9.1E−07	0.06	2.0E−07	0.012	19	4347	Intergenic: ~ 3 kb from 3′ end of *OR10H1*	G-protein coupled receptor/ olfactory receptor activity
cg26107275	–	–	−0.07	2.2E−07	0.012	12	−407	Promoter: *BC04264*	–

Δβ: (Delta beta; absolute mean difference of β values between groups (Nicoyans–non-Nicoyans)

TSS Transcription start site, *Chr* chromosome

three CpG sites that were observed to have a significant between group (Nicoyans and non-Nicoyans) difference in DNAm of ≥5.0% (cg02853387, cg02438481, and cg13979274) (sequences listed in Additional file 6: Table S1). Correlation coefficients between the pyrosequencing and 450k array for each CpG site showed a strong concordance between the two technologies [$r_s = 0.87$ (cg02853387), $r_s = 0.92$ (cg02438481), and $r_s = 0.88$ (cg13979274)]. Bland–Altman plots were generated for each CpG site to illustrate the agreement between the two quantification techniques (Fig. 5). To verify our differential DNAm findings between Nicoyans and

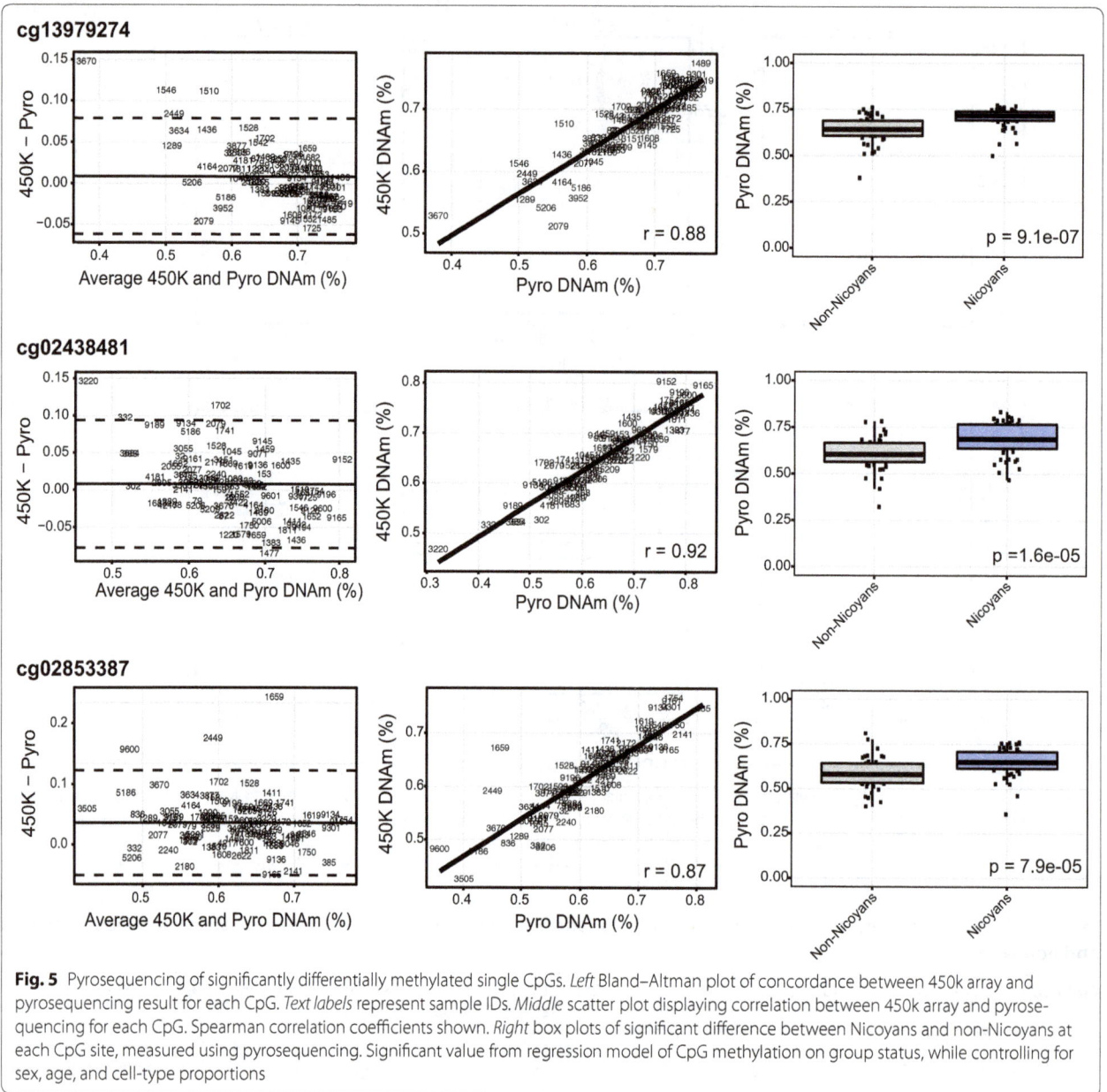

Fig. 5 Pyrosequencing of significantly differentially methylated single CpGs. *Left* Bland–Altman plot of concordance between 450k array and pyrosequencing result for each CpG. *Text labels* represent sample IDs. *Middle* scatter plot displaying correlation between 450k array and pyrosequencing for each CpG. Spearman correlation coefficients shown. *Right* box plots of significant difference between Nicoyans and non-Nicoyans at each CpG site, measured using pyrosequencing. Significant value from regression model of CpG methylation on group status, while controlling for sex, age, and cell-type proportions

non-Nicoyans, we confirmed these associations by statistically regressing DNAm determined by pyrosequencing onto group status, while controlling for age, cell-type proportions, and sex. All three sites remained significantly different between Nicoyans and non-Nicoyans (Table 3; Fig. 5).

Lastly, to determine whether a measure of genetic population structure was confounded with group (Nicoyans vs non-Nicoyans), we performed a post hoc analysis using 'Epistructure' [28]. Principal component analysis was completed on DNAm of CpGs previously

identified as genetically informative loci. The first two principal components generated from this analysis have been proposed to confer composites of genetic structure to be used as covariates in a DNAm analysis. Using this technique, we did not observe a significant difference between the Epistructure principal components of the measures in Nicoyans and non-Nicoyans, while controlling for sex, age, and cell-type proportions (PC1: $p = 0.60$, PC2: $p = 0.93$, Additional file 5: Figure S5); therefore, we did not include these measures in our analyses.

Discussion

In this study, we investigated differential patterns of DNAm in a population with well-characterized high longevity: Nicoya, Costa Rica. We aimed to identify unique patterns of DNAm that may underlie biological pathways associated with the longevity observed in Nicoya. Our study sample was drawn from a nationally representative demographic study of Costa Ricans age 60 years and over, and we randomly sampled individuals from within Nicoya and the rest of Costa Rica who were aged 60–75 and 95 years and above in order to assess age-associated DNAm. We have four primary findings. First, we observed a bioinformatically inferred younger immune profile in Nicoyan individuals compared to those living in the rest of Costa Rica, finding cellular proportion differences in CD8+ naïve and CD8+ memory T cells. Next, we found a lower level of total mean DNAm variation in Nicoyans compared to non-Nicoyans. We found 20 DMRs and four single CpG sites that were differentially methylated between Nicoyans and non-Nicoyans. While any of these differences we observed may be due to genetic differences in the populations, the fact that we show that estimated genetic structure did not differ between Nicoyans and non-Nicoyans suggests that these biological differences are more likely to be the result of environmental differences between these two populations. Lastly, DNAm age was not significantly different between Nicoyans and non-Nicoyans, although Costa Ricans overall had a younger mean DNAm age than the mean chronological age.

Our finding of proportional differences in CD8+ naïve and CD8+ memory T cells is intriguing in the context of previous work from animal models and other human research that has established that these blood cell proportions change as a function of age. Specifically, the naïve T cell response diminishes with time as they are naturally replaced by memory T cells through age-related thymic involution [17, 29]. Therefore, younger immune profiles have been hypothesized to delay the onset of infection vulnerability and extend health span [30]. Interestingly, centenarian offspring have been investigated in this context and show this 'youthful' immune cell phenotype [31]. The fact that our work suggested that Nicoyans had a lower proportion of CD8+ T memory cells and higher CD8+ T naïve cells is interesting in the context of immunoaging and might be suggestive of an age-related immune phenotype in Nicoyans.

Given their lower mortality rate, we were surprised about the lack of differences in DNAm age between Nicoyans non-Nicoyans, especially given that this biological aging measure has been associated with many age-related conditions such as cognitive fitness decline, frailty, and mortality [9, 11, 32]. It is important to note, however, that we only had the statistical power to test for a very large difference in DNAm age, and our 95% confidence interval suggests that Nicoyan individuals could be up to 6 years younger in DNAm age. We calculated epigenetic age in our samples using three published DNAm age predictors and found no significant difference in these measures of biological age between Nicoyans and non-Nicoyans. Previous work found that peripheral blood mononuclear cells from the offspring of semi-super centenarians, in an Italian cohort, appear 5.1 years epigenetically younger than controls. Centenarians also were reportedly 8.6 years younger than their chronological age [33]. This is consistent with our population; when we collapsed the Nicoyan and non-Nicoyan groups, we found centenarians were 12.7 years epigenetically younger. Furthermore, all Costa Ricans were 6.9 years epigenetically younger than their chronological age. It is possible that this might reflect the recently reported epigenetically younger phenotype in Hispanic populations [34].

To examine age-associated DNAm variability, we assessed a measure of variance in Nicoyans and non-Nicoyans independently. We were able to demonstrate the phenomenon of epigenetic drift within each of these groups, as our sample consisted of two age ranges in each group. Additionally, we observed that DNAm variation was lower in Nicoyans, both in the ≤80-year-old and in the >80-year-old age groups, when compared to non-Nicoyans. Given that DNAm variability across individuals has been reported to increase with age, our finding may highlight an epigenetic characteristic of the longevity in Nicoyans [12]. Although the biological pre- and antecedents of increased DNAm variability are poorly understood, our findings suggest lower DNAm variability may be associated with lower mortality. Some explanations for this age-associated DNAm variability have been proposed, such that DNAm variability may result from a functional decline in DNAm maintenance machinery or that the variability is a product of environmental exposures over time [35].

We found 20 genomic regions and four single CpGs that were significantly differentially methylated between Nicoyans and non-Nicoyans. One such DMR contained six CpGs and was located in the promoter region of *NUDT12*, a gene encoding a protein shown in vitro to cleave NADH, NADPH, and NAD+ [36]. Given that NUDT12 may play a role in NAD metabolism, a regulatory process associated with health span and aging, it is consistent with the fact that DNAm may be involved in the regulation of this gene that may have downstream effects on NAD biosynthesis. Nicoyans had a lower level of DNAm in the promoter region of *NUDT12*, a signature often associated with higher gene expression. In addition to our DMR findings, we also found four individual sites to be both statistically and biologically significant, three of which existed in intergenic regions. We further investigated three of these CpGs by quantifying DNAm with pyrosequencing, allowing us to verify both the accuracy of the 450k array and the significant differences between Nicoyans and non-Nicoyans. However, it remains unclear whether differential DNAm of these single CpGs or DMRs, at the observed effect sizes (<10%), are sufficient to yield a biological change. Interpretation of these findings at a biological level will require future mechanistic experiments.

Our findings should be interpreted within the context of several limitations. One limitation was the lack of genotype information for these samples, as genetic variation is considerably associated with DNAm [37]. In order to reduce genetic heterogeneity, we restricted control sampling to areas within Costa Rica, but outside of Nicoya. Nicoyan status is determined as of the time of the survey, i.e., at older ages, not based on birth or life-course residence, but in our sample 44 out of 48 Nicoyan residents have lived there their entire lives. While there are no documented differences between the historical migration patterns of the inhabitants of Nicoya and the rest of Costa Rica, minor differences may exist. Therefore, we implemented a recently published tool to infer genetic information using DNAm data obtained from the 450k array called 'Epistructure,' a tool from the python package GLINT [28]. We found no significant difference in population structure measures between Nicoyans and non-Nicoyans and thus did not include these composite measures in our analyses. Another consideration of our study results is that we used a bioinformatics approach to predict CD8+ T cell proportions, which are relative compositional estimates. As these predictions do not estimate actual cell counts, the abundance of other cell types will affect the proportional estimate of the CD8+ T cells. Ideally, we will need to validate our findings using a quantitative approach, such as fluorescent-activated cell sorting, to obtain actual cell counts. We note, perhaps not surprisingly, that we did not observe a significant difference when we performed an overall compositional analysis of these predictions [38], meaning that the blood composition overall was not different between Nicoyans and non-Nicoyans. However, our findings are supported by using two separate reference-based approaches [6, 14], both of which identified CD8+ T cells as being significantly different between Nicoyans and non-Nicoyans. Furthermore, these bioinformatic cell-type proportion predictions have been well validated in the literature when compared to actual cell counts, and so we were confident that this approach reflected true cell-type proportions [16].

Conclusions
Our findings thus highlight DNAm as a potential factor underlying the unique longevity observed in Nicoya region of Costa Rica. This work also supports the

demographic data on longevity as characterizing this population as unique. The specific differences in immune cell proportions we observed in Nicoyans will lay the framework for a validation study to observe whether cell-sorting experiments yield similar results. Additionally, the differential DNAm findings may provide a candidate list of CpGs to test for differences in other longevity populations. Lastly, upon validating our findings, our work will contribute to narrowing the focus of mechanistic studies to assess whether the DNAm differences we observed are involved in gene regulation that may alter gene expression trajectories.

Methods
Sample preparation and data collection
Whole blood was collected from participants and genomic DNA was extracted at the University of Costa Rica from 2 ml of frozen whole blood using the phenol–chloroform method. DNA was bisulfite converted with the Zymo Research EZ DNA Methylation™ Kit (Irvine, CA, USA). Approximately 160 ng of bisulfite-converted DNA from each sample, with the addition of one technical replicate, was randomized across eight 450k array BeadChips as well as sentrix row and run in one batch according to the manufacturer's protocol (San Diego, CA, USA).

Qiagen Pyromark Assay Design 2.0 software (Hilden, Germany) was used to generate pyrosequencing assays targeted to three 450k array CpGs. Pyrosequencing was performed on the a Qiagen Pyromark™ Q96 (Hilden, Germany) according to manufacturer's instructions. All primer sequences are listed in Additional file 6: Table S1.

Data preprocessing and normalization
Illumina GenomeStudio software (San Diego, CA, USA) was used to subtract background noise and color correct raw data using control probes. Data were extracted in the form of an average β value matrix and imported into R Statistical software for the remainder of data processing. Logit-transformed β values to M values, a less heteroscedastic value, were used for all statistical analyses, whereas β values were used for visualization purposes as they represent percent methylation (0 = no methylation, 1 = fully methylated). We have included a comparison table of β values compared to M values for CpGs identified in the site-specific differential methylation analysis (Additional file 7: Table S2).

All data processing and statistical analyses were implemented in R version 3.2.3. We removed a subset of probes that could potentially lead to erroneous results. These consisted of 65 SNP control probes, probes that were specific to either the X or Y sex chromosomes, probes with missing β values or with a detection $p \geq 0.01$ in 5 or more samples, polymorphic CpG probes, and cross-hybridizing XY probes. The total number of probes post-filtering based on these criteria was 441,109 [25]. No sample outliers were removed, defined as having more than 5% of their total probes fail.

Subset-quantile within array normalization (SWAN) was used to account for type I and II probe differences on the 450k array [39]. Known technical variation (sentrix ID and position) was accounted for with the function 'ComBat' [40]. Confirmation of these corrections was assessed before and after using principal component analysis.

Estimation of blood cell proportions

A validated cellular deconvolution method was used to estimate cell-type proportions in each blood sample, namely CD4+ and CD8+ T cells, natural killer cells, B cells, monocytes and granulocytes [16]. The predicted abundance levels of CD8+ T naïve and CD8+ T memory were obtained from the 'Advanced Blood Analysis' of the online DNAm age predictor [6]. Significance values were generated from performing a Kruskal–Wallis test for each cell-type proportion by group.

Prediction of epigenetic age

Three epigenetic clocks were used to predict biological age. The 'Horvath' and 'Hannum' estimates were computed with the online epigenetic clock software [6, 7]. The 'Weidner' age prediction was generated using the previously described 99 CpG model [41]. We investigated, to the best of our ability, the possibility that this finding was due to a global batch effect influencing all samples by performing the DNAm age calculation on raw data, after SWAN normalization data, and again after ComBat correction. In all cases, the mean DNAm age for all Costa Ricans was younger than the mean chronological age. We chose to proceed with calculating DNAm age using the most corrected data as we expected data that is corrected for technical batch effects, inclusive of probe design and chip–chip variance, to best represent true biological signal.

DNA methylation analysis

The R package 'DMRcate' was used to find DMRs [42]. The DMRcate model contained Nicoya group, chronological age, sex, and estimated cell-type proportions. This tool uses a Gaussian kernel smoothing of DNAm across the genome. Benjamini–Hochberg (BH) method was applied with a threshold of ≤ 0.05 and a β value difference of $\geq 5\%$ [43].

Site-specific differential DNAm analysis was conducted using moderated t-statistics with empirical Bayesian variation estimation using the bioconductor R package 'limma' with chronological age, sex, and cell-type proportions as covariates [44]. M values consisted of log-transformed β values to achieve a measure with uniform variation and decreased heteroscedasticity. Significance values were corrected for multiple testing using the BH method [43].

DNAm variability was calculated using the interquantile range (IQR) across the 90th and 10th percentiles of each group, independently, at each CpG. A significance value was generated by performing a Wilcoxon signed-rank test between groups.

Inferred genetic ancestry

Population structure was inferred using the 'Epistructure' command-line tool GLINT [28]. This method applies principal component analysis on a reference list of genetically informative 450k array probes. This tool suggests the top two principal components can infer genetic structure. Linear regression was used for comparison of each PC and group status, while adjusting for sex, cell-type proportions, and age.

Additional files

Additional file 1: Figure S1. Correlation plot of DNA methylation-based estimated blood cell-type proportions. Colored blocks represent correlation p values below 0.05, red indicates negative correlation and blue indicates positive correlation. Gran = granulocyte, Mono = monocyte, NK = natural killer.

Additional file 2: Figure S2. EEAA (extrinsic epigenetic age acceleration), General Age Acceleration (residuals from a linear model of DNAm age regressed onto chronological age), IEAA (intrinsic age acceleration). All measures were generated from the online epigenetic age software. No significant differences were observed between Nicoyans (blue) and non-Nicoyans (red). Significant values generated from ANOVA statistical tests.

Additional file 3: Figure S3. Continuation of differentially methylated regions between Nicoyans and non-Nicoyans. Remaining 14 of the 20 statistically significant DMRs obtained from DMRcate analysis found by the R package 'DMRcate'. Unadjusted DNA methylation values, shown as percent of cells methylated, are displayed on the y-axis, and genomic distance (bp) to the TSS is plotted on the × axis. Associated genes are based on closest distance to the TSS of each differentially methylation region. Nicoyans are represented by blue points with the mean of each CpG site illustrated by a blue line. Non-Nicoyans are represented with red with each point indicating an individual, with the red line illustrating the mean at each CpG.

Additional file 4: Figure S4. QQ plots of each identified CpG modeled using M values or β values. Linear model included DNA methylation value regressed on group (Nicoya vs non-Nicoya) with sex, age, and estimated cell-type proportions included as covariates.

Additional file 5: Figure S5. Epistructure-derived principal component analysis. Genetically informative 450k array sites were used to estimate genetic population structure in our data. Principal component analysis was performed on a reference CpG list to generate measures (PCs) of genetic structure. No significant difference in these estimates was seen between Nicoyans and non-Nicoyans.

Additional file 6: Table S1. Pyrosequencing primer sequences designed with Qiagen Pyromark Assay Design 2.0 software.

Additional file 7: Table S2. Comparison of M values and β values of each identified CpG.

Abbreviations

450k array: Infinium HumanMethylation450 array; BH: Benjamini and Hochberg; CpG: cytosine–phosphate–guanine dinucleotide; DMR: differentially methylated region; DNAm: DNA methylation; FDR: false discovery rate; IQR: interquantile range; PCA: principal component analysis.

Authors' contributions

DHR, LMM, and MSK designed the research study. LMM performed all data analyses and wrote the manuscript together with DHR and MSK. AMM performed pyrosequencing assays. AMM and JLM performed DNAm microarray experiments. RDE provided statistical advice and assisted with data visualization. WHD and LRB designed the CRELES Study. MJJ helped with data interpretation and analysis plan. All authors read and approved the final manuscript.

Author details

[1] Department of Medical Genetics, Centre for Molecular Medicine and Therapeutics, BC Children's Hospital Research Institute, University of British Columbia, 950 West 28th Ave, Vancouver, Canada. [2] School of Public Health, University of California, Berkeley, Berkeley, CA, USA. [3] Centro Centroamericano de Población, Universidad de Costa Rica, San José, Costa Rica. [4] Division of General Medical Disciplines, Department of Medicine, School of Medicine, Stanford University, 1070 Arastradero Road, Suite 300, Palo Alto, CA 94304, USA.

Acknowledgements

We would like to acknowledge Alexandre Lussier and Nicole Gladish for their editorial contributions.

Competing interests

The authors declare that they have no competing interests.

Funding

The CRELES data were collected by the Central American Population Center of the University of Costa Rica with support from Wellcome Trust Grant 072406. We gratefully acknowledge funding for the DNAm assays from the Center on the Economics and Demography of Aging, UC Berkeley (P30AG012839). We would also like to acknowledge funding from the Canadian Institute of Health Research (CIHR) (F16-00910) and the R. Howard Webster Foundation (F13-00031). DH Rehkopf was supported by a grant from the National Institute of Aging (K01 AG047280). MS Kobor is the Canada Research Chair in Social Epigenetics, Senior Fellow of the Canadian Institute for Advanced Research, and Sunny Hill BC Leadership Chair in Child Development. LM McEwen was supported by a CIHR Frederick Banting and Charles Best Doctoral Research Award (F15-04283).

References

1. Brooks-Wilson AR. Genetics of healthy aging and longevity. Hum Genet. 2013;132(12):1323–38.
2. Passarino G, De Rango F, Montesanto A. Human longevity: genetics or lifestyle? It takes two to tango. Immun Ageing. 2016;13(1):12.
3. Jones MJ, Goodman SJ, Kobor MS. DNA methylation and healthy human aging. Aging Cell. 2015;14(6):924–32.
4. Shoemaker R, Deng J, Wang W, Zhang K. Allele-specific methylation is prevalent and is contributed by CpG-SNPs in the human genome. Genome Res. 2010;20(7):883–9.
5. Farré P, Jones MJ, Meaney MJ, Emberly E, Turecki G, Kobor MS. Concordant and discordant DNA methylation signatures of aging in human blood and brain. Epigenetics Chromatin. 2015;8(1):19.
6. Horvath S. DNA methylation age of human tissues and cell types. Genome Biol. 2013;14(10):R115.
7. Hannum G, Guinney J, Zhao L, Zhang L, Hughes G, Sadda S, et al. Genome-wide methylation profiles reveal quantitative views of human aging rates. Mol Cell. 2013;49(2):359–67.

8. Weidner CI, Lin Q, Koch CM, Eisele L, Beier F, Ziegler P, et al. Aging of blood can be tracked by DNA methylation changes at just three CpG sites. Genome Biol. 2014;15(2):R24.
9. Marioni RE, Shah S, McRae AF, Chen BH, Colicino E, Harris SE, et al. DNA methylation age of blood predicts all-cause mortality in later life. Genome Biol. 2015;16(1):25.
10. Chen BH, Marioni RE, Colicino E, Peters MJ, Ward-Caviness CK, Tsai P-C, et al. DNA methylation-based measures of biological age: meta-analysis predicting time to death. Aging. 2016;8(9):1844–65.
11. Breitling LP, Saum K-U, Perna L, Schöttker B, Holleczek B, Brenner H. Frailty is associated with the epigenetic clock but not with telomere length in a German cohort. Clin Epigenetics. 2016;8(1):21.
12. Martin GM. Epigenetic drift in aging identical twins. Proc Natl Acad Sci. 2005;102(30):10413–4.
13. Jones MJ, Islam SA, Edgar RD, Kobor MS. Adjusting for cell type composition in DNA methylation data using a regression-based approach. Methods Mol Biol. 2015;1589:99–106.
14. Houseman E, Accomando WP, Koestler DC, Christensen BC, Marsit CJ, Nelson HH, et al. DNA methylation arrays as surrogate measures of cell mixture distribution. BMC Bioinformatics. 2012;13(1):86.
15. Guintivano J, Aryee MJ, Kaminsky ZA. A cell epigenotype specific model for the correction of brain cellular heterogeneity bias and its application to age, brain region and major depression. Epigenetics. 2013;8(3):290–302.
16. Koestler DC, Christensen B, Karagas MR, Marsit CJ, Langevin SM, Kelsey KT, et al. Blood-based profiles of DNA methylation predict the underlying distribution of cell types: a validation analysis. Epigenetics. 2013;8(8):816–26.
17. Fagnoni FF, Vescovini R, Passeri G, Bologna G, Pedrazzoni M, Lavagetto G, et al. Shortage of circulating naive CD8+ T cells provides new insights on immunodeficiency in aging. Blood. 2000;95(9):2860–8.
18. Beekman M, Blanché H, Perola M, Hervonen A, Bezrukov V, Sikora E, et al. Genome-wide linkage analysis for human longevity: Genetics of Healthy Aging Study. Aging Cell. 2013;12(2):184–93.
19. Rose G. Sick individuals and sick populations. Int J Epidemiol. 2001;30(3):427–32 **(discussion 433–4)**.
20. Diez Roux AV. Investigating neighborhood and area effects on health. Am J Public Health. 2001;91(11):1783–9.
21. Rosero-Bixby L, Dow WH, Rehkopf DH. The Nicoya region of Costa Rica: a high longevity island for elderly males. Vienna Yearb Popul Res. 2013;11:109–36.
22. Bogin B, Varela-Silva MI. Leg length, body proportion, and health: a review with a note on beauty. Int J Environ Res Public Health. 2010;7(3):1047–75.
23. Rehkopf DH, Dow WH, Rosero-Bixby L, Lin J, Epel ES, Blackburn EH. Longer leukocyte telomere length in Costa Rica's Nicoya Peninsula: a population-based study. Exp Gerontol. 2013;48(11):1266–73.
24. Rosero-Bixby L, Fernández X, Dow W, National Archive of Compuetrized Data on Aging, editors. CRELES: costa rican longevity and healthy aging study, 2005 (Costa Rica Estudio De Longevidad Y Envejecimiento Saludable) (ICPSR 26681). 2010. doi:10.3886/ICPSR26681.v2. http://www.icpsr.umich.edu/icpsrweb/NACDA/studies/26681/documentation%20.
25. Price ME, Cotton AM, Lam LL, Farré P, Emberly E, Brown CJ, et al. Additional annotation enhances potential for biologically-relevant analysis of the Illumina Infinium HumanMethylation450 BeadChip array. Epigenetics Chromatin. 2013;6(1):4.
26. Jaffe AE, Irizarry RA. Accounting for cellular heterogeneity is critical in epigenome-wide association studies. Genome Biol. 2014;15(2):R31.
27. Fagnoni FF, Vescovini R, Mazzola M, Bologna G, Nigro E, Lavagetto G, et al. Expansion of cytotoxic CD8+ CD28− T cells in healthy ageing people, including centenarians. Immunology. 1996;88(4):501–7.
28. Rahmani E, Shenhav L, Schweiger R, Yousefi P, Huen K, Eskenazi B, et al. Genome-wide methylation data mirror ancestry information. Epigenetics Chromatin. 2017;10:1.
29. Wherry EJ, Kurachi M. Molecular and cellular insights into T cell exhaustion. Nat Rev Immunol. 2015;15(8):486–99.
30. Gui J, Mustachio LM, Su D-M, Craig RW. Thymus size and age-related thymic involution: early programming, sexual dimorphism, progenitors and stroma. Aging Dis. 2012;3(3):280–90.
31. Pellicanò M, Buffa S, Goldeck D, Bulati M, Martorana A, Caruso C, et al. Evidence for less marked potential signs of t-cell immunosenescence in centenarian offspring than in the general age-matched population. J Gerontol A Biol Sci Med Sci. 2013. doi:10.1093/gerona/glt120.

32. Marioni RE, Shah S, McRae AF, Ritchie SJ, Muniz-Terrera G, Harris SE, et al. The epigenetic clock is correlated with physical and cognitive fitness in the Lothian Birth Cohort 1936. Int J Epidemiol. 2015. doi:10.1093/ije/dyu277.

33. Horvath S, Pirazzini C, Bacalini MG, Gentilini D, Di Blasio AM, Delledonne M, et al. Decreased epigenetic age of PBMCs from Italian semi-supercentenarians and their offspring. Aging. 2015;7(12):1159–70.

34. Horvath S, Gurven M, Levine ME, Trumble BC, Kaplan H, Allayee H, et al. An epigenetic clock analysis of race/ethnicity, sex, and coronary heart disease. Genome Biol. 2016;17(1):171.

35. Shah S, McRae AF, Marioni RE, Harris SE, Gibson J, Henders AK, et al. Genetic and environmental exposures constrain epigenetic drift over the human life course. Genome Res. 2014;24(11):1725–33.

36. Garten A, Schuster S, Penke M, Gorski T, de Giorgis T, Kiess W. Physiological and pathophysiological roles of NAMPT and NAD metabolism. Nat Rev Endocrinol. 2015;11(9):535–46.

37. Banovich NE, Lan X, McVicker G, van de Geijn B, Degner JF, Blischak JD, et al. Methylation QTLs are associated with coordinated changes in transcription factor binding, histone modifications, and gene expression levels. Reddy TE, editor. PLoS Genet [Internet]. 2014;10(9):e1004663. http://eutils.ncbi.nlm.nih.gov/entrez/eutils/elink.fcgi?dbfrom=pubmed&id=25233095&retmode=ref&cmd=prlinks.

38. van den Boogaart KG, Tolosana-Delgado R. Fundamental concepts of compositional data analysis. In: Gentleman R, Giovanni P, and Hornik, K, editors. Analyzing compositional data with R. Berlin: Springer; 2013. p. 13–50.

39. Maksimovic J, Gordon L, Oshlack A. SWAN: subset-quantile within array normalization for Illumina Infinium HumanMethylation450 BeadChips. Genome Biol. 2012;13(6):R44.

40. Johnson WE, Li C, Rabinovic A. Adjusting batch effects in microarray expression data using empirical Bayes methods. Biostatistics. 2007;8(1):118–27.

41. Lin Q, Weidner CI, Costa IG, Marioni RE, Ferreira MRP, Deary IJ, et al. DNA methylation levels at individual age-associated CpG sites can be indicative for life expectancy. Aging. 2016;8(2):394–401.

42. Peters TJ, Buckley MJ, Statham AL, Pidsley R, Samaras K, Lord RV, et al. De novo identification of differentially methylated regions in the human genome. Epigenetics Chromatin. 2015;8:6.

43. Hochberg Y, Benjamini Y. More powerful procedures for multiple significance testing. Stat Med. 1990;9(7):811–8.

44. Smyth GK. Linear models and empirical Bayes methods for assessing differential expression in microarray experiments. Stat Appl Genet Mol Biol. 2004;3(1):Article 3–25.

45. Rosero-Bixby L, Dow WH. Surprising SES gradients in mortality, health, and biomarkers in a Latin American population of adults. J Gerontol B Psychol Sci Soc Sci. 2009;64(1):105–17.

46. Horvath S, Ritz BR. Increased epigenetic age and granulocyte counts in the blood of Parkinson's disease patients. Aging. 2015;7(12):1130–42.

47. Horvath S, Levine AJ. HIV-1 infection accelerates age according to the epigenetic clock. J Infect Dis. 2015;212(10):1563–73.

Links between DNA methylation and nucleosome occupancy in the human genome

Clayton K. Collings[1] and John N. Anderson[2*]

Abstract

Background: DNA methylation is an epigenetic modification that is enriched in heterochromatin but depleted at active promoters and enhancers. However, the debate on whether or not DNA methylation is a reliable indicator of high nucleosome occupancy has not been settled. For example, the methylation levels of DNA flanking CTCF sites are higher in linker DNA than in nucleosomal DNA, while other studies have shown that the nucleosome core is the preferred site of methylation. In this study, we make progress toward understanding these conflicting phenomena by implementing a bioinformatics approach that combines MNase-seq and NOMe-seq data and by comprehensively profiling DNA methylation and nucleosome occupancy throughout the human genome.

Results: The results demonstrated that increasing methylated CpG density is correlated with nucleosome occupancy in the total genome and within nearly all subgenomic regions. Features with elevated methylated CpG density such as exons, SINE-Alu sequences, H3K36-trimethylated peaks, and methylated CpG islands are among the highest nucleosome occupied elements in the genome, while some of the lowest occupancies are displayed by unmethylated CpG islands and unmethylated transcription factor binding sites. Additionally, outside of CpG islands, the density of CpGs within nucleosomes was shown to be important for the nucleosomal location of DNA methylation with low CpG frequencies favoring linker methylation and high CpG frequencies favoring core particle methylation. Prominent exceptions to the correlations between methylated CpG density and nucleosome occupancy include CpG islands marked by H3K27me3 and CpG-poor heterochromatin marked by H3K9me3, and these modifications, along with DNA methylation, distinguish the major silencing mechanisms of the human epigenome.

Conclusions: Thus, the relationship between DNA methylation and nucleosome occupancy is influenced by the density of methylated CpG dinucleotides and by other epigenomic components in chromatin.

Keywords: DNA methylation, Nucleosome, Epigenetics, CpG, NOMe-seq, MNase-seq

Background

The genomes of eukaryotic organisms are packaged into tightly condensed arrangements of nucleoprotein complexes referred to as chromatin. At the primary level of chromatin compaction, a 147-base-pair segment of DNA spirals nearly twice around an octamer of histone proteins to form a structure known as the nucleosome [1, 2]. The degree of nucleosome occupancy that occurs along

DNA in chromatin is important because it can dictate the accessibility of DNA to the transcriptional machinery and to other proteins involved in genome regulation [3]. With advances in high-throughput sequencing technologies, nucleosome maps have revealed differential nucleosome occupancy patterns over entire genomes for a variety of species and cell types [4]. For example, nucleosome-depleted regions are observed overlapping transcription start sites of active genes, while high nucleosome occupancy is found to encompass the promoters of silent genes [5]. Furthermore, several genome-wide explorations in conjunction with biochemical modifications have

*Correspondence: andersjn@purdue.edu
[2] Department of Biological Sciences, Purdue University, 915 W. State Street, West Lafayette, IN 47907, USA
Full list of author information is available at the end of the article

elucidated mechanisms that have been evoked to explain the differences in nucleosome occupancy detected across intragenic and intergenic chromatin.

In vertebrate cells, the most common mode of DNA methylation entails the addition of a methyl group to a cytosine residue in the context of a CpG dinucleotide. CpG methylation is perhaps the best understood epigenetic mark and is maintained through cell division during DNA replication primarily by the DNA methyltransferase Dnmt1 with some assistance from the de novo methyltransferases, Dnmt3a and Dnmt3b [6]. This modification has been linked to gene silencing and is considered to be an important factor in the formation of constitutive and facultative heterochromatin [7]. Additionally, DNA methylation has also been shown to be essential for normal tissue-specific development. During embryonic stem cell differentiation, select CpGs throughout the genome become methylated by the de novo DNA methyltransferases, and through DNA methylation's epigenetic influence on chromatin structure and gene regulation, the inheritability of diverse cellular phenotypes within higher eukaryotic species is sustained [8, 9].

Although CpG dinucleotides are underrepresented in mammals, they are not randomly distributed, and regions with high CpG density, referred to as CpG islands, are typically unmethylated and cover the promoters of many housekeeping genes [10]. The chromatin architecture of active CpG island promoters is characterized by nucleosome depletion, histone acetylation, H3K4 methylation, but not H3K36 methylation [10]. Additionally, numerous transcription factors bind to CpG islands, and proteins with CXXC domains, which target unmethylated CpGs, are especially enriched in CpG Islands [10]. Some examples include the transcription factor Sp1 [11], the H3K36me2 demethylase Kdm2a [12], and the H3K4me3 methyltransferase subunit Cfp1 [13]. These features of active transcription are thought to protect CpG islands from de novo methylation [10].

Downstream of active promoters, Setd2 catalyzes the methylation of H3K36 with elongating Pol II in the bodies of transcribed genes [5]. H3K36me3 is enriched in exons over introns and has been proposed to be associated with co-transcriptional splicing mechanisms [14]. With the transfer of Pol II, histone H3K4 demethylation is performed by Kdm5 and Kdm1 [15]. In parallel, the de novo DNA methyltransferases preferentially bind to unmethylated H3K4 and to H3K36me3 [16–19], and recently, the presence of H3K36me3 was shown to be linked to the enrichment of binding and de novo methylation by DNMT3b over DNMT3a in gene bodies in vivo [20].

We previously examined the effects of DNA methylation on the stability of a large heterogeneous population of nucleosomes [21]. Specifically, with bacterial artificial chromosomes (BACs), the CpG methyltransferase M. SssI, isolated histones, and micrococcal nuclease, we conducted nucleosome reconstitution experiments in conjunction with high-throughput sequencing on ~1 MB of mammalian DNA that was unmethylated or methylated. The features by which DNA methylation was found to increase the stability of nucleosomes (already positioned by nucleotide sequence) were elevated CpG frequency and a tendency for the minor grooves of CpGs to be rotationally oriented toward the histone surface, and these methylation-sensitive nucleosomes were found to be enriched in exons and in CpG islands [21]. Our in vitro nucleosome data reflected nucleosomal DNA methylation patterns observed in vivo in terms of the co-enrichment of DNA methylation and nucleosome occupancy in exons [22–25], the increased nucleosome occupancy associated with methylated CpG islands [25, 26], and the rotational orientation of methylated CpGs in Arabidopsis nucleosomes [22].

In order to extend our research beyond our in vitro experiments, we sought to understand the relationship between DNA methylation and nucleosome occupancy in the cell. In this study, we first perform an integrated analysis of MNase-seq and NOMe-seq data. Through this approach, we survey chromatin landscapes from the perspective of the nucleosome and find an underlying positive correlation between methylated CpG density and nucleosome occupancy. We also acknowledge exceptions to this pattern that can be linked to the presence or absence of other epigenetic factors. Finally, we extensively characterize the chromatin in CpG islands and at conserved transcription factor binding sites to reveal regulation of DNA methylation and nucleosome occupancy in the vicinities of these genomic landmarks.

Results

DNA methylation and nucleosome occupancy at the genome level

A common strategy used to study chromatin from genome-wide high-throughput sequencing data involves designating boundary elements and then characterizing the markers surrounding these sites [27]. An example of this approach is given in Additional file 1: Figure S1, which displays average DNA methylation and nucleosome occupancy levels amid computationally predicted sites for the transcription factor, CTCF. These CTCF recognition sequences often mark the boundaries of topologically associated domains [28]. The results show that these presumptive CTCF sites are flanked by a series of regularly spaced nucleosomes and that methylation at CpG dinucleotides occurs preferentially in the linker regions of these phased nucleosomes [29–31]. The elucidation of these patterns raises the question of what can constitute a chromatin boundary.

Approximately 15 million nucleosomes are positioned along the haploid human genome. The primary bioinformatics approach utilized in this work treated each nucleosome identified by MNase-seq as a chromatin boundary element. Using the flanking DNA centered on nucleosome midpoints, the sequence content, DNA methylation levels, and nucleosome occupancies were measured for the entire genome and for several subgenomic regions. In order to compare DNA methylation patterns derived from nucleosomes reconstituted in vitro to those formed in the cell, BS-seq [32] and MNase-seq [33], data derived from leukocytes were used (Fig. 1a–f). For the analysis of nucleosome occupancy and DNA methylation, in vivo MNase-seq and NOMe-seq data from fetal lung fibroblasts (IMR90 cells) were used (Fig. 1g–j) [29].

A hallmark of rotationally positioned nucleosomal DNA is the ~10-base-pair periodicity of A/T-rich sequences with minor grooves facing toward the histone octamer alternating with G/C-rich sequences whose minor grooves face away [34–36]. This arrangement is illustrated using the in vitro data presented in Fig. 1a. It was suggested long ago that this pattern facilitates the winding of DNA around the histone octamer. According to this widely accepted model, the minor grooves of CpG dinucleotides should face outwards at positions ±10, ±20, ±30, ±40, ±50, ±60, and ±70 base pairs from the nucleosome midpoint, and this pattern has been observed in *S. cerevisiae* and *C. elegans* which lack CpG methylation [37–39]. To our knowledge, such a clear pattern for *total* CpG occurrence has never been observed in mammalian nucleosomal DNA, and the lack of this periodic pattern is illustrated by the in vitro and in vivo datasets in Fig. 1b. However, a weak ~10-bp periodicity of the methylated CpG fraction is detected in the in vitro data (Fig. 1c), which becomes magnified when the fraction of CpGs that are methylated are quantified in Fig. 1e. This periodicity is indicative of an unusual rotational orientation as the minor grooves of the methylated CpG dinucleotides face the histone surface, in agreement with previous reports in Arabidopsis and human DNA sequences [21, 22, 40]. Note that the frequency of CpGs is ~1.6-fold higher in the in vitro data as compared to the

in vivo data (Fig. 1b). Also note that nucleosomes from CpG islands were excluded from the analysis for Fig. 1 and for all other analyses unless otherwise specified.

Comparisons between the in vitro and in vivo datasets yield striking differences in nucleosomal DNA methylation patterns. Preferential methylation is observed in nucleosome cores from the in vitro dataset (Fig. 1c, e), while a linker preference is found in the in vivo dataset (Fig. 1d, e). This linker preference is extended to nucleosomes that flank the fixed nucleosomes as DNA methylation levels peak between phased nucleosomes (Fig. 1f). On the other hand, nucleosome phasing is not observed in the in vitro data, presumably because the in vitro reconstitutions were carried out in moderate DNA excess. This condition may also explain the enrichment of mCpGs in the nucleosome core from the in vitro dataset since, as shown in Figs. 2 and 3, CpG occurrence is positively correlated with DNA methylation levels and nucleosome occupancy, and as noted earlier, the in vitro data are enriched in CpGs relative to the in vivo data (Fig. 1b). The NOMe-seq data aligned to MNase-seq derived nucleosome midpoints (Fig. 1g–j) mirror the results for the combined BS-seq and MNase-seq data (Fig. 1e, f). Additionally, the nucleosome phasing (Fig. 1j), the enrichment of DNA methylation in linker DNA (Fig. 1h), and the 10-bp periodic patterns of DNA methylation levels with the unique rotational orientation (Fig. 1g) give us confidence that the integration of MNase-seq and NOMe-seq data was precisely executed.

In an attempt to understand the relationship between DNA methylation and nucleosome occupancy, we first divided the MNase-seq derived nucleosome sequences into sublibraries based on the numbers of CpGs. The average mCpG/CpG fractions in the in vitro and in vivo sublibraries from leukocytes were determined, and the results are plotted in panels a and b of Fig. 2, respectively. In both cases, low CpG densities (1–3 CpGs per fragment) correspond to preferred linker methylation, while higher frequencies correspond to preferential methylation in the nucleosome core. These results can explain the preferential linker methylation in the total in vivo data and the preferential core methylation in the total in vitro data since the overall CpG content in the unfractionated in vitro library is 1.6-fold higher than what is observed in

---

(See figure on next page.)

Fig. 1 Genome-wide nucleosomal DNA sequence and methylation patterns in leukocytes and IMR90 cells. **a** For the in vitro MNase-seq data from leukocytes, average occurrences of select dinucleotides were computed from forward and reverse complement sequences aligned to nucleosome midpoints. Using BS-seq and MNase-seq data from leukocytes, average occurrences of CpGs (**b**), methylated and unmethylated CpGs (**c**, **d**) and mCpG/CpG fractions (**e**, **f**) were computed for both the in vitro and in vivo libraries from forward and reverse complement sequences aligned to nucleosome midpoints. Using MNase-seq and NOMe-seq data from IMR90 cells, average mHCG/HCG fractions (**g**, **h**) and uGCH/GCH fractions (**i**, **j**) were computed from forward and reverse complement sequences aligned to MNase-seq derived nucleosome midpoints. Nucleosomes within CpG Islands were excluded from this analysis

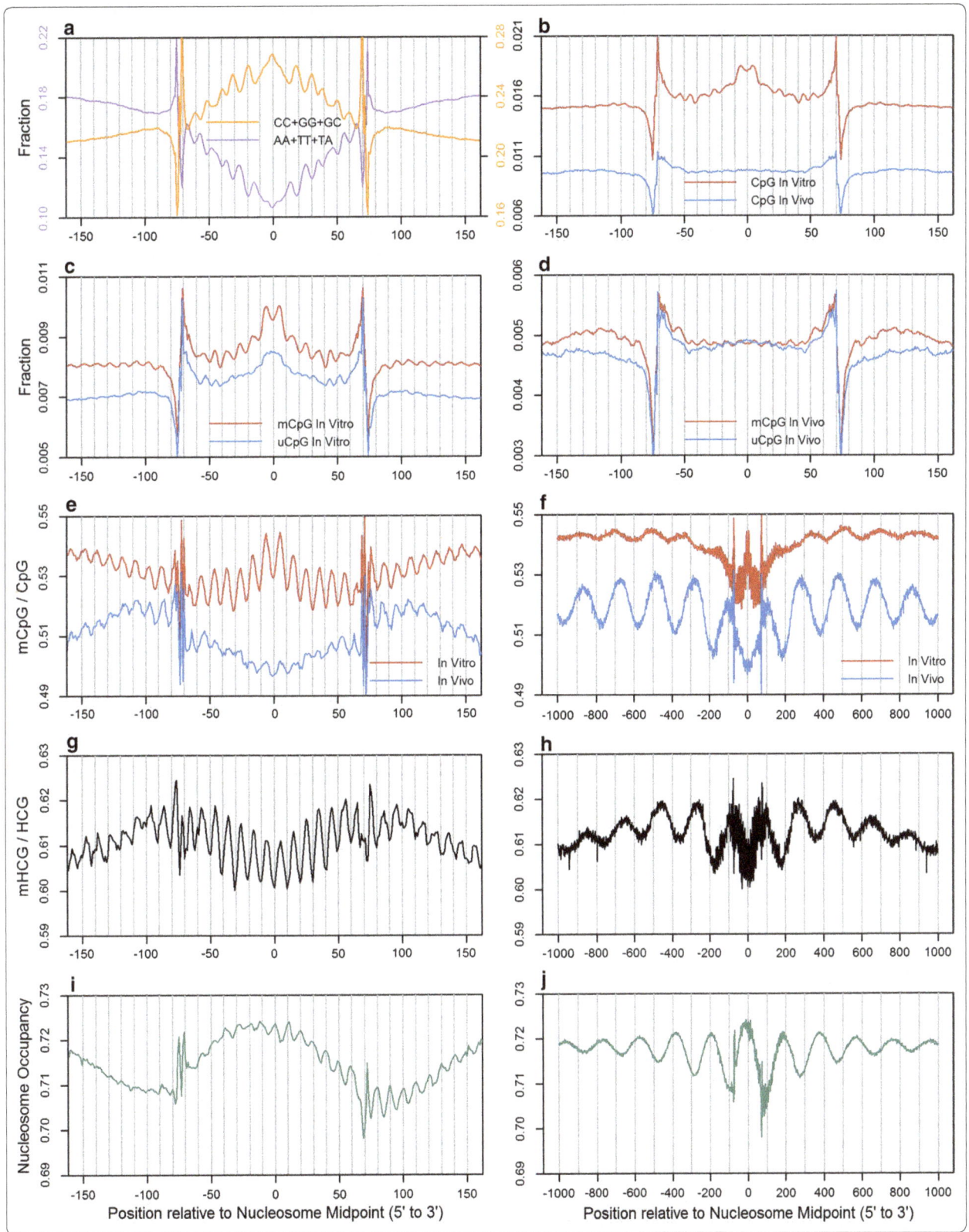

Position relative to Nucleosome Midpoint (5' to 3')

the in vivo data (Fig. 1b, e, f). Figure 2c shows essentially the same results using the MNase-seq derived sublibraries and NOMe-seq data from IMR90 cells, and Fig. 2d shows a corresponding graded increase in nucleosome occupancy as a function of increasing CpG content in the central selected nucleosome, and this increase does not appreciably extend into the two flanking nucleosomes.

To further characterize the relationship between DNA methylation and nucleosome occupancy, boxplots were used to show the distributions of nucleosome occupancy in different sequence contexts using all nucleosomes in the genome excluding those within CpG islands (Fig. 3). See "Methods" section for more details. Figure 3a shows an apparent linear relationship between CpG frequency and nucleosome occupancy, while a cooperative relationship between CpG frequency and DNA methylation is indicated in Fig. 3b [20, 39]. Combining mCpG frequency and nucleosome occupancy data in Fig. 3c yields a stronger correlation compared to total CpG frequency (Fig. 3a), suggesting that the presence of mCpGs has a role in dictating nucleosome occupancy levels. This interpretation is further supported by the negative relationship between uCpGs and occupancy (Fig. 3d). An additional feature that is dependent on CpG content is the ratio of mCpG levels in the core versus linker, such that linker preference gradually gives rise to core preference as a function of increasing CpG content (Fig. 3e).

In order to provide balance for the analyses conducted in Figs. 2 and 3, we also investigated the effects of CpG density in linker DNA on nucleosome occupancy and on DNA methylation. Interestingly, with increasing linker CpG density, average nucleosome occupancies remained constant while DNA methylation percentages, for the most part, appeared to drop in the core (Additional file 1: Figure S2). Moreover, when comparing the effects of increasing CpG density in linker versus nucleosome core DNA on mCpG density and on nucleosome occupancy side by side, it becomes apparent that mCpG density in nucleosomal DNA but not linker DNA is correlated with nucleosome occupancy (Additional file 1: Figure S3).

Many studies have implicated G + C content in the control of nucleosome occupancy [41, 42]. However, these models have shown limited applicability in mammalian cells [32]. It is conceivable that the increase in nucleosome occupancy that we observe upon increasing CpG is actually due to an increase in G + C content since CpG-rich fragments tend to be rich in G + C content as computed in Fig. 3f. However, this increase in G + C content is accompanied by only a marginal increase in nucleosome occupancy (Fig. 3g), and when CpGs are removed from the G + C content, there is essentially no considerable effect on nucleosome occupancy as shown in Fig. 3h, i.

DNA methylation and nucleosome occupancy within various genomic features

The MNase-seq derived nucleosome midpoints from IMR90 cells were annotated by HOMER [43] in order to determine whether the results for the entire genome displayed in Figs. 2 and 3 are observed within 20 different genomic features (Fig. 4; Additional file 1: Figures S4–S11). We first carried out each of the analyses portrayed by Fig. 3a–i using exons and featureless intergenic sequences in place of the sequences used from the total genome, and the results are similar to those presented in Fig. 3 (Additional file 1: Figures S4, S5). Average profiles of CpG occurrence, DNA methylation, and nucleosome occupancy aligned to nucleosomes within these features reveal a diverse set of chromatin landscapes (Additional file 1: Figures S6–S11), but within each of these genomic features, positive correlations are observed when plotting the distributions of nucleosome occupancy as function mCpG frequency as exhibited in Additional file 1: Figure S12. These results provide additional evidence for the dependence of nucleosome occupancy on mCpG density and imply that this phenomenon is a universal or a nearly universal property in the genome.

Among the genomic features with the highest nucleosome occupancies and mCpG density include exons [22–24] and SINE-Alu transposable elements [44]. Characteristics of these elements along with intron sequences are shown in Fig. 4. As compared to flanking and bulk sequences, both exons and SINE-Alu elements are enriched in CpGs, possess high levels of DNA methylation within the nucleosome core, and display an enrichment in nucleosome occupancy. In contrast, intron sequences are similar to bulk DNA, possessing low CpG content, average nucleosome occupancy, and higher methylation levels in linker DNA. The results in Additional file 1: Figure S13 also show that there are sharp increases in CpG occurrence, DNA methylation, and nucleosome occupancy in exons over introns at both the 5' and 3' exon–intron junctions. The higher CpG occurrences in exons as compared to introns are at least partially due to coding constraints. The potential consequences of this effect in terms of the control of cotranscriptional splicing by DNA methylation have been discussed previously [24].

DNA methylation and nucleosome occupancy in domains of selected histone modifications

Nucleosomes containing posttranslationally modified and variant histones are considered to represent key players in the control of the genome [5, 6, 9, 23, 45]. It was therefore of interest to characterize the DNA methylation status of these nucleosomes in order to

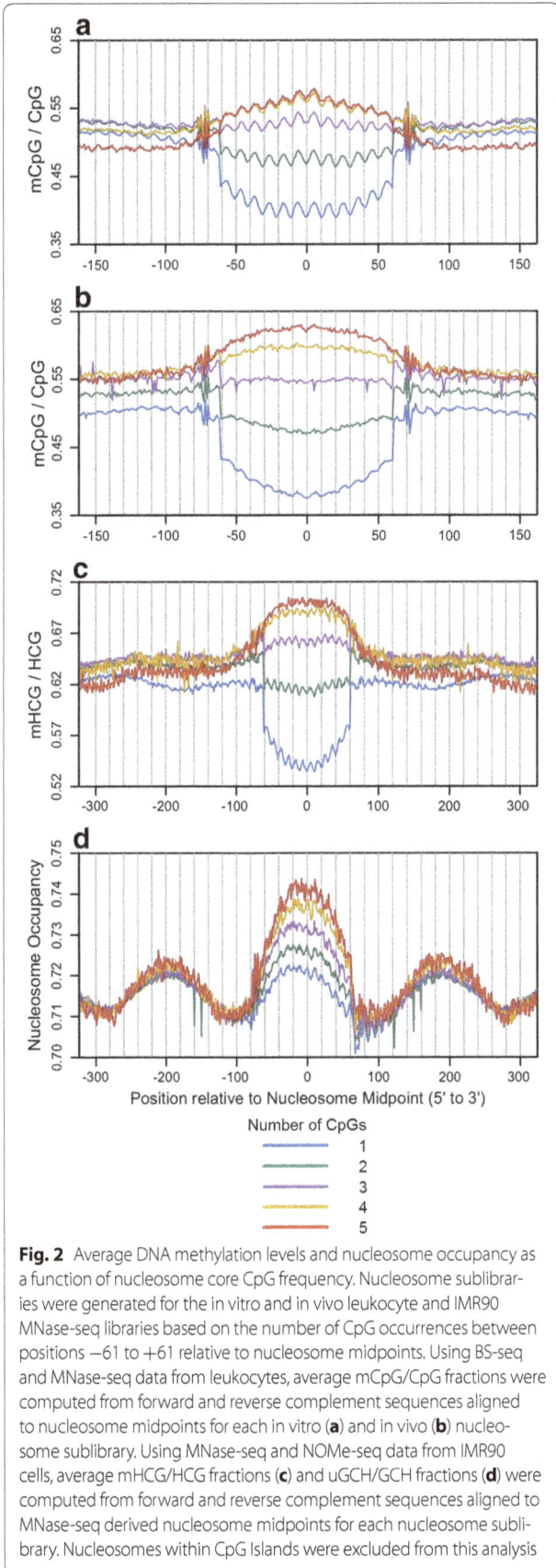

Fig. 2 Average DNA methylation levels and nucleosome occupancy as a function of nucleosome core CpG frequency. Nucleosome sublibraries were generated for the in vitro and in vivo leukocyte and IMR90 MNase-seq libraries based on the number of CpG occurrences between positions −61 to +61 relative to nucleosome midpoints. Using BS-seq and MNase-seq data from leukocytes, average mCpG/CpG fractions were computed from forward and reverse complement sequences aligned to nucleosome midpoints for each in vitro (**a**) and in vivo (**b**) nucleosome sublibrary. Using MNase-seq and NOMe-seq data from IMR90 cells, average mHCG/HCG fractions (**c**) and uGCH/GCH fractions (**d**) were computed from forward and reverse complement sequences aligned to MNase-seq derived nucleosome midpoints for each nucleosome sublibrary. Nucleosomes within CpG Islands were excluded from this analysis

determine whether or not their occupancies are proportional to mCpG density as it is in the bulk of the genome. Before performing this analysis, we first annotated every base pair in the peaks of 12 well-studied histone modifications and the variant H2A.Z using IMR90 ChIP-seq data from the Roadmap Epigenomics Project [46]. The annotation data shown in Additional file 1: Figure S14 show where these modifications are enriched or depleted relative to the entire genome. Figure 5a–c displays the CpG content, DNA methylation levels, and nucleosome occupancy data surrounding nucleosome midpoints that are located within peaks of the histone modifications listed in the key of panel d. For each histone modification, the average nucleosome occupancy values were plotted as a function of the number of mCpGs per nucleosome (Fig. 5d).

Nucleosomes marked by H3K4me3 are highly enriched in promoters, 5'UTRs, and CpG Islands (Additional file 1: Figure S12) and as expected contain high levels of unmethylated CpGs and possess low nucleosome occupancy values (Fig. 5). On the other hand, nucleosomes marked by H3K36me3 are located in gene bodies and are enriched by a factor of 2 in exons over introns (Additional file 1: Figure S12). Accordingly, H3K36me3-modified nucleosomes contain moderately high levels of methylated CpGs are associated with high nucleosome occupancy (Fig. 5). Regardless of histone modification, occupancy levels for most nucleosomes are correlated with their mCpG frequencies and DNA methylation levels in this analysis (Fig. 5d; Additional file 1: Figure S15). The only major exception is observed with nucleosomes marked by H3K9me3, which display very low levels of CpGs and low DNA methylation levels yet possess the highest nucleosome occupancy out of all the examined histone modifications (Fig. 5d). H3K9me3-modified nucleosomes are preferentially associated with constitutive heterochromatin where they play critical roles in DNA silencing [47].

In order to further investigate the relationship between DNA methylation and nucleosome occupancy within modified nucleosomes, we expanded the analysis in Fig. 5a–d by assessing the pairwise data among the 12 histone modifications and the variant H2A.Z (Fig. 5e–g). Comparison of the heatmaps displaying nucleosome occupancy (Fig. 5e) and mCpG density (Fig. 5f) as well as DNA methylation levels (Additional file 1: Figure S15) reveals a strong correspondence for most histone modification pairs and is supported by the high correlation given in Fig. 5g. The major exceptions to this correspondence include pairs marked by H3K9me3 indicated by the blue dots in Fig. 5g, which reflect the findings in Fig. 5d.

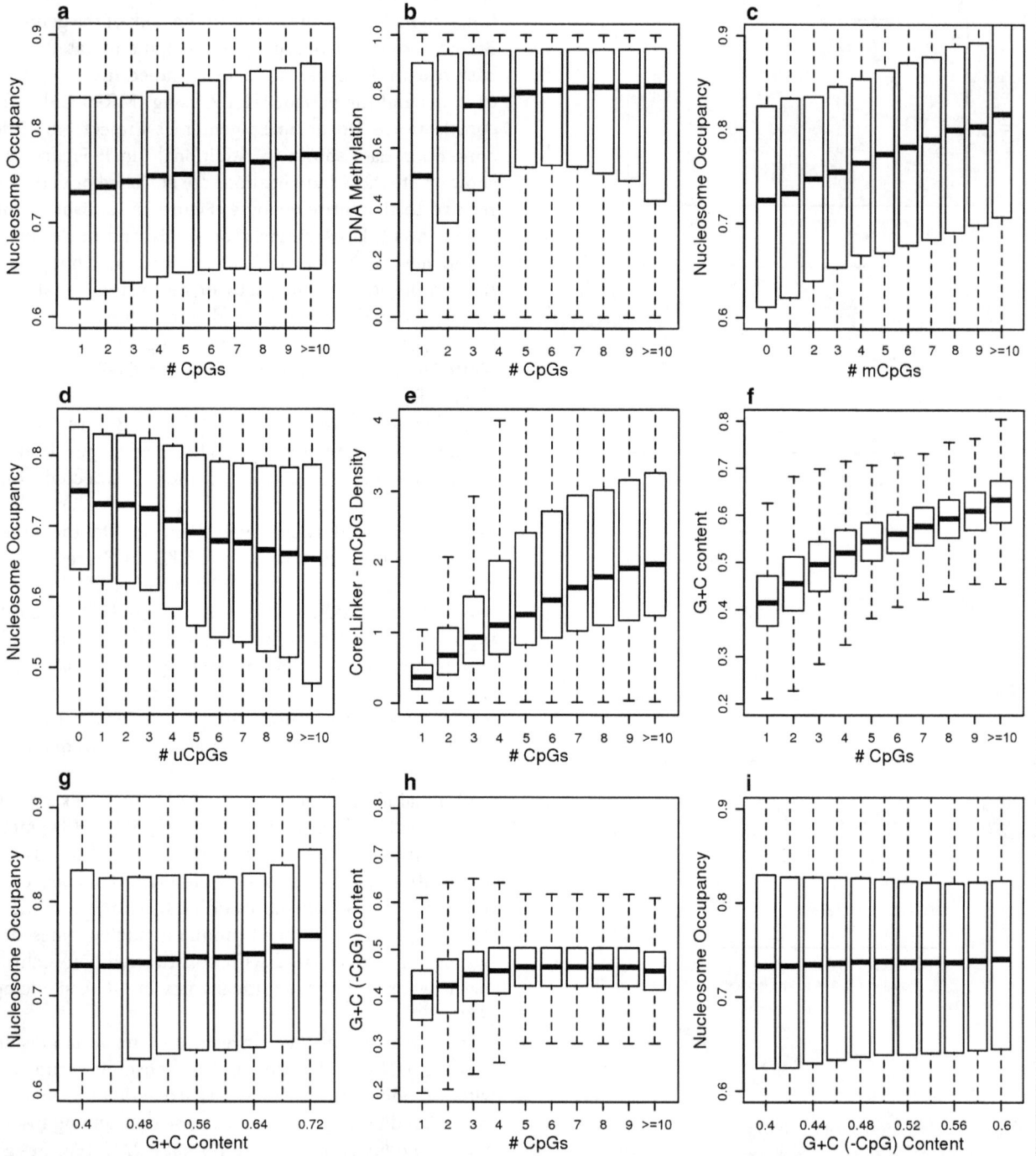

Fig. 3 Effects of increasing methylated CpG density in the nucleosome core on nucleosome occupancy and the ratio of methylated CpG density in core versus linker DNA. *Boxplots* display the distribution of nucleosome occupancy as a function of the number of CpGs (**a**), the number of methylated CpGs (**c**), the number of unmethylated CpGs (**d**), G + C content (**g**), and G + C-CpG content (**i**) for all nucleosomes in the genome outside of CpG islands. *Boxplots* also show the distribution of DNA methylation levels (**b**), G + C content (**f**), and G + C-CpG content (**h**) as a function of the number of CpGs. **e** *Boxplots* show the distribution of the ratios of methylated CpG density per base pair in the core versus linker DNA as a function of the number of CpGs. See "Methods" section for more details

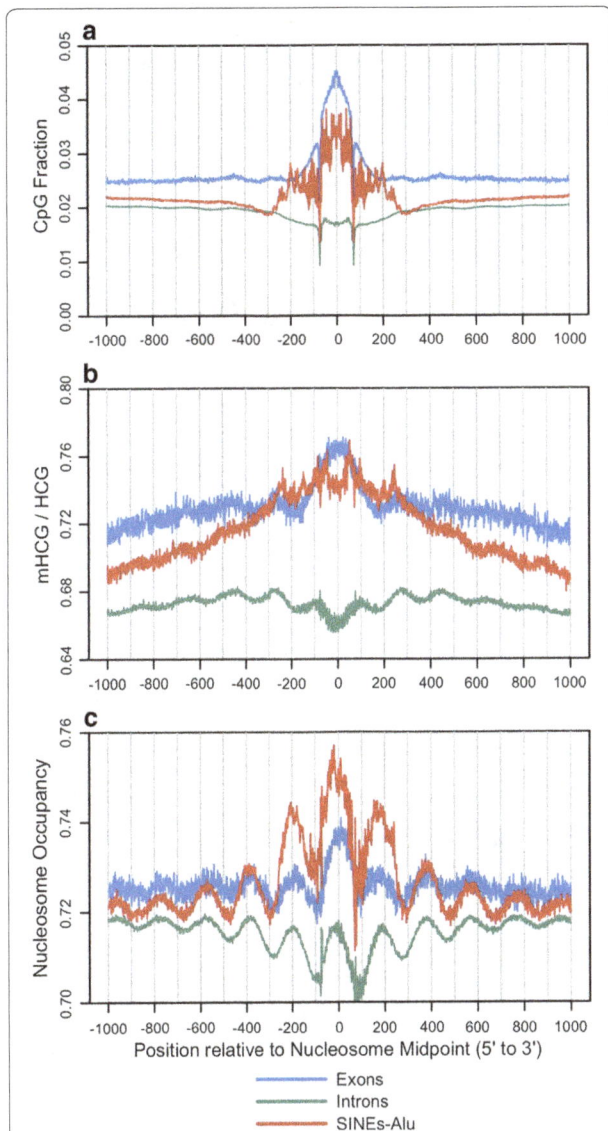

Fig. 4 Frequency profiles of CpGs, DNA methylation levels, and nucleosome occupancy surrounding nucleosomes positioned within exons, introns, and SINE-Alu elements. Using MNase-seq and NOMe-seq data from IMR90 cells, average occurrences of CpGs (**a**), mHCG/HCG fractions (**b**) and uGCH/GCH fractions (**c**) were computed from forward and reverse complement sequences aligned to MNase-seq derived nucleosome midpoints

DNA methylation and nucleosome occupancy in CpG islands

Over half of the promoters in the human genome reside within CpG islands. These CpG- and G + C-rich segments serve as platforms for the assembly of unstable nucleosomes and sites for attracting regulatory proteins leading to regions of chromatin that are permissive for transcriptional activation [10]. We characterized the chromatin in CpG islands in IMR90 cells by analyzing the relationships among nucleosome occupancy, DNA methylation, and the various histone modifications analyzed in Fig. 5. Figure 6a shows that, like the bulk of the genome, the frequencies of mCpG are positively correlated with nucleosome occupancies in CpG islands. The CpG island nucleosomes were then divided into two groups, non-TSS and TSS, based on whether or not their midpoints occurred within 500 bp of a TSS. For each MNase-seq derived nucleosome, nucleosome occupancy and DNA methylation values were plotted in 2D color-coded scatterplots (Fig. 6b, c). These results reveal an exception to the theme observed with bulk DNA in that there are significant numbers of the sequences that display low methylation but high nucleosome occupancy.

The heatmaps representing CpG islands displayed in Fig. 6d, e, which consist of NOMe-seq, DNase-seq, and ChIP-seq data sorted by nucleosome occupancy, were constructed in order to elucidate the link between the low DNA methylation levels and high nucleosome occupancies observed in a subset of CpG island nucleosomes (Fig. 6b, c). Figure 6d displays data aligned to TSSs that are overlapped by CpG islands, and Fig. 6e shows data aligned to CpG island centers in non-TSS regions, which were subdivided into intragenic and intergenic groups. An open chromatin configuration at promoters is signified by the void of nucleosomes and the enrichment in DNaseI hypersensitivity at the TSSs in Fig. 6d. The heatmaps appear to show a strong overall correspondence between nucleosome occupancy and DNA methylation in gene bodies and intergenic regions but not at TSSs with high nucleosome occupancy. Most histone modifications appear to follow the overall positive correlation between DNA methylation and nucleosome occupancy.

(See figure on next page.)
Fig. 5 Analysis of nucleosome occupancy and mCpG density in differentially marked chromatin across the genome. Using MNase-seq, NOMe-seq, and Roadmap ChIP-seq data from IMR90 cells, average occurrences of CpGs (**a**), mHCG/HCG fractions (**b**), and uGCH/GCH fractions (**c**) were computed from forward and reverse complement sequences aligned to MNase-seq derived nucleosome midpoints within several histone modification peaks indicated in the key. The 12 histone modifications (and variant H2A.Z) in the key are ordered by decreasing average DNA methylation at the nucleosome midpoint. **d** Using data between positions −61 to +61 relative to the nucleosome midpoint, the average nucleosome occupancy for each histone modification was *plotted* against the corresponding average number of methylated CpGs in the central 123 base pairs of the nucleosome. **e–g** Nucleosome occupancy (**e**) and mCpGs density (**f**) were weighted by ChIP-seq Z-scores for both single and paired histone modifications (see "Methods" section). **g** *Plots* of nucleosome occupancy and mCpG density from the heatmaps in **e** and **f** show that H3K9me3-modified chromatin (*blue dots*) deviates from the correlation observed in the genome-wide data

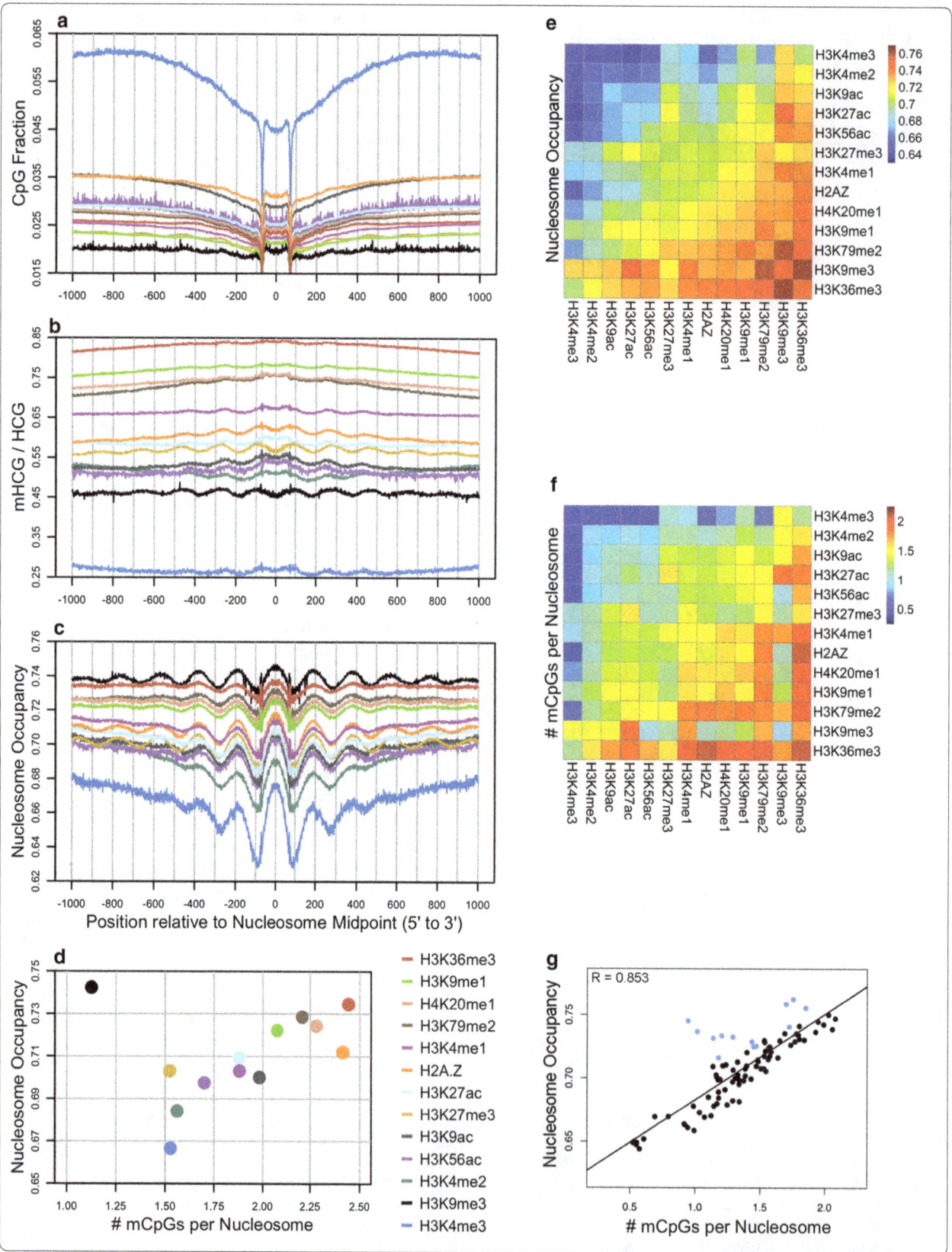

For example, the nucleosomes marked by acetylated histones as well as by H3K4me2/3 are located in unmethylated CpG islands with low nucleosome occupancy, while nucleosomes marked by H3K36me3 are located in methylated CpG islands with moderate to high levels of both occupancy and methylation in gene bodies. One noticeable exception in Fig. 6d is enrichment of H3K27me3, which is an epigenetic mark of polycomb-repressed genes [48]. However, when the heatmaps are sorted by H3K27me3 or H3K9me3 signals, it becomes apparent that both of these epigenetic modifications are characterized by low levels of methylation and moderate to high nucleosome occupancy in CpG islands in all three annotations (Additional file 1: Figures S16, S17) with a stronger anti-correlation associated with the presence of H3K27me3 [49].

DNA methylation and nucleosome occupancy at conserved transcription factor binding sites

In order to further evaluate the control of DNA methylation and nucleosome occupancy at regulatory elements in an unbiased manner, we studied all transcription factor binding sites (TFBSs) that are conserved in mammals and that contain CpG in their recognition sequences [50–52]. In this analysis, we characterized the chromatin at TFBSs in HCT116 cells using NOMe-seq [53], BS-seq [54], and also ChIP-seq data from ENCODE (Fig. 7) [55]. All CpG-containing TFBSs provided by the UCSC genome browser were divided into unmethylated and methylated sites (see "Methods" section), and the data above were aligned to these coordinates and sorted by decreasing nucleosome occupancy (Fig. 7a). Nearly all methylated sites display high nucleosome occupancy, while a large fraction of unmethylated sites are nucleosome depleted, DNaseI hypersensitive, and marked by H3K27ac (Fig. 7a, b), in agreement with several studies that have characterized the chromatin at active enhancers [56–58]. We also examined the occupancy levels of 10 transcription factors from ENCODE at their respective unmethylated and methylated CpG-containing TFBSs (Fig. 7c, d). All ten transcription factors exhibit binding at their unmethylated sites and appear nearly or completely absent at their methylated sites (Fig. 7d). Further analysis of the SP1 transcription factor shows that its occupancy is associated with low nucleosome occupancy,

unmethylated CpGs, DNaseI hypersensitivity and H3K27 acetylation (Fig. 7c).

We extended the characterization of the chromatin shown in Fig. 7a by examining 5 additional histone modifications and the variant H2A.Z using ChIP-seq data from the Jones laboratory (Additional file 1: Figure S18) [53]. For the methylated sites, the enrichment of H3K36 trimethylation and depletion of H3K27 acetylation and H3K4 trimethylation imply that several methylated sites are located in the gene bodies and deficient in active promoters or enhancers (Additional file 1: Figure S18). On the other hand, the unmethylated sites can be divided into two main groups. One group is enriched in CpG islands as well as active promoters and enhancers, which are indicated by low nucleosome occupancy and DNA methylation with high H3K27 acetylation and H3K4 methylation, and the second group is less CpG rich and possesses low to moderate levels of DNA methylation and high nucleosome occupancy (Additional file 1: Figure S18). Thus, the second group represents another exception to the positive correlation between methylated CpGs and nucleosome occupancy. Interestingly, the second group exhibits an enrichment of H3K9 and H3K27 methylation (Additional file 1: Figure S18), which are the same modifications linked to the exceptions observed in CpG-poor genomic regions and in CpG islands (Figs. 5, 6). These modifications along with the presence of H3K4me1 suggest that some of the sites in the second group may signify poised enhancers (Additional file 1: Figure S18) [56–58].

Discussion

Relationships between DNA methylation and nucleosome occupancy in vitro and in vivo

In order to explore links between DNA methylation and nucleosome occupancy, we relied heavily on data derived from NOMe-seq because this methodology, developed by Jones and coworkers, enables the simultaneous measurement of nucleosome occupancy and endogenous methylation for cell populations and single cells [29, 53, 59]. In comparison with NOMe-seq, MNase-seq requires higher coverage and relies on enrichment-based measurements of nucleosome occupancy, and these estimations can be skewed by sequence biases generated by enzyme cutting preferences, extents of digestion, and

(See figure on next page.)
Fig. 6 Characterization of chromatin at CpG islands. **a** *Boxplots* display the distribution of nucleosome occupancy as a function of the number of methylated CpGs. **b, c** Nucleosome with midpoints within CpG islands was divided into two groups, non-TSS and TSS. For each nucleosome, nucleosome occupancy and DNA methylation values were plotted in the 2D color-coded scatterplots. Using NOMe-seq and Roadmap ChIP-seq data from IMR90 cells, nucleosome occupancy, DNA methylation levels, and Z-scores from 12 histone modifications and the histone variant H2A.Z were aligned to TSSs that overlapped CpG Islands (**d**) and to the centers of CpG islands located in gene bodies and intergenic regions (**e**). Heatmap values were computed in 101 21-bp bins surrounding the TSSs and non-TSS CpG island centers, and sites (heatmap rows) were sorted by nucleosome occupancy

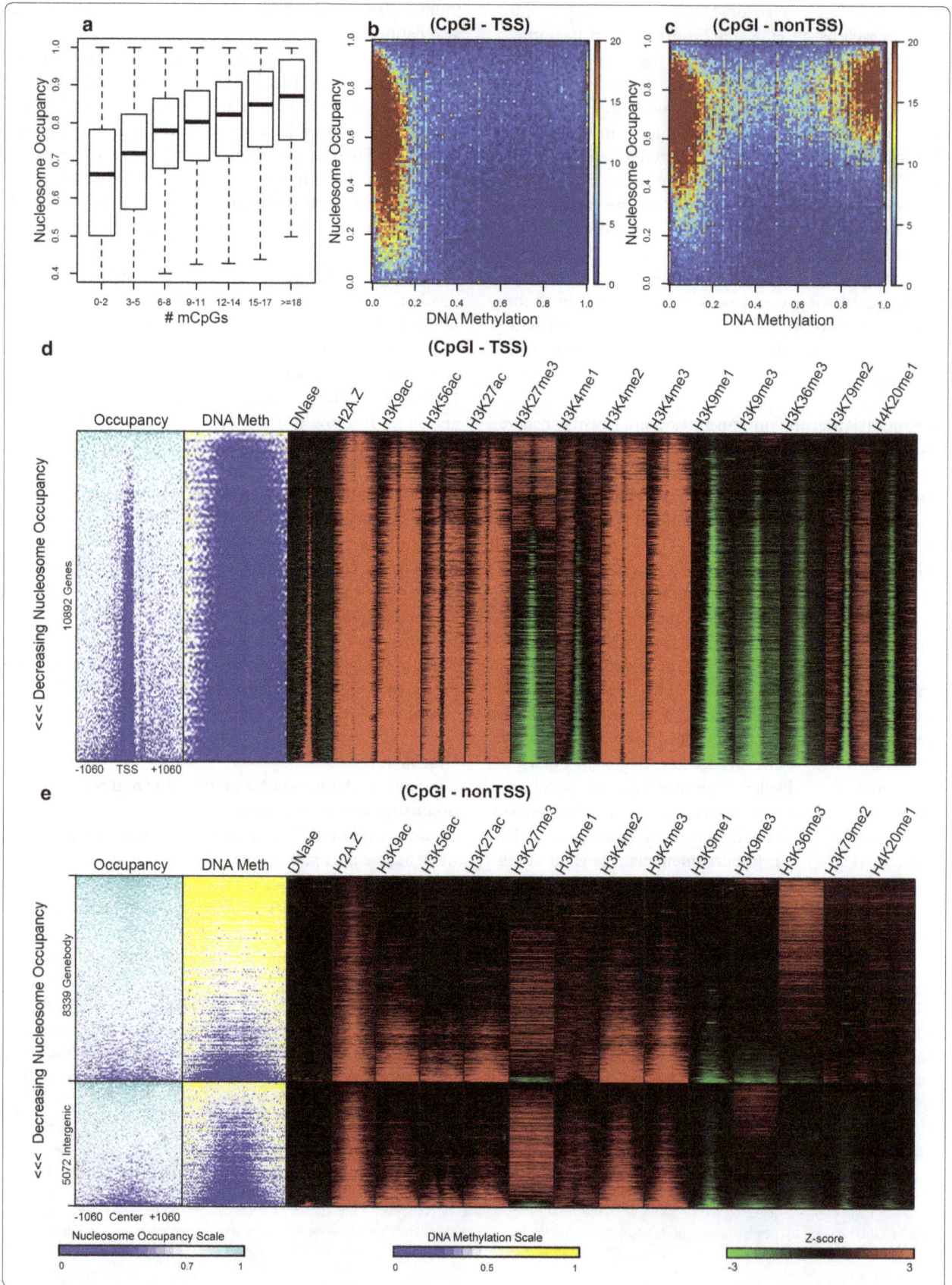

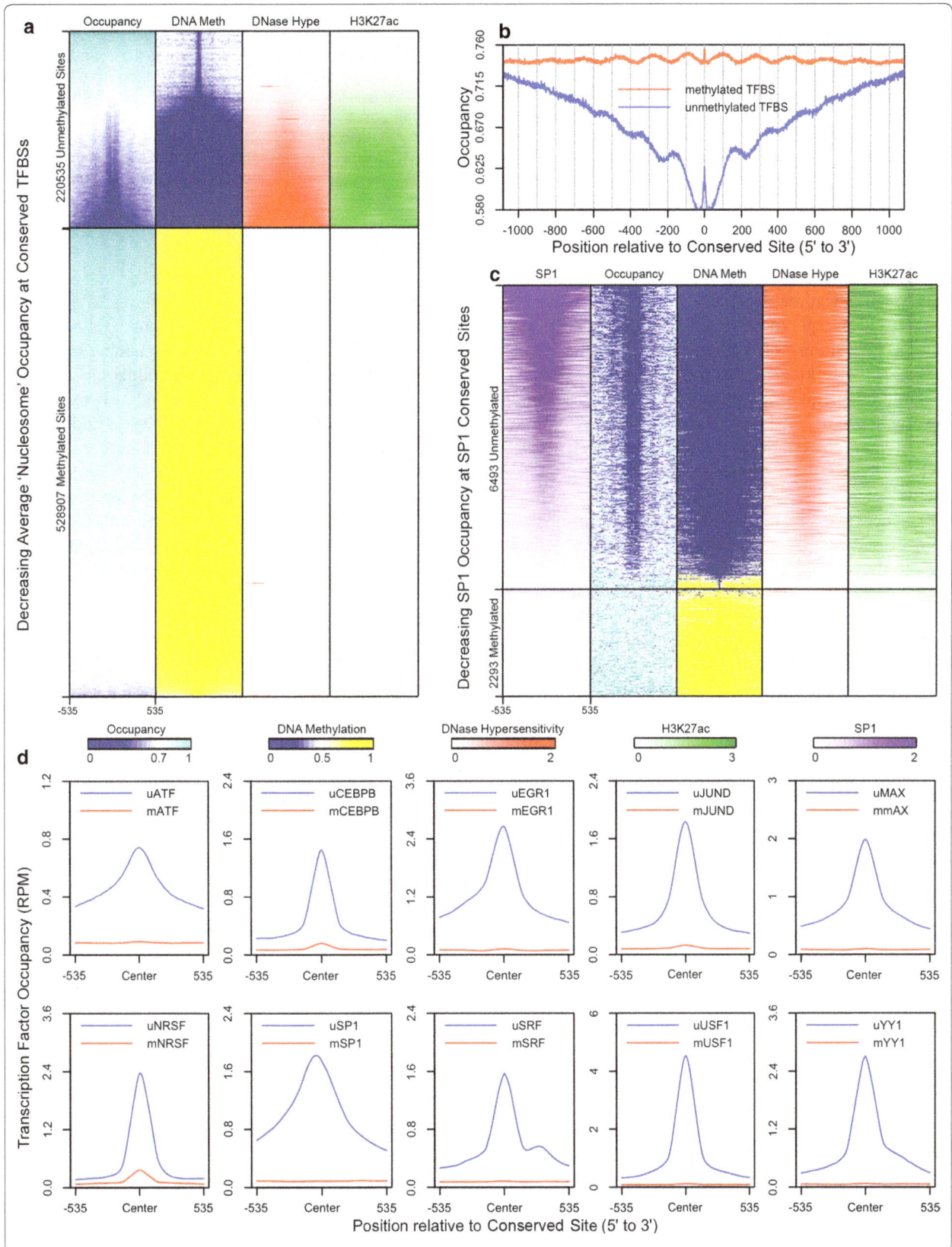

(See figure on previous page.)

Fig. 7 Characterization of chromatin at unmethylated and methylated conserved transcription factor binding sites. All conserved transcription factor binding sites (TFBSs) from the UCSC genome browser were divided into unmethylated and methylated sites (see "Methods" section). **a** Using NOMe-seq, BS-seq, and ENCODE ChIP-seq data from HCT116 cells, nucleosome occupancy, DNA methylation levels, and DNase-seq and H3K27ac signals were aligned to unmethylated and methylated conserved TFBSs. **b** Average uGCH/GCH fractions were also computed from forward and reverse complement sequences aligned to these unmethylated and methylated conserved TFBSs. **c** Similar to **a**, nucleosome occupancy, DNA methylation levels, and DNase-seq, H3K27ac, and SP1 signals were aligned to unmethylated and methylated SP1 conserved TFBSs. **d** Using ENCODE ChIP-seq data for 10 transcription factors, including SP1, average signals were plotted with respect to their corresponding unmethylated and methylated conserved TFBSs. In the heatmaps, average signals were computed in 51 21-bp bins surrounding each conserved site. Scales for DNase-seq, H3K27ac, and the transcription factors represent occupancy levels in reads per million

library amplification steps [60]. However, MNase-seq's primary advantage comes from its capability to determine the positions of nucleosome midpoints at near base-pair resolution, and this precision can be enhanced if paired-end sequencing is applied [21, 60]. By exploiting the strengths of both MNase-seq and NOMe-seq, we were able to confidently examine the effects of DNA methylation on nucleosome occupancy (Figs. 1, 2, 3; Additional file 1: Figures S2, S3).

Epigenetic factors must be maintained during development and cell differentiation, for without this characteristic they would be diluted during cell division. Two major epigenetic factors that satisfy this criterion are DNA methylation and stable posttranslationally modified histones. The metabolically stable histone methylation of H3K9 and H3K27, in contrast to histones modified by acetylation or phosphorylation, has half-lives measured in hours or longer and is commonly viewed as memory markers [45, 61–64]. These stable histone modifications along with DNA methylation are also involved in chromatin silencing, which raises the question as to whether or not their silencing mechanisms display similarities. The results of this investigation provide implications for this question, for the modes of action of these factors, and for epigenetics in general.

Our results suggest that the nucleosome serves as an effector arm of epigenetic mechanisms and that enhanced nucleosome occupancy is causally related to silencing induced by DNA methylation and stable histone modifications. The results in our previous studies demonstrated that DNA methylation enhances the stability of nucleosomes in a fraction of the human genome in vitro that contains multiple CpGs arranged in an unusual rotation with their minor grooves facing toward the histone octamer [21]. The studies described in this report elaborated on these features and extended the findings to the entire genome under in vivo conditions. The results show that nucleosome occupancy is correlated with mCpG frequency in the total genome and in subgenomic regions defined by CpG frequency (Figs. 2, 3; Additional file 1: Figures S2, S3) and by the coordinates of annotated

features (Figs. 4, 6; Additional file 1: Figures S4–S13, S16, S17, S19), histone modification domains (Fig. 5; Additional file 1: Figure S15), and CpG-containing transcription factor binding sites (Fig. 7; Additional file 1: Figure S18). However, our analyses uncovered two exceptions to the generalization between mCpG levels and nucleosome occupancy, which were nucleosomes marked by H3K9me3 and H3K27me3 (Figs. 5, 6; Additional file 1: Figures S16–S18). These nucleosomes displayed low levels of mCpG but high nucleosome occupancy values, which is indicative of some feature, other than DNA methylation, being responsible for their high nucleosome residency.

Modes of action for mCpG

The results in this report raise the question of why increasing methylated CpG density is associated with increasing nucleosome occupancy. One possible answer could be derived from the influential silencing mechanisms of methyl-binding proteins [65, 66]. It is also conceivable that highly methylated CpG-rich DNA could directly enhance nucleosome stability before nucleosomes are assembled as demonstrated in vitro [21, 67, 68] or become more stable after de novo methylation. We proposed in our MNase-seq study with in vitro reconstituted nucleosomes [21] that the hydrophobic, bulky methyl groups in the accessible major groove could cause narrowing of the corresponding minor groove. Indeed, in a recent DNase-seq experiment conducted on naked DNA, cutting frequencies were shown to be influenced by the methylation-induced narrowing of the minor groove [69]. The consequence of this action could strengthen the interactions between positive-charged histone arginines and the negative-charged DNA phosphate backbone, thereby enhancing nucleosome stability and in turn, increasing nucleosome occupancy in the cell [21, 69].

Although previous studies have indicated that the nucleosome core is the preferred site of DNA methylation [21, 22, 70, 71], others have suggested that methylation occurs preferentially in the linker regions [30, 72]. The results in Figs. 2 and 3 provide a possible explanation for this apparent discrepancy since a transition of preferential methylation from linker to core as a function of increasing CpG content was revealed along with a corresponding increase in nucleosome occupancy. For example, nucleosomes in CpG-rich exonic DNA and SINE-Alu sequences display higher mCpG density and nucleosome occupancy levels in the core, while nucleosomes in CpG-poor intronic DNA show selective methylation in linker DNA (Fig. 4). In light of this trend, the in vivo data displayed in Fig. 1d imply that methylated CpG dinucleotides have virtually no effect on nucleosome occupancy or positioning at the unfractionated genome level. However, in increasingly CpG-rich DNA, the effect of methylated CpG density on nucleosome occupancy becomes more apparent, and therefore, the majority of this effect is likely limited to a small fraction of the genome where CpG density is high. In fact, only ~7% of the genome is represented by 4 or more CpGs in 123-base-pair sliding windows.

The transition of preferential methylation from linker to nucleosome core may reflect the cooperative binding and enzymatic activities of the de novo DNA methyltransferases DNMT3a and DNMT3b, which have been shown to increase with CpG density outside of CpG islands [20]. Similar to bulk genomic DNA, we find a similar cooperative relationship between DNA methylation levels and CpG frequency in nucleosomal DNA (Fig. 3b). This cooperative mode of binding and methylation may be due to the heteromeric nature of the DNMT3 complexes in which two DNMT active sites display a spacing equivalent to about 10 nucleotides of DNA [73]. Thus, in the domains of high CpG density, clusters of CpGs would tend to become preferentially methylated, conceivably promoting nucleosome assembly or repositioning to the site. The unusual rotational orientation of mCpGs in a 10-nucleotide period reported previously [20, 21] and shown in Fig. 1 may also facilitate the action of the DNMTs since multiple mCpGs with major grooves facing away from the histone surface should be the most assessable to these enzymes.

It is important to emphasize that in methylated CpG dinucleotides, the methyl groups reside within the major groove, and it is likely that the major grooves of some CpGs along nucleosomal DNA are less accessible to the de novo methyltransferases. Indeed, it was proposed that the rotational orientation observed in methylated CpGs in Arabidopsis nucleosomes is a product of major groove accessibility [22]. The results of the de novo methyltransferase studies conducted by the Schubeler group imply that in domains of low CpG density, de novo methyltransferase binding and activity is reduced, and consequently, the DNA methylation levels in these regions are expected to be less efficiently maintained [20]. Moreover, with decreasing CpG density, the methylation of

nucleosome core DNA may be more effectively inhibited due to the relatively CpG-rich substrate preference of the de novo methyltransferases and due to a decrease in probability of major groove accessibility to CpG dinucleotides. Accordingly, the DNA flanking CTCF sites (Additional file 1: Figure S1), encompassing partially methylated domains (Additional file 1: Figure S19), and in the bulk of the genome (Fig. 1), where DNA methylation levels are higher in linker DNA, are located in CpG-deficient regions. Furthermore, the positive correlation between mCpG density and nucleosome occupancy, as well as the lack of an effect of increasing linker mCpG density on nucleosome occupancy levels, suggests that methylated CpG dinucleotides in adjacent linker DNA are not significantly influencing nucleosome formation (Figs. 2, 3; Additional file 1: Figures S2–S3).

Genome silencing by DNA methylation, H3K9Me3 and H3K27Me3

A central question concerning epigenetic mechanisms is the source and nature of the primary signals for epigenetic silencing. The signals must ultimately reside in the DNA sequence, but their nature is poorly understood. The signals for relating nucleosome occupancy to DNA methylation are the patterns of CpG dinucleotides, which are encoded in the sequence, and the factors that dictate the methylation status of CpGs such as nucleosome core versus linker localization and the rotational orientation of the CpGs in the nucleosome core. Furthermore, it is most often assumed that the initial signal for posttranslational modification of histones originates with specific regulatory proteins like transcription factors that recognize specific sequences in order to elicit a chain of events that lead to the final modification.

An alternative view is that simple DNA sequence patterns are directly responsible for the initial recognition process [74, 75]. For example, the clustering of unmethylated CpGs in G + C-rich regions is thought to serve as a signal for recognition of polycomb group complexes, which in turn results in the methylation of lysine 27 on histone H3 and ultimately chromatin silencing [48, 75]. Likewise, A + T-rich oligonucleotide sequences have been proposed to play a role in the recognition of H3K9 methylases leading to heterochromatization and repression [74, 75]. It has also been proposed that abundant nuclear proteins such as the HMG box proteins recognize these sites, and it is interesting to note that HMG proteins preferentially bind to AT duplex sequences of the form (ATATAT)N as compared to (AAAAAA)N in physiological ionic strength and temperature and that this AT-heteropolymeric specificity is shared by nucleosomes that contain H3K9Me3 [74, 76].

DNA methylation, transcription factors, and enhancer chromatin

Active enhancer chromatin typically contains multiple bound transcription factors, certain histone marks such as H3K27ac, undermethylated DNA and an open structure as evidence by DNaseI hypersensitivity [56–58]. We attempted to simplify our analysis by characterizing evolutionarily conserved transcription factor chromatin that contain at least one CpG in the recognition sequence. There exists apparent heterogeneity in the dataset as seen by the presence of H3K9me3, H3K27me3, partially methylated DNA, and deficiencies in H3K27ac (Additional file 1: Figure S18), which are characteristics of poised enhancers, sequences which are inactive but have potential for activation. The results in Fig. 7d show that 10 out of 10 transcription factors were preferentially associated with unmethylated DNA sequences, which raises the question whether this specificity is reflected in the binding of transcription factors to naked DNA. There are numerous examples where methylation blocks the in vitro binding of transcription factors that have CpG in their binding sites [77–79], but there are also cases including Sp1 where methylation is without effect on transcription factor binding [79, 80]. The molecular complexity of enhancer chromatin also makes it difficult to unravel cause and effect relationships. It is conceivable that the undermethylation of enhancer DNA is responsible for the reduced nucleosome occupancy and more open chromatin structure of enhancer sequences which would be consistent with the results presented in this study. However, there are alternative explanations to this proposal including the ability of transcription factors to induce loss of methylation at CpG sites, bound transcription factors excluding DNA methyltransferases, and passive DNA demethylation by DNA replication in the absence of maintenance methylation [80, 81].

Cancer, DNA methylation, and the nucleosome

A conserved property of cancer cells and tumors is global genomic hypomethylation and the local hypermethylation of some CpG Islands [82–84]. The global hypomethylation of the genome is viewed as a driving force for genomic instability in cancer, which characterizes the disease [83]. The activation of many mobile elements such as the SINE-Alu sequences observed in almost all cancers provides examples of this phenomenon [82, 83]. In fact, it has been suggested that the main carcinogenic effect of global DNA hypomethylation in cancer is mediated by its ability to create genomic instability [84]. The results of this study may be relevant to these observations since a reduction in mCpGs levels by as little as a one CpG per nucleosome results in detectable decreases in nucleosome occupancy. A decrease in the occupancy

or stability of such nucleosomes might be expected to have effects on the positions of these nucleosomes as well as the positions of nucleosome arrays adjacent to these nucleosomes, which are prevalent in the genome. Changes in the stabilities and positioning of nucleosomes in a significant fraction of the chromatin, as predicted from the present study, are expected to have profound, inheritable effects on the expression of the genome.

Conclusions

Previous studies have suggested that DNA methylation directly enhances the stability of nucleosomes in vitro. However, the relationship between DNA methylation and nucleosome occupancy is poorly understood. In this study, we implemented a bioinformatics approach that combines MNase-seq and NOMe-seq data to study links between DNA methylation and nucleosome occupancy throughout the human genome. Using this approach, we demonstrated that increasing mCpG density is correlated with nucleosome occupancy and that in mCpG-rich nucleosomes, methylation levels are greater in the core than in the adjacent linker DNA. These nucleosomal DNA methylation patterns were detected not only in total genomic DNA but also within most subgenomic regions. Prominent exceptions to the positive correlation between mCpG density and nucleosome occupancy included CpG islands marked by H3K27me3 and CpG-poor heterochromatin marked by H3K9me3, and these modifications, along with DNA methylation, characterize the major silencing mechanisms of mammalian chromatin. Thus, the density of methylated CpG dinucleotides may be an important factor in regulating nucleosome occupancy levels the human genome.

Methods

Data acquisition

Previously aligned in vivo and in vitro MNase-seq nucleosome and control data from leukocyte cells (neutrophil granulocytes) were acquired from GEO under accession number GSE25133 (GSM678045-63) [33]. Processed ENCODE BS-seq data from leukocyte cells (GM12878) were obtained from EMBL-EBI and are associated with GEO accession number GSE40832 (GSM1002650) [32]. Previously aligned MNase-seq data and processed NOMe-seq data from IMR90 cells were obtained from GEO under accession numbers GSE21823 (GSM543311) and GSE40770 (GSM1001125), respectively [29]. Previously aligned DNase-seq and ChIP-seq data for 12 histone modifications and the histone variant H2A.Z from IMR90 cells were obtained from the UCSD Human Reference Epigenome Mapping Project (Roadmap, GSE16256) [46]. Processed NOMe-seq data (for Fig. 7; Additional file 1: Figure S16) and ChIP-seq data (for Additional file 1: Figure S17) from HCT116 cells were obtained from GEO (GSE58638) [53]. Processed BS-seq data (for Fig. 7; Additional file 1: Figure S16) and previously aligned ChIP-seq and DNase-seq data (for Fig. 7) from HCT116 cells were obtained from GEO (GSE60106, GSM1465024) [54] and the ENCODE project [55], respectively. Exon, CpG island, and conserved TFBS coordinates were obtained from the UCSC genome browser [85] and/or the HOMER software [43], and computationally predicated CTCF sites were obtained from the CTCF Database 2.0 [86].

Identification and annotation of nucleosome midpoints
Using MNase-seq data from leukocyte and IMR90 cells, nucleosome midpoints were determined by adding or subtracting 73 base pairs from the 5' end of each read that aligned to plus or minus strands, respectively. For the MNase-seq in vitro and in vivo nucleosome and control data from leukocyte cells, coverage was calculated in 147-base-pair windows across the genome, and fractions of coverage between the control and the nucleosome libraries were computed. Subsequently, the numbers of reads at the nucleosome midpoints were normalized by these fractions of coverage in order to subtract background that could be generated by MNase cutting biases. Regardless of whether or not the MNase control data were subtracted from the nucleosome data, the dinucleotide and DNA methylation frequency profiles in Fig. 1a–f appeared nearly identical to ones where control data were not subtracted (data not shown). Nucleosome midpoints from IMR90 cells were annotated using the HOMER software package, and all frequency profiles surrounding nucleosome midpoints were generated using in-house scripts.

Data analysis for Fig. 3
Using MNase-seq and NOMe-seq data from IMR90 cells, the number of CpGs, G + C content, average mHCG/HCG fraction, and average uGCH/GCH fraction were computed for each nucleosome outside of CpG islands between positions −61 and +61 relative to the MNase-seq derived nucleosome midpoint. Methylation data for at least 2 cytosines within HCGs and GCHs regardless of strand were required for inclusion of a nucleosome in the analysis. With these data, boxplots were used to display the distributions presented in the figure. The numbers of methylated and unmethylated CpGs were determined by multiplying the average mHCG/HCG and uHCG/HCG fractions, respectfully, by the number of CpGs and rounding to the nearest integer. For the analysis in panel e, the number of CpGs and the average mHCG/HCG ratio were computed for the linker DNA of each included nucleosome between positions (±85 to ±115) relative to

the nucleosome midpoint. Methylation data for at least 2 cytosines within HCGs in the linker DNA regardless of strand were required for inclusion of a nucleosome in this analysis. With these data, boxplots show the distribution of the ratios of methylated CpG density per base pair in the core versus linker DNA as a function of the number of CpGs. Methylation CpG density per base pair in a nucleosome and its linker DNA were computed by multiplying the average mHCG/HCG ratio by the number of CpGs in the core and linker separately and dividing by 123 and 62 base pairs, respectfully.

Identification and annotation of ChIP-seq peaks
Peaks for ChIP-seq data from IMR90 cells were identified using SICER with default parameters [87]. Each base pair in every peak called by SICER for the 12 histone modifications and the histone variant H2A.Z was annotated by the HOMER software.

Data analysis for Fig. 5
For the analysis represented by panels e–g, the number of CpGs, average mHCG/HCG fraction, and average uGCH/GCH fraction were determined for each nucleosome between positions −61 and +61 relative to each MNase-seq derived nucleosome midpoint. ChIP-seq Z-scores for the 12 histone modifications and the variant H2A.Z were also determined at each midpoint. For each histone modification, averages values were weighted by Z-scores using the following formula.

$$\bar{x}(h) = \frac{\sum_i^n x_i \times z(h)_i}{\sum_i^n z(h)_i} \quad \text{if } z(h)_i > 0$$

For histone modification pairs, average values were weighted by the modification with the smaller Z-score.

$$\text{if } z(h_A)_i \text{ and } z(h_B)_i > 0$$

$$\bar{x}(h_{A,B}) = \frac{\sum_i^n x_i \times z(h_A)_i + \sum_i^n x_i \times z(h_B)_i}{\sum_i^n z(h_A)_i + \sum_i^n z(h_B)_i}$$

$$z(h_A)_i = 0 \quad \text{if } z(h_A)_i > z(h_B)_i$$

$$z(h_B)_i = 0 \quad \text{if } z(h_B)_i > z(h_A)_i$$

Data analysis for Fig. 6
Using MNase-seq and NOMe-seq data from IMR90 cells, the same procedure described for Fig. 3 was carried out for nucleosomes positioned in CpG islands. CpG island nucleosomes were divided into two groups, non-TSS and TSS, based on whether or not their midpoints occurred within 500 bp of a TSS. These information were used to generate the plots in panels a, b, and c. For panel d, if a CpG island overlapped a RefSeq TSS, the CpG island was included. For panel e, if a CpG island

center was annotated as an exon or intron (gene body) or as intergenic by HOMER and if the CpG island was not used in panel d, the CpG island was included. Subsequently, NOMe-seq data in bigwig format were aligned to CpG island TSSs and CpG island gene body and intergenic centers using an unpublished Perl script written by Yaping Liu, and in-house scripts were used to express the occupancy and methylation levels in 101 21-base pair bins. Subsequently, the CpG islands were sorted by decreasing average occupancy across the 101 bins. Using Roadmap ChIP-seq data from IMR90 cells in bed format and another Perl script written by Yaping Liu, RPM values minus input were computed across the genome for each dataset, and these values were transformed into Z-scores. These Z-scores were then aligned to the TSSs and CpG islands centers sorted by decreasing occupancy and binned in the same way as above. Heatmaps were generated using Java Tree View [88].

Data analysis for Fig. 7 and Additional file 1: Figure S18

BS-seq data from HCT116 were used instead of NOMe-seq data to evaluate the DNA methylation levels at conserved TFBSs so that more sites could be analyzed. All hg19 conserved TFBSs from the UCSC genome browser with methylation data for a cytosine in at least one CpG were divided into unmethylated and methylated sites depending on whether or not the average mCpG/CpG fraction was greater than or equal to 0.5. These conserved TFBSs do not include CTCF conserved sites. Alignment of the NOMe-seq occupancy data to conserved TFBSs was executed in the same way as described for Fig. 6. For Fig. 7, ENCODE ChIP-seq and DNase-seq data in bam format were aligned to conserved TFBSs using ngs.plot [89], but RPM levels were not transformed into Z-scores. For Additional file 1: Figure S18, previously generated ChIP-seq Z-scores in bigwig format were aligned to the conserved TFBSs, and these values were expressed in 21-bp bins. All heatmaps were generated using Java Tree View [88].

Supplementary methods

More detailed bioinformatics procedures and in-house scripts used in this study are provided in Additional file 2.

Authors' contributions
CKC and JNA conceived the ideas and analyses conducted for this study. CKC and JNA wrote the manuscript. CKC wrote in-house scripts and generated figures. Both authors read and approved the final manuscript.

Author details
[1] Department of Biochemistry and Molecular Genetics, Northwestern University Feinberg School of Medicine, 320 E. Superior Street, Chicago, IL 60611, USA. [2] Department of Biological Sciences, Purdue University, 915 W. State Street, West Lafayette, IN 47907, USA.

Acknowledgements
We would like to thank the Purdue Bioinformatics Core for the use of their computational resources. We would also like to thank Yaping Lui for sharing his bioinformatics tools, which were used in some analyses.

Competing interests
The authors declare that they have no competing interests.

Funding
This work was funded by the Department of Biological Sciences at Purdue University.

References
1. Kornberg RD. Chromatin structure: a repeating unit of histones and DNA. Science. 1974;184:868–71.
2. Luger K, Mader AW, Richmond RK, Sargent DF, Richmond TJ. Crystal structure of the nucleosome core particle at 2.8 A resolution. Nature. 1997;389:251–60.
3. Li B, Carey M, Workman JL. The role of chromatin during transcription. Cell. 2007;128:707–19.
4. Radman-Livaja M, Rando OJ. Nucleosome positioning: how is it established, and why does it matter? Dev Biol. 2010;339:258–66.
5. Owen-Hughes T, Gkikopoulos T. Making sense of transcribing chromatin. Curr Opin Cell Biol. 2012;24:296–304.
6. Jones PA, Liang G. Rethinking how DNA methylation patterns are maintained. Nat Rev Genet. 2009;10:805–11.
7. Lewis J, Bird A. DNA methylation and chromatin structure. FEBS Lett. 1991;285:155–9.
8. Cedar H, Bergman Y. Programming of DNA methylation patterns. Annu Rev Biochem. 2012;81:97–117.
9. Bird A. DNA methylation patterns and epigenetic memory. Genes Dev. 2002;16:6–21.
10. Deaton AM, Bird A. CpG islands and the regulation of transcription. Genes Dev. 2011;25:1010–22.
11. Brandeis M, Frank D, Keshet I, Siegfried Z, Mendelsohn M, Nemes A, Temper V, Razin A, Cedar H. Sp1 elements protect a CpG island from de novo methylation. Nature. 1994;371:435–8.
12. Tsukada Y, Fang J, Erdjument-Bromage H, Warren ME, Borchers CH, Tempst P, Zhang Y. Histone demethylation by a family of JmjC domain-containing proteins. Nature. 2006;439:811–6.
13. Lee JH, Skalnik DG. CpG-binding protein (CXXC finger protein 1) is a component of the mammalian Set1 histone H3-Lys4 methyltransferase complex, the analogue of the yeast Set1/COMPASS complex. J Biol Chem. 2005;280:41725–31.
14. Kolasinska-Zwierz P, Down T, Latorre I, Liu T, Liu XS, Ahringer J. Differential chromatin marking of introns and expressed exons by H3K36me3. Nat Genet. 2009;41:376–81.
15. Kooistra SM, Helin K. Molecular mechanisms and potential functions of histone demethylases. Nat Rev Mol Cell Biol. 2012;13:297–311.
16. Ooi SK, Qiu C, Bernstein E, Li K, Jia D, Yang Z, Erdjument-Bromage H, Tempst P, Lin SP, Allis CD, et al. DNMT3L connects unmethylated lysine 4 of histone H3 to de novo methylation of DNA. Nature. 2007;448:714–7.
17. Hu JL, Zhou BO, Zhang RR, Zhang KL, Zhou JQ, Xu GL. The N-terminus of histone H3 is required for de novo DNA methylation in chromatin. Proc Natl Acad Sci USA. 2009;106:22187–92.
18. Zhang Y, Jurkowska R, Soeroes S, Rajavelu A, Dhayalan A, Bock I, Rathert P, Brandt O, Reinhardt R, Fischle W, Jeltsch A. Chromatin methylation activity of Dnmt3a and Dnmt3a/3L is guided by interaction of the ADD domain with the histone H3 tail. Nucleic Acids Res. 2010;38:4246–53.
19. Dhayalan A, Rajavelu A, Rathert P, Tamas R, Jurkowska RZ, Ragozin S, Jeltsch A. The Dnmt3a PWWP domain reads histone 3 lysine 36 trimethylation and guides DNA methylation. J Biol Chem. 2010;285:26114–20.
20. Baubec T, Colombo DF, Wirbelauer C, Schmidt J, Burger L, Krebs AR, Akalin A, Schubeler D. Genomic profiling of DNA methyltransferases reveals a role for DNMT3B in genic methylation. Nature. 2015;520:243–7.
21. Collings CK, Waddell PJ, Anderson JN. Effects of DNA methylation on nucleosome stability. Nucleic Acids Res. 2013;41:2918–31.

22. Chodavarapu RK, Feng S, Bernatavichute YV, Chen PY, Stroud H, Yu Y, Hetzel JA, Kuo F, Kim J, Cokus SJ, et al. Relationship between nucleosome positioning and DNA methylation. Nature. 2010;466:388–92.

23. Choi JK. Contrasting chromatin organization of CpG islands and exons in the human genome. Genome Biol. 2010;11:R70.

24. Gelfman S, Cohen N, Yearim A, Ast G. DNA-methylation effect on cotranscriptional splicing is dependent on GC architecture of the exon-intron structure. Genome Res. 2013;23:789–99.

25. Jones PA. Functions of DNA methylation: islands, start sites, gene bodies and beyond. Nat Rev Genet. 2012;13:484–92.

26. Baylin SB, Jones PA. A decade of exploring the cancer epigenome—biological and translational implications. Nat Rev Cancer. 2011;11:726–34.

27. Fu Y, Sinha M, Peterson CL, Weng Z. The insulator binding protein CTCF positions 20 nucleosomes around its binding sites across the human genome. PLoS Genet. 2008;4:e1000138.

28. Dixon JR, Selvaraj S, Yue F, Kim A, Li Y, Shen Y, Hu M, Liu JS, Ren B. Topological domains in mammalian genomes identified by analysis of chromatin interactions. Nature. 2012;485:376–80.

29. Kelly TK, Liu Y, Lay FD, Liang G, Berman BP, Jones PA. Genome-wide mapping of nucleosome positioning and DNA methylation within individual DNA molecules. Genome Res. 2012;22:2497–506.

30. Berman BP, Liu Y, Kelly TK. DNA methylation marks inter-nucleosome linker regions throughout the human genome. PeerJ Preprints. 2013;1:e27v3. doi:10.7287/peerj.preprints.27v3.

31. Teif VB, Beshnova DA, Vainshtein Y, Marth C, Mallm JP, Hofer T, Rippe K. Nucleosome repositioning links DNA (de)methylation and differential CTCF binding during stem cell development. Genome Res. 2014;24:1285–95.

32. Varley KE, Gertz J, Bowling KM, Parker SL, Reddy TE, Pauli-Behn F, Cross MK, Williams BA, Stamatoyannopoulos JA, Crawford GE, et al. Dynamic DNA methylation across diverse human cell lines and tissues. Genome Res. 2013;23:555–67.

33. Valouev A, Johnson SM, Boyd SD, Smith CL, Fire AZ, Sidow A. Determinants of nucleosome organization in primary human cells. Nature. 2011;474:516–20.

34. Trifonov EN, Sussman JL. The pitch of chromatin DNA is reflected in its nucleotide sequence. Proc Natl Acad Sci USA. 1980;77:3816–20.

35. Satchwell SC, Drew HR, Travers AA. Sequence periodicities in chicken nucleosome core DNA. J Mol Biol. 1986;191:659–75.

36. Segal E, Fondufe-Mittendorf Y, Chen L, Thastrom A, Field Y, Moore IK, Wang JP, Widom J. A genomic code for nucleosome positioning. Nature. 2006;442:772–8.

37. Kaplan N, Moore IK, Fondufe-Mittendorf Y, Gossett AJ, Tillo D, Field Y, LeProust EM, Hughes TR, Lieb JD, Widom J, Segal E. The DNA-encoded nucleosome organization of a eukaryotic genome. Nature. 2009;458:362–6.

38. Valouev A, Ichikawa J, Tonthat T, Stuart J, Ranade S, Peckham H, Zeng K, Malek JA, Costa G, McKernan K, et al. A high-resolution, nucleosome position map of C. elegans reveals a lack of universal sequence-dictated positioning. Genome Res. 2008;18:1051–63.

39. Collings CK, Fernandez AG, Pitschka CG, Hawkins TB, Anderson JN. Oligonucleotide sequence motifs as nucleosome positioning signals. PLoS ONE. 2010;5:e10933.

40. Gaidatzis D, Burger L, Murr R, Lerch A, Dessus-Babus S, Schubeler D, Stadler MB. DNA sequence explains seemingly disordered methylation levels in partially methylated domains of Mammalian genomes. PLoS Genet. 2014;10:e1004143.

41. Tillo D, Kaplan N, Moore IK, Fondufe-Mittendorf Y, Gossett AJ, Field Y, Lieb JD, Widom J, Segal E, Hughes TR. High nucleosome occupancy is encoded at human regulatory sequences. PLoS ONE. 2010;5:e9129.

42. Tillo D, Hughes TR. G + C content dominates intrinsic nucleosome occupancy. BMC Bioinformatics. 2009;10:442.

43. Heinz S, Benner C, Spann N, Bertolino E, Lin YC, Laslo P, Cheng JX, Murre C, Singh H, Glass CK. Simple combinations of lineage-determining transcription factors prime cis-regulatory elements required for macrophage and B cell identities. Mol Cell. 2010;38:576–89.

44. Salih F, Salih B, Kogan S, Trifonov EN. Epigenetic nucleosomes. Alu sequences and CG as nucleosome positioning element. J Biomol Struct Dyn. 2008;26:9–16.

45. Bintu L, Yong J, Antebi YE, McCue K, Kazuki Y, Uno N, Oshimura M, Elowitz MB. Dynamics of epigenetic regulation at the single-cell level. Science.

2016;351:720–4.

46. Roadmap Epigenomics C, Kundaje A, Meuleman W, Ernst J, Bilenky M, Yen A, Heravi-Moussavi A, Kheradpour P, Zhang Z, Wang J, et al. Integrative analysis of 111 reference human epigenomes. Nature. 2015;518:317–30.

47. Elgin SC, Grewal SI. Heterochromatin: silence is golden. Curr Biol. 2003;13:R895–8.

48. Wiles ET, Selker EU. H3K27 methylation: a promiscuous repressive chromatin mark. Curr Opin Genet Dev. 2016;43:31–7.

49. Brinkman AB, Gu H, Bartels SJ, Zhang Y, Matarese F, Simmer F, Marks H, Bock C, Gnirke A, Meissner A, Stunnenberg HG. Sequential ChIP-bisulfite sequencing enables direct genome-scale investigation of chromatin and DNA methylation cross-talk. Genome Res. 2012;22:1128–38.

50. Lister R, Pelizzola M, Dowen RH, Hawkins RD, Hon G, Tonti-Filippini J, Nery JR, Lee L, Ye Z, Ngo QM, et al. Human DNA methylomes at base resolution show widespread epigenomic differences. Nature. 2009;462:315–22.

51. Blattler A, Farnham PJ. Cross-talk between site-specific transcription factors and DNA methylation states. J Biol Chem. 2013;288:34287–94.

52. Zhu H, Wang G, Qian J. Transcription factors as readers and effectors of DNA methylation. Nat Rev Genet. 2016;17:551–65.

53. Lay FD, Liu Y, Kelly TK, Witt H, Farnham PJ, Jones PA, Berman BP. The role of DNA methylation in directing the functional organization of the cancer epigenome. Genome Res. 2015;25:467–77.

54. Blattler A, Yao L, Witt H, Guo Y, Nicolet CM, Berman BP, Farnham PJ. Global loss of DNA methylation uncovers intronic enhancers in genes showing expression changes. Genome Biol. 2014;15:469.

55. Consortium EP. A user's guide to the encyclopedia of DNA elements (ENCODE). PLoS Biol. 2011;9:e1001046.

56. Zentner GE, Tesar PJ, Scacheri PC. Epigenetic signatures distinguish multiple classes of enhancers with distinct cellular functions. Genome Res. 2011;21:1273–83.

57. Zhu Y, Sun L, Chen Z, Whitaker JW, Wang T, Wang W. Predicting enhancer transcription and activity from chromatin modifications. Nucleic Acids Res. 2013;41:10032–43.

58. Siggens L, Ekwall K. Epigenetics, chromatin and genome organization: recent advances from the ENCODE project. J Intern Med. 2014;276:201–14.

59. Pott, S. Simultaneous measurement of chromatin accessibility, DNA methylation, and nucleosome phasing in single cells. 2016. doi:10.1101/061739.

60. Zentner GE, Henikoff S. High-resolution digital profiling of the epigenome. Nat Rev Genet. 2014;15:814–27.

61. Byvoet P, Shepherd GR, Hardin JM, Noland BJ. The distribution and turnover of labeled methyl groups in histone fractions of cultured mammalian cells. Arch Biochem Biophys. 1972;148:558–67.

62. Honda BM, Candido PM, Dixon GH. Histone methylation. Its occurrence in different cell types and relation to histone H4 metabolism in developing trout testis. J Biol Chem. 1975;250:8686–9.

63. Kim J, Kim H. Recruitment and biological consequences of histone modification of H3K27me3 and H3K9me3. ILAR J. 2012;53:232–9.

64. Cheung P, Lau P. Epigenetic regulation by histone methylation and histone variants. Mol Endocrinol. 2005;19:563–73.

65. Boyes J, Bird A. Repression of genes by DNA methylation depends on CpG density and promoter strength: evidence for involvement of a methyl-CpG binding protein. EMBO J. 1992;11:327–33.

66. Du Q, Luu PL, Stirzaker C, Clark SJ. Methyl-CpG-binding domain proteins: readers of the epigenome. Epigenomics. 2015;7:1051–73.

67. Lee JY, Lee J, Yue H, Lee TH. Dynamics of nucleosome assembly and effects of DNA methylation. J Biol Chem. 2015;290:4291–303.

68. Kaur P, Plochberger B, Costa P, Cope SM, Vaiana SM, Lindsay S. Hydrophobicity of methylated DNA as a possible mechanism for gene silencing. Phys Biol. 2012;9:065001.

69. Lazarovici A, Zhou T, Shafer A, Dantas Machado AC, Riley TR, Sandstrom R, Sabo PJ, Lu Y, Rohs R, Stamatoyannopoulos JA, Bussemaker HJ. Probing DNA shape and methylation state on a genomic scale with DNase I. Proc Natl Acad Sci USA. 2013;110:6376–81.

70. Razin A, Cedar H. Distribution of 5-methylcytosine in chromatin. Proc Natl Acad Sci USA. 1977;74:2725–8.

71. Solage A, Cedar H. Organization of 5-methylcytosine in chromosomal DNA. Biochemistry. 1978;17:2934–8.

72. Felle M, Hoffmeister H, Rothammer J, Fuchs A, Exler JH, Langst G. Nucleosomes protect DNA from DNA methylation in vivo and in vitro. Nucleic Acids Res. 2011;39:6956–69.

73. Jia D, Jurkowska RZ, Zhang X, Jeltsch A, Cheng X. Structure of Dnmt3a bound to Dnmt3L suggests a model for de novo DNA methylation. Nature. 2007;449:248–51.

74. Wang Z, Willard HF. Evidence for sequence biases associated with patterns of histone methylation. BMC Genomics. 2012;13:367.

75. Quante T, Bird A. Do short, frequent DNA sequence motifs mould the epigenome? Nat Rev Mol Cell Biol. 2016;17:257–62.

76. Brown JW, Anderson JA. The binding of the chromosomal protein HMG-2a to DNA regions of reduced stabilities. J Biol Chem. 1986;261:1349–54.

77. Tate PH, Bird AP. Effects of DNA methylation on DNA-binding proteins and gene expression. Curr Opin Genet Dev. 1993;3:226–31.

78. Perini G, Diolaiti D, Porro A, Della Valle G. In vivo transcriptional regulation of N-Myc target genes is controlled by E-box methylation. Proc Natl Acad Sci USA. 2005;102:12117–22.

79. Medvedeva YA, Khamis AM, Kulakovskiy IV, Ba-Alawi W, Bhuyan MS, Kawaji H, Lassmann T, Harbers M, Forrest AR, Bajic VB. The FANTOM consortium. Effects of cytosine methylation on transcription factor binding sites. BMC Genomics. 2014;15:119.

80. Brandeis M, Frank D, Keshet I, Siegfried Z, Mendelsohn M, Nemes A, Temper V, Razin A, Cedar H. Sp1 elements protect a CpG island from de novo methylation. Nature. 1994;371(6496):435–8.

81. Mummaneni P, Yates P, Simpson J, Rose J, Turker MS. The primary function of a redundant Sp1 binding site in the mouse aprt gene promoter is to block epigenetic gene inactivation. Nucleic Acids Res. 1998;26:5163–9.

82. Feinberg AP, Vogelstein B. A technique for radiolabeling DNA restriction endonuclease fragments to high specific activity. Anal Biochem. 1983;132:6–13.

83. Belancio VP, Roy-Engel AM, Deininger PL. All y'all need to know 'bout retroelements in cancer. Semin Cancer Biol. 2010;20:200–10.

84. Gronbaek K, Hother C, Jones PA. Epigenetic changes in cancer. APMIS. 2007;115:1039–59.

85. Karolchik D, Hinrichs AS, Furey TS, Roskin KM, Sugnet CW, Haussler D, Kent WJ. The UCSC Table Browser data retrieval tool. Nucleic Acids Res. 2004;32:D493–6.

86. Ziebarth JD, Bhattacharya A, Cui Y. CTCFBSDB 2.0: a database for CTCF-binding sites and genome organization. Nucleic Acids Res. 2013;41:D188–94.

87. Xu S, Grullon S, Ge K, Peng W. Spatial clustering for identification of ChIP-enriched regions (SICER) to map regions of histone methylation patterns in embryonic stem cells. Methods Mol Biol. 2014;1150:97–111.

88. Saldanha AJ. Java Treeview—extensible visualization of microarray data. Bioinformatics. 2004;20:3246–8.

89. Shen L, Shao N, Liu X, Nestler E. ngs.plot: quick mining and visualization of next-generation sequencing data by integrating genomic databases. BMC Genomics. 2014;15:284.

Identification of epigenetic signature associated with alpha thalassemia/mental retardation X-linked syndrome

Laila C. Schenkel[1], Kristin D. Kernohan[2], Arran McBride[2], Ditta Reina[8,9], Amanda Hodge[8,9], Peter J. Ainsworth[1,3,4,5,6,7], David I. Rodenhiser[4,5,6,7], Guillaume Pare[8,9], Nathalie G. Bérubé[4,5,6,7], Cindy Skinner[10], Kym M. Boycott[2], Charles Schwartz[10] and Bekim Sadikovic[1,3,7*]

Abstract

Background: Alpha thalassemia/mental retardation X-linked syndrome (ATR-X) is caused by a mutation at the chromatin regulator gene *ATRX*. The mechanisms involved in the ATR-X pathology are not completely understood, but may involve epigenetic modifications. ATRX has been linked to the regulation of histone H3 and DNA methylation, while mutations in the *ATRX* gene may lead to the downstream epigenetic and transcriptional effects. Elucidating the underlying epigenetic mechanisms altered in ATR-X will provide a better understanding about the pathobiology of this disease, as well as provide novel diagnostic biomarkers.

Results: We performed genome-wide DNA methylation assessment of the peripheral blood samples from 18 patients with ATR-X and compared it to 210 controls. We demonstrated the evidence of a unique and highly specific DNA methylation "epi-signature" in the peripheral blood of ATRX patients, which was corroborated by targeted bisulfite sequencing experiments. Although genomically represented, differentially methylated regions showed evidence of preferential clustering in pericentromeric and telometric chromosomal regions, areas where ATRX has multiple functions related to maintenance of heterochromatin and genomic integrity.

Conclusion: Most significant methylation changes in the 14 genomic loci provide a unique epigenetic signature for this syndrome that may be used as a highly sensitive and specific diagnostic biomarker to support the diagnosis of ATR-X, particularly in patients with phenotypic complexity and in patients with *ATRX* gene sequence variants of unknown significance.

Keywords: ATRX, DNA methylation, Epi-signature, Intellectual disability, Biomarker

Background

An emerging development in the field of medical genetics has been the identification of Mendelian disorders involving genes encoding the writers, erasers, readers and remodelers of the epigenetic machinery [1]. Building on several decades of evidence regarding the functions of covalent DNA methylation [2, 3] and histone modifications [4] in regulating gene transcription, it is evident that mutations in the proteins responsible for creating, interpreting or removing the broad arrays of epigenetic marks can be linked to genetic conditions including cancer [5], imprinting disorders and/or single-gene disorders [6].

Along with these discoveries came the opportunity, not only for the elucidation of underlying molecular mechanisms altered in these disorders, but also for the identification of epi-signatures that can be diagnostically useful, specifically where patients express a subset of clinical manifestations associated with a phenotypic spectrum shared across more than one syndrome, making a specific clinical diagnosis difficult to reach.

*Correspondence: Bekim.Sadikovic@lhsc.on.ca
3 Molecular Genetics Laboratory, Victoria Hospital, London Health Sciences Center, 800 Commissioner's Road E, B10-104, London, ON N6A 5W9, Canada
Full list of author information is available at the end of the article

Among the rapidly expanding number of proteins responsible for chromatin maintenance and remodeling related to transcription is alpha thalassemia/mental retardation X-linked (ATRX; NG_008838.2). Mutations in the *ATRX* gene cause alpha thalassemia/mental retardation X-linked syndrome (ATR-X, OMIM 301040), a disorder characterized by moderate to severe intellectual disability, expressive language disorder, characteristic facial gestalt during infancy, often associated with hematological signs of alpha thalassemia [7].

The ATRX protein functions as an agent of ATP-dependent chromatin remodeling and is a member of the SWI/SNF superfamily of proteins. The latter can have a wide variety of cellular functions, as described in detail in several recent reviews [8–10]. Briefly, ATRX protein is involved in cellular processes such as meiosis, mitosis, DNA repair and regulation of transcription through an effect on chromatin [11–15]. Disruption of these activities may contribute to developmental abnormalities associated with the ATR-X syndrome.

Within the ATRX protein, a histone-binding ATRX–DNMT3–DNMT3L (ADD) domain can sense the methylation modifications of both H3K4 and H3K9 [16], essentially acting as an interpreter of these histone states. ATRX is also known to associate with the transcription cofactor DAXX. ATRX–DAXX complex is responsible for deposition of histone H3.3 at the telomeric and pericentromeric heterochromatic regions within chromosomes [17]. Loss of ATRX in ES cells leads to the loss of histone H3.3 at imprinting control regions and telomeric regions, along with the concurrent loss of H3K9me3 [18, 19]. ATRX has also been linked to DNA methylation regulation, as mutations at the *ATRX* gene result in DNA methylation changes at subtelomeric and repetitive regions [20].

The role of ATRX as a regulator of heterochromatin dynamics raises the possibility that mutations in *ATRX* may lead to downstream transcriptional effects across the complex of genes or repetitive regions involved in the global context of the disorder, in addition to explaining phenotypical differences in these patients. For example, *ATRX* mutations affect the expression of α-globin gene cluster, causing α-thalassemia [21]. Mechanistically, α-globin cluster, among other genes, has G-rich tandem repeats (TRs) sites, which have been reported to bind ATRX resulting in H3.3 deposition and gene expression regulation. In addition, it was suggested that the differences in size of these TRs among ATR-X patients contribute to the ranges in severity of the syndrome [22].

The orchestrated regulation of epigenetic mechanisms, including associations between ATRX and DNA methylation [11, 12], is essential for tissue homeostasis, cell identity and proper human development. Here, we describe the findings of a genome-wide DNA methylation array (GWMA) performed on peripheral blood samples from patients with ATR-X and show the genome-wide changes in DNA methylation that occur in patients with this epigenetic syndrome. We have identified a specific epi-signature of differentially hypo- and hypermethylated genes in patients clinically diagnosed with ATR-X syndrome. Our study shows the preponderance of differentially methylated genes within, or adjacent to, pericentromeric or telomeric chromosomal regions, suggesting a major role of heterochromatin in the pathophysiology of ATR-X, linked to the disruption of ATRX function in the context of its role as a regulator of heterochromatin dynamics.

Results

The epi-signature identified in blood samples from ATR-X patients

The genome-wide DNA methylation array of 20 blood samples obtained from ATR-X patients was compared with a reference cohort (controls). Various methylation changes at a single-probe level were identified across the genome, consisting of both hypo- and hypermethylation (estimate value > ±0.15) (Fig. 1a). Hierarchical clustering of significant probes ($p < 0.01$) clearly demonstrated a unique methylation profile and subclustering for these patients compared with our large laboratory reference cohort (Fig. 1c). The global methylation analysis revealed an increase in methylation at low methylation value regions (0.1–0.2 methylation value) in patients relative to controls (Fig. 1b). This pattern suggests that increased methylation is taking place in normally unmethylated regions, majority of which are normally located in promoters and CpG islands.

Statistical filtering to identify regions with most robust methylation changes, using multiple parameters including p value <0.01, F value >50, number of consecutive probes >4 and methylation difference > ±20%, revealed 16 genetic regions with significant statistical difference between ATR-X and controls (Table 1). Of these, 13 regions showed hypermethylation (methylation difference higher than +0.2) and three regions showed hypomethylation (methylation difference lower than −0.2). These regions were distributed across the genome both outside ($n = 5$) and within CpG islands ($n = 11$), including seven regions at gene promoter CpG islands and two intragenic CpG islands (Table 1). This epi-signature was specific to ATR-X patients and did not correlate with the type of mutation at the *ATRX* gene locus.

We then performed a single-patient analysis to identify possible patient-specific, as opposed to patient cohort-specific, recurring methylation changes, using statistical parameters of $p < 0.01$, methylation difference > ±15%,

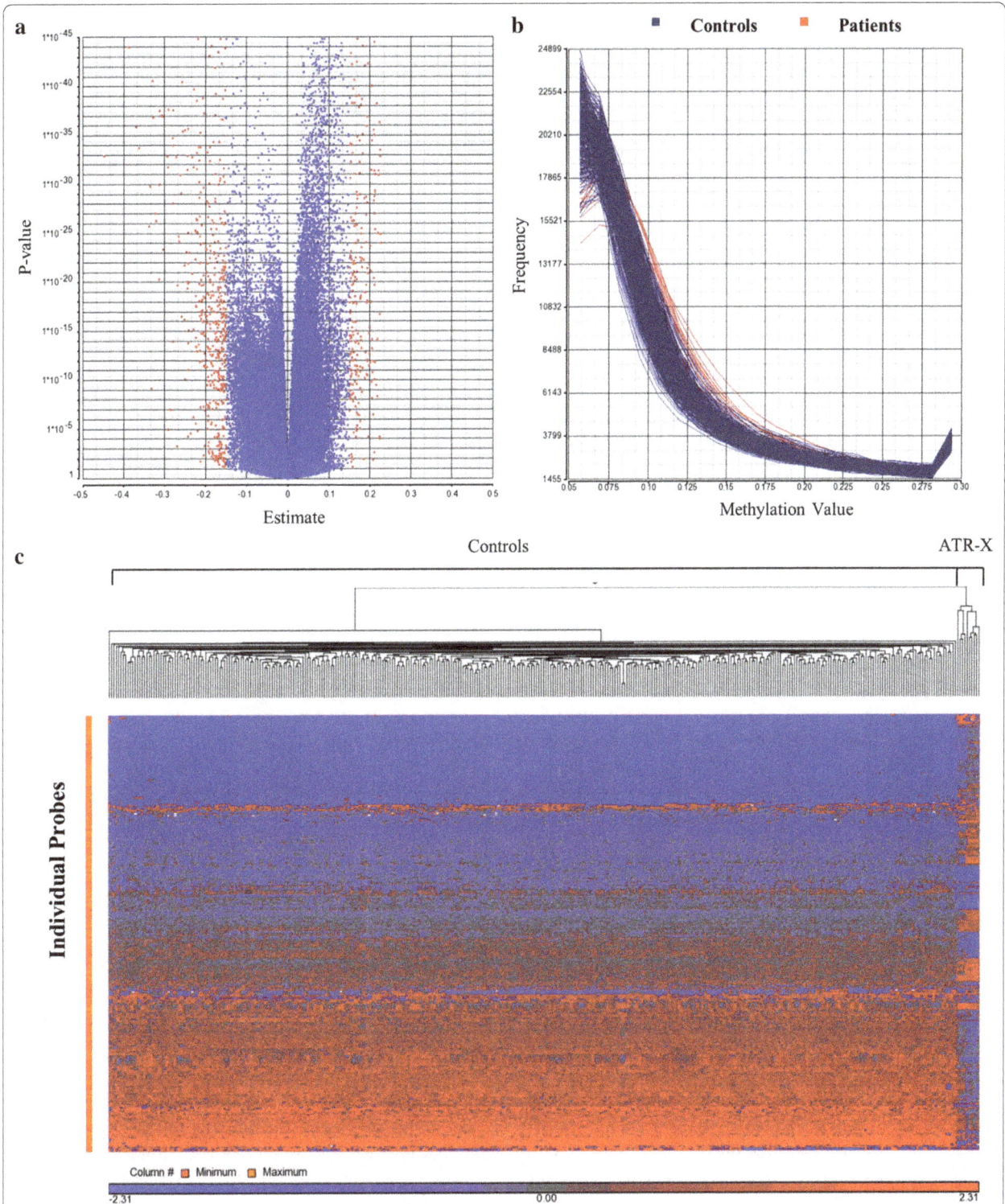

Fig. 1 **a** Volcano plot of methylation difference (estimate) between ATR-X and controls versus statistical significance (log *p* value) of individual probes. In *red* are highlighted probes with estimate value higher than 0.15. Positive estimate value = hypermethylation; negative estimate value = hypomethylation. **b** Histogram of all samples showing methylation value (*X-axis*) versus frequency of methylation levels across the genome (*Y-axis*) in ATR-X patients (*red*) and controls (*blue*). Patients with ATR-X showed a higher frequency of methylation at low methylation value regions (0.1–0.2 methylation value). Low methylation value regions are characteristic of promoter CpG islands. **c** Hierarchical clustering of probes differentially methylated (*p* < 0.01) between ATR-X and controls demonstrating marked asymmetry of the two groups. Cases are represented in the columns and probes in the rows

Table 1 ATR-X methylation signature: significant regions detected by methylation array in ATR-X patients ($n = 17$) compared with controls ($n = 210$)

Chr	Region start	Region stop	# Probes	Nearest gene	Distance to gene (bp)	Within CpG island	Gene promoter/intragenic
chr1	228291476	228291715	5	C1orf35	453	No	–
chr3	107810351	107810801	6	CD47	415	Yes	–
chr3	109056339	109056907	4	DPPA4	0	No	Promoter
chr4	124222	125122	5	ZNF718	0	Yes	Intragenic/promoter
chr5	23506728	23507762	13	PRDM9	0	No	promoter
chr5	134073420	134073589	5	CAMLG	581	No	–
chr5	150284292	150284806	9	ZNF300	0	Yes	Promoter
chr5	150325572	150326872	13	ZNF300P1	0	Yes	Promoter
chr6	168045258	168045898	5	LOC401286	21624	No	–
chr6	34498908	34499514	5	PACSIN1	0	Yes	Intragenic
chr8	43131250	43132517	9	POTEA	15068	Yes	–
chr10	38069509	38069955	4	ZNF248	0	Yes	Intragenic
chr16	10479552	10480299	6	ATF7IP2	0	Yes	Promoter
chr19	12305543	12306507	7	LOC100289333	0	Yes	Promoter
chr20	623079	623431	4	SRXN1	3837	Yes	–
chrY	21664286	21665031	4	BCORP1	0	Yes	Promoter

Significant regions: Human reference genome Hg19. Probes >4, estimate (net methylation difference) > ± 20%, F value >50, p value <0.01. Gene promoters were defined as any sequence immediately surrounding the annotated 5′ end of the gene

across four consecutive probes. First we observed that the cohort-specific epi-signature was absent in one of the patients (patient #12). A follow-up assessment showed that although this patient has previously been identified to carry a possible mutation in the *ATRX* gene, more recent data demonstrated that c.5579A>G; p.N1860S in the *ATRX* gene is indeed a benign polymorphism and hence this patent did not have the ATR-X syndrome. The remaining 17 patients with molecular diagnosis of ATR-X, using the above statistical cutoffs, showed an average of 9.8 significant loci from the epi-signature per individual, with the minimum of four significant loci observed in two patients (#13 and #15). To evaluate the specificity of this assay, we applied the ATR-X epi-signature in randomly selected 15 individuals, which included normal controls that were not part of the discovery cohort, as well as individuals with Fragile X syndrome, Prader–Willi syndrome, Angelman syndrome and Beckwith–Wiedemann syndrome. The majority of these individuals did not present any statistical significant changes at the ATRX epi-signature loci. Six control individuals showed significant changes at 1 or 2 ATRX epi-signature loci (*POTEA* and *PACSIN1*). These two genes had slightly higher level of variable DNA methylation and were as a result removed from the final ATRX epi-signature.

Biological interpretation

A more comprehensive gene list with methylation changes in the ATR-X group passing the following criteria, minimum of three consecutive probes with $p < 0.01$, F value across the region >50 and methylation difference > ±15% (i.e., at least a 15% methylation difference), showed an overrepresentation of genes involved in biosynthetic processes, nucleic acid metabolic processes and methylation process (Table 2). Many of the genes are involved in transcriptional regulation: *PRDM9* encodes a histone H3 lysine-4 trimethyltransferase [23]; CTDP1 functions in recruiting RNA polymerase to DNA promoters and is an OMIM gene for congenital cataracts, facial dysmorphism and neuropathy [24]; TFB2M regulates mtDNA transcription and maintenance [25]; also ZNF300, ZNF274 and ATF7IP2 are transcriptional regulators [26–28]. Another gene, *QKI*, regulates RNA splicing, export of target RNAs from the nucleus, translation of proteins and RNA stability [29]. In addition, three genes encode proteins associated with methylation process, including the histone H3 lysine-4 trimethyltransferase, a known target of ATRX and the betaine–homocysteine methyltransferase 2-BHMT2, which catalyzes the methylation of homocysteine. Evidences suggest that ATRX functions as a high-affinity RNA-binding protein and may regulate RNA stability of function [30]. These findings further support previous evidences, suggesting that ATRX has a role in regulating DNA and RNA metabolism and stability, as well as in the epigenetic regulation. Mutation at the *ATRX* gene may result in transcriptional deregulation of several genes across the genome and consequently

Table 2 Biological pathways identified by pathway analysis of the differentially methylated genes in ATR-X

Biological pathway	Biosynthetic process		Nucleic acid metabolic process		Methylation process
Number of genes in group	14		12		3
Fisher's exact enrichment score −ln(p value)	5.9		4.6		4.6
Fisher's exact right-tail p value	<0.01		0.01		0.01
Genes	DPPA4	PPAP2C	ZNF274	ATF7IP2	PRDM9
	ZNF718	ZNF486	CTDP1	ZNF486	BHMT2
	PRDM9	ZNF274	TFB2M	ZNF300	TFB2M
	BHMT2	TFB2M	QKI	DPPA4	
	ZNF300	ATF7IP2	ZNF718	ZNF248	
	QKI	CTDP1	UBE2U	PRDM9	
	ZNF248	ALG10B			

Pathway analysis was performed with differentially methylated regions using cutoff of estimate > ±0.15, $p < 0.01$, $F > 50$ and probes >3

neurodevelopmental problems associated with the disorder.

Technical validation of the methylation array

To technically confirm the methylation array findings, we performed bisulfite mutagenesis/sequencing for two regions identified by the array: CD47 and ZNF300P1 (Fig. 2; Additional file 1). The methylation array at these regions showed hypermethylation in ATR-X patients compared with controls (Fig. 2), with an average methylation value for the 6 CD47 probes of 0.55 and 0.8 in controls and patients (1.45 fold increase), respectively, and average methylation for the 12 probes at ZNF300P1 of 0.3 and 0.54 in controls and patients (1.8 fold increase), respectively. Using bisulfite mutagenesis/sequencing, we detected an overall increase in methylation at these regions in patients as compared to controls (Fig. 2). The bisulfite analysis showed average methylation at CD47 gene of 0.19 and 0.31 in controls and patients (1.63 fold increase), respectively, and at ZNF300P1 gene of 0.16 and 0.25 in controls and patients (1.56 fold increase), respectively. These findings confirm the accuracy and specificity of the DNA methylation array data.

Uneven distribution of altered methylation sites across the genome

ATRX has multiple functions related to maintenance of heterochromatin and genomic integrity that are essential during mammalian development. Hence, we hypothesized that genes in the ATRX epi-signature may be clustered in highly heterochromatinized pericentromeric and telomeric regions of the chromosomes. Using a statistical cutoff of $p < 0.01$, estimate > ±0.15, F value >50 and probes >3, the top 40 genes were localized using Karyogram view (Partek GS). We found that 17/40 (42.5%) of these genes in our signature mapped to telomeric and subtelomeric regions (defined as the last full cyto-band region on the chromosome) and 8/40 (20%) to pericentromeric regions (defined as the full cyto-band region on the centromere of the chromosome), with the remaining 15 (37.5%) scattered throughout the genome (Fig. 3). These findings suggest that ATRX dysfunction has a functional consequence to genes localized within heterochromatinized regions of the genome, with potentially inappropriate gene expression of these genes/epi-signature secondary to loss of ATRX as a regulator of chromatin integrity.

Discussion

The interplay between ATRX and DNA methylation has been evidenced by early studies in EBV-transformed cells from patients with ATR-X and controls, showing that mutations at the ATRX gene cause changes in the pattern of methylation at subtelomeric and rDNA sequences [20]. Furthermore, loss of ATRX expression has been linked to extensive epigenomic alteration including CPG island hypermethylation observed in astrocytic tumors [31, 32]. The involvement of ATRX in the regulation of DNA methylation was further supported by the discovery that ATRX interacts with MeCP2 and cohesion subunits in the brain [33, 34]. MeCP2 is a methyl-CpG-binding domain protein with affinity for GC-rich sequences and methylated DNA which in turn facilitates the recruitment of histone modifiers and chromatin remodeling complexes [35]. Similar to ATRX, MeCP2 is essential for neurodevelopment and mutations or duplications of the *MeCP2* gene cause Rett syndrome, a neurodevelopmental disorder [36]. In addition, cohesin proteins play a role in the regulation of chromosome organization and gene expression by binding to unmethylated CTCF-associated regions and mutations at cohesin genes are associated with the developmental defects seen in patients with Cornelia de Lange syndrome [37]. MeCP2 was shown to recruit the helicase domain of ATRX to heterochromatic

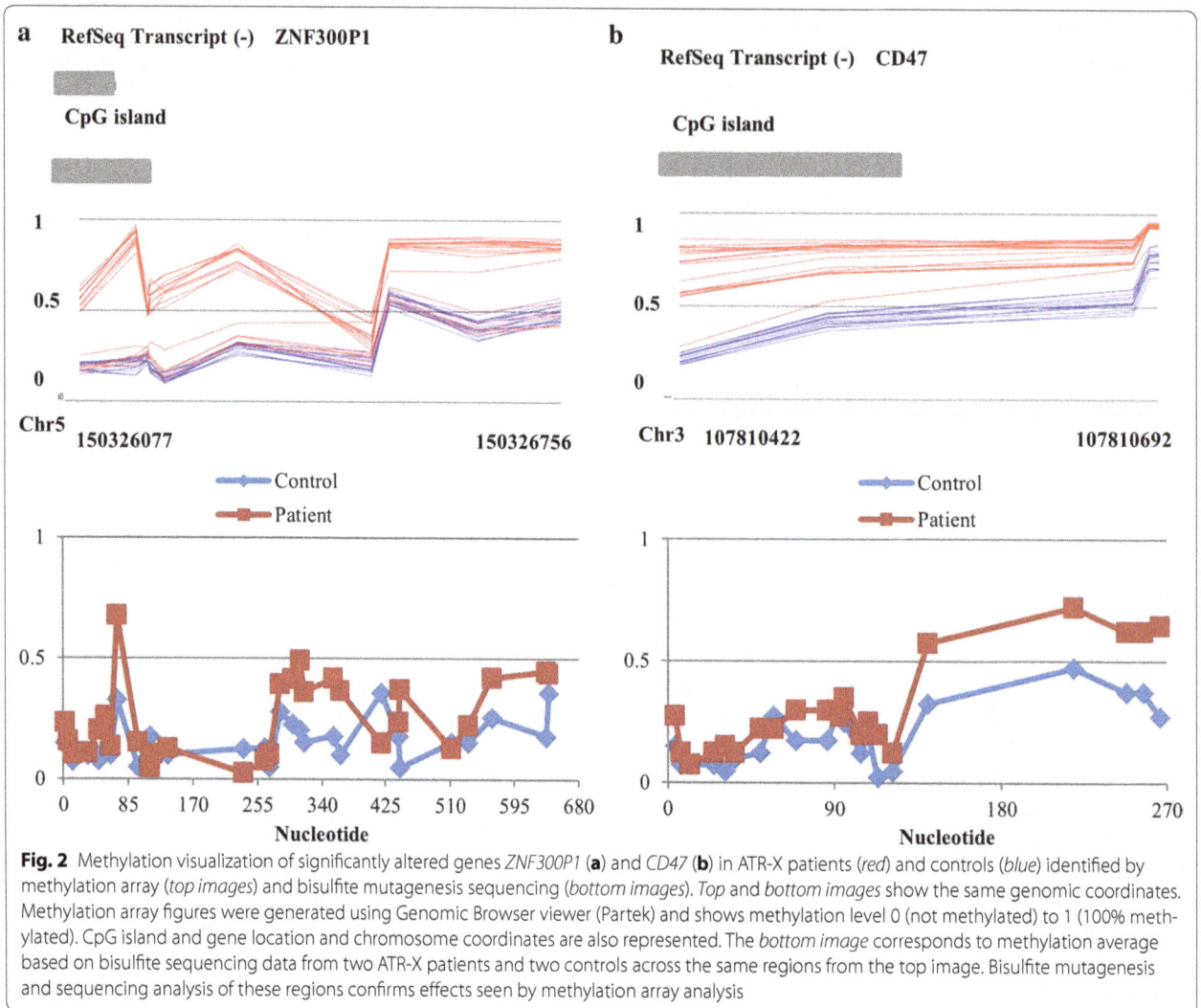

Fig. 2 Methylation visualization of significantly altered genes *ZNF300P1* (**a**) and *CD47* (**b**) in ATR-X patients (*red*) and controls (*blue*) identified by methylation array (*top images*) and bisulfite mutagenesis sequencing (*bottom images*). *Top* and *bottom images* show the same genomic coordinates. Methylation array figures were generated using Genomic Browser viewer (Partek) and shows methylation level 0 (not methylated) to 1 (100% methylated). CpG island and gene location and chromosome coordinates are also represented. The *bottom image* corresponds to methylation average based on bisulfite sequencing data from two ATR-X patients and two controls across the same regions from the top image. Bisulfite mutagenesis and sequencing analysis of these regions confirms effects seen by methylation array analysis

regions in a DNA methylation-dependent manner [33]. In addition, MeCP2 has been reported to interact with DNA methyltransferase 1 in order to perform maintenance methylation in vivo [38], as well as with histone H3 lysine 9 methyltransferase enzymes, to reinforce a repressive chromatin state by bridging DNA methylation and histone methylation [39].

Both ATRX protein and de novo DNA methyltransferases DNMT3A/B/L contain a histone-binding domain (ADD) that has been shown to play a role in the establishment and maintenance of DNA methylation patterns. The ADD domain interacts with specific methylation modifications of histone lysine 4 and 9 (H3K4 and H3K9). The H3K4 methylation is associated with gene transcription and promoters DNA hypomethylation, whereas methylation of H3K9 is a heterochromatin-associated mark associated with transcriptional repression and DNA

hypermethylation [16, 17]. H3K4 and H3K9 methylation is proposed to act as chromatin-based signals for regulation of DNA methylation, while ATRX–heterochromatin interaction depends on these histone methylation markers [16]. ATRX ADD domain binds to the methylated H3K9 (H3K9me3) in conjunction with unmodified H3K4 which are commonly seen in the repressed repeat elements. Therefore, mutations that functionally disrupt the ATRX protein and result in "mis-targeting" of ATRX–heterochromatin interaction may provide a mechanism for abnormal DNA methylation patterns in patients with the ATR-X syndrome. While ATRX does not contain a DNA methyltransferase domain, we and others [20] have clearly shown association between ATRX mutation and abnormal patterns of DNA methylation.

The mechanism for ATRX induction of DNA methylation aberrations is not well known. Evidence has shown

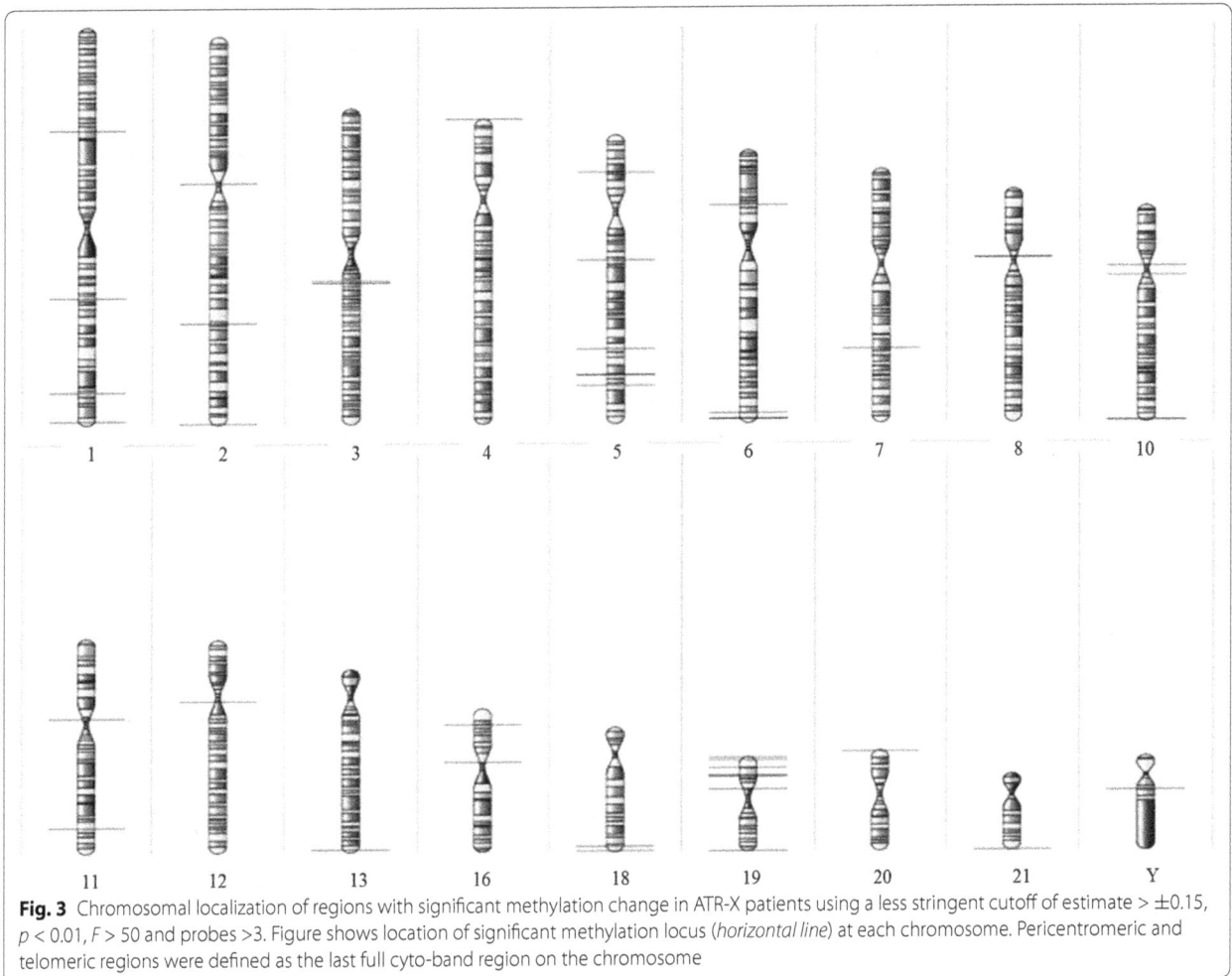

Fig. 3 Chromosomal localization of regions with significant methylation change in ATR-X patients using a less stringent cutoff of estimate > ±0.15, $p < 0.01$, $F > 50$ and probes >3. Figure shows location of significant methylation locus (*horizontal line*) at each chromosome. Pericentromeric and telomeric regions were defined as the last full cyto-band region on the chromosome

an overlapping function of ATRX and ATF7IP2. A genome-wide promoter DNA methylation study, using methylation-dependent immunoprecipitation–Chip assay, has demonstrated hypermethylation at ATF7IP2 gene in patients with ATRX mutation [40]. ATF7IP2, also known as MBD1-containing chromatin-associated factor 2, is known to bind to the transcription repression domain of the methylated cytosine-binding protein MBD1, as well as to interact with the H3K9 methyltransferase SETDB1 [28]. The overlapping protein interaction of ATRX–ATF7IP2 suggests that they form part of the same repressive chromatin complex, which involves ATF7IP2-induced H3K9me3 and ATRX binding to H3K9me3. There is also evidence for ATF7IP2 and ATRX transcriptional activation role, through SP1 and DAXX interaction, in promyelocytic leukemia nuclear bodies [28, 41, 42]. These data suggest that ATRX and ATF7IP2 have overlapping repressive/activating chromatin remodeling properties and potentially function in overlapping gene regulation pathways.

Most studies assessing the regulation of DNA methylation by ATRX have been focused on repetitive sequences and gene-specific methylation analysis. A recent study using methylation-sensitive restriction endonuclease has shown that *ATRX* mutations are associated with alterations in the DNA methylation profiles in highly repetitive sequences [20]. Another study using bisulfite mutagenesis analysis in mice model has demonstrated that specific gene activation at ancestral pseudoautosomal regions, which are repetitive sequences regulated by ATRX, does not involve gene-specific changes on DNA methylation, but relies on the ATRX-dependent H3.3 deposition mechanism [43]. However, none of these studies have analyzed global DNA methylation and/or specific gene CpG islands methylation in non-repetitive sequences. By using a high-resolution methylation array technique and a large reference cohort (controls), our study has clearly shown the existence of a pattern of DNA methylation changes, including in promoter CpG islands, telomeric and pericentromeric regions, in patients with ATR-X.

Accordingly, in our study, most of the DNA methylation changes observed in patients with an *ATRX* gene mutation were localized at telomeric and pericentromeric regions. How the epigenetic consequences of ATRX mutations actually result in the disease phenotype is not well understood. It is possible that the methylation alterations could result in differences in transcriptional regulation. For example, hypomethylation in a gene promoter CpG island may result in increased transcription, whereas hypermethylation may result in decreased transcriptional activity [6]. Gene pathway analysis showed that many of the genes identified in the ATR-X epi-signature are associated with DNA and RNA metabolic process, which may be involved in the regulation of specific gene expression and corroborate to the cardinal developmental processes disrupted in this rare disease; however, further research involving integrative analysis of gene expression and DNA methylation profiling to investigate the relationship between these DNA methylation changes and gene expression is warranted.

Here, we propose that the most significant and recurrent regions altered in the genomic DNA of patients with ATR-X, consisting of 14 loci, provide an epigenetic signature for this syndrome which may be used as a high sensitive and specific diagnostic biomarker to support the diagnosis of ATR-X, particularly in patients with phenotypical complexity and/or with *ATRX* gene sequence variants of unknown significance. Previous findings have demonstrated evidence of loss of DNA methylation in the repetitive elements [20]. While theoretically, it would be possible to use repetitive element methylation patterns as part of a unique ATRX mutation epi-signature, routine analysis of the repetitive elements DNA methylation pattern can be challenging due to the lack of specificity for assays designed for assessment of methylation of genomic repeats. Furthermore, most array or sequencing-based bisulfate protocols are limited to targeting unique genomic sequences. Therefore, identification of a robust unique epigenetic signature across a large number of unique genetic sequences that we describe in this manuscript presents an opportunity for utilization of these findings in routine clinical diagnostics.

In addition to the ATR-X epi-signature described here, our group has recently demonstrated unique DNA methylation signatures in patients with two other conditions, including Floating–Harbor syndrome, which is caused by mutation in the *SRCAP* gene, as well as cerebellar ataxia, deafness and narcolepsy syndrome, which is caused by mutations in the *DNMT1* gene [44, 45]. Other groups have also identified epi-signatures in patients with Sotos syndrome [46], and the X-linked intellectual disability

caused by the *KDM5C* gene [47]. Taken together, these studies demonstrate the ability of genome-wide methylation array to accurately diagnose multiple epigenetic disorders. Utilization of this technology in routine clinical practice will enable the discovery of new epigenetic biomarkers and will serve to enhance our understanding of human disease etiology. However, the identification of epigenetic variants of unknown clinical significance (E-VUS) will require delivery of testing to be performed in regulated clinical laboratories along with an adequate control cohort of normal samples, together with the development and implementation of clinical and laboratory testing guidelines, and availability and integration with pre- and posttest genetic counseling.

Conclusion

In conclusion, the observation of genome-wide epigenetic defects in ATR-X patients expands our understanding of the pathology of this disease, in which specific DNA methylation changes could lead directly to an aberrant expression of a number of genes in *ATRX*-deficient patients, particularly, but not restricted to telomeric and pericentromeric regions, thus contributing to the phenotypes associated with ATR-X syndrome. In addition, the unique epi-signature identified for ATR-X syndrome can now be used as an epigenetic biomarker to support the diagnosis of new patients using a sensitive, specific and cost-effective GWMA testing protocol.

Methods
Sample collection, DNA extraction and genotyping
Peripheral blood samples from patients referred for genetic testing at the Greenwood Genetic Center were collected for methylation study. All patients were consented and counseled for ATRX testing as part of their clinical referral. Ethical approval was consented by the Self Regional Healthcare Institutional Review Board (IRB #26). Genomic DNA was extracted from peripheral blood leukocyte using standard techniques. Patients presenting with the alpha thalassemia/mental retardation X-linked syndrome (ATR-X) underwent molecular diagnostic confirmation (ATRX gene analysis) and were selected for the methylation study. Table 3 shows the molecular (mutations) characteristics of these patients. The ATR-X panel of patients is composed of 18 males with average age of 12.2 years (ranging from 8 m to 27 years).

The methylation array of these patients was compared with a reference cohort composed of controls and individuals previously referred for microarray with no significant methylation alteration. The reference cohort (controls) is composed of 210 male controls with average age of 7.3 years (2 m–53 years).

Table 3 Clinical and molecular characteristics of ATR-X patients referred for methylation study

ATR-X patient no.	Mutation
1	c.109C>T; p.R37X
2	c.109C>T; p.R37X
3	c.109C>T; p.R37X
4	c.109C>T; p.R37X
5	c.109C>T; p.R37X
6	c.730A>C; p.I244L
7	c.758T>C; p.L253S
8	c.736T>C; p.R246C
9	c.736C>T; P.R246C
10	c.952G>T; p.G249C
11	c.4817G>A; p.S1606N
12	c.5579A>G; p.N1860S
13	c.5786A>G; p.K1929R
14	c.6254G>A; p.R2085H
15	c.6593A>G; p.H2198R
16	c.7156C>T; p.R2386X
17	c.7366_7367 InsA; p.M2456Nfs X42
18	Deletion of exon 2

Methylation array and data analysis

The DNA methylation array was performed using the Infinium HumanMethylation450 BeadChip (Illumina) according to standard protocol at the Genetic and Molecular Epidemiology Laboratory at McMaster University. The array coverage includes >485,000 individual methylation sites, 99% of RefSeq genes and 96% of CpG islands. Beta and intensity values for methylation were generated using the Illumina Genome Studio Software, and .idat files were imported to Partek Genomic Suite software (Partek GS). The patient cohort was compared with the laboratory reference cohort. Statistical analysis was performed to compare ATR-X patients versus the control cohort using the ANOVA test to generate probe-level statistics, including p value (t test), F value (signal to noise) and estimate value (net methylation difference). The cutoff of estimate > ± 0.20, $p < 0.01$, $F > 50$ and probes >4 was used to select the top significant regions to be included in the epi-signature. A less stringent cutoff of estimate > ± 0.15, $p < 0.01$, $F > 50$ and probes > 3 was used for pathway analysis and karyogram view in order to include a larger number of regions in those analysis. Significant regions were mapped against the CpG islands and gene promoter regions. Genomic visualization of the data was performed using the karyogram view toll (Partek GS) for chromosome distribution of differentially methylated regions, and the Genomic Browser Wizard (Partek GS) for locus-specific methylation levels.

Pathway analysis

The top 45 differentially methylated genes identified using a less stringent cutoffs (Additional file 2) were assessed using the pathway analysis tool in the Partek Genomics Suite software. Briefly, statistical analysis included Fisher's exact test and was restricted to functional groups at least two genes. Results show the enrichment p value (p value of the Fisher exact test reflective of the number of the genes in versus not in the list or functional group) and the enrichment score (negative log of the enrichment p value; a high score indicates that the genes in the functional group are overrepresented in the gene list).

Bisulfite mutagenesis

Genomic DNA isolated from blood of ATR-X patients ($n = 2$) and controls ($n = 2$) was sodium bisulfite treated using the EZ DNA Methylation-Direct Kit (Zymo Research) according to manufacturer's instructions. DNA was amplified by nested PCR and the resulting products ligated into the pGEM-T Easy vector using a TOPO-TA cloning kit (Invitrogen). Positive clones were sequenced with Applied Biosystem 3730xl DNA Analyzer technology (Center for Applied Genomics, McGill University). Clones were accepted at $\geq 95\%$ conversion. Non-converted cytosine residues and mismatched base pairs were used to ensure all clones originated from unique template DNA.

Additional files

Additional file 1: Figure S1. Methylation string diagrams of significantly altered regions in ATR-X patients and controls. Bisulfite mutagenesis and sequencing analysis was performed in approximately 20 alleles from each sample, and individual alleles are represented as a string of CpGs. The total average methylation for each sample is indicated. Unmethylated CpGs are represented as empty circles, and methylated CpGs as filled circles.

Additional file 2: Table S1. ATR-X methylation; significant regions detected by methylation array in ATR-X patients (n = 17) compared with controls (n = 210) using cutoff of probes >3, estimate >15%, F value >50, p value <0.01.

Abbreviations
ATRX: alpha thalassemia/mental retardation X-linked protein; ATR-X: alpha thalassemia/mental retardation X-linked syndrome; DAXX: death domain-associated protein; DNMT: DNA methyltransferase; ES cells: embryonic stem cells; E-VUS: epigenetic variants of unknown clinical significance; GWMA: genome-wide DNA methylation array; H3K4: histone 3 lysine 4; H3K9: histone 3 lysine 9; KDM5C: lysine (K)-specific demethylase 5C; MeCP2: methyl CpG-binding protein 2; Partek GS: Partek Genomic Suite software; PCR: polymerase chain reaction; SRCAP: Snf2-related CREBBP activator protein; SWI/SNF: SWItch/sucrose non-fermentable family; TR: tandem repeats.

Authors' contributions
LCS analyzed the methylation data and was a major contributor in writing the manuscript. KDK and AM performed bisulfite methylation analysis. DR and AH performed the methylation microarray. PJA, NGB and DIR contributed to data interpretation and manuscript writing. KMB, GP and BS supervised the

methylation study and contributed to manuscript writing. C Schwartz and C Skinner provided the samples and clinical data for the study. All authors read and approved the final manuscript.

Author details
[1] Department of Pathology and Lab Medicine, Western University, London, ON, Canada. [2] Children's Hospital of Eastern Ontario Research Institute, University of Ottawa, Ottawa, ON, Canada. [3] Molecular Genetics Laboratory, Victoria Hospital, London Health Sciences Center, 800 Commissioner's Road E, B10-104, London, ON N6A 5W9, Canada. [4] Department of Paediatrics, Western University, London, ON, Canada. [5] Department of Biochemistry, Western University, London, ON, Canada. [6] Department of Oncology, Western University, London, ON, Canada. [7] Children's Health Research Institute, London, ON, Canada. [8] Department of Pathology and Molecular Medicine, McMaster University, Hamilton, ON, Canada. [9] Department of Clinical Epidemiology and Biostatistics, McMaster University, Hamilton, ON, Canada. [10] Center for Molecular Studies, J.C. Self Research Institute of Human Genetics, Greenwood Genetic Center, Greenwood, SC, USA.

Acknowledgements
We thank the patients and their families for their participation in studies conducted by the Greenwood Genetic Center. Dedicated to the memory of Ethan Francis Schwartz (1996–1998).

Competing interests
The authors declare that they have no competing interests.

Funding
This project was supported in part by a Grant from the South Carolina Department of Disabilities and Special Needs (SC DDSN). This work was also supported by the Care4Rare Canada Consortium (Enhanced Care for Rare Genetic Diseases in Canada) funded by Genome Canada, the Canadian Institutes of Health Research, the Ontario Genomics Institute, Ontario Research Fund, Genome Quebec and Children's Hospital of Eastern Ontario Foundation.

References
1. Bjornsson HT. The Mendelian disorders of the epigenetic machinery. Genome Res. 2015;25:1473–81. doi:10.1101/gr.190629.115.
2. Rodenhiser D, Mann M. Epigenetics and human disease: translating basic biology into clinical applications. CMAJ. 2006;174:341–8. doi:10.1503/cmaj.050774.
3. Jones PA. Functions of DNA methylation: islands, start sites, gene bodies and beyond. Nat Rev Genet. 2012;13:484–92. doi:10.1038/nrg3230.
4. Tan M, Luo H, Lee S, Jin F, Yang JS, Montellier E, Buchou T, Cheng Z, Rousseaux S, Rajagopal N, et al. Identification of 67 histone marks and histone lysine crotonylation as a new type of histone modification. Cell. 2011;146:1016–28. doi:10.1016/j.cell.2011.08.008.
5. Shen H, Laird PW. Interplay between the cancer genome and epigenome. Cell. 2013;153:38–55. doi:10.1016/j.cell.2013.03.008.
6. Schenkel LC, Rodenhiser DI, Ainsworth PJ, Pare G, Sadikovic B. DNA methylation analysis in constitutional disorders: clinical implications of the epigenome. Crit Rev Clin Lab Sci. 2016;53:147–65. doi:10.3109/10408363.2015.1113496.
7. Gibbons R. Alpha thalassaemia-mental retardation, X linked. Orphanet J Rare Dis. 2006;1:15. doi:10.1186/1750-1172-1-15.
8. Voon HP, Gibbons RJ. Maintaining memory of silencing at imprinted differentially methylated regions. CMLS. 2016;73:1871–9. doi:10.1007/s00018-016-2157-6.
9. Watson LA, Goldberg H, Berube NG. Emerging roles of ATRX in cancer. Epigenomics. 2015;7:1365–78. doi:10.2217/epi.15.82.
10. Pickett HA, Reddel RR. Molecular mechanisms of activity and derepression of alternative lengthening of telomeres. Nat Struct Mol Biol. 2015;22:875–80. doi:10.1038/nsmb.3106.
11. Watson LA, Solomon LA, Li JR, Jiang Y, Edwards M, Shin-ya K, Beier F, Berube NG. Atrx deficiency induces telomere dysfunction, endocrine defects, and reduced life span. J Clin Investig. 2013;123:2049–63. doi:10.1172/JCI65634.
12. Ritchie K, Seah C, Moulin J, Isaac C, Dick F, Berube NG. Loss of ATRX leads to chromosome cohesion and congression defects. J Cell Biol. 2008;180:315–24. doi:10.1083/jcb.200706083.
13. Kernohan KD, Vernimmen D, Gloor GB, Berube NG. Analysis of neonatal brain lacking ATRX or MeCP2 reveals changes in nucleosome density, CTCF binding and chromatin looping. Nucleic Acids Res. 2014;42:8356–68. doi:10.1093/nar/gku564.
14. De La Fuente R, Viveiros MM, Wigglesworth K, Eppig JJ. ATRX, a member of the SNF2 family of helicase/ATPases, is required for chromosome alignment and meiotic spindle organization in metaphase II stage mouse oocytes. Dev Biol. 2004;272:1–14. doi:10.1016/j.ydbio.2003.12.012.
15. Leung JW, Ghosal G, Wang W, Shen X, Wang J, Li L, Chen J. Alpha thalassemia/mental retardation syndrome X-linked gene product ATRX is required for proper replication restart and cellular resistance to replication stress. J Biol Chem. 2013;288:6342–50. doi:10.1074/jbc.M112.411603.
16. Noh KM, Allis CD, Li H. Reading between the lines: "ADD"-ing histone and DNA methylation marks toward a new epigenetic "Sum". ACS Chem Biol. 2016;11:554–63. doi:10.1021/acschembio.5b00830.
17. Goldberg AD, Banaszynski LA, Noh KM, Lewis PW, Elsaesser SJ, Stadler S, Dewell S, Law M, Guo X, Li X, et al. Distinct factors control histone variant H3.3 localization at specific genomic regions. Cell. 2010;140:678–91. doi:10.1016/j.cell.2010.01.003.
18. Voon HP, Hughes JR, Rode C, De La Rosa-Velazquez IA, Jenuwein T, Feil R, Higgs DR, Gibbons RJ. ATRX plays a key role in maintaining silencing at interstitial heterochromatic loci and imprinted genes. Cell Rep. 2015;11:405–18. doi:10.1016/j.celrep.2015.03.036.
19. Udugama M, Chang FTM, Chan FL, Tang MC, Pickett HA, McGhie JD, Mayne L, Collas P, Mann JR, Wong LH. Histone variant H3.3 provides the heterochromatic H3 lysine 9 tri-methylation mark at telomeres. Nucleic Acids Res. 2015;43:10227–37. doi:10.1093/nar/gkv847.
20. Gibbons RJ, McDowell TL, Raman S, O'Rourke DM, Garrick D, Ayyub H, Higgs DR. Mutations in ATRX, encoding a SWI/SNF-like protein, cause diverse changes in the pattern of DNA methylation. Nat Genet. 2000;24:368–71. doi:10.1038/74191.
21. Law MJ, Lower KM, Voon HP, Hughes JR, Garrick D, Viprakasit V, Mitson M, De Gobbi M, Marra M, Morris A, et al. ATR-X syndrome protein targets tandem repeats and influences allele-specific expression in a size-dependent manner. Cell. 2010;143:367–78. doi:10.1016/j.cell.2010.09.023.
22. Ratnakumar K, Bernstein E. ATRX: the case of a peculiar chromatin remodeler. Epigenetics. 2013;8:3–9. doi:10.4161/epi.23271.
23. Mihola O, Trachtulec Z, Vlcek C, Schimenti JC, Forejt J. A mouse speciation gene encodes a meiotic histone H3 methyltransferase. Science. 2009;323:373–5. doi:10.1126/science.1163601.
24. Archambault J, Pan G, Dahmus GK, Cartier M, Marshall N, Zhang S, Dahmus ME, Greenblatt J. FCP1, the RAP74-interacting subunit of a human protein phosphatase that dephosphorylates the carboxyl-terminal domain of RNA polymerase IIO. J Biol Chem. 1998;273:27593–601.
25. Cotney J, McKay SE, Shadel GS. Elucidation of separate, but collaborative functions of the rRNA methyltransferase-related human mitochondrial transcription factors B1 and B2 in mitochondrial biogenesis reveals new insight into maternally inherited deafness. Hum Mol Genet. 2009;18:2670–82. doi:10.1093/hmg/ddp208.
26. Gou D, Wang J, Gao L, Sun Y, Peng X, Huang J, Li W. Identification and functional analysis of a novel human KRAB/C2H2 zinc finger gene ZNF300. Biochim Biophys Acta. 2004;1676:203–9.
27. Yano K, Ueki N, Oda T, Seki N, Masuho Y, Muramatsu M. Identification and characterization of human ZNF274 cDNA, which encodes a novel Kruppel-type zinc-finger protein having nucleolar targeting ability. Genomics. 2000;65:75–80. doi:10.1006/geno.2000.6140.
28. Ichimura T, Watanabe S, Sakamoto Y, Aoto T, Fujita N, Nakao M. Transcriptional repression and heterochromatin formation by MBD1 and MCAF/AM family proteins. J Biol Chem. 2005;280:13928–35. doi:10.1074/jbc.M413654200.
29. Lauriat TL, Shiue L, Haroutunian V, Verbitsky M, Ares M Jr, Ospina L, McInnes LA. Developmental expression profile of quaking, a candidate gene for schizophrenia, and its target genes in human prefrontal cortex and hippocampus shows regional specificity. J Neurosci Res. 2008;86:785–96. doi:10.1002/jnr.21534.

30. Sarma K, Cifuentes-Rojas C, Ergun A, Del Rosario A, Jeon Y, White F, Sadreyev R, Lee JT. ATRX directs binding of PRC2 to Xist RNA and Polycomb targets. Cell. 2014;159:869–83. doi:10.1016/j.cell.2014.10.019.

31. Turcan S, Rohle D, Goenka A, Walsh LA, Fang F, Yilmaz E, Campos C, Fabius AW, Lu C, Ward PS, et al. IDH1 mutation is sufficient to establish the glioma hypermethylator phenotype. Nature. 2012;483:479–83. doi:10.1038/nature10866.

32. Cai J, Yang P, Zhang C, Zhang W, Liu Y, Bao Z, Liu X, Du W, Wang H, Jiang T, Jiang C. ATRX mRNA expression combined with IDH1/2 mutational status and Ki-67 expression refines the molecular classification of astrocytic tumors: evidence from the whole transcriptome sequencing of 169 samples samples. Oncotarget. 2014;5:2551–61. doi:10.18632/oncotarget.1838.

33. Nan X, Hou J, Maclean A, Nasir J, Lafuente MJ, Shu X, Kriaucionis S, Bird A. Interaction between chromatin proteins MECP2 and ATRX is disrupted by mutations that cause inherited mental retardation. Proc Natl Acad Sci USA. 2007;104:2709–14. doi:10.1073/pnas.0608056104.

34. Kernohan KD, Jiang Y, Tremblay DC, Bonvissuto AC, Eubanks JH, Mann MR, Berube NG. ATRX partners with cohesin and MeCP2 and contributes to developmental silencing of imprinted genes in the brain. Dev Cell. 2010;18:191–202. doi:10.1016/j.devcel.2009.12.017.

35. Portela A, Esteller M. Epigenetic modifications and human disease. Nat Biotechnol. 2010;28:1057–68. doi:10.1038/nbt.1685.

36. LaSalle JM, Yasui DH. Evolving role of MeCP2 in Rett syndrome and autism. Epigenomics. 2009;1:119–30. doi:10.2217/epi.09.13.

37. Watrin E, Kaiser FJ, Wendt KS. Gene regulation and chromatin organization: relevance of cohesin mutations to human disease. Curr Opin Genet Dev. 2016;37:59–66. doi:10.1016/j.gde.2015.12.004.

38. Kimura H, Shiota K. Methyl-CpG-binding protein, MeCP2, is a target molecule for maintenance DNA methyltransferase, Dnmt1. J Biol Chem. 2003;278:4806–12. doi:10.1074/jbc.M209923200.

39. Fuks F, Hurd PJ, Wolf D, Nan X, Bird AP, Kouzarides T. The methyl-CpG-binding protein MeCP2 links DNA methylation to histone methylation. J Biol Chem. 2003;278:4035–40. doi:10.1074/jbc.M210256200.

40. Carvill GaS A. Genome-wide DNA methylation analysis in patients with familial ATR-X mental retardation syndrome. In: Appasani K, editor. Epigenomics from chromatin biology to therapeutics, vol. 1. New York: Cambridge University Press; 2012. p. 434–46.

41. Xue Y, Gibbons R, Yan Z, Yang D, McDowell TL, Sechi S, Qin J, Zhou S, Higgs D, Wang W. The ATRX syndrome protein forms a chromatin-remodeling complex with Daxx and localizes in promyelocytic leukemia nuclear bodies. Proc Natl Acad Sci USA. 2003;100:10635–40. doi:10.1073/pnas.1937626100.

42. Vallian S, Chin KV, Chang KS. The promyelocytic leukemia protein interacts with Sp1 and inhibits its transactivation of the epidermal growth factor receptor promoter. Mol Cell Biol. 1998;18:7147–56.

43. Levy MA, Kernohan KD, Jiang Y, Berube NG. ATRX promotes gene expression by facilitating transcriptional elongation through guanine-rich coding regions. Hum Mol Genet. 2015;24:1824–35. doi:10.1093/hmg/ddu596.

44. Hood RL, Schenkel LC, Nikkel SM, Ainsworth PJ, Pare G, Boycott KM, Bulman DE, Sadikovic B. The defining DNA methylation signature of Floating-Harbor Syndrome. Sci Rep. 2016;6:38803. doi:10.1038/srep38803.

45. Kernohan KD, Cigana Schenkel L, Huang L, Smith A, Pare G, Ainsworth P, Boycott KM, Warman-Chardon J, Sadikovic B. Identification of a methylation profile for DNMT1-associated autosomal dominant cerebellar ataxia, deafness, and narcolepsy. Clin Epigenetics. 2016;8:91. doi:10.1186/s13148-016-0254-x.

46. Choufani S, Cytrynbaum C, Chung BH, Turinsky AL, Grafodatskaya D, Chen YA, Cohen AS, Dupuis L, Butcher DT, Siu MT, et al. NSD1 mutations generate a genome-wide DNA methylation signature. Nat Commun. 2015;6:10207. doi:10.1038/ncomms10207.

47. Grafodatskaya D, Chung BH, Butcher DT, Turinsky AL, Goodman SJ, Choufani S, Chen YA, Lou Y, Zhao C, Rajendram R, et al. Multilocus loss of DNA methylation in individuals with mutations in the histone H3 lysine 4 demethylase KDM5C. BMC Med Genomics. 2013;6:1. doi:10.1186/1755-8794-6-1.

PERMISSIONS

LIST OF CONTRIBUTORS

Zac Chatterton, Natalia Mendelev, Sean Chen and Fatemeh Haghighi
Friedman Brain Institute, Icahn School of Medicine at Mount Sinai, 1425 Madison Avenue, New York, NY 10029, USA
Department of Neuroscience, Icahn School of Medicine at Mount Sinai, 1425 Madison Ave, Floor 10, Room 10-70D, New York, NY 10029, USA
Medical Epigenetics,James J. Peters VA Medical Center, Bronx, NY 10468, USA

Brigham J. Hartley, Man-Ho Seok and Kristen Brennand
Friedman Brain Institute, Icahn School of Medicine at Mount Sinai, 1425 Madison Avenue, New York, NY 10029, USA
Department of Neuroscience, Icahn School of Medicine at Mount Sinai, 1425 Madison Ave, Floor 10, Room 10-70D, New York, NY 10029, USA
Department of Psychiatry, Icahn School of Medicine at Mount Sinai, 1425 Madison Avenue, New York, NY 10029, USA

Maria Milekic
Department of Psychiatry, Columbia University, New York, NY 10032, USA

Gorazd Rosoklija
Department of Psychiatry, Columbia University, New York, NY 10032, USA
Macedonian Academy of Sciences and Arts, Skopje, Macedonia
School of Medicine, Skopje, Macedonia

Aleksandar Stankov
School of Medicine, Skopje, Macedonia. 10 Psychiatric Hospital Skopje, Skopje, Macedonia

Iskra Trencevsja-Ivanovska
Psychiatric Hospital Skopje, Skopje, Macedonia.

Yongchao Ge
Department of Neurology, Icahn School of Medicine at Mount Sinai, 1425 Madison Avenue, New York, NY 10029, USA

Andrew J. Dwork
Department of Psychiatry, Columbia University, New York, NY 10032, USA

Department of Pathology and Cell Biology, Columbia University, New York, NY 10032, USA
Macedonian Academy of Sciences and Arts, Skopje, Macedonia

Megan Guntrum, Ekaterina Vlasova and Tamara L. Davis
Department of Biology, Bryn Mawr College, 101 N. Merion Avenue, Bryn Mawr, PA 19010-2899, USA

Arbel Moshe and Tommy Kaplan
School of Computer Science and Engineering, The Hebrew University of Jerusalem, Jerusalem 91904, Israel

Kevin T. Ebata, Kathryn Mesh and Miguel Ramalho-Santos
Eli and Edythe Broad Center of Regeneration Medicine and Stem Cell Research, University of California, San Francisco, San Francisco, CA, USA

Shichong Liu and Benjamin A. Garcia
Epigenetics Program, Department of Biochemistry and Biophysics, Perelman School of Medicine, University of Pennsylvania, Philadelphia, PA, USA

Misha Bilenky
Canada's Michael Smith Genome Sciences Centre, BC Cancer Agency, Vancouver,BC, Canada

Alexander Fekete and Michael G. Acker
Novartis Institutes for Biomedical Research, Cambridge, MA, USA

Martin Hirst
Canada's Michael Smith Genome Sciences Centre, BC Cancer Agency, Vancouver, BC, Canada
Department of Microbiology and Immunology, Centre for High-Throughput Biology, University of British Columbia, Vancouver, BC, Canada

Christopher Schröder and Sven Rahmann
Genome Informatics, Institute of Human Genetics, University of Duisburg- Essen, University Hospital Essen, Essen, Germany

Elsa Leitão and Bernhard Horsthemke
Institute of Human Genetics,University of Duisburg-Essen, University Hospital Essen, Hufelandstraße 55, 45147 Essen, Germany

Stefan Wallner and Gerd Schmitz
Institute for Clinical Chemistry and Laboratory Medicine, University Hospital Regensburg, Regensburg, Germany

Ludger Klein-Hitpass
Institute of Cell Biology, University Hospital Essen, Essen, Germany.

Anupam Sinha
Institute of Clinical Molecular Biology, Kiel University, University Hospital, Kiel, Germany

Karl-Heinz Jöckel
Institute of Medical Informatics, Biometry and Epidemiology, University Hospital Essen, Essen, Germany

Stefanie Heilmann-Heimbach and Markus M. Nöthen
Institute of Human Genetics, School of Medicine, University Hospital of Bonn, University of Bonn, Bonn, Germany
Department of Genomics, Life and Brain Center, University of Bonn, Bonn, Germany

Per Hoffmann
Institute of Human Genetics, School of Medicine, University Hospital of Bonn, University of Bonn, Bonn, Germany
Department of Genomics, Life and Brain Center, University of Bonn, Bonn, Germany
Institute of Medical Genetics and Pathology, University Hospital Basel, Basel, Switzerland
Human Genomics Research Group, Department of Biomedicine,University of Basel, Basel, Switzerland.

Michael Steffens
Research Division, Federal Institute for Drugs and Medical Devices (BfArM), Bonn, Germany

Peter Ebert
Max Planck Institute for Informatics, Saarland Informatics Campus, Saarbrücken, Germany
Saarbrücken Graduate School of Computer Science, Saarland Informatics Campus, Saarbrücken, Germany

Taosui Li, Jacob W. Hodgson and Hugh W. Brock
Department of Zoology, Life Sciences Institute, University of British Columbia, 2350 Health Science Mall, Vancouver, BC V6T 1Z4, Canada

Svetlana Petruk and Alexander Mazo
Department of Biochemistry and Molecular Biology, Thomas Jefferson University, Philadelphia, PA 19107, USA

Ryohei Nakamura, Ayako Uno, Masahiko Kumagai and Hiroyuki Takeda
Department of Biological Sciences, Graduate School of Science, The University of Tokyo, 7-3-1 Hongo, Bunkyo-ku, Tokyo 113-0033, Japan

Shinichi Morishita
Department of Computational Biology and Medical Sciences, Graduate School of Frontier Sciences, The University of Tokyo, 5-1-5 Kashiwanoha, Kashiwa 277-8562, Japan

Saikat Bhattacharya
Epigenetics and Chromatin Biology Group, Gupta Lab, Cancer Research Institute, Advanced Centre for Treatment, Research and Education in Cancer (ACTREC), Tata Memorial Centre, Kharghar, Navi Mumbai, MH 410210, India
Homi Bhabha National Institute, Training School Complex, Anushakti Nagar, Mumbai, MH 400085, India.
Stowers Institute for Medical Research,Kansas City, MO 64110, USA.

Divya Reddy, Sanket Shah and Sanjay Gupta
Epigenetics and Chromatin Biology Group, Gupta Lab, Cancer Research Institute, Advanced Centre for Treatment, Research and Education in Cancer (ACTREC), Tata Memorial Centre, Kharghar, Navi Mumbai, MH 410210, India
Homi Bhabha National Institute, Training School Complex, Anushakti Nagar, Mumbai, MH 400085, India

Vinod Jani, Uddhavesh Sonavane and Rajendra Joshi
Bioinformatics Group, Centre for Development of Advanced Computing (C-DAC), University of Pune Campus, Pune, MH 411007, India

Nikhil Gadewal
BTIS, Cancer Research Institute, Advanced Centre for Treatment, Research and Education in Cancer (ACTREC), Tata Memorial Centre, Kharghar, Navi Mumbai, MH 410210, India

Raja Reddy and Kakoli Bose
Integrated Biophysics and Structural Biology Lab, Cancer Research Institute, Advanced Centre for Treatment, Research and Education in Cancer (ACTREC), Tata Memorial Centre, Kharghar, Navi Mumbai, MH 410210, India
Homi Bhabha National Institute, Training School Complex, Anushakti Nagar, Mumbai, MH 400085, India

Yiqin Ma and Laura Buttitta
Department of Molecular, Cellular and Developmental Biology, University of Michigan, Ann Arbor, MI 48109, USA

Ruiqi Liao
Department of Cell and Developmental Biology, University of Illinois at Urbana Champaign, B107 Chemistry and Life Sciences Building, MC-123 601 S. Goodwin Ave., Urbana, IL 61801, USA

Craig A. Mizzen
Department of Cell and Developmental Biology, University of Illinois at Urbana Champaign, B107 Chemistry and Life Sciences Building, MC-123 601 S. Goodwin Ave., Urbana, IL 61801, USA
Institute for Genomic Biology, University of Illinois at Urbana Champaign, Urbana, IL 61801, USA

Lisa M. McEwen, Alexander M. Morin, Rachel D. Edgar, Julia L. MacIsaac, Meaghan J. Jones and Michael S. Kobor
Department of Medical Genetics, Centre for Molecular Medicine and Therapeutics, BC Children's Hospital Research Institute, University of British Columbia, 950 West 28th Ave, Vancouver, Canada

William H. Dow
School of Public Health, University of California, Berkeley, Berkeley, CA, USA

Luis Rosero-Bixby
Centro Centroamericano de Población, Universidad de Costa Rica, San José, Costa Rica

David H. Rehkopf
Division of General Medical Disciplines, Department of Medicine, School of Medicine, Stanford University, 1070 Arastradero Road, Suite 300, Palo Alto, CA 94304, USA

Clayton K. Collings
Department of Biochemistry and Molecular Genetics, Northwestern University Feinberg School of Medicine, 320 E. Superior Street, Chicago, IL 60611, USA

John N. Anderson
Department of Biological Sciences, Purdue University, 915 W. State Street, West Lafayette, IN 47907, USA

Laila C. Schenkel
Department of Pathology and Lab Medicine, Western University, London,ON, Canada

Kristin D. Kernohan, Arran McBride and Kym M. Boycott
Children's Hospital of Eastern Ontario Research Institute,University of Ottawa, Ottawa, ON, Canada

Ditta Reina, Amanda Hodge and Guillaume Pare
Department of Pathology and Molecular Medicine, McMaster University, Hamilton, ON, Canada
Department of Clinical Epidemiology and Biostatistics,McMaster University, Hamilton, ON, Canada

Peter J. Ainsworth
Department of Pathology and Lab Medicine, Western University, London, ON, Canada
Molecular Genetics Laboratory, Victoria Hospital, London Health Sciences Center, 800 Commissioner's Road E, B10-104, London, ON N6A 5W9, Canada
Department of Paediatrics, WesternUniversity, London, ON, Canada
Department of Biochemistry, Western University, London, ON, Canada
Department of Oncology, Western University, London, ON, Canada
Children's Health Research Institute, London, ON, Canada

David I. Rodenhiser and Nathalie G. Bérubé
Department of Paediatrics, Western University, London, ON, Canada
Department of Biochemistry, Western University, London, ON, Canada
Department of Oncology, Western University, London, ON, Canada
Children's Health Research Institute, London, ON, Canada

Cindy Skinner and Charles Schwartz
Center for Molecular Studies, J.C. Self Research
Institute of Human Genetics, Greenwood Genetic
Center, Greenwood, SC, USA

Bekim Sadikovic
Department of Pathology and Lab Medicine,
Western University, London, ON, Canada
Molecular Genetics Laboratory, Victoria Hospital,
London Health Sciences Center, 800 Commissioner's
Road E, B10-104, London, ON N6A 5W9, Canada
Children's Health Research Institute, London, ON,
Canada

Index

A

Additional Sex Combs, 70, 85-86

Allele-specific Methylation, 52, 59, 61-62, 64

Atrx, 193-194, 196-203

B

Biodemography, 160

Biomarker, 117, 161, 193, 200

Brain, 1-2, 4, 7-11, 21, 47, 51, 67-68, 102, 104, 113, 117, 172, 197, 202-203

C

Cancer, 2, 10-11, 24, 50, 85-86, 100-101, 103, 112-114, 117, 136-137, 156, 158, 160, 188, 191-192, 202

Cdk2, 120, 129, 135, 137, 139-141, 146-147, 155, 158-159

Cdk7, 139, 141, 146-149, 151, 155-156, 158

Cdk9, 71, 139-142, 146-149, 151-153, 155-159

Cell Cycle Exit, 119-120, 125-126, 129, 131, 133, 135-137

Cell Deconvolution, 9

Cell Differentiation, 2, 7-8, 95, 139-140, 142, 144, 149, 153-154, 156, 158-159, 175, 186

Chromatin, 11, 23, 36, 41, 48, 50, 58, 63, 68, 71, 77, 81, 87, 90, 102, 106, 109, 113, 117, 122, 129, 133, 137, 142, 144, 151, 160, 166, 176, 178, 181, 183, 188, 194, 199, 203

Chromatin Search, 26

Cpg, 1, 3-6, 9, 11-15, 18, 20-24, 52, 54-55, 59-68, 89, 95, 97, 160-162, 164-165, 167-169, 171-176, 178-181, 183, 186-188, 190-192, 194-201, 203

Cpg Dyads, 12-14, 20-22, 24

Cyclin-dependent Kinase (CDK), 139

D

Differentially Methylated Regions, 1, 4-6, 11-12, 14, 52-53, 63, 66, 68, 164, 167, 173, 193, 197, 201-202

Differentiation, 1-3, 7-9, 11, 40, 68, 95, 100-102, 119-120, 122, 125, 129, 131, 133, 135-137, 139-144, 149, 151, 153-154, 156, 158-159, 175, 186

Dna Methylation, 1-4, 6-18, 24, 41, 44-47, 50, 53, 56, 58-63, 65-68, 87, 89, 94-95, 97-99, 160, 162-165, 167, 170-172, 174-176, 181, 183, 188, 190-194, 196-203

Dna Sequence, 53, 87-88, 90, 92, 95-96, 98, 115, 161, 176, 187, 191-192

Dorsolateral Prefrontal Cortex (DLPFC), 1

Drosophila, 26-28, 32-35, 38-39, 70-71, 73-77, 80-81, 84-86, 98, 119-120, 122, 125-126, 129, 131, 133, 136-137

E

E2f, 119, 125-126, 129, 131, 133, 135-137

Embryonic Stem Cell, 51, 139, 157-159

Enhancer Of Zeste, 70, 85, 119, 136

Enhancers, 24, 26-27, 31, 33, 35, 38, 56, 62, 70, 87, 98, 129, 137, 154, 156, 174, 183, 188, 191

Epi-signature, 193-194, 196-197, 200-201

Epigenetic Age, 160-162, 164, 169, 171, 173

Epigenetics, 1, 11-12, 24, 40, 50, 100, 117, 159-160, 172-174, 186, 191, 202-203

F

Fetal, 1-11, 176

Fetal Samples, 4, 6-7, 9

G

Gaga Factor, 26, 34-35, 71

Gametic Dmr, 12

Genomic Imprinting, 12, 23-24, 52

Gtl2, 12-17, 19-24

H

H3k9me2, 40-49, 51

Haplotype, 52, 66-67

Hemimethylation, 12-22

Heterochromatin Binding Proteins, 119, 129

Histone, 12, 26, 33, 36, 42, 47, 52, 54, 56, 58, 63, 68, 71, 74, 76, 78, 88, 90, 102, 108, 114, 122, 126, 129, 131, 133, 137, 139, 143, 146, 153, 160, 176, 181, 183, 188, 198, 203

Histone H1, 100, 139, 157-159

Histone Lysine Demethylase, 40

Histone Methylation, 40-41, 44, 47, 50, 186, 191-192, 198, 203

Histone Modifications, 12, 26-30, 32-33, 35-36, 38, 56, 58, 61-62, 65-66, 81, 87-88, 98, 119, 126, 129, 131, 133, 135-136, 160, 173, 178-179, 181, 183, 186, 188

Histone Trimethylation, 70

Homomethylation, 12, 14

Hsp70 Transcriptional Elongation, 70, 83

I

Immune Aging, 160

Immunofluorescence, 3, 40-42, 47, 49, 122

In Utero, 1-2, 8-10
Intellectual Disability, 193, 200

L
Longevity, 160-162, 164, 169-170, 172

M
Maternal-to-zygotic Transition, 26-28, 38-39
Methylation Array, 11, 52, 194, 196-201
Mnase-seq, 88-90, 92, 97, 174-176, 178-179, 181, 183, 186, 188

N
Neurodevelopment, 1, 8, 197
Neuron, 1, 8
Nicotine, 1-3, 8-11
Nome-seq, 174-176, 178-179, 181, 183, 186, 188, 190
Nucleosome, 26-28, 38-39, 70-71, 81, 85, 87-90, 92-98, 100, 106, 108-113, 117, 174-176, 178-181, 183, 186-188, 190-192, 202
Nucleosome Positioning, 27, 87-88, 90, 92, 94-96, 98, 190-191

P
P-tefb, 71, 139, 146-147, 151, 155-159

Phosphorylation, 26, 71, 136-137, 139-144, 146-147, 149, 151, 153-159, 186
Pluripotency Factors, 139, 143

S
Set Domain, 70, 72, 74-75, 78, 80, 84-85
Smoking, 1-2, 4-11, 161-162
Snp Genotyping, 52, 64
Somatic Dmr, 12
Spectral Clustering, 26, 30-33, 38-39

T
Tobacco, 1-2, 7, 9-11
Trithorax, 34, 70, 85-86

V
Vertebrate, 87-88, 95-96, 99, 175
Vitamin C, 10, 40-47, 49-50

W
Whole Genome Bisulfite Sequencing, 52, 61, 63, 67

Z
Zelda Peaks, 26, 28-36, 38

www.ingramcontent.com/pod-product-compliance
Lightning Source LLC
Chambersburg PA
CBHW082033190326
41458CB00010B/3354